Mathematical Physics Studies

More information about this series at http://www.springer.com/series/6316

Alexander Cardona · Pedro Morales
Hernán Ocampo · Sylvie Paycha
Andrés F. Reyes Lega
Editors

Quantization, Geometry and Noncommutative Structures in Mathematics and Physics

Springer

Editors
Alexander Cardona
Departamento de Matemáticas
Universidad de los Andes
Bogotá
Colombia

Sylvie Paycha
Institut für Mathematik
Universität Potsdam
Potsdam
Germany

Pedro Morales
Department of Mathematics
The University of Texas
Austin, TX
USA

Andrés F. Reyes Lega
Departamento de Física
Universidad de los Andes
Bogotá
Colombia

Hernán Ocampo
Departamento de Física
Universidad del Valle
Cali
Colombia

ISSN 0921-3767 ISSN 2352-3905 (electronic)
Mathematical Physics Studies
ISBN 978-3-319-65426-3 ISBN 978-3-319-65427-0 (eBook)
DOI 10.1007/978-3-319-65427-0

Library of Congress Control Number: 2017948622

Printed on acid-free paper

This Springer imprint is published by Springer Nature
The registered company is Springer International Publishing AG
The registered company address is: Gewerbestrasse 11, 6330 Cham, Switzerland

Preface

The Summer School *Geometric, Algebraic and Topological Methods for Quantum Field Theory* has been taking place in Villa de Leyva (Colombia) every second year since 1999. In a world of ever-increasing academic specialization, the organizers of this school have sought over the years to provide an environment where students and researchers working in various areas of pure mathematics and theoretical physics may find a real chance to communicate with each other and to learn about subjects which at first sight would appear either irrelevant to their field of study or simply too remote in terms of prerequisites and language.

Not only does E. Wigner's remark[1] (1960) about the "*the unreasonable effectiveness of mathematics in physics*" remain valid, more so, it has gained in significance in a much broader context: Following M. Atiyah, R. Dijkgraaf and N. Hitchin[2] (2010), one may be tempted to exchange the role of the words *physics and mathematics* in the previous sentence, and it would still make sense! It is also in this spirit that the 9th edition of the school was planned. It comprised six courses, delivered by experts, as well as several contributed research talks, all of which revolved around one (or more) of the broad themes *Quantization, Geometry and Noncommutativity*.

Following the tradition of the previous editions to publish the lectures delivered during the school, this volume presents contributions by the lecturers of the school but it differs from traditional proceedings in so far as it comprises chapters on related topics based on the lectures, with a special effort put in achieving a pedagogical presentation. The present volume, we hope, will reflect the effort the authors made during the school to adapt the lectures to the needs of the participants.

The book is therefore intended for curious readers eager to get acquainted with one of the above-listed topics, namely *Quantization, Geometry, and Noncommutative Structures*, and the relations between them. It is addressed to a broad readership ranging from master students to advanced researchers, who will

---

[1]Wigner, Eugene P., *Communications on Pure and Applied Mathematics* (1960) 13, 1–14.

[2]Atiyah, M., Dijkgraaf, R. and Hitchin, N., *Phil. Trans. R. Soc. A* (2010) 368, 913–926.

find in this volume both the necessary prerequisites and the openings to active areas of research as well as links between these closely related fields of mathematical physics. We hope this volume will be of interest to both physicists and mathematicians, and that it will serve as an introduction to ongoing research in very active areas of mathematics and physics at the border line between geometry, topology, algebra, and quantum field theory. In spite of the diversity of the topics, the reader will find a coherent and homogeneous presentation, reflecting the strong ties between the various topics. For the lay reader, it should serve as a smooth first encounter with some of them.

We are indebted to various institutions for their financial support for this school. Let us first of all thank *Universidad de los Andes* in Colombia, which has been from the very beginning our main source of financial support. We also warmly thank Universidad Nacional de Colombia, Pontificia Universidad Javeriana, Universidad Sergio Arboleda, and Institut de Recherche Mathématique Avancée de Strasbourg for their financial support and their contribution to the success of this school.

Special thanks to the administrative staff at Universidad de los Andes, particularly to Silvia Restrepo (Vice Rector for Research), Ferney Rodríguez (Dean of the Faculty of Sciences), Adolfo Queiroz (Director of the Mathematics Department), and Gabriel Téllez (Director of the Physics Department), for their constant encouragement and support. We also would like to thank Leidy Castillo and Luisa Fernanda Amarillo, who did a great job for the practical organization of the school, the quality of which was very much appreciated by participants and lecturers. We are also very indebted to Marbel Galindo and Paola Pardo for their help in various essential tasks needed for the successful development of the school.

We also want to express our gratitude to the editorial staff of Springer. Aldo Rampioni guided us during the initial stages of this project. His advice was decisive and helped us achieve our initial editorial project. Kirsten Theunissen guided us throughout the whole editorial process. We thank them both for their excellent job and also for their patience.

Without the people named here, all of whom helped in the organization in some way or another, before, during, and after the school, this scientific event would not have left such vivid memories in the lecturers' and participants' minds. Last but not least, thanks to all the participants who gave us all, lecturers and editors, the impulse to prepare this volume through the enthusiasm they showed during the school, and thanks to all the contributors and referees for their participation in the realization of this volume.

Bogotá, Colombia	Alexander Cardona
Austin, USA	Pedro Morales
Cali, Colombia	Hernán Ocampo
Potsdam, Germany	Sylvie Paycha
Bogotá, Colombia	Andrés F. Reyes Lega

Contents

Contributors

Nicolás Andruskiewitsch FaMAF-CIEM (CONICET), Universidad Nacional de Córdoba, Córdoba, República Argentina

Nathan Berkovits ICTP South American Institute for Fundamental Research, Instituto de Física Teórica UNESP-Universidade Estadual Paulista, Sao Paulo, SP, Brazil

Alexander Cardona Departamento de Matemáticas, Universidad de los Andes, Bogotá, Colombia

Alberto S. Cattaneo Institut für Mathematik, Universität Zürich, Zürich, Switzerland

Pierre J. Clavier Institut für Mathematik, Potsdam Universität, Potsdam, Deutschland

Mboyo Esole Department of Mathematics, Northeastern University, Boston, MA, USA

Kevin Falk Centre de Physique Théorique, CNRS, CPT, Aix-Marseille Université, Marseille, France; et Université de Toulon, La Garde, France

Humberto Gomez Facultad de Ciencias Básicas, Universidad Santiago de Cali, Cali, Valle, Colombia

Simone Gutt Université Libre de Bruxelles, Brussels, Belgium

Christian Kassel Institut de Recherche Mathématique Avancée, CNRS & Université de Strasbourg, Strasbourg, France

Pavel Mnev Max-Planck-Institut für Mathematik, Bonn, Germany; St. Petersburg Department of V.A., Steklov Institute of Mathematics of the Russian Academy of Sciences, St. Petersburg, Russia

Viet Dang Nguyen Institut Camille Jordan, Université Claude Bernard Lyon 1, Lyon, France; Institut Camille Jordan, CNRS, Lyon, France

Sylvie Paycha Institut für Mathematik, Universität Potsdam, Potsdam, Germany

Andrés F. Reyes Lega Departamento de Física, Universidad de los Andes, Bogotá, Colombia

Konstantin Wernli Institut für Mathematik, Universität Zürich, Zürich, Switzerland

Chapter 1
Prelude: A General Overview

Alexander Cardona, Sylvie Paycha and Andrés F. Reyes Lega

Abstract This chapter provides the reader with a general overview of the various topics discussed in this volume, emphasizing the deep relations existing between them. Following a brief historical account of the emergence of the concept of *"quantization"* both in physics and mathematics, a description of the main concepts and tools appearing in subsequent chapters is presented.

1.1 Introduction

This volume presents various ongoing approaches to the vast topic of quantization, namely to the process of forming a quantum mechanical system starting from a classical one and discusses their numerous fruitful interactions with mathematics.

In its early years, quantum theory was understood in terms of a set of empirical rules that would allow to make sense—to a certain extent—of experimental results. Thus, for instance, in the old quantum theory, an electron would still orbit the nucleus obeying the laws of classical dynamics, but an additional condition, the Bohr–Sommerfeld quantization condition, had to be fulfilled. This reduced the set of allowed orbits to a discrete one, providing a way to explain the quantization of energy levels. The subsequent development of wave mechanics by de Broglie, the

A. Cardona (✉)
Departamento de Matemáticas, Universidad de los Andes, Carrera 1 18A-10,
Edificio H, Bogotá, Colombia
e-mail: acardona@uniandes.edu.co

S. Paycha
Institut für Mathematik, Universität Potsdam, Am Neuen Palais 10,
14469 Potsdam, Germany
e-mail: paycha@math.uni-potsdam.de

A.F. Reyes Lega
Departamento de Física, Universidad de los Andes, Carrera 1 18A-10,
Bloque Ip, Bogotá, Colombia
e-mail: anreyes@uniandes.edu.co

A. Cardona et al. (eds.), *Quantization, Geometry and Noncommutative Structures in Mathematics and Physics*, Mathematical Physics Studies,
DOI 10.1007/978-3-319-65427-0_1

introduction of Schrödinger's equation and the development of Heisenberg's "matrix mechanics", eventually led to the formulation based on operators in Hilbert space, as presented by Dirac and von Neumann.

In particular, Dirac emphasized that quantum observables—described by operators acting on a Hilbert space—can be obtained by replacing classical observables (i.e., smooth functions on phase space) by self-adjoint operators, in such a way that the Poisson bracket of two classical observables becomes, up to a constant, the commutator of the corresponding quantum observables. Thus, the quantum analogue of the classical Poisson bracket $\{x, p\} = 1$ of classical mechanics is given by the canonical commutation relations $[\hat{x}, \hat{p}] = i\hbar$, with $\hbar$ the Planck constant. So quantization brings in non-commuting operator algebras due to the presence of the parameter $\hbar$.

A first approach to quantization presented in this volume, called deformation quantization, an approach initiated by M. Flato, A. Lichnerowicz, and D. Sternheimer, in viewing the Planck constant $\hbar$ as a small parameter, provides a deformation of the structure of the algebra of classical observables rather than a radical change in the nature of the observables. It is defined in terms of a star product viewed as a formal deformation in the parameter $\hbar$ of the algebraic structure of the space of smooth functions on a Poisson manifold. When symmetries come into play, deformation quantization needs to be merged with group actions, the topic of Chap. 2, by Simone Gutt.

The non-commutativity arising from quantization is the main concern of non-commutative geometry, which has become an autonomous area of research under the impulse of A. Connes. His and Chamseddine's spectral action principle applied to an appropriate non-commutative space yields the standard model action coupled to Einstein and Weyl gravity. Allowing for the presence of symmetries requires working with principal fiber bundles in a non-commutative setup, the topic of Chap. 3, by Christian Kassel. Non-commutativity is central to N. Andruskiewitsch's contribution which presents Nichols algebras that provide a unifying concept for various viewpoints on the quantized enveloping algebra of a simple finite-dimensional Lie algebra $\mathfrak{g}$ at a generic parameter q.

An alternative quantization procedure which claims to encompass gravity was born in the late 1960s and early 1970s under the name of string theory. Indeed, one of the many vibrational states of the string is supposed to correspond to the graviton, a quantum mechanical particle that carries gravitational force. It went through a first golden age in the late 1980s and early 1990s known as the first string theory revolution, and a revival around the concept of duality in the late 1990s and early 2000, known as the second string theory revolution. In Chap. 6, N. Berkovits and H. Gomez present its supersymmetric version which encodes both the bosons and the fermions. Superstrings have drawn the attention of many a mathematician, due to its various fruitful interactions with algebraic geometry, some of which are described here by M. Esole.

The quantization of gauge theories entails many subtleties, in great part due to the presence of gauge invariance. From the point of view of classical dynamics, in gauge theories we are faced with the problem that the theory, initially defined in terms of a Lagrangian density, cannot be described in a Hamiltonian setting without taking into account the presence of constraints. An appropriate treatment of the quantum problem leads to BRST symmetry, as illustrated in the example of the superstring in Berkovits' lectures. The quantization of a field theory can also be performed in a Lagrangian setting, making use of path integrals. In the case of gauge theories, the problems reappear in the form of the Gribov ambiguity [1]. A very general approach devised to properly dealing with the gauge-fixing problem is the Batalin–Vilkovisky formalism, which is the topic of Chaps. 8 and 9.

Reflecting the deep relations between the various topics discussed in the lectures to follow are the many common mathematical or physical concepts and tools they bring into play. Let us name a few transversal concepts to various lectures that can serve as guiding threads for the reader:

- **Group actions** which arise wherever there are symmetries, so in any quantization procedure which claim to take symmetries into account, such as deformation quantization in a G-equivariant setup in Simone Gutt's contribution. In Ch. Kassel's lectures, group actions are generalized to the non-commutative world in the form of comodule algebras over a Hopf algebra. In N. Berkovits' lectures, which uses the BRST formalism, the local symmetries are fixed and ghost and antighost parameters (parameters with inverse statistics) are introduced, thus giving rise to global symmetries and an associated conserved charge, the BRST charge.
- **Hopf algebras**, the dual counterparts of groups, that correspond to structures encoding simultaneously an (unital associative) algebra and a (counital coassociative) coalgebra, with compatibility conditions between these structures together with an antiautomorphism satisfying a certain property. Hopf algebras naturally occur in algebraic topology, in group theory (via the concept of a group ring), quantum groups as can be seen from the lectures by Ch. Kassel where they are used to quantize homogeneous spaces and in the context of Nichols algebras presented by N. Andruskiewitsch, that play a crucial role in the classification program of Hopf algebras. They also have diverse applications ranging from condensed-matter physics and quantum field theory to string theory.
- **Fibrations** that arise wherever quantization meets geometry, here in the form of (i) elliptic fibrations, describing an elliptic curve moving along a variety, the topic of M. Esole's lectures, whose physical background lies in the realm of strings where elliptic curves arise naturally via conformal field theory, (ii) the non-commutative principal fiber bundles discussed in Ch. Kassel's lectures, a non-commutative generalization of ordinary principal fiber bundles that developed with gauge theory, (iii) the Weyl bundle, a bundle used in S. Gutt's lectures, whose fibers are modeled on the Weyl algebra, and on whose flat sections one builds a star product, (iv) as an instance of the more general concept of foliation arising in A. Ashtekar's lectures as globally hyperbolic space-time in the context of quantum field theory on curved space-time.
- **Supersymmetry** which takes different forms depending on the context, e.g., that of a supersymmetric action in N. Berkovits' lectures. Supersymmetry is a key ingredient in string theory; there are various string theories in ten dimensions related by dualities which give rise to challenging questions in mathematics requiring sophisticated tools such as the elliptic fibrations of M. Esole's lectures.

- **Quantization**, a deep and rich concept which is a unifying thread throughout these lectures where it comes up in various disguises, in the form of BRST quantization in Berkovits' lectures, in that of functional quantization used to quantize the strings that serve as one of the motivations for M. Esole's study of elliptic fibrations, as a deformation quantization in S. Gutt's lectures, in the form of unitary representations of the Weyl algebra of an infinite-dimensional symplectic vector space discussed in A. Ashtekar's lectures.
- **Non-commutativity and deformation** inherent in quantization procedures that typically bring—possibly deformed—non-commuting operators into the scene, is reflected in the canonical commutation relations obeyed by the annihilation and creation operators in A. Astekar's lectures and lies behind the operator product expansions in conformal field theory used in N. Berkovits' presentation. In the framework of quantization by deformation discussed in S. Gutt's lectures, Poisson brackets are substituted by $\hbar$-deformed operator brackets, $\hbar$ being the Planck constant. Similarly, in Ch. Kassel's lectures, the coordinate algebra $\mathbb{C}[X, Y]$ of the complex plane is deformed to the q-deformed "coordinate algebra" $\mathbb{C}_q[X, Y]$ of a hypothetical quantum space and symmetry groups such as $SL(2)$ are deformed to quantum groups $SL_q(2)$. Such quantum groups relate to Nichols algebras central to N. Andruskiewitsch's lectures and that appear as the invariant part of Woronowicz's non-commutative differential calculus.

In view of their importance in this volume, the concept of "quantization" and the related concept of non-commutativity deserve further explanations.

The word *quantization* is commonly used to describe a procedure to link the "classical" description of a dynamical system with its "quantum" description. In some cases, such a quantization can be reached exploiting geometric features of the system, but approaches involving rather algebraic or analytical tools are also used when the "quantization rules" can be read of the classical description of the system in algebraic or analytic terms. There is by far no unified approach to quantization, even when only very simple dynamical systems are considered, and in general it is not clear either that such procedure may exist. In any case, the quest for a bridge between the mathematical structures used to describe classical dynamical systems and those used to come up with a quantum description of them gave rise to many deep and interesting ideas in mathematical physics and, in particular, to new mathematical theories.

From the point of view of mathematics, classical dynamics can be achieved using tools borrowed from differential equations, classical analysis, and differential geometry and whenever symmetries are involved, group theory comes into play in more or less sophisticated ways (from special functions and representation theory to the geometry of Lie groups and fiber bundles). Quantum descriptions of dynamics involve functional analysis in an essential way, but they also use non-commutative algebras and shed light on the role of topology for systems sensitive to such type of constraints. In addition, in recent times, new mathematical tools arise from theories inspired by the principles and rules of quantum physics, and by the heuristics of what one expects of a mathematical quantization of the classical structures. Among many others, theories motivated by quantization are non-commutative geometry, quantum groups, and algebraic deformation theory discussed in this volume.

1.2 Poisson Geometry and Classical Dynamics

The basic objects in the commonly used geometric approaches to classical dynamics are smooth manifolds equipped with 2-tensors in terms of which a Lie algebra structure (compatible with differentiation) can be given to the space of functions on the manifold. Alternatively, the starting point can be the operation

$$\{\cdot,\cdot\} : C^\infty(M) \times C^\infty(M) \to C^\infty(M)$$

providing the space of smooth functions (here, we consider real-valued functions on the manifold M, although complex-valued functions can also be considered as observables, see, e.g., Simone Gutt's lectures) with a Lie algebra structure. In other words, the bracket $\{\cdot,\cdot\}$ enjoys the following properties

1. Linearity,
$$\{\alpha f + \beta g, h\} = \alpha\{f, h\} + \beta\{g, h\},$$
2. Antisymmetry,
$$\{f, g\} = -\{g, f\},$$
3. Jacobi identity,
$$\{f, \{g, h\}\} + \{g, \{h, f\}\} + \{h, \{f, g\}\} = 0$$

for all f, g and $h \in C^\infty(M)$, and any scalars α and β, to which we add the compatibility with the usual product of functions, i.e.

4. Leibniz rule,
$$\{f, gh\} = g\{f, h\} + \{f, g\}h.$$

These four identities define the a *Poisson bracket* on $C^\infty(M)$, and we call M equipped with such a bracket a *Poisson manifold*.

Symplectic manifolds, which are Poisson manifolds for which the Poisson tensor is non-degenerate, are the most popular ground used to model dynamical systems. A symplectic manifold is a pair (M, ω), where ω is a closed and non-degenerate differential 2-form on M (in the context of Poisson geometry, the dual of the Poisson 2-tensor). For example, cotangent bundles are symplectic manifolds particularly well adapted to model phase spaces: If Q is a smooth manifold with local coordinates $(q_1, \ldots, q_n)$, its cotangent bundle T^*Q is a $2n$-dimensional symplectic manifold with local coordinates $(q_1, \ldots, q_n, p_1, \ldots, p_n)$ whose first n coordinates define the position in the configuration space Q and last n coordinates correspond to their associated generalized momenta. The symplectic structure in this case is canonical, since the cotangent bundle projection onto the configuration space $T^*Q \xrightarrow{\pi} Q$ defines a 1-form θ in terms of which the symplectic form can be written $\omega = d\theta$. This 1-form is called the *symplectic potential*, and in local coordinates it has the form

$$\theta = \sum_{i=1}^{n} p_i dq_i.$$

On a general symplectic manifold (M, ω), given that ω is closed, by the Poincaré lemma such a symplectic potential exists locally (and it is not unique in general). However, Darboux's theorem shows that every symplectic manifold locally has the structure of a cotangent bundle, so that any two symplectic manifolds with the same dimension are locally diffeomorphic since, locally, every symplectic 2-form looks like

$$\omega = \sum_{i=1}^{n} dp_i \wedge dq_i.$$

To illustrate how the symplectic structure can be used to model the classical dynamics of a physical system, let us consider a system whose phase space is the symplectic manifold $(M = T^*Q, \omega)$. A physical observable is, by definition, any real-valued smooth function $f \in C^\infty(M)$; examples are usual physical quantities—energy, momentum, etc. Since the symplectic 2-form ω is non-degenerate, there is a natural linear isomorphism

$$i_. : T_.M \to T_.^*M$$

given by contraction $i(X) = i_X\omega = \omega(X, \cdot)$. This isomorphism can be used to identify tangent and cotangent vectors and in particular, to associate to each smooth function $f \in C^\infty(M)$ a vector field X_f on M by the relation

$$i_{X_f}\omega = -df.$$

Such a vector field X_f is called the *Hamiltonian vector field* associated with f, in terms of which the Poisson bracket operation is given by

$$\{f, g\} = \omega(X_f, X_g).$$

Since the exterior derivative of f can locally be written as

$$df = \sum_{i=1}^{n} \frac{\partial f}{\partial q_i} dq_i + \frac{\partial f}{\partial p_i} dp_i,$$

the Hamiltonian vector field defined by this function is the one given in local coordinates by

$$X_f = \sum_{i=1}^{n} \frac{\partial f}{\partial p_i}\frac{\partial}{\partial q_i} - \frac{\partial f}{\partial q_i}\frac{\partial}{\partial p_i}. \tag{1.1}$$

In local coordinates, the Poisson bracket of two functions $f, g \in C^\infty(M)$ is the smooth function defined by

$$\{f, g\} = \sum_{i=1}^{n} \frac{\partial f}{\partial q_i}\frac{\partial g}{\partial p_i} - \frac{\partial g}{\partial q_i}\frac{\partial f}{\partial p_i}, \tag{1.2}$$

where $2n$ is the dimension of M, which is the usual expression for such an operation used in physics.

Poisson brackets are useful to describe the dynamics of physical systems because, given a Hamiltonian function H for the system, the evolution of classical observables is given by their bracket with the corresponding Hamiltonian [2], i.e.

$$\frac{df}{dt} = \{f, H\}, \tag{1.3}$$

for any smooth function $f \in C^\infty(M)$. Notice that, if $\gamma(t) = (p_i(t), q_i(t))$ is an integral curve of the Hamiltonian vector field (1.1) associated with a function H, the time evolution of the canonical variables on the symplectic manifold is given by

$$-\frac{\partial H}{\partial q_i} = \dot{p}_i = \{p_i, H\} \quad , \quad \frac{\partial H}{\partial p_i} = \dot{q}_i = \{q_i, H\},$$

which are precisely the Hamilton equations in the case in which H is a Hamiltonian for the system. Thus, once a Hamiltonian function is given, dynamics follows directly from the Poisson bracket defined by the 2-form ω in (1.2).

Remark. All the facts illustrated here in the context of symplectic manifolds hold in the more general context of Poisson manifolds, where the expressions before in terms of the 2-form ω must be replaced by their counterparts in terms of the Poisson tensor (see, e.g., Simone Gutt's lectures).

Some years after the birth of quantum mechanics, Paul Dirac realized that the Hamiltonian description of the dynamics, and in particular, the algebraic structure defined by the Poisson bracket $\{\cdot, \cdot\}$ on the algebra of classical observables, is crucial to understand the relationship between classical and quantum dynamics. One of the main features of the quantum description of a physical system is the use of self-adjoint operators acting on Hilbert spaces as quantum observables, highlighting the non-commutative nature of this algebra of observables. Since, with respect to the Poisson bracket operation, classical observables as position and momenta already satisfy commutation relations of the form $\{q_i, p_j\} = \delta_{ij}$, Dirac noticed that to a certain extent, the non-commutativity of quantum observables was already present in the classical setting and, as a consequence, the quantization process should be understood as a morphism between similar algebraic structures in very different contexts. On the one hand the differential-geometric approach of the dynamics in terms of smooth functions on a manifold (as classical observables) a Hamiltonian and a Poisson bracket and, on the other hand, the functional-analytic approach in terms of self-adjoint operators (as quantum observables) acting on a Hilbert space (of "wave functions") with the usual commutator of operators as natural bracket [3]. In this sense, one can think of Poisson manifolds as "maximal noncommutative spaces" between the world of classical physics (commutative algebras of smooth functions on smooth manifolds) and the quantum world of non-commutative algebras, the triple $(C^\infty(M), \{\cdot, \cdot\}, H)$—the algebra of observables plus a distinguished object in terms of which the evolution can be given, see (1.3), often called *dynamical system*— being the starting point of any quantization model.

1.3 Geometric and Deformation Quantization

Quantizing a dynamical system $(C^\infty(M), \{\cdot, \cdot\}, H)$ corresponds to a rule which assigns to the system a representation $f \mapsto \widehat{f}$ of the algebra of classical observables in the algebra of self-adjoint operators $\mathscr{A}$ acting on certain Hilbert space $\mathscr{H}$. How to build the Hilbert space and the representation itself can vary according to the physical system or the mathematical purpose, and in some cases a "complete" quantization cannot be achieved. From the point of view of mathematics, there are two methods which have been successfully studied and given rise to very stimulating ideas beyond their relationship with physics, *geometric quantization* and *deformation quantization*. Both methods strive to fulfill the Dirac quantization conditions [3]:

1. The application $f \mapsto \widehat{f}$ must be linear
2. If f is constant then $\widehat{f}$ must be the multiplication (by the constant f) operator
3. If, for three classical observables, $\{f, g\} = h$ then

$$[\widehat{f}, \widehat{g}] = -i\hbar\widehat{h} \tag{1.4}$$

must be verified by their quantum counterparts.

1.3.1 Geometric Quantization

The goal of geometric quantization is to build both a Hilbert space and a representation of observables from the geometry and the topology of the dynamical system one started from. If ones starts from a symplectic manifold (M, ω), which models the classical phase space for a dynamical system, to quantize *geometrically* such system means finding a map

$$C^{\infty}(M) \times \Gamma(\mathscr{L}) \to \Gamma(\mathscr{L})$$
$$(f, \psi) \mapsto \hat{f}\psi,$$

where $\Gamma(\mathscr{L})$ denotes the space of sections of a Hermitian line bundle $\mathscr{L} \to M$, modeling "wave functions", satisfying the Dirac quantization conditions. The idea goes back to Kostant and Souriau [4, 5], for whom the "prequantization bundle" $\mathscr{L}$ is a complex line bundle over M, endowed with a connection ∇ with curvature prescribed by the symplectic form, namely equal to $\hbar^{-1}\omega$. Such a bundle exists if and only if the class of $(2\pi\hbar)^{-1}\omega$ in $H^2(M, \mathbb{R})$ is in the image of $H^2(M, \mathbb{Z})$ under the inclusion (see, e.g., [6]) and, if this integrality condition is verified, the Hilbert space of prequantization $\mathscr{H}(M, \mathscr{L})$ is the completion of the space of square integrable sections $s : M \to \mathscr{L}$, denoted by $\Gamma(\mathscr{L})$, with inner product

$$\left(s, s'\right) = \int_M \left\langle s, s'\right\rangle \varepsilon$$

where $\varepsilon = \frac{1}{2\pi\hbar} dp_1 \wedge \cdots \wedge dp_n \wedge dq_1 \wedge \cdots \wedge dq_n$ is the volume element defined by the symplectic form on the manifold M.

Beyond the obvious theoretical importance of this construction, a very relevant feature of this approach is the integrality condition on the symplectic form it involves, namely the topological restriction $[(2\pi\hbar)^{-1}] \in H^2(M, \mathbb{Z})$, which can be used to explain the quantization of certain numbers associated with elementary physical systems (the so-called quantum numbers, see, e.g., [6]). Regarding the representation of observables, in this setting, to each smooth function $f \in C^{\infty}(M)$, we associate an Hermitian operator according to the Konstant–Souriau representation

$$\hat{f} = f - i\hbar\nabla_{X_f},$$

where X_f denotes the Hamiltonian vector field generated by f. Both the Hilbert space and the representation of observables are determined by the symplectic form; from this point of view, in this quantization, the quantum dynamics is completely determined by the classical dynamics of the system, via a topological condition.

In order to illustrate how this construction works, let us compute a simple example, namely the operators corresponding to position q_i and momentum p_i in the phase space $M = T^*\mathbb{R}^n$ with canonical symplectic form. In this case, the correspondent line bundle associated is $M \times \mathbb{C}$ and the representation corresponding to the observables gives $\hat{f} = f - i\hbar X_f - (p_i dq_i)(X_f)$ so that, since $X_{p_i} = \frac{\partial}{\partial q_i}$ and $X_{q_i} = -\frac{\partial}{\partial p_i}$,

$$\hat{p}_i = p_i - i\hbar\left(\frac{\partial}{\partial q_i}\right) - p_i = -i\hbar\frac{\partial}{\partial q_i}$$

and

$$\hat{q}_i = q_i - i\hbar\left(-\frac{\partial}{\partial p_i}\right) = q_i + i\hbar\frac{\partial}{\partial p_i}.$$

This result disagrees with the usual rules of quantum mechanics (Scrhödinger's version) that read $\hat{p}_i = -i\hbar\frac{\partial}{\partial q_i}$ and $\hat{q}_i = q_i$, and this is why we use the name "prequantization" at this stage for this procedure, which should be promoted to a quantization by means of a polarization (see [6] for details). Both, the geometric prequantization and the polarization procedures, can be carried out on Poisson manifolds, see [7].

1.3.2 Deformation Quantization

The appearance of the Planck constant $\hbar$ in the course of the last few paragraphs is completely incidental, and more related with the wish of recovering the usual commutation rules of quantum mechanics as an output of the quantization process. In contrast, for the deformation theory of quantization, it is the main parameter (actually it is, in this context, everything but a constant), the one in terms of which the algebra of quantum observables will be built as a deformation of the algebra of classic observables $\mathscr{A}_o = (C^\infty(M), \{,\})$.

The algebra $\mathscr{A}_o = C^\infty(M)$ of classical observables is replaced by $\mathscr{A}_\hbar = C^\infty(M)[\![\hbar]\!]$, the algebra of formal power series in $\hbar$ of elements in $\mathscr{A}_o$, whose elements have the form $\mathsf{f} = \sum_{k=0}^\infty \hbar^k f_k$. Viewing this formal power series as analogues of symbols in the theory of pseudo-differential operators gives an idea of the composition formula of the corresponding elements in $\mathscr{A}_\hbar$ (Weyl's quantization). A formal deformation quantization of a Poisson manifold M is a couple $(\mathscr{A}_\hbar = C^\infty(M)[\![\hbar]\!], *)$, where

$$* : \mathscr{A}_\hbar \otimes \mathscr{A}_\hbar \to \mathscr{A}_\hbar$$

denotes a star product defined on the algebra of formal power series of elements in $\mathscr{A}_o$ such that, for any $f, g \in \mathscr{A}_o$,

$$f * g = \sum_{l \geq 0} \hbar^k C_k(f, g),$$

where the C_k are defined by bidifferential operators (and define Hochschild 2-cochains on $\mathscr{A}_o$) satisfying $(f * g) * h = f * (g * h)$. Thus, for any $\mathsf{f} = \sum_{k=0}^\infty \hbar^k f_k$ and $\mathsf{g} = \sum_{k=0}^\infty \hbar^k g_k$ in $\mathscr{A}_\hbar$, with $f_k, g_k \in \mathscr{A}_o$ for all k,

$$\mathsf{f} * \mathsf{g} = \sum_{l+i+j \geq 0} \hbar^l C_l(f_i, g_j), \tag{1.5}$$

where it is assumed that the first two cochains satisfy $C_0(f, g) = fg$ (the usual commutative product of smooth functions) and $C_1(f, g) - C_1(g, f) = \{f, g\}$, so that

$$f * g = fg + O(\hbar)$$

and

$$f * g - g * f = -i\hbar\{f, g\} + O(\hbar^2),$$

a week version of (1.4).

This theory was initiated in the 1970s by F. Bayen, M. Flato, C. Fronsdal, A. Lichnerowicz, and D. Sternheimer [8], in the context of symplectic manifolds, and revisited by B. Fedosov in the 1980s [9] in the same context but in a much richer geometric approach. Since then many aspects of the theory have been studied (e.g., classification issues, generalizations to Poisson manifolds and more general differential-geometric/algebraic structures, index theory, etc.) giving rise to very important advances in different areas of mathematical physics. Chapter 2 offers a complete exposition of these and other aspects of deformation quantization by Simone Gutt, a leading expert in the subject who contributed with the theory from an early stage, reaching important developments of the theory such as Kontsevich's formality theorem, the concept of reduction in the formal deformation setting and convergence issues in the deformation quantization programme. Professor Gutt's lectures on deformation quantization, aimed at graduate students in physics or mathematics, are self-contained and contain a very complete list of references to the abundant literature on the subject; we refer the reader to that chapter for more on this interesting point of view on quantization.

1.4 Non-commutative Geometry and Quantum Groups

As mentioned before, the starting point for a description of the dynamics of a classical system is a triple $(\mathscr{A}_o, \{\cdot, \cdot\}, H)$, where $\mathscr{A}_o = C^\infty(M)$ denotes the algebra of classical observables and $\{\cdot, \cdot\}$ the Poisson bracket of smooth functions. Instead of a "constructive" quantization of such dynamical system by a deformation as indicated before, or an explicit construction of the quantum algebra of observables from geometric data, there are methods involving mathematical objects supposed to represent the quantum counterparts of classical dynamical systems without explicit mention to some particular quantization process. From these points of view, the non-commutative algebras involved in the description of the quantum dynamics of a system must, in some appropriate limit, give back the classic algebraic setting of classical dynamics, but they must not necessarily be built from them. Among these theories, we want to mention Alain Connes' *non-commutative differential geometry* and the theory of *quantum groups*.

The basic object in non-commutative geometry is a *spectral triple* $(\mathscr{A}, \mathscr{H}, D)$ (also called unbounded Fredholm module), and it involves an involutive algebra $\mathscr{A}$ represented in a Hilbert space $\mathscr{H}$, together with a self-adjoint operator D with compact resolvent in $\mathscr{H}$ such that $[D, a]$ is bounded for any $a \in \mathscr{A}$ [10]. This triple is the non-commutative generalization of the natural triple $(C^\infty(M), L^2(S, M), D)$ of classical differential spin geometry, where the algebra is the one of smooth functions on a (spin) manifold M, which is commutative with respect to the usual product of functions, the Hilbert space is the one of L^2-spinors (sections of the spin bundle $S \to M$ over M) and D is the classical Dirac operator (the square root of the Laplacian). Thus, the algebra $\mathscr{A}$ models the algebra of functions of a "noncommutative space" which we only see through the spectral properties of the operator D the same way as, for example, the Riemannian metric on M is encoded in the operator spectrum of D in classical global analysis. Conditions can be imposed on a spectral triple to generalize many important features of the usual spectral theory of pseudo-differential operators on manifolds to these non-commutative spaces, obtaining in addition to the usual notions of differential geometry (distances, scalar curvature, etc.) more involved constructions as index theory (see B. Iochum's lectures in [11], and references therein).

In physics, non-commutative spectral triples have been used to describe elementary particle models over non-commutative space-times, conformal field theories, and dualities among many other uses (see M. Marcolli's lectures in [11]). Many interesting examples of non-commutative spaces in mathematics come from the theory of *quantum groups*, objects which are deformations of (algebras of functions on) groups, but still have a very similar representation theory. The notion of quantum group comes from the one of Hopf algebras, which are algebraic structures often used to describe deformations of the function algebras on semisimple Lie groups or enveloping algebras of semisimple Lie algebras (see Christian Kassel's lectures in this volume). These deformations are commonly parametrized by a parameter q which, for some authors, is related to $\hbar$ as $q = \exp(c\hbar)$ for some appropriate scalar c and is used to exhibit explicit deformations of their classical counterparts. For example, the algebra of the quantum group $SU_q(2)$ is the polynomial algebra generated by four elements a, b, c and d satisfying the following relations, for a parameter $0 < q < 1$,

$$
\begin{aligned}
ba &= qab, \quad ca = qac, \\
db &= qbd, \quad dc = qcd, \\
bc &= cb; \quad ad - q^{-1}bc = da - qbc = 1,
\end{aligned}
$$

so that the case $q = 1$ would correspond to the classical matrix representation $\begin{pmatrix} a & b \\ c & d \end{pmatrix}$ of elements of the Lie group $SU(2)$ in terms of the (commuting) coordinates a, b, c and d.

It is interesting to notice that, if we forget the group-like features of these objects, it has been possible to use the representation theory of many classes of quantum groups to define appropriate *Dirac operators* and, as a consequence, it is possible to realize them as a class of non-commutative spaces in the context of spectral triples. Although the spectral triples $(\mathscr{A}_q, \mathscr{H}_q, D)$ associated with such classes of quantum groups often use the classical Dirac operator on the corresponding classical group, they have interesting properties with potential applications both in mathematics and theoretical physics (see, e.g., [12, 13]).

In classical field theory, principal fiber bundles play a very important role to model gauge symmetries, i.e., internal symmetries of classical systems modeled by the fiber (a Lie group) of a fibration over the space-time manifold; when these classical objects are replaced by their quantum analogues, we obtain different types of objects which appear often in the following pages. First, in Fedosov's approach to deformation quantization, fibrations of non-commutative algebras over symplectic manifolds appear in a natural way (the Weyl bundle) and their geometry is used to build up start products as explained in Simone Gutt's lecture notes. Fibrations in the context of non-commutative geometry play an important role in applications in physics and come in very different flavors which can be used in different situations: classical fibrations on non-commutative spaces (i.e., classical fibers on non-commutative base manifolds), parametrized families of non-commutative spaces or fibrations with quantum fibers on non-commutative spaces. The role of quantum groups in equivariant non-commutative algebraic geometry, in particular the notion of non-commutative principal bundle, or Hopf–Galois extension, will be discussed by Christian Kassel in this volume. Many other examples of fibrations involving Hopf algebras can be studied from the spectral point of view of non-commutative differential geometry; let us just quote the case of non-commutative Hopf fibrations considered by Giovanni Landi and Walter van Suijlekom in [14] and the non-commutative homogeneous spaces studied by Joseph Várilly in [15].

1.5 Quantum Fields

Quantization of a classical field theory brings new features, such as the existence of inequivalent representations of the algebra generated by creation and annihilation operators. This is due to the fact that (by definition) such a theory is a dynamical system with an infinite number of degrees of freedom.

As an example, let us consider the Klein–Gordon equation

$$(\partial_\mu \partial^\mu + m^2)\varphi(x) = 0, \tag{1.6}$$

which is the simplest one compatible with the Poincaré symmetry of Minkowski space $\mathscr{M}$. This equation can be obtained from a Lagrangian density $\mathscr{L}(\varphi, \partial_\mu \varphi)$ as a solution of the corresponding Euler–Lagrange equations, but the dynamics can also be described in terms of a symplectic structure that is naturally associated with the differential equation (1.6). In fact, let V denote the space of real smooth solutions of the Klein–Gordon equation, in a suitable topology. Then, given a choice of a space-like hypersurface Σ, we can define a symplectic form on V,

$$\sigma(\varphi_1, \varphi_2) := \int_\Sigma (\varphi_1 \nabla_\alpha \varphi_2 - \varphi_2 \nabla_\alpha \varphi_1)\, n^\alpha d\mathrm{vol}_\Sigma,$$

which is independent of the choice of Σ. Let $\mathbb{E}^{\pm} : C_0^\infty(\mathcal{M}) \to C^\infty(\mathcal{M})$ denote the retarded/advanced fundamental solutions of (1.6). Then, for any $f \in C_0^\infty(\mathcal{M})$, it is easy to see that $\mathbb{E}f$ is a solution to the field Eq. (1.6), where $\mathbb{E} = \mathbb{E}^- - \mathbb{E}^+$. From this, we obtain an isomorphism $V \cong C_0^\infty(\mathcal{M})/\ker(\mathbb{E})$. Under this isomorphism, the symplectic form can be written as follows:

$$\sigma([f],[g]) = \int_{\mathcal{M}} f(x)(\mathbb{E}g)(x)d^4x.$$

The quantized field corresponding to this dynamical system can be described in terms of a unital $*$-algebra generated by symbols $\Phi(f)$ (with f in the complexification of $C_0^\infty(\mathcal{M})$), that are subject to the following relations:

$$\Phi(\bar{f}) = \Phi(f)^*, \tag{1.7}$$

$$\Phi((\partial_\mu\partial^\mu + m^2)f) = 0, \tag{1.8}$$

$$[\Phi(f), \Phi(g)] = i\sigma([f],[g]). \tag{1.9}$$

Physically, the generators $\Phi(f)$ can be regarded as "smeared" field operators. In terms of the more familiar operator-valued distribution $\hat{\varphi}(x)$ ("the quantum field") we have, at least formally,

$$\Phi(f) = \int \hat{\varphi}(x)f(x)d^4x.$$

Thus, Eq. (1.8) expresses the idea that the quantized field is still a solution of the field equation, whereas (1.9), when written in terms of the quantum field $\hat{\varphi}(x)$, takes the more familiar form of the canonical commutation relations (CCR):

$$[\hat{\varphi}(x), \hat{\varphi}(y)] = i\Delta(x,y). \tag{1.10}$$

Here, $\Delta(x,y)$ denotes the *Pauli–Jordan function*, a distributional solution of (1.6) with causal support [16]. The relation between (1.9) and (1.10) is due to the fact that $\Delta(x,y)$ is also the kernel of the (integral) operator $\mathbb{E}$.

An alternative point of view consists in starting with the symplectic vector space (V,σ) and constructing the corresponding Weyl algebra. The commutation relations obeyed by the generators of the Weyl algebra can be understood as the exponentiated form of the CCR (1.10)

One of the main differences between (standard) quantum mechanics and quantum field theory comes from the Stone–von Neumann theorem, which asserts that, up to unitary equivalence, there is only one irreducible representation of the CCR,

$$[\hat{q}_i, \hat{p}_j] = i\hbar\delta_{i,j}. \tag{1.11}$$

In this case, the symplectic space is $T^*\mathbb{R}^n$, a finite dimensional symplectic vector space. These two assumptions (that the symplectic manifold is a vector space and of finite dimensionality) are essential for the proof of the Stone–von Neumann theorem. Its failure in the case of finite dimensional symplectic manifolds leads to the richness of interplay between topology and symplectic geometry, as discussed previously. In the case of a (free, scalar) quantum field, we are still working with a symplectic vector space, but now of infinite dimensionality.

For the example of the scalar field discussed here, the Hilbert space where the CCR are represented is a *bosonic* Fock space. It can be described in terms of the symmetric tensor algebra of V.

On the other hand, quantization of fermionic fields (such as the one described by the Dirac equation) differs from its bosonic counterpart for commutation relations have to be substituted by *anticommutation* relations due to the spin-statistics connection. The Fock space is then accordingly related to the *exterior algebra* of the space of solutions of the classical equation [17]. As mentioned in Sect. 1.1, quantization of a *gauge theory* entails new difficulties, since the Lagrangian describing such a theory is *singular*, meaning that there are constraints that have to be dealt with in a proper way. Examples of such theories and their quantization are the subject of Chap. 6 (in the context of string theory) and Chaps. 8 and 9 (dealing with different aspects of the Batalin–Vilkovisky formalism).

From the point of view of both mathematics and physics, the appearance of *renormalization* is perhaps one of the most intriguing, as well as interesting, aspects of quantum field theory. Although not discussed in this volume, it is convenient to observe that, at the core of renormalization calculations arising in perturbative quantum field theory, there is a Hopf algebra structure, known as the Connes–Kreimer Hopf algebra [17, 18], which provides an algebraic interpretation of the mechanisms underlying the "forest formula" used by physicists. Another point of view, stemming from the algebraic approach to quantum field theory, uses ideas from deformation quantization to study perturbative renormalization [19]. These two examples provide further illustrations as to how deeply interconnected are the topics discussed in this volume.

References

1. V.N. Gribov, Quantization of non-abelian gauge theories. Nuclear Phys. B **139**(1), 1–19 (1978)
2. V.I. Arnold, *Mathematical Methods of Classical Mechanics* (Springer, 1989)
3. P.A.M. Dirac, *The Principles of Quantum Mechanics* (Clarendon Press, 1982)
4. B. Kostant, Quantization and unitary representations, in *Lectures in Modern Analysis and Applications III*, vol. 170, Lecture Notes in Mathematics (Springer, 1970), pp. 87–208
5. J.-M. Souriau, *Structure des Systèmes Dynamiques* (Dunod, Paris, 1970)
6. N.M.J. Woodhouse, *Geometric Quantization* (Clarendon Press, 1991)
7. I. Vaisman, *Lectures on the Geometry of Poisson Manifolds* (Birkhauser, 1991)
8. F. Bayen, M. Flato, C. Fronsdal, A. Lichnerowicz, D. Sternheimer, *Deformation Theory and Quantization. I. Deformations of Symplectic Structures* (Annals of Physics, 111, 1978)

9. B. Fedosov, A simple geometrical construction of deformation quantization. J. Differ. Geom. **40**(2), 213–238 (1994)
10. A. Connes, *Noncommutative Geometry* (Academic press, 1995)
11. A. Cardona, C. Neira-Jiménez, H. Ocampo, S. Paycha, A.F. Reyes-Lega, (eds.) *Geometric, Algebraic and Topological Methods for Quantum Field Theory* (World Scientific, 2013)
12. A. Connes, Cyclic cohomology, quantum group symmetries and the local index formula for $SU_q(2)$. J. Inst. Math. Jussieu **3**, 203–225 (2004)
13. L. Dabrowski, G. Landi, A. Sitarz, W. van Suijlekom, J.C. Varilly, The Dirac operator on $SU_q(2)$. Commun. Math. Phys. **259**, 729–759 (2005)
14. G. Landi, W. van Suijlekom, Principal fibrations from noncommutative spheres. Commun. Math. Phys. **260**, 729–759 (2005)
15. A. Cardona, H. Ocampo, S. Paycha (eds.), Geometric and topological methods for quantum field theory, in *Proceedings of the Summer School held in Villa de Leyva, July 9–27, 2001* (World Scientific Publishing Co., Inc., River Edge, 2003)
16. F. Scheck, *Quantum Physics* (Springer, 2007)
17. J.M. Gracia-Bondía, J.C. Vărilly, H. Figueroa, *Elements of Noncommutative Geometry* (Springer Science & Business Media, 2013)
18. D. Manchon. Hopf algebras, from basics to applications to renormalization (2004), arXiv:math/0408405
19. R. Brunetti, C. Dappiaggi, K. Fredenhagen, J. Yngvason, *Advances in Algebraic Quantum Field Theory* (Springer, 2015)

Chapter 2
Deformation Quantization and Group Actions

Simone Gutt

Abstract This set of notes corresponds to a mini-course given in Villa de Leyva in July 2015. It does not contain any new result and is meant to be an elementary first introduction to formal Deformation Quantization, hoping it will be an incentive to learn more advanced topics in the subject. Quantization of a classical system is a way to pass from classical to quantum results. There exist several mathematical attempts to formulate possible quantization methods. Formal deformation quantization was introduced in the seventies by Flato et al. and understands quantization as a deformation (called a star product) of the structure of the algebra of classical observables. After an introduction to the concept of quantization in Sect. 2.1, we introduce formal deformation quantization in Sect. 2.2, the description of Fedosov's construction of a star product on a symplectic manifold in Sect. 2.3, an introduction to classifications of star products in Sect. 2.4 and a brief introduction to the notion of formality and its link with star products on a Poisson manifold in Sect. 2.5. Various notions of group actions in the context of deformation quantization are given in Sect. 2.6, along with the study of the invariance of a Fedosov's star product, and classifications of invariant star products on a manifold endowed with an invariant connection. We present in Sect. 2.7 the concept of reduction in the formal deformation quantization setting, and show how quantization commutes with reduction, considering here only the simplest form of reduction and following a simplified version of Bordemann–Waldmann's approach. We conclude by briefly mentioning in Sect. 2.8 convergence issues in the deformation quantization programme.

S. Gutt—Also on leave from Université de Lorraine, Ile du Saulcy, 57045 Metz Cedex 01, France.

S. Gutt (✉)
Université Libre de Bruxelles, Campus Plaine, CP 218, bd du Triomphe,
1050 Brussels, Belgium
e-mail: sgutt@ulb.ac.be

A. Cardona et al. (eds.), *Quantization, Geometry and Noncommutative Structures in Mathematics and Physics*, Mathematical Physics Studies,
DOI 10.1007/978-3-319-65427-0_2

Foreword A rather large, though not complete, bibliography is included for those who want to go beyond this introduction. Some references are directly linked to this introduction:

- the book [167] by S. Waldmann is an introduction to deformation quantization (in German);
- concerning Sect. 2.3, the book [82] by B.V. Fedosov presents his construction of star products along with many of their properties, and introduces index theorems for deformation quantization on symplectic manifolds;
- extending Sect. 2.4, the expository paper [104], joint with J. Rawnsley, gives an introduction to Deligne's Čech cohomology classes, associated to star products on a symplectic manifold;
- to develop Sect. 2.5, the expository paper [59] by A. Cattaneo and D. Indelicato introduces formality and star products and the paper [64] by A. Cattaneo, G. Felder and L. Tomassini gives the globalization of a star product on a Poisson manifold; see also the original paper by Kontsevich [122];
- Section 2.7 is taken from [106]; the reduction presented is a special case of a reduction procedure introduced by M. Bordemann, C. Herbig and S. Waldmann in [40].

There are many important aspects of deformation quantization which are not addressed in these notes; some of them are mentioned with corresponding references.

Possible connexions with other classes given at the school appear in the text. There is in Sect. 2.1 a mention of the lectures of Abhay Ashtekar when quantum field theory is alluded to, and the lectures of Nathan Berkovits are referred to concerning superstring theory. In Sect. 2.6, one mentions the lectures of Christian Kassel when one speaks about quantum groups, and in Sect. 2.7 there is again a link to the lectures of Nathan Berkovits concerning BRST formalism.

2.1 What Do We Mean by Quantization?

Quantum theory provides a description of nature which is more fundamental than classical theory. It is necessary to describe atomic or subatomic physics (and it is also needed to describe some macroscopical phenomena such as superconductors and superfluids). It incorporates phenomena which can not be accounted for by classical physics like the quantization of certain physical properties, the uncertainty principle, etc.

We shall consider here quantum mechanics which provides a non relativistic description (i.e. the speed is far less than 3×10^8 m/s) of a finite number of particles with a finite number of degrees of freedom.

Remark 2.1 To go beyond this, quantum field theory provides a description which incorporates higher velocities, for instance to describe a system including photons, or a system with a varying number of particles; it merges quantum principles and special relativity. In this realm quantum electrodynamics provides a description of electromagnetic interactions, quantum chromodynamics of strong nuclear forces, electroweak theory of electromagnetic force and weak nuclear forces, and the standard model of particles unifies the three type of interactions. However, it has been proven difficult to build quantum theories of gravity (the remaining fundamental force); string theory is a candidate for such a theory. Quantum field theory on curved space time is the object of the lectures given by Abhay Ashtekar and an introduction to superstring theory is given in the lectures of Nathan Berkovits.

By quantization of a classical system, we mean a way to pass from classical to quantum results. One could wonder why we are interested in quantization, since it could appear to be an artificial problem, nature being quantum. A first motivation lies in the difficulty of directly providing a quantum description of a physical system, and the classical description is often easier to obtain; hence one often uses the classical description as a starting point to find a quantum description. Furthermore, a given physical theory remaining valid within a range of measurements, any modified theory should give the same results in the initial range. The description of a system by classical mechanics is adequate in the macroscopic non relativistic world, for size much larger than 10^{-9} m and speed far less than 3×10^8 m/s.

Guidelines as how to pass from a classical description to a quantum one are based on the precepts that there exists a classical limit, and that to any classical observable there corresponds a quantum one.

2.1.1 Classical Mechanics

Classical mechanics, in its Hamiltonian formulation on the motion space, can be described in the framework of symplectic manifolds (or more generally Poisson manifolds). The motion space is in general the quotient of the evolution space by the motion; it can often be identified with a space of possible initial values for positions and momenta. Observables are families $f_\bullet = \{f_t \mid t \in \mathbb{R}\}$ of smooth functions on that manifold M. The dynamics is defined in terms of a Hamiltonian $H \in C^\infty(M)$ and the time evolution of an observable $f_\bullet$ is governed by the equation:

$$\frac{d}{dt} f_t = -\{H, f_t\}.$$

For instance a particle of mass m moving in $\mathbb{R}^3$ subject to a force F which is the gradient of a potential $F = -\nabla V$, has a position determined by the 3 coordinates q^1, q^2, q^3 whose evolution in time is governed by Newton's equations

$$m\frac{d^2q^i}{dt^2} = -\frac{\partial V}{\partial q^i}.$$

Introducing the momenta $p_i = m\frac{dq^i}{dt}$, and the Hamiltonian $H' := \frac{p^2}{2m} + V$ with $p^2 := \sum_{j=1}^3 p_j^2$, the motion of the system in the evolution space $\mathscr{E} := \mathbb{R} \times T^*\mathbb{R}^3$ with coordinates $(t, q^1, q^2, q^3, p_1, p_2, p_3)$ is given by the flow of the vector field

$$\frac{\partial}{\partial t} + \sum_{i=1}^{3} \left(\frac{\partial H'}{\partial p_i}\frac{\partial}{\partial q^i} - \frac{\partial H'}{\partial q^i}\frac{\partial}{\partial p_i} \right).$$

Let $\pi = \mathscr{E} \to \mathscr{E}/\!\sim =: M$ be the projection on the space of motions (two points in $\mathscr{E}$ being equivalent when they belong to the same orbit under the flow). The 2-form $\Omega := \sum_{j=1}^3 dp_j \wedge dq^j - dH' \wedge dt$ on $\mathscr{E}$ is the pullback under π of a symplectic form ω on M. The Hamiltonian H' is the pullback of a function H on M. An observable given by a time independent function f on $\mathscr{E}$ (i.e. the pullback by the projection on the second factor of a function on the phase space $T^*\mathbb{R}^3$, that is a function of the positions and momenta), is now represented by a collection of functions $\{f_t\}$ on the motion space M, the function f_t evaluated at a point $m \in M$ being the value of f at time t in the corresponding motion. Then

$$\frac{d}{dt} f_t = \sum_{i=1}^{3} \left(\frac{\partial H'}{\partial p_i}\frac{\partial f}{\partial q^i} - \frac{\partial H'}{\partial q^i}\frac{\partial f}{\partial p_i} \right)(t) = -\{H, f_t\}.$$

An Incursion in Poisson and Symplectic Manifolds

Definition 2.1 A **Poisson bracket**, defined on the space of complex valued smooth functions on a manifold M, is a $\mathbb{C}$-bilinear map $(u, v) \mapsto \{u, v\}$ on $C^\infty(M) := \mathbb{C}^\infty(M, \mathbb{C})$, such that, for any $u, v, w \in C^\infty(M)$:

1. $\overline{\{u, v\}} = \{\overline{u}, \overline{v}\}$ (reality)
2. $\{u, v\} = -\{v, u\}$ (skew-symmetry);
3. $\{\{u, v\}, w\} + \{\{v, w\}, u\} + \{\{w, u\}, v\} = 0$ (Jacobi's identity);
4. $\{u, vw\} = \{u, v\}w + \{u, w\}v$ (Leibniz rule).

Exercise 2.1 The Leibniz rule is equivalent to the bracketing with a function u being a derivation of the associative algebra of smooth functions on M.

Bracketing with a function u is therefore given by a vector field X_u on M, which is called the **Hamiltonian vector field** associated to the function u:

$$\{u, v\} = X_u v \tag{2.1}$$

By skew-symmetry, a Poisson bracket is thus given in terms of a contravariant skew-symmetric 2-tensor P on M, called the **Poisson tensor**, by

$$\{u, v\} = P(du \wedge dv) \qquad \left(\text{in local coordinates} \quad \{u, v\} = \sum_{i,j=1}^{m=\dim M} P^{ij}\frac{\partial u}{\partial y_i}\frac{\partial v}{\partial y_j} \right). \tag{2.2}$$

Exercise 2.2 The Jacobi identity for the Poisson bracket is equivalent to the vanishing of the Schouten bracket:

$$[P, P] = 0 \quad \left(\text{locally} \sum_{r=1}^{\dim M} \left(P^{ir}\frac{\partial}{\partial y_r}P^{jk} + P^{jr}\frac{\partial}{\partial y_r}P^{ki} + P^{kr}\frac{\partial}{\partial y_r}P^{ij}\right) = 0\right).$$

The Schouten bracket is the extension -as a graded derivation for the exterior product- of the bracket of vector fields to skew-symmetric contravariant tensor fields; it will be developed in Sect. 2.5.2.

A **Poisson manifold**, denoted (M, P), is a manifold M with a Poisson bracket defined by the Poisson tensor P.

A **first example** is $\mathbb{R}^{2n}$ with coordinates $\{q^i, p_i; 1 \le i \le n\}$ and the **canonical Poisson bracket**

$$\{u, v\} = \sum_{j=1}^{n}\left(\frac{\partial u}{\partial q^j}\frac{\partial v}{\partial p_j} - \frac{\partial u}{\partial p_j}\frac{\partial v}{\partial q^j}\right). \tag{2.3}$$

More generally, on $\mathbb{R}^m$ with coordinates $\{x^i; 1 \le i \le m\}$, any constant real skew-symmetric contravariant 2-tensor P defines a Poisson structure with Poisson bracket

$$\{u, v\} = \sum_{j,k=1}^{m} P^{jk}\frac{\partial u}{\partial x^j}\frac{\partial v}{\partial x^k}.$$

A particular class of Poisson manifolds, essential in classical mechanics, is the class of **symplectic manifolds**. If (M, ω) is a symplectic manifold (i.e. ω is a closed nondegenerate 2-form on M) and if $u, v \in C^\infty(M)$, the Poisson bracket of u and v is defined by

$$\{u, v\} := X_u(v) = \omega(X_v, X_u),$$

where X_u denotes the Hamiltonian vector field corresponding to the function u, given by $i(X_u)\omega = du$.

Exercise 2.3 In coordinates the components of the corresponding Poisson tensor P^{ij} form the inverse matrix of the components ω_{ij} of ω. Symplectic manifolds are exactly Poisson manifolds for which the Poisson tensor is non degenerate at each point.

Amongst the symplectic manifolds, there is the cotangent bundle $T^*N \xrightarrow{\pi} N$ to a manifold N, endowed with the symplectic form $d\Theta$ where Θ is the Liouville 1-form on T^*N:

$$\Theta_\eta(X) := \eta(\pi_* X) \qquad \text{for any } \eta \in T^*N \text{ and } X \in T_\eta T^*N.$$

This appears as the phase space of a classical system with configuration space N.

Duals of Lie algebras form the class of linear Poisson manifolds. If $\mathfrak{g}$ is a Lie algebra then its dual $\mathfrak{g}^*$ is endowed with the Poisson tensor P defined by

$$P_\xi(X, Y) := \xi([X, Y])$$

for $X, Y \in \mathfrak{g} = (\mathfrak{g}^*)^* \sim (T_\xi \mathfrak{g}^*)^*$. If $\{ X^1, \dots, X^m \}$ is a basis of $\mathfrak{g}$ and x^j the corresponding linear coordinates on $\mathfrak{g}^*$

$$x^j : \mathfrak{g}^* \to \mathbb{R} : \xi \mapsto \xi(X^j),$$

and if c_k^{ij} denote the stucture constants $[\, X^i, X^j \,] = \sum_k c_k^{ij} X^k$, this bracket writes

$$\{u, v\} = \sum_{i,j,k} c_k^{ij} x^k \frac{\partial u}{\partial x^i} \frac{\partial v}{\partial x^j}. \tag{2.4}$$

2.1.2 *Quantum Mechanics*

Quantum mechanics, in its usual Heisenberg formulation, takes place in the framework of Hilbert spaces (states are rays in such a space). Observables are families $A_\bullet = \{A_t, t \in \mathbb{R}\}$ of self-adjoint operators on the Hilbert space. The dynamics is defined in terms of a Hamiltonian H, which is a self-adjoint operator, and the time evolution of an observable $A_\bullet$ is governed by the equation:

$$\frac{dA_t}{dt} = \frac{i}{\hbar}[H, A_t]$$

where $\hbar$ is the reduced Planck constant $\hbar = \frac{h}{2\pi} \simeq 10^{-34}$ J.s.

A natural suggestion for quantization is a correspondence $\mathscr{Q}\colon f \mapsto \mathscr{Q}(f)$ mapping a function f to a self-adjoint operator $\mathscr{Q}(f)$ on a Hilbert space $\mathscr{H}$ in such a way that $\mathscr{Q}(1) = \mathrm{Id}$ and

$$[\mathscr{Q}(f), \mathscr{Q}(g)] = i\hbar \mathscr{Q}(\{f, g\}) + O(\hbar^2).$$

There is no correspondence defined on all smooth functions on M so that

$$[\mathscr{Q}(f), \mathscr{Q}(g)] = i\hbar \mathscr{Q}(\{f, g\}),$$

when one puts an irreducibility requirement which is necessary not to violate Heisenberg's uncertainty principle. More precisely, Van Hove [165] proved that there is no irreducible representation of the Heisenberg algebra, viewed as the algebra of constants and linear functions on $\mathbb{R}^{2n}$ endowed with the Poisson braket, which extends to a representation of the algebra of polynomials on $\mathbb{R}^{2n}$.

We shall now describe commonly used quantizations of $\mathbb{R}^{2n}$ endowed with its **canonical Poisson bracket** as defined in Eq. (2.3):
in coordinates $\{q^i, p_i; 1 \le i \le n\}$ $\{u, v\} = \sum_{j=1}^{n} \left(\frac{\partial u}{\partial q^j}\frac{\partial v}{\partial p_j} - \frac{\partial u}{\partial p_j}\frac{\partial v}{\partial q^j} \right)$.

1. Standard Ordering

The standard ordering yields a bijection $\mathcal{Q}_{st}$ between (complex valued) polynomials on $\mathbb{R}^{2n}$, $\mathbb{C}[p_i, q^j]$, and the space of differential operators on $\mathbb{R}^n$ with polynomial coefficients $D_{(polyn)}(\mathbb{R}^n)$. It assigns to the constant function 1, the operator $\mathcal{Q}_{st}(1) = \mathrm{Id}$, to the classical observables q^i the quantum operators of multiplication by q^i, $\mathcal{Q}_{st}(q^i) := Q^i := q^i\cdot$, and to p_i the differential operators of order 1 involving derivation with respect to q^i, $\mathcal{Q}_{st}(p_i) := P_i := -i\hbar\frac{\partial}{\partial q^i}$. One has to specify what is associated to other classical observables given by polynomials in q^i and p_j since Q^j and P_j no longer commute. For the standard ordering, one defines

$$\mathcal{Q}_{st}({q^1}^{i_1} \dots {q^n}^{i_n} p_1^{j_1} \dots p_n^{j_n}) := {Q^1}^{i_1} \dots {Q^n}^{i_n} P_1^{j_1} \dots P_n^{j_n}.$$

Equivalently, for any $f, g \in \mathbb{C}[p_i, q^j]$ and any $\phi \in C^\infty(\mathbb{R}^n, \mathbb{C})$:

$$\mathcal{Q}_{st}(f)\phi = \sum_{r=0}^{\infty} \sum_{r_1+\ldots+r_n=r} \frac{(\hbar/i)^r}{r!} \frac{\partial^r f}{\partial p_1^{r_1} \dots p_n^{r_n}}_{|p=0} \frac{\partial^r \phi}{\partial {q^1}^{r_1} \dots {q^n}^{r_n}},$$

so that the deformed product on $\mathbb{C}[p_i, q^j]$ corresponding to the composition of operators in $D_{(polyn)}(\mathbb{R}^n)$ via the bijection $\mathcal{Q}_{st}$ is given by

$$\begin{aligned} f *_{Std}\, g :&= \mathcal{Q}_{st}^{-1}(\mathcal{Q}_{st}(f) \circ \mathcal{Q}_{st}(g)) \\ &= \sum_{r=0}^{\infty} \sum_{r_1+\cdots+r_n=r} \frac{(\hbar/i)^r}{r!} \frac{\partial^r f}{\partial p_1^{r_1} \dots \partial p_n^{r_n}} \frac{\partial^r g}{\partial {q^1}^{r_1} \dots \partial {q^n}^{r_n}}. \end{aligned} \tag{2.5}$$

Its classical limit is: $f *_{Std}\, g = fg + (\hbar/i)\sum_j \frac{\partial f}{\partial p_j}\frac{\partial g}{\partial q^j} + O(\hbar^2)$.

Remark 2.2 Consider the space $C_c(\mathbb{R}^n, \mathbb{C})$ of compactly supported smooth functions endowed with the Hermitian scalar product

$$< \phi, \psi > := \int \overline{\phi(q^1, \dots, q^n)}\psi(q^1, \dots, q^n) dq^1 \dots dq^n.$$

Then $< \phi, A\psi > = < A\phi, \psi >$ for $A = P_i$ or $A = Q^j$ but the property is not true for all polynomials; for instance, it is not true for the operator associated to the real function $p_i q^i$; indeed, the adjoint of $Q^i P_i$ is $P_i Q^i = Q^i P_i - i\hbar \,\mathrm{Id}$. Hence the operator $\mathcal{Q}_{st}(p_i q^i)$ cannot be extended to a self-adjoint operator in the Hilbert completion $L^2(\mathbb{R}^n, dq)$ of $(C_c(\mathbb{R}^n, \mathbb{C}), < .,. >)$.

2. Weyl Ordering

The Weyl ordering is again a bijection $\mathcal{Q}_{Weyl}$ between the polynomials $\mathbb{C}[p_i, q^j]$ and the space of differential operators $D_{(polyn)}(\mathbb{R}^n)$. It assigns to the constant function 1, the operator $\mathcal{Q}_{Weyl}(1) = \mathrm{Id}$, to the classical observables q^i the quantum operators $\mathcal{Q}_{Weyl}(q^i) := Q^i := q^i\cdot$ of multiplication by q^i, and to p_i the differential operators of order 1 $\mathcal{Q}_{Weyl}(p_i) := P_i$ and to a polynomial in $p's$ and $q's$ the corresponding totally symmetrized polynomial in Q^i and P_j, e.g.

$$\mathcal{Q}_{Weyl}(q^1(p^1)^2) = \frac{1}{3}(Q^1(P^1)^2 + P^1Q^1P^1 + (P^1)^2Q^1).$$

Exercise 2.4 $\mathcal{Q}_{Weyl}(\exp(aq + bp)) = \exp(aQ + bP)$ and $\mathcal{Q}_{st}(\exp(aq + bp)) = \exp aQ \exp bP$ for a, b formal parameters (i.e. when one expands in powers of a and b the equality is true for any power of a and b); now $\exp(aQ + bP) = e^{\frac{\hbar ab}{2i}} \exp aQ \exp bP$, so that

$$\mathcal{Q}_{Weyl}(f) = \mathcal{Q}_{st}(\tilde{T} f)$$

for $\tilde{T} = e^{\frac{\hbar}{2i}\sum_j \frac{\partial^2}{\partial q^j \partial p_j}}$. Then the deformed product on $\mathbb{C}[p_i, q^j]$ corresponding to the composition of operators in $D_{(polyn)}(\mathbb{R}^n)$ via the bijection $\mathcal{Q}_{Weyl}$ is

$$\begin{aligned} f *_{Weyl} g : &= \mathcal{Q}_{Weyl}^{-1}\left(\mathcal{Q}_{Weyl}(f) \circ \mathcal{Q}_{Weyl}(g)\right) = \tilde{T}^{-1}\left((\tilde{T} f) *_{Std} (\tilde{T} g)\right) \quad (2.6)\\ &= \sum_{r=0}^{\infty}\sum_{s=0}^{r} \frac{r!}{s!(r-s)!}(-1)^{r-s}\frac{(\hbar/i)^r}{r!}\frac{\partial^r f}{\partial q^s \partial p^{r-s}}\frac{\partial^r g}{\partial p^s \partial q^{r-s}}\\ &= f.g + \frac{\hbar}{2i}\{f, g\} + O(\hbar^2) \end{aligned}$$

(using multi indices or working in dimension 1).

3. Wick Ordering

Set $z = q + ip$ (we present here the complex dimension 1 case; the formulas are analogous in dimension n with multiindices) and let $\mathcal{O}(\mathbb{C})$ be the set of antiholomorphic functions on $\mathbb{C}$ with hermitian scalar product defined by $< \phi, \psi > := \frac{1}{2\pi\hbar}\int \overline{\phi(\overline{z})}\psi(\overline{z})e^{\frac{-|z|^2}{2\hbar}}dzd\overline{z}$ which may diverge. Let

$$\mathcal{H} := \{\phi \in \mathcal{O}(\mathbb{C}) \mid < \phi, \phi > < \infty\}.$$

The set of polynomials in $\overline{z}$ is dense in $\mathcal{H}$. The Wick ordering assigns to the constant function 1, the operator $\mathcal{Q}_{Wick}(1) = \mathrm{Id}$, to the function z the quantum operators $\mathcal{Q}_{Wick}(z) := 2\hbar\frac{\partial}{\partial \overline{z}}$, to $\overline{z}$ the multiplication by $\overline{z}$, $\mathcal{Q}_{Wick}(\overline{z}) := \overline{z}\cdot$ and to any polynomial

$$\mathscr{Q}_{Wick}(\overline{z}^n z^m) := (2\hbar)^m \overline{z}^n \left(\frac{\partial}{\partial \overline{z}}\right)^m \quad \text{i.e.}$$

$$\mathscr{Q}_{Wick}(f)\phi = \sum_{r=0}^{\infty} \frac{(2\hbar)^r}{r!} \frac{\partial^r f}{\partial z^r}_{|z=0} \frac{\partial^r \phi}{\partial \overline{z}^r} \quad \forall f \in \mathbb{C}[p,q], \phi \in \mathbb{C}[\overline{z}].$$

Exercise 2.5 The deformed product corresponding to the composition of operators is given by

$$\begin{aligned} f *_{Wick} g &:= \mathscr{Q}^{-1}_{Wick}\left(\mathscr{Q}_{Wick}(f) \circ \mathscr{Q}_{Wick}(g)\right) \\ &= \sum_{r=0}^{\infty} \frac{(2\hbar)^r}{r!} \frac{\partial^r f}{\partial z^r} \frac{\partial^r g}{\partial \overline{z}^r} = fg + 2\hbar \frac{\partial f}{\partial z} \frac{\partial g}{\partial \overline{z}} + O(\hbar^2). \end{aligned} \tag{2.7}$$

This Wick product satisfies the hermitian property: $\overline{f *_{Wick} g} = \overline{g} *_{Wick} \overline{f}$. Setting $\Delta' := \frac{\partial^2}{\partial q^2} + \frac{\partial^2}{\partial p^2}$ and $\tilde{T}' := \exp \frac{\hbar}{4}\Delta'$, one gets

$$\tilde{T}'\left(f *_{Weyl} g\right) = \tilde{T}' f *_{Wick} \tilde{T}' g \qquad \forall f, g \in \mathbb{C}[p,q]. \tag{2.8}$$

Remark 2.3 Formulas (2.5), (2.6) and (2.7) do not converge in general if we replace polynomials by smooth functions. To make them well defined, a way to proceed is to replace the purely imaginary complex number $\frac{\hbar}{i}$ by a formal parameter ν and to consider formal power series in that parameter. This will lead to the definition of formal deformation quantization (see next section).

Remark 2.4 Other mathematical formulations of quantization exist, such as
- Geometric Quantisation of Kostant and Souriau [158] which proceeds in two steps. Prequantization of a symplectic manifold (M, ω) where one builds, if it exists, a prequantum bundle which is a Hermitian line bundle with a connection $(L \to M, h, \nabla)$ such that the curvature is $\frac{\omega}{i\hbar}$; if $\mathscr{H}$ denotes the Hilbert space of L^2 sections of the bundle L, one defines a correspondence $Q : C^\infty(M) \to (Op)(\mathscr{H})$, with values in operators acting on $\mathscr{H}$, by $Q(f) := i\hbar\nabla_{X_f} + f$. Clearly $[Q(f), Q(g)] = i\hbar Q(\{f, g\})$ and $Q(1) = \text{id}$ but there is no irreducibility. In a second step, one introduces the concept of polarization to "cut down the number of variables".
- In the case where the symplectic manifold is compact Kähler and admits a prequantization line bundle, one can use the framework of geometric quantization to define the Toeplitz quantization (see, for instance, [38]) which acts on holomorphic sections of this line bundle. A function f acts on a holomorphic section s by projecting fs on the space of holomorphic sections.
- Closely related is Berezin's quantisation [19, 20] where one builds on a particular class of Kähler manifolds a family of associative algebras using a symbolic calculus. Examples of deformation quantization have been constructed using asymptotic expansions of these quantizations (see, for instance, [38, 51, 117, 147]).

2.2 Deformation Quantization

Observe that the two mathematical frameworks for classical and quantum mechanics are very different. This makes it difficult to see classical mechanics as a limit of quantum mechanics. Deformation Quantization was introduced by Flato, Lichnerowicz and Sternheimer in [93], and developed in [15]: they

> suggest that quantization be understood as a deformation of the structure of the algebra of classical observables rather than a radical change in the nature of the observables.

This deformation approach to quantization is part of a "deformation approach" to the developments of physics which was one of the seminal ideas stressed by Moshe Flato: one looks at some (new) level of a theory in physics as a deformation of a former one [92].

One stresses here the fundamental aspect of the space of observables rather than the set of states; observables behave indeed in a nice way when one deals with composite systems: both in the classical and in the quantum picture, the space of observables for combined systems is the tensor product of the spaces of observables.

The algebraic structure of classical observables that one deforms is the algebraic structure of the space of smooth functions on a Poisson manifold: the associative structure given by the usual product of functions and the Lie structure given by the Poisson bracket. Formal deformation quantisation is defined in terms of a star product which is a formal deformation of that structure.

2.2.1 *Definition and Examples of Star Products*

Definition 2.2 (*Bayen et al.* [15]) A **star product** on a Poisson manifold (M, P) is a bilinear map

$$N \times N \to N[\![\nu]\!], \qquad (u, v) \mapsto u * v = u *_\nu v := \sum_{r \geq 0} \nu^r C_r(u, v)$$

where $N = C^\infty(M)$ [we consider in general complex valued functions] such that

1. when the map is extended ν-linearly (and continuously in the ν-adic topology) to $N[\![\nu]\!] \times N[\![\nu]\!]$ it is formally associative $(u * v) * w = u * (v * w)$;
2. (a) $C_0(u, v) = uv$, (b) $C_1(u, v) - C_1(v, u) = \{u, v\}$ (c) $1 * u = u * 1 = u$;
3. the C_r's are bidifferential operators on M, i.e. given in any local chart (U, φ) with local coordinates $\{x^i : 1 \leq i \leq m = \dim M\}$ by

$$C_r(u, v)_{|U} = \sum_{k \leq K, k' \leq K'} \sum_{i_1,\dots,i_k; j_1,\dots,j_{k'}} p^{i_1,\dots,i_k; j_1,\dots,j_{k'}} \frac{\partial^{|k|} u}{\partial x^{i_1} \dots \partial x^{i_k}} \frac{\partial^{|k'|} v}{\partial x^{j_1} \dots \partial x^{j_{k'}}}$$

(it is then more precisely a **differential star product**).

When each C_r is of order maximum r in each argument, one speaks of a **natural star product**.
If $\overline{f * g} = \overline{g} * \overline{f}$ for any $\nu = i\lambda,\ \lambda \in /R$, the star product is called **Hermitian**.

If there were a quantization in the usual sense, i.e. a correspondence between functions on the Poisson manifold (M, P) and algebras A_h of operators on a Hilbert space (depending on a parameter h related to the Plank's constant), one could look at the deformed products $*_h$ of two functions as corresponding to the composition of the corresponding operators in A_h. One can think of a star product as the expansion in the parameter h of such deformed products. In particular, one can define the star products on $\mathbb{R}^{2n}$ (with its canonical Poisson structure) coming from the quantization of polynomial functions given by the standard, the Weyl and the Wick orderings.

Exercise 2.6 The standard ordering (see Eq. (2.5)) yields:

$$f *_{st} g := \sum_{r=0}^{\infty} \frac{\nu^r}{r!} \sum_{r_1+\ldots+r_n=r} \frac{\partial^r f}{\partial p_1^{r_1} \ldots p_n^{r_n}} \frac{\partial^r g}{\partial q^{1 r_1} \ldots q^{n r_n}}, \tag{2.9}$$

the Weyl ordering (see Eq. (2.6)) yields in coordinates $\{x^1 = p_1, \ldots, x^n = p_n, x^{n+1} = q^1, \ldots x^{2n} = q^n\}$

$$f *_{weyl} g = \sum_{r=0}^{\infty} \frac{\nu^r}{r!} \sum_{i_1,\ldots,i_r,j_1,\ldots,j_r} P_0^{i_1 j_1} \ldots P_0^{i_r j_r} \frac{\partial^r f}{\partial x^{i_1} .. \partial x^{i_r}} \frac{\partial^r g}{\partial x^{j_1} .. \partial x^{j_r}} \tag{2.10}$$

with $P_0 = \begin{pmatrix} 0 & -I \\ I & 0 \end{pmatrix}$, and the Wick ordering (see Eq. (2.7)) yields

$$f *_{wick} g := \sum_{r=0}^{\infty} \nu^r \sum_{i_1,\ldots,i_r} \frac{(2i)^r}{r!} \frac{\partial^r f}{\partial z^{i_1} .. \partial z^{i_r}} \frac{\partial^r g}{\partial \overline{z}^{i_1} .. \partial \overline{z}^{i_r}} \tag{2.11}$$

Those three star products are natural; the ones corresponding to Weyl and Wick orderings are Hermitian.

Remark 2.5 A star product can also be defined not on the whole of $C^\infty(M)$ but on any subspace N of it which is stable under pointwise multiplication and Poisson bracket.
In (b) we require the skew-symmetric part of C_1 to be $\frac{1}{2}\{\ ,\ \}$; one finds in the literature other normalisations; originally it was $\{\ ,\ \}$ and often it is $\frac{i}{2}\{\ ,\ \}$; all these amount to a rescaling of the parameter.
By (b) the centre of the deformed algebra $(C^\infty(M)[\![\nu]\!], *)$ consists of series whose terms Poisson commute with all functions, so elements of $\mathbb{R}[\![\nu]\!]$ when M is symplectic and connected.
Properties (a) and (b) imply that the **star commutator** defined by $[u, v]_* = u * v - v * u$, which obviously makes $C^\infty(M)[\![\nu]\!]$ into a Lie algebra, has the form $[u, v]_* =$

$\nu\{u, v\} + \ldots$ so that repeated bracketing leads to higher and higher order terms. We denote $ad_* u\ (v) := [u, v]_*$.

Example 2.1 (The Moyal star product) The simplest example of a deformation quantisation is the Moyal product for a constant Poisson structure P on a vector space $V = \mathbb{R}^m$:

$$P = \sum_{i,j} P^{ij} \partial_i \wedge \partial_j, \quad P^{ij} = -P^{ji} \in \mathbb{R}$$

where $\partial_i = \partial/\partial x^i$ is the partial derivative in the direction of the coordinate x^i, $i = 1, \ldots, n$. The formula for **the (formal) Moyal product associated to** P is

$$(u *_{M(P)} v)(z) = \exp\left(\frac{\nu}{2} P^{rs} \partial_{x^r} \partial_{y^s}\right) (u(x)v(y))\Big|_{x=y=z}. \tag{2.12}$$

Associativity of $*_{M(P)}$ follows from the fact that

$$\partial_{t^k}(u *_{M(P)} v)(t) = (\partial_{x^k} + \partial_{y^k}) \exp\left(\frac{\nu}{2} P^{rs} \partial_{x^r} \partial_{y^s}\right) (u(x)v(y))\Big|_{x=y=t}.$$

Indeed,

$$\begin{aligned}
((u *_{M(P)} v) *_{M(P)} w)(x') &= \exp\left(\frac{\nu}{2} P^{rs} \partial_{t^r} \partial_{z^s}\right) ((u *_{M(P)} v)(t) w(z))\Big|_{t=z=x'} \\
&= \exp\left(\frac{\nu}{2} P^{rs} (\partial_{x^r} + \partial_{y^r}) \partial_{z^s}\right) \exp\left(\frac{\nu}{2} P^{r's'} \partial_{x^{r'}} \partial_{y^{s'}}\right) ((u(x)v(y))w(z))\Big|_{x=y=z=x'} \\
&= \exp\left(\frac{\nu}{2} P^{rs} (\partial_{x^r} \partial_{z^s} + \partial_{y^r} \partial_{z^s} + \partial_{x^r} \partial_{y^s})\right) ((u(x)v(y))w(z))\Big|_{x=y=z=x'} \\
&= (u *_{M(P)} (v *_{M(P)} w))(x').
\end{aligned}$$

The **(formal) Moyal product** $*_M$ is the one associated to a non degenerate P on $\mathbb{R}^{2n}$.

Exercise 2.7 Writing $P_0 = \begin{pmatrix} 0 & -I \\ I & 0 \end{pmatrix}$, and using (see Eq. (2.10)), show that

$$f *_M g = f *_{weyl} g. \tag{2.13}$$

Definition 2.3 When P is non degenerate $\left(\text{i.e.} V = \mathbb{R}^{2n},\ P_0 = \begin{pmatrix} 0 & -I \\ I & 0 \end{pmatrix}\right)$, the space of polynomials in ν whose coefficients are polynomials on V with Moyal product is called **the Weyl algebra** $(S(V^*)[\nu], *_M)$.

Remark 2.6 Moyal star product is the star product (see Eq. (2.13)) coming from the quantization of polynomials on $\mathbb{R}^{2n}$ with Weyl's ordering. Moyal used in 1949 the deformed bracket which corresponds to the commutator of operators to study quantum statistical mechanics and the Moyal product first appeared in Groenewold [99]. Weyl quantization can be extended beyond polynomials; heuristically one would like to write

$$\text{“}\mathcal{Q}_{Weyl}(F)\text{”} = \left(\frac{1}{2\pi}\right)^{2n} \int_{\mathbb{R}^n}\int_{\mathbb{R}^n} \hat{F}(u,v)e^{i(uQ+vP)}dudv,$$

where $\hat{F}$ is the Fourier transform $\hat{F}(u,v) = \int_{\mathbb{R}^n}\int_{\mathbb{R}^n} F(q,p)e^{-i(uq+vp)}dqdp$.

Exercise 2.8 If one develops the above formally, using the fact that on a nice test function ϕ, $(e^{iuQ}\phi)(x) = e^{iu.x}\phi(x)$, $(e^{ivP}\phi)(x) = \phi(x+\hbar v)$ and $e^{i(uQ+vP)} = e^{-\frac{i}{2}\hbar u.v}e^{iuQ}\circ e^{ivP}$, one gets the formula

$$(\mathcal{Q}_{Weyl}(F)(\phi))(x) := \int_{\mathbb{R}^n}\left(\int_{\mathbb{R}^n} F\left(\frac{x+y}{2}, 2\pi\hbar\xi\right)e^{-2\pi i(y-x)\xi}\phi(y)dy\right)d\xi;$$

which one takes as a definition of $\mathcal{Q}_{Weyl}(F)$; it is well defined for a test function ϕ in the Schwartz space when F satisfies weak regularity bounds (there exists a constant $C > 0$ and constants $C_{i,j} > 0$, such that $\forall i, j \geq 0$ and for all x, p, one has $|\nabla_x^i\nabla_p^i F(x,p)| \leq C_{i,j}(1+|x|+|p|)^C$).

The above formula coincides with the previous one when F is a polynomial. The map $\mathcal{Q}_{Weyl}$ gives an isometry between the space $L^2(\mathbb{R}^{2n})$ and the space of Hilbert Schmidt operators on $L^2(\mathbb{R}^n)$, associating a self-adjoint operator to a real function.

Exercise 2.9 If F and G are two Schwartz functions, then the composition

$$\mathcal{Q}_{Weyl}(F)\circ\mathcal{Q}_{Weyl}(G)$$

is equal to $\mathcal{Q}_{Weyl}(F\times_\hbar G)$ where $F\times_\hbar G$ is the function defined by

$$(F\times_\hbar G)(u) := \left(\frac{1}{\pi\hbar}\right)^{2n}\int_{\mathbb{R}^{2n}}\int_{\mathbb{R}^{2n}} e^{\frac{2i}{\hbar}\Omega(v,w)}F(u+v)G(u+w)dvdw \tag{2.14}$$

$$= \left(\frac{1}{\pi\hbar}\right)^{2n}\int_{\mathbb{R}^{2n}}\int_{\mathbb{R}^{2n}} e^{\frac{2i}{\hbar}(\Omega(u,v)+\Omega(v,w)+\Omega(w,u))}F(v)G(w)dvdw. \tag{2.15}$$

with $\Omega = \begin{pmatrix} 0 & I \\ -I & 0\end{pmatrix}$.

The result is a Schwartz function; hence $\times_\hbar$ gives an associative product on the space of Schwartz functions, called the convergent Moyal star product. The (formal) Moyal star product introduced before can be seen as an asymptoptic expansion in $\nu = \hbar/i$ of this composition law.

*Example 2.2 (The standard *-product on* $\mathfrak{g}^*$*)* Let $\mathfrak{g}^*$ be the dual of a Lie algebra $\mathfrak{g}$. The algebra of polynomials on $\mathfrak{g}^*$ is identified with the symmetric algebra $S(\mathfrak{g})$. One defines a new associative law on this algebra by a transfer of the product $\circ$ in the universal enveloping algebra $U(\mathfrak{g})$, via the bijection between $S(\mathfrak{g})$ and $U(\mathfrak{g})$ given by the total symmetrization σ:

$$\sigma : S(\mathfrak{g}) \to U(\mathfrak{g}) \quad X_1 \dots X_k \mapsto \frac{1}{k!} \sum_{\rho \in S_k} X_{\rho(1)} \circ \dots \circ X_{\rho(k)}.$$

Then $U(\mathfrak{g}) = \oplus_{n \geq 0} U_n$ where $U_n := \sigma(S^n(\mathfrak{g}))$ and we decompose an element $u \in U(\mathfrak{g})$ accordingly $u = \sum u_n$. We define for $P \in S^p(\mathfrak{g})$ and $Q \in S^q(\mathfrak{g})$

$$P * Q = \sum_{n \geq 0} (\nu)^n \sigma^{-1}((\sigma(P) \circ \sigma(Q))_{p+q-n}). \tag{2.16}$$

This yields a differential star product on $\mathfrak{g}^*$ [102]; it is characterised by

$$X * (X_1 \dots X_k) = X X_1 \dots X_k + \sum_{j=1}^{k} \frac{(-1)^j}{j!} \nu^j Bj[[[X, X_{r_1}], \dots], X_{r_j}] X_1 \dots \widehat{X_{r_1}} \dots \widehat{X_{r_j}} \dots X_k$$

where B_j are the Bernouilli numbers. For $\nu = 2\pi i$, this star product writes [79]:

$$u * v(\xi) = \int_{\mathfrak{g} \times \mathfrak{g}} \hat{u}(X) \hat{v}(Y) e^{2i\pi \langle \xi, CBH(X,Y) \rangle} dX dY$$

where $\hat{u}(X) = \int_{\mathfrak{g}^*} u(\eta) e^{-2i\pi \langle \eta, X \rangle}$ and where CBH denotes Campbell-Baker-Hausdorff formula for the product of elements in the group in a logarithmic chart ($\exp X \exp Y = \exp CBH(X, Y) \quad \forall X, Y \in \mathfrak{g}$).

Remark 2.7 The standard star product on $\mathfrak{g}^*$ does not always restrict to orbits (except for the Heisenberg group) so other algebraic constructions of star products on $S(\mathfrak{g})$ were considered (for instance in [9, 10, 50, 90]). When $\mathfrak{g}$ is semisimple, if $\mathcal{H}$ is the space of harmonic polynomials and if $I_1, \dots I_r$ are generators of the space of invariant polynomials, then any polynomial $P \in S(\mathfrak{g})$ writes uniquely as a sum $P = \sum_{a_1 \dots a_r} I_1^{a_1} \dots I_r^{a_r} h_{a_1 \dots a_r}$ where $h_{a_1 \dots a_r} \in \mathcal{H}$. One considers the linear isomorphism σ' between $S(\mathfrak{g})$ and $U(\mathfrak{g})$ induced by this decomposition

$$\sigma'(P) = \sum_{a_1 \dots a_r} (\sigma(I_1)\circ)^{a_1} \dots (\sigma(I_r)\circ)^{a_r} \circ \sigma(h_{a_1 \dots a_r}).$$

The associative composition law in $U(\mathfrak{g})$, pulled back by this isomorphism σ', gives a star product on $S(\mathfrak{g})$ which is not defined by differential operators. With Cahen and Rawnsley, we proved [56] that if $\mathfrak{g}$ is semisimple, there is no differential star product on any neighbourhood of 0 in $\mathfrak{g}^*$ such that $C * u = Cu$ for the quadratic invariant polynomial $C \in S(\mathfrak{g})$ and all $u \in S(\mathfrak{g})$ (thus no differential star product which is tangential to the orbits).

2.2.2 Existence of Star Products

In 1983, De Wilde and Lecomte proved [68] that on any symplectic manifold there exists a differential star product. Their technique works to prove the existence of a differential star product on a regular Poisson manifold [129]. In 1985, but appearing only in the West in the nineties [83], Fedosov gave a recursive construction of a star product on a symplectic manifold (M, ω). In 1994, he extended this result to give a recursive construction in the context of regular Poisson manifold [82]. Independently, also using the framework of Weyl bundles, Omori, Maeda and Yoshioka [140] gave an other proof of existence of a differential star product on a symplectic manifold, gluing local Moyal star products.

In 1997, Kontsevich [122] gave a proof of the existence of a star product on any Poisson manifold and gave an explicit formula for a star product for any Poisson structure on $V = \mathbb{R}^m$. This appeared as a consequence of the proof of his formality theorem. Tamarkin [162] gave a version of the proof in the framework of the theory of operads.

2.2.3 The Notion of States

The star product model gives a quantization model for the algebra of observables, so here an algebra over formal power series $\mathbb{C}[[\nu]]$. In the usual presentation of quantum mechanics, observables are operators on a Hilbert and states are rays in that Hilbert space. Model algebras of quantum observables are complex algebras of bounded linear operators on a complex Hilbert space. These are prototypes of C^*-algebras. Recall that a C^*-algebra is a Banach algebra over $\mathbb{C}$ endowed with a $*$ involution (i.e. an involutive semilinear antiautomorphism) such that $\|a\| = \|a^*\|$ and $\|aa^*\| = \|a\|^2$ for each element a in the algebra. Recall that if $\mathscr{A} = \mathscr{B}(\mathscr{H})$ is the algebra of bounded linear operators on a Hilbert space $\mathscr{H}$ and if ψ is a non vanishing element of $\mathscr{H}$, the ray it generates defines the linear functional

$$\omega : \mathscr{A} \to \mathbb{C} : A \mapsto \omega(A) := \frac{\langle \psi, A\psi \rangle}{\langle \psi, \psi \rangle}$$

which is positive in the sense that $\omega(A^*A) \geq 0$. This lead to define a state in the theory of C^* algebras as a positive linear functional. Bordemann, Römer and Waldmann [39] give the following intrinsic description of the notion of states for formal star products, generalizing the notion of a state to the framework of $*$-algebras.

Definition 2.4 (1) An associative commutative unital ring R is said to be **ordered** with positive elements P if the product of two elements in P is in P, the sum of two elements in P is in P, and R is the disjoint union $R = P \cup \{0\} \cup -P$.(Examples are given by $\mathbb{Z}$, $\mathbb{Q}$, $\mathbb{R}$, $\mathbb{R}[[\lambda]]$; in the case of $\mathbb{R}[[\lambda]]$, a series $a = \sum_{r=r_0}^{\infty} a_r \lambda^r$ is positive if its lowest order non vanishing term is positive ($a_{r_0} > 0$).)

Let $C = R(i)$ be the ring extension by a square root i of -1 of an ordered ring. For instance $C = \mathbb{C}$ for $R = \mathbb{R}$ or, for our use here in deformation quantization, $C = \mathbb{C}[[\lambda]]$ for $R = \mathbb{R}[[\lambda]]$ with $\nu = i\lambda$.

(2) An associative algebra $\mathscr{A}$ over C is called a ***-algebra** if it has an involutive antilinear antiautomorphism $^* : \mathscr{A} \to \mathscr{A}$ called the *-involution. (Examples: any C^* algebra is a *-algebra over $\mathbb{C}$, in particular the *-algebra over $\mathbb{C}$ of bounded linear operators on a Hilbert space with the involution given by taking the adjoint; also the deformed algebra $(C^\infty(M)[[\nu = i\lambda]], *)$ with a Hermitian star product and conjugaison is a *-algebra over $\mathbb{C}[[\lambda]]$).

(3) A linear functional $\omega : \mathscr{A} \to C$ over a *-algebra over C is called **positive** if

$$\omega(A^*A) \geq 0 \qquad \text{for any } A \in \mathscr{A}.$$

(4) A **state** for a *-algebra $\mathscr{A}$ with unit over C is a positive linear functional which satisfies $\omega(1) = 1$.

Remark 2.8 The positive linear functionals on $C^\infty(M)$ are the compactly supported Borel measures.

The δ-functional on $\mathbb{R}^{2n}$ is not positive with respect to the Moyal star product: if $H := \frac{1}{2m}p^2 + kq^2$, $\left(H *_{Moyal} H\right)(0,0) = \frac{k\nu^2}{2m} = \frac{-k\lambda^2}{2m} < 0$.

Bursztyn and Waldmann prove in [45] that for a Hermitian star product, any classical state ω_0 on $C^\infty(M)$ can be deformed into a state for the deformed algebra, $\omega = \sum_{r=0}^{\infty} \lambda^r \omega_r$.

2.3 Fedosov's Star Products on a Symplectic Manifold

Fedosov gives a construction [83] of a star product on a symplectic manifold (M, ω), when one has chosen a symplectic connection and a sequence of closed 2-forms on M. One obtains the star product by identifying the space $C^\infty(M)[[\nu]]$ with an algebra of flat sections of an associative algebra bundle, the so-called Weyl bundle, endowed with a flat connection.

2.3.1 *The Weyl Bundle*

Let (V, Ω) be a symplectic vector space and consider the space of polynomials in ν whose coefficients are polynomials on V with Moyal star product; this is the Weyl algebra $S(V^*)[\nu]$.

Exercise 2.10 Show that the Weyl algebra $S(V^*)[\nu]$ is isomorphic to the universal enveloping algebra $U(\mathfrak{h})$ of the Heisenberg Lie algebra $\mathfrak{h} = V^* \oplus \mathbb{R}\nu$ with Lie bracket

$$\left[y^i, y^j\right] = (\Omega^{-1})^{ij}\nu.$$

Indeed both are associative algebras generated by V^* and ν and the map sending an element of $V^* \subset \mathfrak{h}$ to the corresponding element in $V^* \subset S(V^*)$ viewed as a linear function on V and mapping $\nu \in \mathfrak{h}$ on $\nu \in \mathbb{R}[\nu] \subset S(V^*)[\nu]$ satisfies: $\xi *_M \xi' - \xi' *_M \xi = [\xi, \xi']$ for all $\xi, \xi' \in \mathfrak{h}$ so extends to a morphism of associative algebras.

There is a grading on $U(\mathfrak{h})$ assigning the degree 1 to the y^i's and the degree 2 to the element ν. The **formal Weyl algebra** W is the completion in that grading of the above algebra. An element of the formal Weyl algebra is of the form

$$a(y,\nu) = \sum_{m=0}^{\infty} \Bigg(\sum_{2k+l=m} a_{k,i_1,\ldots,i_l} \nu^k y^{i_1} \ldots y^{i_l} \Bigg).$$

The product in $U(\mathfrak{h})$ is given by the Moyal star product

$$(a \circ b)(y,\nu) = \left(\exp\left(\frac{\nu}{2} (\Omega^{-1})^{ij} \frac{\partial}{\partial y^i} \frac{\partial}{\partial z^j} \right) a(y,\nu) b(z,\nu) \right) \Bigg|_{y=z}$$

and the same formula also defines the product in W.

Definition 2.5 The symplectic group $Sp(V, \Omega)$ of the symplectic vector space (V, Ω) consists of all invertible linear transformations A of V with $\Omega(Au, Av) = \Omega(u, v)$, for all $u, v \in V$. $Sp(V, \Omega)$ acts as automorphisms of $\mathfrak{h}$ by $A \cdot f = f \circ A^{-1}$ for $f \in V^*$ and $A \cdot \nu = 0$. This action extends to both $U(\mathfrak{h})$ and W and on the latter is denoted by ρ. It satisfies $\rho(A)(a \circ b) = \rho(A)(a) \circ \rho(A)(b)$. Explicitely: $\rho(A)(\sum_{2k+l=m} a_{k,i_1,\ldots,i_l} \nu^k y^{i_1} \ldots y^{i_l}) = \sum_{2k+l=m} a_{k,i_1,\ldots,i_l} \nu^k (A^{-1})^{i_1}_{j_1} \ldots (A^{-1})^{i_l}_{j_l} y^{j_1} \ldots y^{j_l}$.

To any element B in the Lie algebra $sp(V, \Omega)$ of the symplectic group, we associate the quadratic element $\overline{B}$ in W defined by

$$\overline{B} = \frac{1}{2} \sum_{ijr} \Omega_{ri} B^r_j y^i y^j.$$

This is an identification since the condition to be in $sp(V, \Omega)$ is that $\sum_r \Omega_{ri} B^r_j$ is symmetric in i and j.

Exercise 2.11 Show that the natural action $\rho_*(B)$ is given by:

$$\rho_*(B) y^l = \frac{-1}{\nu} [\overline{B}, y^l]$$

where $[a, b] := (a \circ b) - (b \circ a)$ for any $a, b \in W$.

Since both sides act as derivations this extends to all of W as

$$\rho_*(B) a = \frac{-1}{\nu} [\overline{B}, a]. \tag{2.17}$$

Definition 2.6 If (M, ω) is a symplectic manifold, we can form its bundle $F(M)$ of symplectic frames. A symplectic frame at the point $x \in M$ is a linear symplectic isomorphism $\xi_x : (V, \Omega) \to (T_x M, \omega_x)$. The bundle $F(M)$ is a principal $Sp(V, \Omega)$-bundle over M (the action on the right of an element $A \in Sp(V, \Omega)$ on a frame ξ_x is given by $\xi_x \circ A$).

The associated bundle $\mathcal{W} = F(M) \times_{Sp(V,\Omega),\rho} W$ is a bundle of algebras on M called the bundle of formal Weyl algebras, or, more simply, **the Weyl bundle**. Its **sections** are formal series

$$a(x, y, \nu) = \sum_{2k+l \geq 0} \nu^k a_{k,i_1,\ldots,i_l}(x) y^{i_1} \cdots y^{i_l} \tag{2.18}$$

where the coefficients $a_{k,i_1,\ldots,i_l}$ define (in the i's) symmetric covariant l–tensor fields on M. So $\mathcal{W} \simeq \oplus_p \mathbb{C} \otimes S^p(T^*M)[[\nu]]$. We denote by $\Gamma(\mathcal{W})$ the space of those sections. The pointwise product of two sections makes $\Gamma(\mathcal{W})$ into an algebra, and **the multiplication** has the form

$$(a \circ b)(x, y, \nu) = \left(\exp\left(\frac{\nu}{2} P^{ij} \frac{\partial}{\partial y^i} \frac{\partial}{\partial z^j}\right) a(x, y, \nu) b(x, z, \nu)\right)\Bigg|_{y=z}, \tag{2.19}$$

where P is the Poisson tensor associated to the symplectic structure (thus $\sum_j P^{ij}$ $\omega_{jk} = \delta^i_k$). The center of this algebra coincide with $C^\infty(M)[[\nu]]$.

2.3.2 *Flat Connections on the Weyl Bundle*

Let (M, ω) be a symplectic manifold. A **symplectic connection** on M is a connection ∇ on TM which is torsion-free and satisfies $\nabla_X \omega = 0$. Such connections always exist but, unlike the Riemannian case, are not unique.

Exercise 2.12 To see the existence, take any torsion-free connection ∇' and define S by $\omega(S(X, Y), Z) = \frac{1}{3}\big((\nabla'_X \omega)(Y, Z) + (\nabla'_Y \omega)(X, Z)\big)$. Check that $\nabla_X Y = \nabla'_X Y + S(X, Y)$ defines a symplectic connection.

Remark 2.9 ([105]) Any natural star product $* = \sum_{r \geq 0} C_r$ on a symplectic manifold defines a unique symplectic connection ∇ such that

$$C_1(u, v) = \tfrac{1}{2}\{u, v\} + [\{u, Ev\} + \{Eu, v\} - E(\{u, v\})]$$

with E a differential operator of order 2 and

$$C_2(u, v) + C_2(v, u) = \tfrac{1}{4} P^{ij} P^{i'j'} \nabla^2_{ii'} u \, \nabla^2_{jj'} v + ((\mathrm{ad}\, E)^2 m)(u, v) + A_2(u, v)$$

with $((\mathrm{ad}\, E)^2 m)(u, v) = E^2(uv) + 2Eu.Ev - E^2u.v - u.E^2v - 2E(Eu.v + u.Ev)$ and A_2 a differential operator of order 1 in each argument.

A symplectic connection defines a connection 1-form in the symplectic frame bundle and so a connection in all associated bundles (i.e. a covariant derivative of sections); we denote by ∂ the connection in $\mathcal{W}$. For any vector field X on M, the covariant derivative ∂_X is a derivation of the algebra $\Gamma(\mathcal{W})$. We consider $\mathcal{W}$-valued q-forms on M to express the connection and its curvature; these are sections of the bundle $\mathcal{W} \otimes \Lambda^q T^*M \simeq \mathbb{C} \otimes (\oplus_p S^p(T^*M) \otimes \Lambda^q(T^*M))[[\nu]]$ and locally have the form

$$\sum_{2k+p\geq 0} \nu^k a_{k,i_1,\dots,i_l,j_1,\dots,j_q}(x) y^{i_1} \dots y^{i_p} dx^{j_1} \wedge \dots \wedge dx^{j_q}$$

where the coefficients, symmetric in $i_1, \dots, i_p$ and anti-symmetric in $j_1, \dots, j_q$, are covariant tensors. Such sections can be multiplied using the product in $\mathcal{W}$ and simultaneously exterior multiplication $a \otimes \omega \circ b \otimes \omega' = (a \circ b) \otimes (\omega \wedge \omega')$. The space of $\mathcal{W}$-valued forms $\Gamma(\mathcal{W} \otimes \Lambda^*)$ is then a graded Lie algebra with respect to the bracket

$$[s, s'] = s \circ s' - (-1)^{q_1 q_2} s' \circ s \qquad \text{for } s_i \in \Gamma(\mathcal{W} \otimes \Lambda^{q_i}).$$

The connection ∂ in $\mathcal{W}$ is given by

$$\partial \colon \Gamma(\mathcal{W}) \to \Gamma(\mathcal{W} \otimes \Lambda^1) \qquad \partial a = da - \frac{1}{\nu}[\overline{\Gamma}, a] \text{ with } \overline{\Gamma} = \frac{1}{2} \sum_{ijkr} \omega_{ki} \Gamma^k_{rj} y^i y^j dx^r,$$

where Γ^i_{kl} are the Christoffel symbols of ∇ in TM (which define an element of the symplectic Lie algebra with respect to the il indices). As usual, the connection ∂ in $\mathcal{W}$ extends to a covariant exterior derivative on $\Gamma(\mathcal{W} \otimes \Lambda^*)$, also denoted by ∂, by using the Leibnitz rule:

$$\partial(a \otimes \omega) = \partial(a) \wedge \omega + a \otimes d\omega.$$

The curvature of ∂ is then given by $\partial_\circ \partial$ which is a 2-form with values in End($\mathcal{W}$).

Exercise 2.13 The curvature of ∂ admits a simple expression in terms of the curvature R of the symplectic connection ∇:

$$\partial \circ \partial a = \frac{1}{\nu}[\overline{R}, a] \quad \text{where } \overline{R} = \frac{1}{4} \sum_{ijklr} \omega_{rl} R^l_{ijk} y^r y^k dx^i \wedge dx^j. \tag{2.20}$$

The idea is to try to modify ∂ to have zero curvature. In order to do so we use a further technical tool, coming from Koszul's long exact sequence. Given any finite dimensional vector space V', the Koszul long exact sequence is:

$$0 \to S^q(V') \xrightarrow{\delta'} V' \otimes S^{q-1}(V') \xrightarrow{\delta'} \Lambda^2 V' \otimes S^{q-2}(V') \xrightarrow{\delta'} \dots \xrightarrow{\delta'} \Lambda^{q-1}(V') \otimes V' \xrightarrow{\delta'} \Lambda^q(V') \longrightarrow 0$$

where δ' is the skew-symmetrisation operator:

$$\delta'(v^1 \wedge \dots \wedge v^q \otimes w^1 \dots w^p) = \sum_{i=1}^{p} v^1 \wedge \dots \wedge v^q \wedge w^i \otimes w^1 \dots w^{i-1} w^{i+1} \dots w^p.$$

The symmetrisation operator reads:

$$s(v^1 \wedge \ldots \wedge v^q \otimes w^1 \ldots w^p) \sum_{i=1}^{q} (-1)^{q-i} v^1 \wedge \ldots v^{i-1} \wedge v^{i+1} \ldots \wedge v^q \otimes v^i \cdot w^1 \ldots w^p.$$

They satisfy $(\delta')^2 = 0,\ s^2 = 0,\ (\delta' \circ s + s \circ \delta')_{|\Lambda^q V' \otimes S^p(V')} = (p+q)\,\mathrm{Id}$.
For any $a \in \Gamma(\mathscr{W} \otimes \Lambda^q)$, we write

$$a = \sum_{p \geq 0, q \geq 0} a_{pq} = \sum_{2k+p \geq 0, q \geq 0} \nu^k a_{k,i_1,\ldots,i_p,j_1,\ldots,j_q} y^{i_1} \ldots y^{i_p} dx^{j_1} \wedge \ldots \wedge dx^{j_q};$$

in particular $a_{00} = \sum_k \nu^k a_k$ with $a_k \in C^\infty(M)$; and we define

$$\delta(a) := \sum_k dx^k \wedge \frac{\partial a}{\partial y^k}, \quad \delta^{\sim 1}(a_{pq}) := \begin{cases} \frac{1}{p+q} \sum_k y^k i(\frac{\partial}{\partial x^k}) a_{pq} \text{ if } \mathrm{p+q} > 0; \\ 0 \text{ if } \mathrm{p+q} = 0. \end{cases} \tag{2.21}$$

Exercise 2.14 Show that

$$\delta^2 = 0, \qquad (\delta^{\sim 1})^2 = 0, \qquad (\delta \circ \delta^{\sim 1} + \delta^{\sim 1} \circ \delta)(a) = a - a_{00};$$

and that δ can be written in terms of the algebra structure by

$$\delta(a) = \frac{1}{\nu} \left[\sum_{ij} -\omega_{ij} y^i dx^j, a \right],$$

hence δ is a graded derivation of $\Gamma(\mathscr{W} \otimes \Lambda^*)$. Verify that $\partial\delta + \delta\partial = 0$.

We now look for a connection D on $\mathscr{W}$, so that D_X is a derivation af the algebra $\Gamma(\mathscr{W})$ for any vectorfield X on M, and so that D is flat in the sense that $D \circ D = 0$. Such a connection can be written as a sum of ∂ and a $\mathrm{End}(\mathscr{W})$-valued 1-form. The latter is taken in a particular form:

$$Da = \partial a - \delta(a) - \frac{1}{\nu}[r, a]. \tag{2.22}$$

Exercise 2.15 Show that

$$D_\circ Da = \frac{1}{\nu} \left[\overline{R} - \partial r + \delta r + \frac{1}{2\nu}[r, r], a \right]$$

with $\overline{R}$ defined by (2.20), and that $[r, r] = 2r \circ r$.

The connection D is flat provided the first term in the bracket is a central 2-form.

Theorem 2.1 (Fedosov [83]) *For any given series of closed 2-forms on* M, $\tilde{\Omega} = \sum_{i\geq 1} \nu^i \omega_i$, *the equation*

$$\delta r = -\overline{R} + \partial r - \frac{1}{\nu} r^2 + \tilde{\Omega} \tag{2.23}$$

has a unique solution $r \in \Gamma(\mathscr{W} \otimes \Lambda^1)$ *satisfying the normalization condition* $\delta^{\sim 1} r = 0$ *and such that the* $\mathscr{W}$*-degree of the leading term of* r *is at least 3.*

Proof We apply $\delta^{\sim 1}$ to the Eq. (2.23) using the fact that r is a 1-form and thus $r_{00} = 0$. Then r, if it exists, must satisfy

$$r = \delta^{\sim 1}\delta r = -\delta^{\sim 1}\overline{R} + \delta^{\sim 1}\partial r - \frac{1}{\nu}\delta^{\sim 1} r^2 + \delta^{-1}\tilde{\Omega}. \tag{2.24}$$

Two solutions of this equation will have a difference which satisfies the same equation but without the $\overline{R}$ term and the $\tilde{\Omega}$ term. If the first non-zero term of the difference has finite degree m, then the leading term of $\delta^{\sim 1}\partial r$ has degree $m+1$ and of $\delta^{\sim 1}(r^2/h)$ has degree $2m-1$. Since both of these are larger than m for $m \geq 2$, such a term cannot exist so the difference must be zero. Hence the solution is unique. Existence is very similar. We observe that the above argument shows that the Eq. (2.24) for r determines the homogeneous components of r recursively. So it is enough to show that such a solution satisfies both conditions of the theorem. Obviously $\delta^{\sim 1} r = 0$. Let $A = \delta r + \overline{R} - \partial r + \frac{1}{\nu} r^2 - \tilde{\Omega} \in \Gamma(\mathscr{W} \otimes \Lambda^2)$. Then $\delta^{\sim 1} A = \delta^{\sim 1}\delta r + \delta^{\sim 1}(\overline{R} - \partial r + \frac{1}{\nu} r^2 - \tilde{\Omega}) = r - r = 0$. Also $DA = \partial A - \delta A - \frac{1}{\nu}[r, A] = 0$. We can now apply a similar argument to that which proved uniqueness. Since $A_{00} = 0$, $\delta^{\sim 1} A = 0$ and $DA = 0$ we have $A = \delta^{\sim 1}\delta A = \delta^{\sim 1}(\partial A - \frac{1}{\nu}[r, A])$ and recursively we can see that each homogeneous component of A must vanish, which shows that (2.23) holds and the theorem is proved. □

Carrying out the recursion (2.24) to determine r explicitly, one easily sees [21] that: r_m only depends on ω_i for $2i + 1 \leq m$ and the first term in r which involves ω_k is:

$$r_{2k+1} = \delta^{-1}(\nu^k \omega_k) + \tilde{r}_{2k+1} \tag{2.25}$$

where the last term does not involve ω_k.

2.3.3 Fedosov's Star Products

Given a series of closed 2-forms on M, $\tilde{\Omega} = \sum_{i\geq 1} h^i \omega_i$, we consider the flat connection D on the Weyl bundle constructed as above, corresponding to r in $\Gamma(\mathscr{W} \otimes \Lambda^1)$ given inductively by (2.24). Since D_X acts as a derivation of the pointwise multiplication of sections, the space $\mathscr{W}_D$ of flat sections is a subalgebra of the space of sections of $\mathscr{W}$:

$$\mathscr{W}_D = \{a \in \Gamma(\mathscr{W}) | Da = 0\}.$$

Theorem 2.2 ([83]) *For any $a_\circ \in C^\infty(M)[[\nu]]$ there is a unique $a \in \mathscr{W}_D$ such that $a(x, 0, \nu) = a_\circ(x, \nu)$.*

Proof This is very much like the above argument. We have $Da = 0 \Leftrightarrow \delta a = \partial a - \frac{1}{\nu}[r, a]$. Since a is a 0-form, $\delta^{\sim 1} a = 0$; we apply $\delta^{\sim 1}$ and get:

$$a = \delta^{\sim 1}\delta a + a_\circ = \delta^{\sim 1}\left(\partial a - \frac{1}{\nu}[r, a]\right) + a_\circ. \tag{2.26}$$

We solve this equation recursively for a, so $a(x, 0, \nu) = a_\circ(x, \nu)$. The fact that $A = Da$ vanishes follows as before by showing that $\delta^{\sim 1} A = 0$ and $DA = D^2 a = 0$. The uniqueness of the element a follows by an induction argument for the difference of two solutions.

Definition 2.7 Define **the symbol map** $\sigma : \Gamma(\mathscr{W}) \to C^\infty(M)[[\nu]]$, by

$$\sigma(a) = a(x, 0, \nu). \tag{2.27}$$

Theorem 2.2 tells us that σ is a linear isomorphism when restricted to $\mathscr{W}_D$; it is used to transport the algebra structure of $\mathscr{W}_D$ to $C^\infty(M)[[\nu]]$.

$$a * b := \sigma(\sigma^{-1}(a) \circ \sigma^{-1}(b)), \quad a, b \in C^\infty(M)[[\nu]]. \tag{2.28}$$

Exercise 2.16 Check that this defines a $*$-product on $C^\infty(M)$. If the curvature of ∇ vanishes and $\tilde{\Omega} = 0$, show that one gets back the Moyal $*$-product.

This $*$-product is called **the Fedosov star product**; its construction depends only on the choice of a symplectic connection ∇ and the choice of a series $\tilde{\Omega}$ of closed 2-forms on M so can be denoted $*_{\nabla,\tilde{\Omega}}$. The Fedosov star product $*_{\nabla,\tilde{\Omega}}$ is natural and the connection associated to it (see Remark 2.9) is ∇. Writing $u *_{\nabla,\tilde{\Omega}} v = \sum_{i \geq 0} \nu^r C_r^{\nabla,\tilde{\Omega}}(u, v)$, we have [21] that, for any r, $C_r^{\nabla,\tilde{\Omega}}$ only depends on ω_i for $i < r$ and

$$C_{r+1}^{\nabla\tilde{\Omega}}(u, v) = \omega_r(X_u, X_v) + \tilde{C}_{r+1}(u, v) \tag{2.29}$$

where the last term does not depend on ω_r.

2.4 Classification of Poisson Deformations and Star Products

2.4.1 Hochschild Cohomology

Star products on a manifold M are examples of deformations -in the sense of Gerstenhaber [98]- of associative algebras. Their study uses the Hochschild cohomology [111] of the algebra, here $C^\infty(M)$, where p-cochains are p-linear maps from $(C^\infty(M))^p$ to $C^\infty(M)$ and where the **Hochschild coboundary operator** maps the p-cochain C to the $p+1$-cochain

$$(\partial C)(u_0,\ldots,u_p) = u_0 C(u_1,\ldots,u_p) + \sum_{r=1}^{p}(-1)^r C(u_0,\ldots,u_{r-1}u_r,\ldots,u_p) \\ + (-1)^{p+1} C(u_0,\ldots,u_{p-1})u_p.$$

For differential star products, we consider differential cochains, i.e. given by differential operators on each argument.

Exercise 2.17 The associativity condition for a star product at order k in the parameter ν reads

$$(\partial C_k)(u,v,w) = \sum_{r+s=k,r,s>0} \left(C_r(C_s(u,v),w) - C_r(u,C_s(v,w)) \right).$$

If one has cochains C_j, $j<k$ such that the star product they define is associative to order $k-1$, then the right hand side above is a cocycle ($\partial(\text{RHS}) = 0$) and one can extend the star product to order k if it is a coboundary ($\text{RHS} = \partial(C_k)$).

Theorem 2.3 (Vey [166]) *Every differential p-cocycle C on a manifold M is the sum of the coboundary of a differential (p-1)-cochain and a* 1*-differential skew-symmetric p-cocycle A:*

$$C = \partial B + A. \tag{2.30}$$

In particular, a cocycle is a coboundary if and only if its total skew-symmetri–zation, which is automatically 1*-differential in each argument, vanishes. Also*

$$H^p_{\text{diff}}(C^\infty(M), C^\infty(M)) = \Gamma(\Lambda^p TM).$$

Furthermore [53], given a connection ∇ on M, B can be defined from C by universal formulas.

By universal, we mean the following: any p-differential operator D of order maximum k in each argument can be written

$$D(u_1,\ldots,u_p) = \sum_{|\alpha_1|<k\ldots|\alpha_p|<k} D^{\alpha_1\ldots\alpha_p}_{|\alpha_1|,\ldots,|\alpha_p|} \nabla_{\alpha_1}u_1\ldots\nabla_{\alpha_p}u_p \tag{2.31}$$

where α's are multiindices, $D_{|\alpha_1|,\ldots,|\alpha_p|}$ are tensors (symmetric in each of the p groups of indices) and $\nabla_\alpha u = (\nabla\ldots(\nabla u))\left(\frac{\partial}{\partial x^{i_1}},\ldots,\frac{\partial}{\partial x^{i_q}}\right)$ when $\alpha = (i_1,\ldots,i_q)$. We claim that there is a B such that the tensors defining B are universally defined as linear combinations of the tensors defining C, universally meaning in a way which is independent of the form of C. An elementary proof of the above theorem can be found in [104]. Note that requiring differentiability of the cochains is essentially the same as requiring them to be local [54], local meaning that $C_r(u, v)(x) = 0$ as soon as u (or v) vanishes in a neighborhood of x.

2.4.2 Equivalence of Star Products

Definition 2.8 Two star products $*$ and $*'$ on (M, P) are said to be **equivalent** if there is a series $T = \mathrm{Id} + \sum_{r=1}^{\infty} \nu^r T_r$ of linear operators T_r on $C^\infty(M)$, such that

$$T(f * g) = Tf *' Tg. \qquad \text{We then write} \quad *' = T \cdot *. \tag{2.32}$$

One can write $T = \exp A$ where A is a series of linear operators on $C^\infty(M)$.

The T_r automatically vanish on constants since 1 is a unit for $*$ and for $*'$. Using linear operators which do not necessarily vanish on constants, one can pass from any associative deformation of the product of functions on a Poisson manifold (M, P) to another such deformation with 1 being a unit.

Exercise 2.18 Show that, on $\mathbb{R}^{2n}$, the Wick, the Standard and Moyal star products are all equivalent, in view of Eqs. (2.6) and (2.8); for instance, $*_{Wick} = T' \cdot *_{Weyl}$ for $T' = \exp \frac{i\nu}{4} \left(\frac{\partial^2}{\partial q^2} + \frac{\partial^2}{\partial p^2} \right)$.

Proposition 2.1 (Lichnerowicz [126], Deligne [72]) *If $*$ and $*'$ are differential star products and $T(u) = u + \sum_{r \geq 1} \nu^r T_r(u)$ is an equivalence so that $*' = T \cdot *$, then the T_r are differential operators.*

Proof If $T = \mathrm{Id} + \nu^k T_k + \cdots$, then $\partial T_k = C'_k - C_k$ is differential, so $C'_k - C_k$ is a differential 2-cocycle with vanishing skew-symmetric part. Thus, using Vey's formula, it is the coboundary of a differential 1-cochain E and $T_k - E$, being a 1-cocycle, is a vector field, hence T_k is differential. One then proceeds by induction, considering $T' = (\mathrm{Id} + \nu^k T_k)^{-1} \circ T = \mathrm{Id} + \nu^{k+1} T'_{k+1} + \cdots$ and the two differential star products $*$ and $*''$, where $*'' = (\mathrm{Id} + \nu^k T_k)^{-1} \cdot *'$, which are differential and equivalent through T' ($*'' = T' \cdot *$).

A differential star product is equivalent to one with linear term in ν given by $\frac{1}{2}\{u, v\}$. Indeed $C_1(u, v)$ is a Hochschild cocycle with antisymmetric part given by $\frac{1}{2}\{u, v\}$ so $C_1 = \frac{1}{2} P + \partial B$ for a differential 1-cochain B. If $T(u) := u + \nu B(u)$ then $*' = T \cdot *$ has the required form.

Proposition 2.2 ([128]) *Let $*$ and $*'$ be two differential star products on (M, ω) and suppose that $H^2(M; \mathbb{R}) = 0$. Then there exists a series $T = \mathrm{Id} + \sum_{k \geq 1} \nu^k T_k$ on $C^\infty(M)[[\nu]]$ such that $*' = T \cdot *$.*

Proof Let us suppose that, modulo some equivalence, the two star products $*$ and $*'$ coincide up to order k. Then associativity at order k shows that $C_k - C'_k$

is a Hochschild 2-cocycle and so by (2.3) can be written as $(C_k - C'_k)(u, v) = (\partial B)(u, v) + A(X_u, X_v)$ for a 2-form A. The total skew-symmetrization of the associativity relation at order $k+1$ shows that A is a closed 2-form. Since the second cohomology vanishes, A is exact, $A = dF$. Transforming by the equivalence defined by $Tu = u + \nu^{k-1} 2F(X_u)$, we can assume that the skew-symmetric part of $C_k - C'_k$ vanishes. Then $C_k - C'_k = \partial B$ where B is a differential operator. Using the equivalence defined by $T = I + \nu^k B$ we can assume that the star products coincide, modulo an equivalence, up to order $k+1$ and the result follows from induction.

In 1994, Fedosov proved the recursive construction explained in Sect. 2.3 and showed that two star products constructed with cohomologous series of 2-forms are equivalent. Following an induction reasoning as above, and using formula (2.29), it is easy [21] to show that any differential star product on a symplectic manifold (M, ω) is equivalent to a Fedosov star product. Hence the equivalence classes of star products on a symplectic manifold are parametrised by elements in $H^2(M; \mathbb{R})[[\nu]]$. This parametrization is also proven by Nest and Tsygan [135], and Deligne [72].

Definition 2.9 A **Poisson deformation** of the Poisson bracket on a Poisson manifold (M, P) is a Lie algebra deformation of $(C^\infty(M), \{\ ,\ \})$ which is a derivation in each argument, i.e. of the form

$$\{u, v\}_\nu = P_\nu(du, dv) \tag{2.33}$$

where $P_\nu = P + \sum \nu^k P_k$ is a series of skew-symmetric contravariant 2-tensors on M (such that $[P_\nu, P_\nu] = 0$). Two Poisson deformations P_ν and P'_ν of the Poisson bracket P on a Poisson manifold (M, P) are **equivalent** if there exists a formal path in the diffeomorphism group of M, starting at the identity, i.e. a series

$$T = \exp D = \mathrm{Id} + \sum_j \frac{1}{j!} D^j \text{ for } D = \sum_{r \geq 1} \nu^r D_r, \tag{2.34}$$

where the D_r are vector fields on M, such that

$$T\{u, v\}_\nu = \{Tu, Tv\}'_\nu \tag{2.35}$$

where $\{u, v\}_\nu = P_\nu(du, dv)$ and $\{u, v\}'_\nu = P'_\nu(du, dv)$.

Flato, Lichnerowicz and Sternheimer studied in [93] 1-differential deformations of the Poisson bracket on symplectic manifolds; one gets.

Proposition 2.3 *On a symplectic manifold (M, ω), the equivalence classes of Poisson deformations of the Poisson bracket P are parametrised by $H^2(M; \mathbb{R})[[\nu]]$.*

One first shows by induction that any Poisson deformation P_ν of the Poisson bracket P on a symplectic manifold (M, ω) is of the form P^Ω for a series $\Omega = \omega + \sum_{k \geq 1} \nu^k \omega_k$ where the ω_k are closed 2-forms, and $P^\Omega(du, dv) = -\Omega(X_u^\Omega, X_v^\Omega)$ where $X_u^\Omega = X_u + \nu(\dots) \in \Gamma(TM)[[\nu]]$ is the element defined by $i(X_u^\Omega)\Omega = du$.

One then shows that two Poisson deformations P^{Ω} and $P^{\Omega'}$ are equivalent if and only if ω_k and ω'_k are cohomologous for all $k \geq 1$. In fact

$$T P^{\Omega}(du, dv) = P^{\Omega'}(d(Tu), d(Tv))$$

with $T = \exp D$ for $D = \sum_{r\geq 1} \nu^r D_r$ iff $\Omega' = \exp(\mathcal{L}_D)\Omega$ so iff $\Omega' - \Omega = d\alpha$ for $\alpha = \sum_{k>0} \nu^k \alpha_k$ with

$$d\alpha = (\exp(\mathcal{L}_D) - \mathrm{Id})\Omega = d\left(\sum_{k\geq 0} \frac{1}{(k+1)!} i(D)(\mathcal{L}_D)^k \Omega\right).$$

In 1997, Kontsevich proved that the coincidence of the set of equivalence classes of star products and Poisson deformations is true for general Poisson manifolds:

Theorem 2.4 ([122]) *The set of equivalence classes of differential star products on a Poisson manifold (M, P) can be naturally identified with the set of equivalence classes of Poisson deformations of P.*

Parametrization of equivalence classes of special star products are known; in particular for star products on pseudo Kähler manifolds with "separation of variables" (i.e. such that $f * u = fu$ and $u * g = u$ whenever f is holomorphic or g antiholomorphic), Karabegov [115] showed that one has even a parametrization of all such star products by series of closed (1, 1)-forms.

Remark 2.10 Although the definition of equivalence is mathematically beautiful, it has drawbacks; a given classical polynomial function on $\mathbb{R}^{2n}$, when quantized relatively to two different orderings, does not lead to operators with the same spectrum. Hence equivalence is too broad to give isospectrality for a given classical observable (provided one could define a good notion of spectrum!). On the other hand, if one considers the whole deformed algebras, one likes to know when two deformed algebras have equivalent sets of representations. This enters the realm of Morita equivalence. The theory of representations of $*$-algebras (in the sense of Definition 2.4) is introduced by Bordemann and Waldmann in [37, 168] extending classical constructions existing for C^*-algebras; the Morita equivalence of star products is studied in [47–49].

Remark 2.11 Deligne [72] defines cohomological classes associated to differential star products on a symplectic manifold and this leads to an intrinsic way to define a characteristic class $c(*)$ of a star product $*$, which parametrizes its equivalence class; the methods do not extend to general Poisson manifolds. A selfcontained presentation of these classes is given in [104]. This allows to characterize isomorphisms in the following way: two differential star products $*$ on (M, ω) and $*'$ on (M', ω') are isomorphic if and only if there exist $f(\nu) = \sum_{r\geq 1} \nu^r f_r \in \mathbb{R}[[\nu]]$ with $f_1 \neq 0$ and $\psi\colon M' \to M$, a symplectomorphism, such that $(\psi^{-1})^* c(*')(f(\nu)) = c(*)(\nu)$. In particular [101]: if $H^2(M; \mathbb{R}) = \mathbb{R}[\omega]$ then there is only one star product up to equivalence and change of parameter. A symplectomorphism ψ of a symplectic manifold can be extended to a ν-linear automorphism of a given differential star product on (M, ω) if and only if $(\psi)^*c(*) = c(*)$. Notice that this is always the case if ψ can be connected to the identity by a path of symplectomorphisms (and this result is in Fedosov [82]). Homomorphisms of star products have been studied by Bordemann in [36].

2.5 Star Products on Poisson Manifolds and Formality

Kontsevich proved that the set of equivalence classes of star products is in bijection with the set of equivalence classes of formal Poisson structures on a general Poisson manifold in [122], as a consequence of his formality theorem. A differential star product on M is defined by a series of bidifferential operators satisfying some identities; a formal Poisson structure on a manifold M is defined by a series of bivector fields (i.e. contravariant skew-symmetric 2-tensors) P satisfying certain properties; to describe a correspondence between these objects, one considers algebras they belong to.

Definition 2.10 A **graded Lie algebra** is a $\mathbb{Z}$-graded vector space $\mathfrak{g} = \bigoplus_{i\in\mathbb{Z}} \mathfrak{g}^i$ endowed with a bilinear operation

$$[\ ,\ ]\colon \mathfrak{g}\otimes\mathfrak{g}\to\mathfrak{g}$$

which is is graded ($[\,a\,,\,b\,]\subset\mathfrak{g}^{|a|+|b|}$), graded skew-symmetric,

$$[\,a\,,\,b\,] = -(-1)^{|a|\,|b|}[\,b\,,\,a\,]$$

and satisfies the graded Jacobi identity

$$[\,a\,,\,[\,b\,,\,c\,]\,] = [\,[\,a\,,\,b\,]\,,\,c\,] + (-1)^{|a|\,|b|}[\,b\,,\,[\,a\,,\,c\,]\,] \qquad (a\in\mathfrak{g}^{|a|}, b\in\mathfrak{g}^{|b|}).$$

Any Lie algebra is a graded Lie algebra concentrated in degree 0, and the degree zero part $\mathfrak{g}^0$ and the even part $\mathfrak{g}^{even} := \bigoplus_{i\in\mathbb{Z}} \mathfrak{g}^{2i}$ of any graded Lie algebra are Lie algebras in the usual sense.

Definition 2.11 A **differential graded Lie algebra** (briefly DGLA) is a graded Lie algebra $\mathfrak{g}$ endowed with a differential, $d\colon \mathfrak{g}\to\mathfrak{g}$, i.e. a linear operator of degree 1 ($d\colon \mathfrak{g}^i\to\mathfrak{g}^{i+1}$) which squares to zero ($d\circ d = 0$) and satisfies the compatibility condition (Leibniz rule)

$$d[\,a\,,\,b\,] = [\,d\,a\,,\,b\,] + (-1)^{\alpha}[\,a\,,\,d\,b\,] \qquad a\in\mathfrak{g}^{\alpha}, b\in\mathfrak{g}^{\beta}.$$

The natural notions of morphisms of graded and differential graded Lie algebras are graded linear maps which commute with the differentials and the brackets (a graded linear map $\phi\colon \mathfrak{g} \to \mathfrak{h}$ of degree k is a linear map such that $\phi(\mathfrak{g}^i) \subset \mathfrak{h}^{i+k}\ \forall i \in \mathbb{Z}$). Remark that a morphism of DGLA's has to be a degree 0 in order to commute with the other structures.

Any DGLA has a cohomology complex defined by

$$\mathscr{H}^i(\mathfrak{g}) := \mathrm{Ker}(d\colon \mathfrak{g}^i \to \mathfrak{g}^{i+1}) \Big/ \mathrm{Im}\,(d\colon \mathfrak{g}^{i-1} \to \mathfrak{g}^i).$$

The set $\mathscr{H} := \bigoplus_i \mathscr{H}^i(\mathfrak{g})$ has a natural structure of graded vector space and inherits the structure of a graded Lie algebra, defined by $[\,|a|\,,\,|b|\,]_{\mathscr{H}} := \big|[\,a\,,\,b\,]_{\mathfrak{g}}\big|$ where $|a| \in \mathscr{H}$ denote the equivalence class of a closed element $a \in \mathfrak{g}$. The cohomology of a DGLA is itself a DGLA with zero differential.

Any morphism $\phi\colon \mathfrak{g}_1 \to \mathfrak{g}_2$ of DGLA's induces a morphism $(\phi)\colon \mathscr{H}_1 \to \mathscr{H}_2$. A morphism of DGLA's inducing an isomorphism in cohomology is called a **quasi-isomorphism**.

2.5.1 The DGLA of Polydifferential Operators

Let A be an associative algebra with unit on a field $\mathbb{K}$; consider the complex of multilinear maps from A to itself:

$$\mathscr{C} := \sum_{i=-1}^{\infty} \mathscr{C}^i \qquad \mathscr{C}^i := \mathrm{Hom}_{\mathbb{K}}(A^{\otimes(i+1)}, A)$$

(we shifted the degree by one; the degree $|A|$ of a $(p+1)$–linear map A is equal to p). The Lie bracket of linear operators is the skew-symmetrization of the composition of linear operators. This notion is extended to multilinear operators: for $A_1 \in \mathscr{C}^{m_1}$, $A_2 \in \mathscr{C}^{m_2}$, one defines:

$$(A_1 \circ A_2)(f_1, \ldots, f_{m_1+m_2+1}) :=$$
$$\sum_{j=1}^{m_1} (-1)^{(m_2)(j-1)} A_1(f_1, \ldots, f_{j-1}, A_2(f_j, \ldots, f_{j+m_2}), f_{j+m_2+1}, \ldots, f_{m_1+m_2+1})$$

for any $(m_1 + m_2 + 1)$- tuple of elements of A and the **Gerstenhaber bracket** is defined by

$$[A_1, A_2]_G := A_1 \circ A_2 - (-1)^{m_1 m_2} A_2 \circ A_1.$$

It gives $\mathscr{C}$ the structure of a graded Lie algebra. The differential d_D is defined by

$$d_D A = -[\mu, A] = -\mu \circ A + (-1)^{|A|} A \circ \mu$$

where μ is the usual product in the algebra A. Hence $dA = (-1)^{|A|+1}\delta A$ where δ is the Hochschild coboundary. The graded Lie algebra $\mathscr{C}$ with the differential d_D is a differential graded Lie algebra.

Here the algebra A is $C^\infty(M)$, and we consider the subalgebra of $\mathscr{C}$ consisting of multidifferential operators $\mathscr{D}_{poly}(M) := \bigoplus \mathscr{D}^i_{poly}(M)$ with $\mathscr{D}^i_{poly}(M)$ the space of multi differential operators acting on $i+1$ smooth functions on M and vanishing on constants. Clearly $\mathscr{D}_{poly}(M)$ is closed under the Gerstenhaber bracket and under the differential d_D, so that it is a DGLA.

Proposition 2.4 *An element C in $\nu\mathscr{D}^1_{poly}(M)[[\nu]]$, i.e. a series of bidifferential operator, yields a deformation of the usual associative pointwise product μ of functions, $* = \mu + C$, which defines a differential star product on M if and only if $d_D C - \frac{1}{2}[C, C]_G = 0$.*

2.5.2 The DGLA of Multivector Fields

A k **-multivector field** is a section of the k-th exterior power $\Lambda^k TM$ of the tangent space TM; the **Schouten-Nijenhuis bracket** is the bracket of multivectorfields defined by extending the usual Lie bracket of vector fields

$$[X_1 \wedge \ldots \wedge X_k, Y_1 \wedge \ldots \wedge Y_l]_S =$$
$$\sum_{r=1}^{k}\sum_{s=1}^{l}(-1)^{r+s}[X_r, X_s]X_1 \wedge \ldots \widehat{X_r} \wedge \ldots \wedge X_k \wedge Y_1 \wedge \ldots \widehat{Y_s} \wedge \ldots \wedge Y_l.$$

Since the bracket of an r- and an s- multivector fields on M is an $r+s-1$- multivector field, we define a structure of graded Lie algebra on the space $\mathscr{T}_{poly}(M)$ of multivector fields on M by setting $\mathscr{T}^i_{poly}(M)$ to be the set of skew-symmetric contravariant $i+1$-tensorfields on M (remark again a shift in the grading). The graded Lie algebra $\mathscr{T}_{poly}(M)$ is a differential graded Lie algebra choosing the differential d_T to be identically zero.

Proposition 2.5 *An element $P \in \nu\mathscr{T}^1_{poly}(M)[[\nu]]$ (i.e. a series of bivectorfields on the manifold M) defines a formal Poisson structure on M if and only if $d_T P - \frac{1}{2}[P, P]_S = 0$.*

If one could construct an isomorphism of DGLA between the algebra $\mathscr{T}_{poly}(M)$ of multivector fields and the algebra $\mathscr{D}_{poly}(M)$ of multidifferential operators, this would give a correspondence between a formal Poisson tensor on M and a formal differential star product on M. By Theorem 2.3 the cohomology of the algebra of multidifferential operators is given by multivector fields

$$\mathscr{H}^i(\mathscr{D}_{poly}(M)) \simeq \mathscr{T}^i_{poly}(M).$$

This bijection is induced by the natural map $U_1 \colon \mathcal{T}^i_{poly}(M) \longrightarrow \mathcal{D}^i_{poly}(M)$ which extends the usual identification between vector fields and first order differential operators, and is defined by:

$$U_1(X_0 \wedge \ldots \wedge X_n)(f_0, \ldots, f_n) = \frac{1}{(n+1)!} \sum_{\sigma \in S_{n+1}} \varepsilon(\sigma)\; X_0(f_{\sigma(0)}) \ldots X_n(f_{\sigma(n)}). \tag{2.36}$$

Exercise 2.19 Compute at order 2 to show that this map fails to preserve the Lie structure.

One extends the notion of morphism between two DGLA's to construct a morphism whose first order approximation is this isomorphism (2.36). To do so one introduces the notion of L_∞-morphism.

2.5.3 L_∞-Algebras, L_∞-Morphism and Formality

A toy picture of our situation (finding a correspondence between a formal Poisson tensor P on M and a formal differential star product $* = \mu + C$ on M) is the following. If C and P were elements in neighborhoods of zero of finite dimensional vector spaces V_1 and V_2, one could consider analytic vector fields X_1 on V_1, X_2 on V_2, vanishing at zero, given by $(X_1)_C = d_D C - \frac{1}{2}[C, C]_G$, $(X_2)_P = d_T P - \frac{1}{2}[P, P]_S$ and one would be interested in finding a correspondence between zeros of X_2 and zeros of X_1. An idea would be to construct an analytic map $\phi : V_2 \to V_1$ such that $\phi(0) = 0$ and $\phi_* X_2 = X_1$. Such a map can be viewed as an algebra morphism $\phi^* : A_1 \to A_2$ where A_i is the algebra of analytic functions on V_i vanishing at zero. The vector field X_i can be seen as a derivation of the algebra A_i. A real analytic function being determined by its Taylor expansion at zero, one can look at $C(V_i) := \sum_{n\geq 1} S^n(V_i)$ as the dual space to A_i; it is a coalgebra. One views the derivation of A_i corresponding to the vector field X_i dually as a coderivation Q_i of $C(V_i)$. One is then looking for a coalgebra morphism $F : C(V_2) \to C(V_1)$ so that $F \circ Q_2 = Q_1 \circ F$. This is generalized to the framework of graded algebras with the notion of L_∞-morphism between L_∞-algebras.

Definition 2.12 A **graded coalgebra** on the base ring $\mathbb{K}$ is a $\mathbb{Z}$–graded vector space $C = \bigoplus_{i\in\mathbb{Z}} C^i$ with a comultiplication, i.e. a graded linear map $\Delta \colon C \to C \otimes C$ such that $\Delta(C^i) \subset \bigoplus_{j+k=i} C^j \otimes C^k$ and such that one has coassociativity, i.e. $(\Delta \otimes \mathrm{id})\Delta(x) = (\mathrm{id} \otimes \Delta)\Delta(x)$ for every $x \in C$.

A counit (if it exists) is a morphism $e \colon C \to \mathbb{K}$ such that $e(C^i) = 0$ for any $i > 0$ and such that $(e \otimes \mathrm{id})\Delta = (\mathrm{id} \otimes e)\Delta = \mathrm{id}$.

The coalgebra is **cocommutative** if $T \circ \Delta = \Delta$ where $T \colon C \otimes C \to C \otimes C$ is the twisting map: $T(x \otimes y) := (-1)^{|x||y|}\, y \otimes x$ for x, y homogeneous elements of degree respectively $|x|$ and $|y|$.

Additional structures that can be put on an algebra can be dualized to give dual versions on coalgebras.

Example 2.3 (*The coalgebra* $C(V)$) If $V = \bigoplus_{i\in\mathbb{Z}} V^i$ is a graded vector space over $\mathbb{K}$, one defines the tensor algebra $T(V) = \bigoplus_{n=0}^{\infty} V^{\otimes n}$ with $V^{\otimes 0} = \mathbb{K}$, and the symmetric algebra $S(V) = T(V)/ < x \otimes y - (-1)^{|x||y|} y \otimes x >$ which is a naturally graded associative algebra. It has a structure of coalgebra with comultiplication Δ defined by $\Delta v := 1 \otimes v + v \otimes 1$ for a homogeneous element $v \in V$ and extended as algebra homomorphism.

The reduced symmetric space is $C(V) : =S^+(V) := \bigoplus_{n>0} S^n(V)$; it is the cofree cocommutative coalgebra without counit constructed on V. (Remark that Δv=0 iff $v\in V$.)

Definition 2.13 A **coderivation** of degree d on a graded coalgebra C is a graded linear map $\delta\colon C^i \to C^{i+d}$ which satisfies the (co–)Leibniz identity $\Delta\delta(v) = \delta v' \otimes v'' + (-1)^{d|v'|} v' \otimes \delta v''$ if $\Delta v = \sum v' \otimes v''$. This can be rewritten with the usual Koszul sign conventions $\Delta\delta = (\delta \otimes \mathrm{id} + \mathrm{id} \otimes \delta)\Delta$.

Definition 2.14 A L_∞**–algebra** is a graded vector space V over $\mathbb{K}$ and a degree 1 coderivation Q defined on the reduced symmetric space $C(V[1])$ so that $Q \circ Q = 0$. (Given any graded vector space V, we can obtain a new graded vector space $V[k]$ by shifting the grading of the elements of V by k, i.e. $V[k] = \bigoplus_{i\in\mathbb{Z}} V[k]^i$ where $V[k]^i := V^{i+k}$.)

Definition 2.15 A L_∞**–morphism** between two L_∞–algebras, $F\colon (V, Q) \to (V', Q')$, is a morphism $F\colon C(V[1]) \longrightarrow C(V'[1])$ of graded coalgebras, so that $F \circ Q = Q' \circ F$.

Any algebra morphism from $S^+(V)$ to $S^+(V')$ is uniquely determined by its restriction to V and any derivation of $S^+(V)$ is determined by its restriction to V. In a dual way, a coalgebra–morphism F from the coalgebra $C(V)$ to the coalgebra $C(V')$ is uniquely determined by the composition of F and the projection on $\pi' : C(V') \to V'$. Similarly, any coderivation Q of $C(V)$ is determined by the composition $F \circ \pi$ where π is the projection of $C(V)$ on V.

Definition 2.16 We call **Taylor coefficients of a coalgebra-morphism** $F\colon C(V) \to C(V')$ the sequence of maps $F_n\colon S^n(V) \to V'$ and **Taylor coefficients of a coderivation** Q of $C(V)$ the sequence of maps $Q_n\colon S^n(V) \to V$.

Proposition 2.6 *Given V and V' two graded vector spaces, any sequence of linear maps $F_n\colon S^n(V) \to V'$ of degree zero determines a unique coalgebra morphism $F\colon C(V) \to C(V')$ for which the F_n are the Taylor coefficients. Similarly, if V is a graded vector space, any sequence $Q_n\colon S^n(V) \to V, n \geq 1$ of linear maps of degree i determines a unique coderivation Q of $C(V)$ of degree i whose Taylor coefficients are the Q_n.*

The Taylor coefficients of a coderivation Q of $C(V[1])$ of degree 1 are the linear maps

$$Q_n : S^n(V[1]) \to V[2].$$

Proposition 2.7 *Any L_∞–algebra (V, Q) such that all the Taylor coefficients Q_n of Q vanish for $n > 2$ yields a differential graded Lie algebra and vice versa.*

A morphism of graded coalgebras between $C(V[1])$ and $C(V'[1])$ is equivalent to a sequence of linear maps (the Taylor coefficients)

$$F_n : S^n(V[1]) \to V'[1];$$

it defines a L_∞–morphism between two L_∞–algebras (V, Q) and (V', Q') iff $F \circ Q = Q' \circ F$. For DGLA's, there exist L_∞–morphisms between two DGLA's which are not DGLA–morphisms.

Definition 2.17 Given a L_∞ algebra (V, Q) over a field of characteristic zero, and given $\mathfrak{m} = \nu\mathbb{R}[[\nu]]$, a $\mathfrak{m}$–**point** is an element $p \in \nu C(V)[[\nu]]$ such that $\Delta p = p \otimes p$ or, equivalently, it is an element

$$p = e^v - 1 = v + \frac{v^2}{2} + \cdots \tag{2.37}$$

where v is an even element in $V[1] \otimes \mathfrak{m} = \nu V[1][[\nu]]$.

A solution of the generalized Maurer–Cartan equation is a $\mathfrak{m}$–point p at which Q vanishes; equivalently, it is an odd element $v \in \nu V[[\nu]]$ such that $Q_1(v) + \frac{1}{2}$ $Q_2(v \cdot v) + \cdots = 0$. If $\mathfrak{g}$ is a DGLA, it is thus an element $v \in \nu\mathfrak{g}[[\nu]]$ such that $dv - \frac{1}{2}[v, v] = 0$.

Exercise 2.20 The image under a L_∞ morphism of a solution of the generalised Maurer–Cartan equation is again such a solution. In particular, if one builds a L_∞ morphism $F : \mathscr{T}_{poly}(M) \to \mathscr{D}_{poly}(M)$ between the two DGLA's we have defined, the image under F of the point $e^\alpha - 1$ corresponding to a formal Poisson tensor,

$$\alpha \in \nu\mathscr{T}^1_{poly}(M)[[\nu]] \text{ such that } [\alpha, \alpha]_S = 0, \tag{2.38}$$

yields a star product on M,

$$* = \mu + \sum_n F_n(\alpha^n). \tag{2.39}$$

Definition 2.18 Two L_∞–algebras (V, Q) and (V', Q') are **quasi-isomorphic** if there is a L_∞–morphism F such that $F_1 : V \to V'$ induces an isomorphism in cohomology. Such a F is called a quasi-isomorphism.

Kontsevich has proven that if F is a L_∞–morphism between two L_∞–algebras (V, Q) and (V', Q') so that $F_1 : V \to V'$ induces an isomorphism in cohomology, then there exists a L_∞–morphism G between (V', Q') and (V, Q) so that $G_1 : V' \to V$ is a quasi inverse for F_1 .

Definition 2.19 Kontsevich's **formality** is a quasi isomorphism between the (L_∞–algebra structure associated to the) DGLA of multidifferential operators, $\mathscr{D}_{poly}(M)$, and its cohomology, the DGLA of multivector fields $\mathscr{T}_{poly}(M)$.

2.5.4 *Formality for* $\mathbb{R}^d$

Kontsevich [122] gave an explicit formula for the Taylor coefficients of a formality for $\mathbb{R}^d$, i.e. the Taylor coefficients F_n of an L_∞–morphism $F : (\mathscr{T}_{poly}(\mathbb{R}^d), Q) \to (\mathscr{D}_{poly}(\mathbb{R}^d), Q')$ where Q corresponds to the DGLA $(\mathscr{T}_{poly}(\mathbb{R}^d),\ [\ ,\ ]_S,\ D_T = 0)$ and Q' to the DGLA $(\mathscr{D}_{poly}(\mathbb{R}^d),\ [\ ,\ ]_G,\ d_D)$ with $F_1 : \mathscr{T}_{poly}(\mathbb{R}^d) \to \mathscr{D}_{poly}(\mathbb{R}^d)$ given by U_1 as in Eq. (2.36). The formula writes

$$F_n = \sum_{m\geq 0} \sum_{\Gamma \in G_{n,m}} \mathscr{W}_\Gamma B_\Gamma$$

- where $G_{n,m}$ is a set of oriented admissible graphs;
  An admissible graph $\Gamma \in G_{n,m}$ has n aerial vertices labelled $p_1, \ldots, p_n$, has m ground vertices labelled $q_1, \ldots, q_m$. From each aerial vertex p_i, a numer k_i of arrows are issued; each of them can end on any vertex except p_i but there can not be multiple arrows. There are no arrows issued from the ground vertices. One gives an order to the vertices:$(p_1, \ldots, p_n, q_1, \ldots, q_m)$ and one gives a compatible order to the arrows, labeling those issued from p_i with $(k_1 + \cdots + k_{i-1} + 1, \cdots, k_1 + \cdots + k_{i-1} + k_i)$. The arrows issued from p_i are named $\mathrm{Star}(p_i) = \{\overrightarrow{p_i a_1}, \ldots, \overrightarrow{p_i a_{k_i}}\}$ with $\overrightarrow{v_{k_1+\cdots+k_{i-1}+j}} = \overrightarrow{p_i a_j}$.

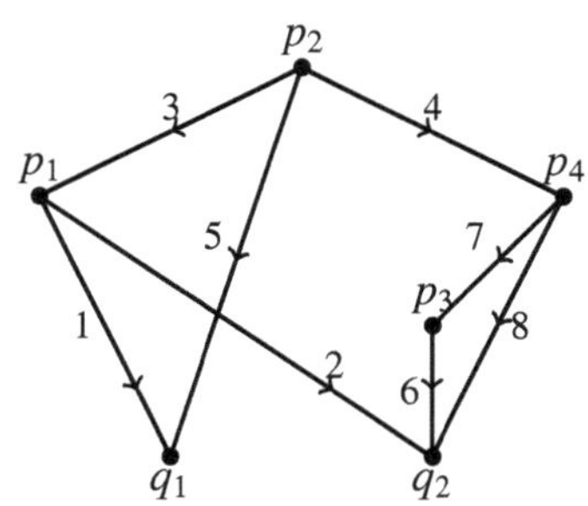

An example of graph Γ_1 in $G_{3,2}$

- where B_Γ associates a m–differential operator to an n–tuple of multivectorfields;
  Given a graph $\Gamma \in G_{n,m}$ and given n multivectorfields $(\alpha_1, \ldots, \alpha_n)$ on $\mathbb{R}^d$, one defines a m– differential operator $B_\Gamma(\alpha_1 \cdot \ldots \cdot \alpha_n)$; it vanishes unless α_1 is a k_1–tensor, α_2 is a k_2–tensor,..., α_n is a k_n–tensor and in that case it is given by:

$$B_\Gamma(\alpha_1 \cdot \ldots \cdot \alpha_n)(f_1, \ldots, f_n) = \sum_{i_1,\ldots,i_K} D_{p_1}\alpha_1^{i_1\cdots i_{k_1}} D_{p_2}\alpha_2^{i_{k_1+1}\cdots i_{k_1+k_2}} \ldots D_{p_n}\alpha_n^{i_{k_1+\cdots+k_{n-1}+1}\cdots i_K} D_{q_1} f_1 \ldots D_{q_m} f_m$$

where $K := k_1 + \cdots + k_n$ and where

$$D_a := \Pi_{j\mid \vec{v_j} = \cdot \vec{a}}\, \partial_{i_j}.$$

For the graph $\boldsymbol{\Gamma}_1$ as above, B_{Γ_1} associates a bidifferential operator [since $n=2$] of order 2 in the first variable [since two arrows arrive at the first ground vertex q_1] and of order 3 in the second variable [since three arrows arrive at the second ground vertex q_2] to a quadruple of multivectorfields $(\alpha_1, \dots, \alpha_4)$ on $\mathbb{R}^d$ [since $m=4$]. For this operator not to vanish, α_1 is a 2–tensor [since two arrows start from the first aerial vertex p_1], α_2 is a 3–tensor [since three arrows start from the second aerial vertex p_2], α_3 is a 1–tensor (a vector field) and α_4 is a 2–tensor; we have then

$$B_{\Gamma_1}(\alpha_1,\alpha_2,\alpha_3,\alpha_4)(f,g) = \sum_{i_1,\dots,i_8} \frac{\partial \alpha_1^{i_1 i_2}}{\partial x^{i_3}} \alpha_2^{i_3 i_4 i_5} \frac{\partial \alpha_3^{i_6}}{\partial x^{i_7}} \frac{\partial \alpha_4^{i_7 i_8}}{\partial x^{i_4}} \frac{\partial^2 f}{\partial x^{i_1} \partial x^{i_5}} \frac{\partial^3 g}{\partial x^{i_2} \partial x^{i_6} \partial x^{i_8}}.$$

- where $\mathscr{W}_\Gamma$ is the integral of a form ω_Γ over the compactification of a configuration space $C^+_{\{p_1,\dots,p_n\}\{q_1,\dots,q_m\}}$.
Let $\mathscr{H}$ denote the upper half plane $\mathscr{H} = \{z \in \mathbb{C} | Im(z) > 0\}$. We define

$$Conf^+_{\{z_1,\dots,z_n\}\{t_1,\dots,t_m\}} := \left\{z_1,\dots,z_n,t_1,\dots,t_m \;\middle|\; \begin{array}{l} z_j \in \mathscr{H};\, z_i \neq z_j \text{ for } i \neq j; \\ t_j \in \mathbb{R};\, t_1 < t_2 \cdots < t_m \end{array}\right\}$$

and $C^+_{\{p_1,\dots,p_n\}\{q_1,\dots,q_m\}}$ to be the quotient of this space by the action of the 2-dimensional group G of all transformations of the form $z_j \mapsto az_j + b \quad t_i \mapsto at_i + b \quad a > 0, b \in \mathbb{R}$. The configuration space $C^+_{\{p_1,\dots,p_n\}\{q_1,\dots,q_m\}}$ has dimension $2n+m-2$ and has an orientation induced on the quotient by $\Omega_{\{z_1,\dots,z_n;t_1,\dots,t_m\}} = dx_1 \wedge dy_1 \wedge \dots dx_n \wedge dy_n \wedge dt_1 \wedge \dots \wedge dt_m$ if $z_j = x_j + iy_j$.
The compactification $\overline{C^+_{\{p_1,\dots,p_n\}\{q_1,\dots,q_m\}}}$ is defined as the closure of the image of the configuration space $C^+_{\{p_1,\dots,p_n\}\{q_1,\dots,q_m\}}$ into the product of a torus and the product of real projective spaces $P^2(\mathbb{R})$ under the map Ψ induced from a map ψ defined on $Conf^+_{\{z_1,\dots,z_n\}\{t_1,\dots,t_m\}}$ in the following way: to any pair of distinct points A, B taken amongst the $\{z_j, \overline{z}_j, t_k\}$, ψ associates the angle $\arg(B-A)$ and to any triple of distinct points A, B, C in that set, ψ associates the element of $P^2(\mathbb{R})$ which is the equivalence class of the triple of real numbers $(|A-B|, |B-C|, |C-A|)$.
Given a graph $\boldsymbol{\Gamma} \in G_{n,m}$, one defines a form on $\overline{C^+_{\{p_1,\dots,p_n\}\{q_1,\dots,q_m\}}}$ induced by

$$\omega_\Gamma = \frac{1}{(2\pi)^{k_1+\dots+k_n}(k_1)!\dots(k_n)!} d\Phi_{\vec{v_1}} \wedge \dots \wedge d\Phi_{\vec{v_K}}$$

where $\Phi_{\overrightarrow{p_j a}} = \mathrm{Arg}(\frac{a-p_j}{a-\overline{p}_j})$.

For a detailed proof of this formality, we refer the reader to [12, 44]. This formality for $\mathbb{R}^d$ associates a star product on $C^\infty(\mathbb{R}^d)$ to a formal Poisson tensor on R^d and gives:

Theorem 2.5 ([122]) *Let α be a Poisson tensor on $\mathbb{R}^d$ (thus $\alpha \in \mathscr{T}^1_{poly}(\mathbb{R}^d)$ and $[\alpha,\alpha]_S = 0$), let X be a vector field on $\mathbb{R}^d$, let $f, g \in C^\infty(\mathbb{R}^d)$. Then the series of* *bidifferential operators*

$$P(\alpha) := \mu + C(\alpha) := \mu + \sum_{j=1}^{\infty} \frac{\nu^j}{j!} F_j(\alpha \cdot \quad \cdot \alpha) \tag{2.40}$$

defines a star product $*$ *on* $\mathbb{R}^d$ *and* $A(X, \alpha) = \sum_{j=0}^{\infty} \frac{\nu^j}{j!} F_{j+1}(X \cdot \alpha \cdot \quad \cdot \alpha)$ *is a series of differential operators yielding the relation*

$$A(X, \alpha) f * g + f * A(X, \alpha) g - A(X, \alpha)(f * g) = \frac{d}{dt}_{|0} P(\Phi^X_{t_*}\alpha)(f, g) \tag{2.41}$$

where Φ_t^X *is the flow of* X.

Kontsevich builds a formality for any manifold M. Cattaneo, Felder and Tomassini give in [64] a globalization on a Poisson manifold of Kontsevich local formula for a star product given above. Using similar techniques, Dolgushev [77] gave a globalisation of Kontsevich's formality, using a torsion free connection on the manifold. In particular this proves the existence of a universal star product when one has chosen a torsion free connection ∇ (universal meaning whose corresponding tensors -see formula (2.31)- are polynomials in the Poisson tensor, the curvature tensor and their covariant derivatives).

Remark 2.12 Tamarkin [162] gave another formulation to the quantization of Poisson manifolds, in the language of operads and Drinfeld's associators. Starting with Kontsevich's and Tamarkin's approaches, formality theory has rapidly evolved and now enters into many fields of research in mathematics (see, for instance [42, 118, 157]). In particular, the general pattern in non commutative geometries is that commutative rings of functions on classical spaces are replaced by more general non-commutative variants, regardless of whether there is still an actual space of points corresponding to this. Deformation theoretic ideas have been important to give classes of examples (see, for instance, [164]) . A nice description of formality and its links with representation theory is given in the book [44]. In that area, recent results give new associators built using the formality, and a new proof of the Kashiwara Vergne conjecture by Alekseev and Torossian [2, 3].

2.6 Group Actions in Deformation Quantization

2.6.1 In a Classical Setting

Definition 2.20 Let (M, P) be a Poisson manifold and consider a smooth left action of a Lie group G on the manifold M,

$$G \times M \to M : (g, p) \mapsto \rho(g)p = g \cdot p.$$

The group acts by **Poisson diffeomorphisms** if and only if

$$\{\rho(g)^* u, \rho(g)^* v\} = \rho(g)^*(\{u, v\}) \qquad \forall u, v \in C^\infty(M), \forall g \in G, \tag{2.42}$$

or, equivalently, if and only if $\rho(g)_* P = P$ for all $g \in G$.

Exercise 2.21 When the Poisson structure is associated to a symplectic structure (M, ω), condition (2.42) is equivalent to $\rho(g)^*\omega = \omega$ for all $g \in G$.

When G acts by Poisson diffeomorphisms, it is a symmetry group for our classical system. Any element X in the Lie algebra $\mathfrak{g}$ of G gives rise to a fundamental vector field X^{*M} defined by

$$X^{*M}_p = \frac{d}{dt}_{|0} \rho(\exp -tX)p$$

(the minus sign is used to have a Lie algebra homomorphism, $\mathfrak{g} \to \chi(M)$ into the Lie algebra of smooth vector fields $[X^{*M}, Y^{*M}] = [X, Y]^{*M}$, $\forall X, Y \in \mathfrak{g}$) and we have an infinitesimal Poisson action of the Lie algebra $\mathfrak{g}$

$$\mathscr{L}_{X^{*M}}\{u, v\} = \{\mathscr{L}_{X^{*M}} u, v\} + \{u, \mathscr{L}_{X^{*M}} v\} \tag{2.43}$$

or equivalently $\mathscr{L}_{X^{*M}} P = 0$; or, in the symplectic case, $\mathscr{L}_{X^{*M}}\omega = 0$ i.e. $\iota(X^{*M})\omega$ is a closed 1-form.

The action of the Lie group is completely determined by the action of its Lie algebra when the Lie group G is connected.

Of particular importance in physics is the case of a so called **(almost) Hamiltonian action** where each fundamental vector field is Hamiltonian, i.e. when for each $X \in \mathfrak{g}$ there exists a function f_X on M such that

$$X^{*M} u = \{f_X, u\} \qquad \forall u \in C^\infty(M). \tag{2.44}$$

In the symplectic case this amounts to say that $\iota(X^{*M})\omega = df_X$.

Indeed, when the Hamiltonian governing the dynamics on (M, P) is invariant under the action of G, any of those functions f_X is a constant of the motion. One can always assume, when all the fundamental vector fields are Hamiltonian, that $X \to f_X$ is linear.

A further assumption is to require that the fundamental vector fields are Hamiltonian by means of a G equivariant map from M into the dual of the Lie algebra (G acting on $\mathfrak{g}^*$ by Ad^*)

$$J : M \to \mathfrak{g}^* \tag{2.45}$$

i.e. $X^{*M} u = \{J(X), u\}$, $\forall u \in C^\infty(M)$ with $J(X) \in C^\infty(M)$ defined by $J(X)(p) :=< J(p), X >$, $< .,. >$ denoting the pairing between $\mathfrak{g}$ and its dual. One says then that the action possesses **a G equivariant moment map** J. Equivariance means that the Hamiltonian functions satisfy

$$J(X)\rho((g)p) = (J(Adg^{-1}X))(p) \text{ and thus } \{ J(X), J(Y) \} = J([X, Y]) \tag{2.46}$$

An action so that each fundamental vector field is Hamiltonian and so that the correspondence $X \mapsto f_X$ can be chosen to be a homomorphism of Lie algebras is also called a **strongly Hamiltonian action**. When the group G acting on M is connected, it is equivalent to the existence of a G equivariant moment map.

2.6.2 In the Deformation Quantization Setting

The action of a Lie group on the classical Hilbert space framework of quantum mechanics is described by a unitary representation of the group on the Hilbert space.

In the setting of deformation quantization, the classical action of a group G on a Poisson manifold extends to the algebra of observables $C^\infty(M)[[\nu]]$ and one can define in this way different notions of invariance of the deformation quantization under the action of a Lie group.

Definition 2.21 Assume (M, P) is a Poisson manifold and G is a Lie group acting on M; as before $G \times M \to M : (g, p) \mapsto \rho(g)p = g \cdot p$. Let $(\mathbb{C}^\infty(M)[[\nu]], *)$ be a deformation quantization of (M, P). The star product is said to be **geometrically invariant** if, for any $g \in G$ and all $u, v \in C^\infty(M)$, one has

$$\rho(g)^* (u * v) = \rho(g)^* u * \rho(g)^* v. \tag{2.47}$$

Exercise 2.22 Show that geometric invariance implies (looking at the skew symmetric part of order 1 in the parameter ν) that

$$\rho(g)^* (\{u, v\}) = \{\rho(g)^* u, \rho(g)^* v\}$$

so that G acts by Poisson diffeomorphisms. Any fundamental vector field X^{*M} is then a derivation of the star product

$$X^{*M} (u * v) = (X^{*M} u) * v + u * (X^{*M} v). \tag{2.48}$$

More generally, symmetries in quantum theories are automorphisms of the algebra of observables. Thus we define a symmetry σ of a star product $* = \sum_r \nu^r C_r$ as an automorphism of the $\mathbb{C}[\![\nu]\!]$-algebra $C^\infty(M)[\![\nu]\!]$ with multiplication given by $*$:

$$\sigma(u * v) = \sigma(u) * \sigma(v), \qquad \sigma(1) = 1,$$

where σ is determined by a formal series $\sigma(u) = \sum_{r \geq 0} \nu^r \sigma_r(u)$ of linear maps. Any such automorphism σ of a star product on a Poisson manifold (M, P) can be written $\sigma(u) = T(u \circ \tau)$ where τ is a Poisson diffeomorphism of (M, P) and $T = \mathrm{Id} + \sum_{r \geq 1} \nu^r T_r$ is a formal series of differential maps.

A Lie group G acts as **symmetries of a deformed algebra** $(C^\infty(M)[[\nu]], *)$ if there is a homomorphism

$$\sigma : G \to \mathrm{Aut}(M, *).$$

In that case, one can write

$$\sigma(g)u = T(g)(\tau(g)^* u) \quad \text{for any } u \in C^\infty(M)$$

and $\tau : G \times M \to M$ defines a Poisson action of G on (M, P).

At the level of the Lie algebra, **an action of the Lie algebra** $\mathfrak{g}$ **on the deformed algebra**, is a homomorphism

$$D : \mathfrak{g} \to \mathrm{Der}(M, *)$$

into the space of derivations of the star product.

Now a derivation D of the star product is said to be **essentially inner** or **Hamiltonian** if $D = \frac{1}{\nu}\,\mathrm{ad}_*\, u$ for some $u \in C^\infty(M)[\![\nu]\!]$. We denote by $\mathrm{Inn}(M, *)$ the essentially inner derivations of $*$. It is a linear subspace of $\mathrm{Der}(M, *)$ and is the quantum analogue of the Hamiltonian vector fields. By analogy with the classical case, we call an action of a Lie algebra (or of a Lie group) on a deformed algebra **almost $*$-Hamiltonian** if each $D(X)$, for any $X \in \mathfrak{g}$, is essentially inner, and we call (quantum) Hamiltonian a linear choice of functions $\tilde{f}_X$ satisfying

$$D(X) = \tfrac{1}{\nu}\,\mathrm{ad}_*\, \tilde{f}_X, \qquad X \in \mathfrak{g}.$$

We say the action is **$*$-Hamiltonian** if u_X can be chosen to make the map

$$\mathfrak{g} \to C^\infty(M)[\![\nu]\!] : X \mapsto \tilde{f}_X$$

a homomorphism of Lie algebras.

When the deformed algebra is invariant by a classical (undeformed) Poisson action of a Lie group G on M, if the action of the Lie algebra $\mathfrak{g}$ defined by the fundamental vector fields ($D(X) = X^{*M}$) is $*$-Hamiltonian, a map $\tilde{J} : \mathfrak{g} \to C^\infty(M)[\![\nu]\!]$ is called **a quantum moment map** [172]. Thus it is a homomorphism of algebras

$$\mathfrak{g} \to C^\infty(M)[\![\nu]\!] : X \mapsto \tilde{J}_X \qquad \text{such that} \quad X^{*M} = \tfrac{1}{\nu}\,\mathrm{ad}_*\, \tilde{J}_X \qquad \forall X \in \mathfrak{g}.$$

In [8], they called quantization such a homomorphism of Lie algebras.

When there is a strongly Hamiltonian action of Lie group G on a Poisson manifold (M, P) a star product is said to be **covariant** under G if

$$f_X * f_Y - f_Y * f_X = \nu f_{[X,Y]} \qquad \forall X, Y \in \mathfrak{g}$$

where $f : \mathfrak{g} \to \mathbb{C}^\infty(M)$ is the homomorphism of Lie algebras describing the fundamental vector fields as Hamiltonian vector fields ($X^{*M}u = \{f_X, u\}$) and it is called **strongly invariant** if it is both geometrically invariant and covariant. In that case, f is a quantum moment map.

Exercise 2.23 Check that the Moyal star product on $\mathbb{R}^{2n}$ endowed with the canonical Poisson bracket $P_0 = \begin{pmatrix} 0 & -I \\ I & 0 \end{pmatrix}$ is strongly invariant under the natural action of the symplectic group on $\mathbb{R}^{2n}$.

Remark 2.13 One can go further and look at actions of a deformed group on a star product [81]. One way to deform a group is to deform in the Hopf category a Hopf algebra associated to the group. One enters into the realm of quantum groups [79]; these are introduced in the lectures of Christian Kassel. Links between quantum groups and deformation quantization appear in [23, 30, 31].

2.6.3 Classification of Invariant Star Products

When a Lie group G acts on the symplectic manifold (M, ω) and is a group of symmetries of a natural $*$ product, then (see Remark 2.9) there is a symplectic connection on (M, ω) which is invariant under G. We shall say that two star products which are invariant under G are G-invariantly equivalent if there is an equivalence $T = \mathrm{id} + \sum_{j=1}^{\infty} \nu^j T_j$ between them which commutes with the action of G. Using the results stated before, one can prove.

Proposition 2.8 [22] *Let G be a Lie group which acts symplectically on (M, ω). Suppose $*$ is a star product which is invariant under G and assume there is a symplectic connection which is invariant under G. Then, there exists a series of G-invariant closed 2-form $\Omega \in Z^2(M; \mathbb{R})^{G-inv}[[\nu]]$ such that $*$ is G-invariantly equivalent to the Fedosov star product constructed from the invariant connection ∇ and Ω, i.e. there exists a series $T = \mathrm{id} + \sum_{j=1}^{\infty} \nu^j T_j$ of G-invariant differential operators such that $* = T \cdot *_{\nabla,\Omega}$.*

Furthermore, two G-invariant star products $*_{\nabla,\Omega}$ and $*_{\nabla,\Omega'}$ are G-invariantly equivalent if and only if $\Omega - \Omega'$ is the boundary of a series of G-invariant 1-forms on M.

Hence there is a bijection between the G-invariant equivalence classes of G-invariant $*$-products on (M, ω) and the space of formal series of elements in the second space of invariant cohomology of M, $H^2(M, \mathbb{R})^{G-inv}[[\nu]]$.

Remark 2.14 On a Poisson manifold (MP) endowed with a $\mathfrak{g}$-action, if there exists a $\mathfrak{g}$-invariant connection, one can use Dolgushev's formality [77] to build also in this case a correspondence between $\mathfrak{g}$-invariant equivalence classes of $\mathfrak{g}$-invariant Poisson deformations of P and $\mathfrak{g}$-invariant equivalence classes of $\mathfrak{g}$ invariant star products.

2.6.4 Invariance of Fedosov's Star Product

We shall denote by $*_{\nabla,\Omega}$ the star product on a symplectic manifold (M,ω) obtained by Fedosov's construction using the symplectic connection ∇ and the series of closed 2-forms Ω.

Lemma 2.1 *Any diffeomorphism ϕ of (M,ω) is a symmetry of $*_{\nabla,\Omega}$ iff it preserves the symplectic 2-form ω, the connection ∇ and the series of closed 2-forms Ω. A vector field X is a derivation of $*_{\nabla,\Omega}$ if and only if $\mathscr{L}_X\omega=0$, $\mathscr{L}_X\Omega=0$, and $\mathscr{L}_X\nabla=0$.*

Exercise 2.24 Prove this Lemma, using the fact that the star product $*_{\nabla,\Omega}$ is natural and the associated connection (see Remark 2.9) is ∇. Hence invariance of ∇ is a necessary condition for the invariance of $*_{\nabla,\Omega}$ by a diffeomorphism of M. Use also the characterization, given in Eq. 2.29, of the 2-forms appearing in Ω.

Many authors have studied whether such a derivation is Hamiltonian for the star product (see, for instance [105, 123, 132]). We give here the proof obtained with J. Rawnsley.

Theorem 2.6 [105] *A vector field X is an inner derivation of $*=*_{\nabla,\Omega}$ if and only if $\mathscr{L}_X\nabla=0$ and there exists a series of functions λ_X such that*

$$i(X)\omega - i(X)\Omega = d\lambda_X.$$

In that case $X(u)=\frac{1}{\nu}(\mathrm{ad}_*\,\lambda_X)(u)$.

Proof With the same notation as above, for any smooth vector field X on M, one has:

$$\delta\circ i(X)+i(X)\circ\delta=\tfrac{1}{\nu}\,\mathrm{ad}_*(\omega_{ij}X^iy^j)\qquad \mathrm{ad}_*\,r\circ i(X)+i(X)\circ\mathrm{ad}_*\,r=\mathrm{ad}_*(i(X)r)$$

and $\partial\circ i(X)+i(X)\circ\partial=\mathscr{L}_X-(\nabla_iX)^jy^i\partial_{y^j}$ which can be rewritten as

$$\partial\circ i(X)+i(X)\circ\partial=\mathscr{L}_X+\tfrac{1}{\nu}\,\mathrm{ad}_*\left(-\tfrac{1}{2}(\nabla_i(i(X)\omega))_jy^iy^j\right)+\tfrac{1}{2}(di(X)\omega)_{ip}y^iP^{jp}\partial_{y^j}.$$

This gives the generalised Cartan formula first given by Neumaier:

$$\begin{aligned}\mathscr{L}_X &= D\circ i(X)+i(X)\circ D+\tfrac{1}{\nu}\,\mathrm{ad}_*(\omega_{ij}X^iy^j)+\tfrac{1}{\nu}\,\mathrm{ad}_*(i(X)r)\\ &\quad+\tfrac{1}{\nu}\,\mathrm{ad}_*\left(\tfrac{1}{2}(\nabla_i(i(X)\omega))_jy^iy^j\right)-\tfrac{1}{2}(di(X)\omega)_{ip}y^iP^{jp}\partial_{y^j}.\end{aligned}\qquad(2.49)$$

The last term obviously drops out when X is a symplectic vector field.

We now assume that X is a symplectic vector field preserving the connection and preserving the series of 2-forms Ω, then $\mathscr{L}_Xr=0$ so

$$-Di(X)r=i(X)Dr+\tfrac{1}{\nu}\left[\omega_{ij}X^iy^j+\tfrac{1}{2}(\nabla_i(i(X)\omega))_jy^iy^j+i(X)r,r\right].$$

Using the definition of r, this gives $-Di(X)r = i(X)\overline{R} - i(X)\Omega + \frac{1}{\nu}\left[\omega_{ij}X^i y^j + \frac{1}{2}(\nabla_i(i(X)\omega))_j y^i y^j, r\right]$. On the other hand, using the fact that $Da = \partial a - \delta(a) - \frac{1}{\nu}[r, a]$ one has

$$D(\omega_{ij}X^i y^j) = -i(X)\omega + \partial(\omega_{ij}X^i y^j) + \tfrac{1}{\nu}[\omega_{ij}X^i y^j, r],$$

$D\left(\frac{1}{2}(\nabla_i(i(X)\omega))_j y^i y^j\right) = -\nabla_i(i(X)\omega))_j dx^i y^j + \partial\left(\frac{1}{2}(\nabla_i(i(X)\omega))_j y^i y^j\right) + \frac{1}{\nu}\left[\frac{1}{2}(\nabla_i(i(X)\omega))_j y^i y^j, r\right]$. Since X is an affine vector field, one has $(i(X)R)(Y)Z = (\nabla^2 X)(Y, Z)$ so that

$$\partial\left(\frac{1}{2}(\nabla_i(i(X)\omega))_j y^i y^j\right) = -\frac{1}{2}((\nabla^2 X)^p_{ki}\omega)_{jp} y^i y^j dx^k = i(X)\overline{R}.$$

Hence $D\left(-i(X)r - \omega_{ij}X^i y^j - \frac{1}{2}(\nabla_i(i(X)\omega))_j y^i y^j\right) = i(X)\omega - i(X)\Omega$. So, for any vector field X so that $\mathcal{L}_X\omega = 0$, $\mathcal{L}_X\Omega = 0$ and $\mathcal{L}_X\nabla = 0$, one has

$$\mathcal{L}_X = D \circ i(X) + i(X) \circ D + \tfrac{1}{\nu}\,\mathrm{ad}_*(T(X))$$

with $T(X) = i(X)r + \omega_{ij}X^i y^j + \frac{1}{2}(\nabla_i(i(X)\omega))_j y^i y^j$ and $DT(X) = -i(X)\omega + i(X)\Omega$.

In particular, if there exists a series of smooth functions λ_X such that

$$i(X)\omega - i(X)\Omega = d\lambda_X \tag{2.50}$$

one can write $\mathcal{L}_X = D \circ i(X) + i(X) \circ D + \frac{1}{\nu}\,\mathrm{ad}_*(\lambda_X + T(X))$ with $D(\lambda_X + T(X)) = 0$.

Thus $\lambda_X + T(X)$ is the flat section associated to the series of smooth function on M obtained by taking the part of $\lambda_X + T(X)$ with no y terms hence λ_X (notice that $i(X)r$ has no terms without a y from the construction of r). If Q denotes the quantisation map associating a flat section to a series in ν of smooth functions, the above yields

$$\mathcal{L}_X = D \circ i(X) + i(X) \circ D + \tfrac{1}{\nu}\,\mathrm{ad}_*(Q(\lambda_X)).$$

Since in those assumptions the map Q commutes with $\mathcal{L}_X$ one has

$$Q(Xf) = \mathcal{L}_X Q(f) = \tfrac{1}{\nu}[Q(\lambda_X), Q(f)]$$

so that for any smooth function f, one has

$$Xf = \tfrac{1}{\nu}(\mathrm{ad}_*\,\lambda_X)(f).$$

This was first stated by Kravchenko (Proposition 4.3 of [123]).

We have seen above that such a vector field X is an inner derivation if $i(X)(\omega - \Omega)$ is exact. We shall show now that this is also a necessary condition.

Assume X is a vector field on M such that there exists a series of smooth functions λ_X with

$$X(u) = \tfrac{1}{\nu}(\mathrm{ad}_* \lambda_X)(u) \tag{2.51}$$

for every smooth function u on M. Then X is a derivation of $*$ so $\mathscr{L}_X\omega = 0$, $\mathscr{L}_X\Omega = 0$, $\mathscr{L}_X\nabla = 0$ and

$$Q(Xf) = \mathscr{L}_X Q(f) = \tfrac{1}{\nu}[T(X), Q(f)]$$

with $T(X) = i(X)r + \omega_{ij}X^i y^j + \frac{1}{2}(\nabla_i(i(X)\omega))_j y^i y^j$ and $DT(X) = -i(X)\omega + i(X)\Omega$.
Taking a contractible open set U in M, there exists a series of smooth locally defined functions λ_X^U on U so that $(i(X)\omega - i(X)\Omega)|_U = d\lambda_X^U$ and, everything being local, we have on U

$$D(\lambda_X^U + T(X))|_U = 0,$$

thus $\lambda_X^U + T(X)$ is the flat section on U associated to the series of smooth functions on U obtained by taking the part of $\lambda_X^U + T(X)$ with no y terms (which is λ_X^U) and

$$Q(X(u))|_U = \mathscr{L}_X Q(u)|_U = \tfrac{1}{\nu}[Q(\lambda_X^U), Q(u)]\big|_U$$

so that $X(u)|_U = \frac{1}{\nu}(\mathrm{ad}_{*_{\nabla,\Omega}} \lambda_X^U)(u)\big|_U$ for any smooth function u. Comparing this with Eq. (2.51) shows that $\lambda_X^U - \lambda_X$ is a constant on U, hence $i(X)\omega - i(X)\Omega = d\lambda_X$.

A direct corollary of the above theorem tells us whether a Fedosov star product which is invariant under the action of a Lie algebra admits a quantum moment map:

Proposition 2.9 *A $\mathfrak{g}$-invariant Fedosov star product for (M, ω) is obtained from a $\mathfrak{g}$-invariant connexion and a $\mathfrak{g}$-invariant series of closed 2-forms Ω. It admits a quantum Hamiltonian if and only if there is a linear map*

$$\hat{J} : \mathfrak{g} \to C^\infty(M)[[\nu]]$$

such that

$$d(\hat{J}(X)) = \iota(X^{*M})\omega - \iota(X^{*M})\Omega \qquad \forall X \in \mathfrak{g}.$$

We then have $X^{*M}u = \frac{1}{\nu}\,\mathrm{ad}_*\, \hat{J}(X)\, u$. It admits a quantum moment map if and only it is Hamiltonian and the linear map $\tilde{J} : \mathfrak{g} \to C^\infty(M)[[\nu]]$ such that $d(\tilde{J}(X)) = \iota(X^{*M})\omega - \iota(X^{*M})\Omega$ can be chosen so that

$$\tilde{J}([X, Y]) = -\omega(X^{*M}, Y^{*M}) + \Omega(X^{*M}, Y^{*M}) \qquad \forall X, Y \in \mathfrak{g}.$$

In a recent preprint [148], Reichert and Waldmann give a characterization of equivalence classes of $\mathfrak{g}$ invariant star products admitting a quantum moment map J, for $\mathfrak{g}$-invariant equivalences intertwining the quantum moments maps, by series in the second $\mathfrak{g}$-equivariant cohomology.

2.7 Reduction in Deformation Quantization

An important classical tool to "reduce the number of variables", i.e. to start from a "big" Poisson manifold M with real Poisson tensor P and construct a smaller one M_{red}, is given by reduction: one considers an embedded coisotropic submanifold in the Poisson manifold,

$$\iota : C \hookrightarrow M.$$

Recall that a submanifold of a Poisson manifold is called **coisotropic** iff the vanishing ideal

$$\mathscr{I}_C = \{f \in C^\infty(M) \mid \iota^* f = 0\} = \ker \iota^*.$$

is closed under Poisson bracket. This is equivalent to say that

$$P^\sharp(N^*C) \subset TC \quad \text{with} \quad N^*C(x) = \{\alpha_x \in T_x^*M \mid \alpha_x(X) = 0\, \forall X \in T_xC\},$$

where $P^\sharp : T^*M \to TM$ is induced by P through $\beta(P^\sharp(\alpha) := P(\alpha, \beta)$.

In the symplectic case $P^\sharp(N^*C) = TC^\perp$ is the orthogonal with respect to the symplectic 2-form ω of the tangent space to C, so that coisotropy means

$$TC_x^\perp := \{Y \in T_x^*M \mid \omega_x(X, Y) = 0 \quad \forall u \in T_xC\} \subset TC.$$

The distribution defined by $P^\sharp(N^*C)$, called the **characteristic distribution**, is involutive. It is spanned at each point by the Hamiltonian vector fields corresponding to functions which are locally in $\mathscr{I}_C$.

We assume that the canonical foliation has a nice leaf space M_{red} (a structure of smooth manifold such that the canonical projection $\pi : C \longrightarrow M_{\text{red}}$ is a submersion). In this case one can show that M_{red} is a Poisson manifold in a canonical way: one defines the normalizer of the vanishing ideal

$$\mathscr{B}_C = \left\{f \in C^\infty(M) \mid \{f, \mathscr{I}_C\} \subseteq \mathscr{I}_C\right\},$$

and

$$\mathscr{B}_C \big/ \mathscr{I}_C \ni [f] \mapsto \iota^* f \in \pi^* C^\infty(M_{\text{red}}) = \mathscr{A}_{red} \tag{2.52}$$

induces an isomorphism of Poisson algebras. We prove this in a simple context in Sect. 2.7.1.

Passing to a deformation quantized version of phase space reduction, one starts with a formal star product $\star$ on M. The associative algebra $\mathscr{A} = (\mathbf{C}^\infty(\mathbf{M})[[\nu]], \star)$ plays the role of the quantized observables of the big system. A good analog of the vanishing ideal $\mathscr{I}_C$ will be a left ideal $\mathscr{I}_\mathrm{C} \subseteq \mathbf{C}^\infty(\mathbf{M})[[\nu]]$ such that the quotient $C^\infty(M)[[\nu]] \big/ \mathscr{I}_\mathrm{C}$ is in $\mathbb{C}[[\nu]]$-linear bijection to the functions $C^\infty(C)[[\nu]]$ on C. We then define the normalizer of $\mathscr{I}_\mathrm{C}$ with respect to the commutator Lie bracket of $\mathscr{A}$,

$$\mathscr{B}_C = \{\mathbf{a} \in \mathscr{A} \mid [\mathbf{a}, \mathscr{J}_C] \subseteq \mathscr{J}_C\},$$

and consider the associative algebra $\mathscr{B}_C / \mathscr{J}_C$ as the reduced algebra $\mathscr{A}_{\mathbf{red}}$.

Of course, we need then to show that $\mathscr{B}_C / \mathscr{J}_C$ is in $\mathbb{C}[[\nu]]$-linear bijection to $C^\infty(M_{\text{red}})[[\nu]]$ in such a way, that the isomorphism induces a star product $\star_{\text{red}}$ on M_{red}. Starting from a strongly invariant star product on M, we describe below the method used in [106] with S. Waldmann to construct a good left ideal inspired by the BRST approach in [40] but simpler as we only need the deformation of the Koszul part of the BRST complex. Other approaches to reduction in deformation quantization appear in Fedosov [82] and in Cattaneo–Felder [62].

Remark 2.15 BRST formalism is a differential geometric approach to quantize a field theory with a gauge symmetry. We refer to the lectures of Nathan Berkovits. The mathematical background of BRST construction describes the space of functions on some reduced spaces as the 0-cohomology space of a complex.

We present only the particular case of the Marsden–Weinstein reduction : consider a smooth left action $G \times M \longrightarrow M : (g, p) \mapsto \rho(g)p$ of a connected Lie group G on M by Poisson diffeomorphisms and assume we have an ad*-equivariant momentum map J. The constraint manifold C is chosen to be the level surface of J for momentum $0 \in \mathfrak{g}^*$ (thus we assume, for simplicity, that 0 is a regular value). Then $C = J^{-1}(\{0\})$ is an embedded submanifold which is coisotropic. The group G acts on C and the reduced space is the orbit space of this group action of G on C. To guarantee a good quotient we assume that G acts freely and properly and we assume that G acts properly not only on C but on all of M. In this case there exists an open neighbourhood $M_{\text{nice}} \subseteq M$ of C with a G-equivariant diffeomorphism

$$\Phi : M_{\text{nice}} \longrightarrow U_{\text{nice}} \subseteq C \times \mathfrak{g}^* \tag{2.53}$$

onto an open neighbourhood U_{nice} of $C \times \{0\}$, where the G-action on $C \times \mathfrak{g}^*$ is the product action of the one on C and Ad*, such that for each $p \in C$ the subset $U_{\text{nice}} \cap (\{p\} \times \mathfrak{g}^*)$ is star-shaped around the origin $\{p\} \times \{0\}$, and the momentum map J is given by the projection onto the second factor, i.e. $J_{|M_{\text{nice}}} = \text{pr}_2 \circ \Phi$. For a proof of this see for instance [40, Lemma 3].

2.7.1 The Classical Koszul Resolution

We consider $C^\infty(M, \Lambda^\bullet_{\mathbb{C}}\mathfrak{g}) = C^\infty(M) \otimes \Lambda^\bullet_{\mathbb{C}}\mathfrak{g}$ with the canonical free $C^\infty(M)$-module structure. The group G acts on $C^\infty(M, \Lambda^\bullet_{\mathbb{C}}\mathfrak{g})$ by the combined action of G on the manifold and the adjoint action on $\mathfrak{g}$ extended to the exterior algebra $\Lambda^\bullet_{\mathbb{C}}\mathfrak{g}$ by automorphisms of the exterior product. We denote this G-action and the corresponding $\mathfrak{g}$-action by ρ. The **Koszul differential** is defined to be

$$\partial x = \iota(J)x, \tag{2.54}$$

where $x \in C^\infty(M, \Lambda^\bullet_{\mathbb{C}}\mathfrak{g})$ and $\iota(J)$ denotes the insertion of J at the first position in the $\Lambda^\bullet_{\mathbb{C}}\mathfrak{g}$-part of x. If $e_1, \ldots, e_N \in \mathfrak{g}$ denotes a basis with dual basis $e^1, \ldots, e^N \in \mathfrak{g}^*$ then we can write $J = \sum_a J_a e^a$ with scalar functions $J_a \in C^\infty(M)$ and $\partial x = J_a \iota(e^a)x$. The map ∂ is a graded derivation of the standard wedge product on $C^\infty(M, \Lambda^\bullet_{\mathbb{C}}\mathfrak{g})$ of degree -1, it is $C^\infty(M)$-linear and $\partial^2 = 0$; we have thus a complex of free $C^\infty(M)$-modules. We write ∂_k for the restriction of ∂ to the antisymmetric degree $k \geq 1$.

We use the particular tubular neighbourhood M_{nice} of C to define a **prolongation map**

$$\text{prol} : C^\infty(C) \ni \phi \mapsto \text{prol}(\phi) = (\text{pr}_1 \circ \Phi)^*\phi \in C^\infty(M_{\text{nice}}). \tag{2.55}$$

This prolongation is G-equivariant: $\rho(g)^*\text{prol}(\phi) = \text{prol}(\rho(g)^*\phi)$. It deserves its name since, for all $\phi \in C^\infty(C)$, we have $\iota^*\text{prol}(\phi) = \phi$.

We define a **homotopy**, on M_{nice} for convenience: let $x \in C^\infty(M_{\text{nice}}, \Lambda^k_{\mathbb{C}}\mathfrak{g})$; since U_{nice} is star-shaped, we set

$$(h_k x)(p) = e_a \wedge \int_0^1 t^k \frac{\partial(x \circ \Phi^{-1})}{\partial \mu_a}(c, t\mu)dt, \tag{2.56}$$

where $\Phi(p) = (c, \mu)$ for $p \in M_{\text{nice}}$ and μ_a denote the linear coordinates on $\mathfrak{g}^*$ with respect to the basis $e^1, \ldots, e^N$. The collection of all these maps h_k gives a map $h : C^\infty(M_{\text{nice}}, \Lambda^\bullet_{\mathbb{C}}\mathfrak{g}) \longrightarrow C^\infty(M_{\text{nice}}, \Lambda^{\bullet+1}_{\mathbb{C}}\mathfrak{g})$.

Proposition 2.10 [40, Lemmas 5 and 6] *The Koszul complex $(C^\infty(M_{\text{nice}}, \Lambda^\bullet_{\mathbb{C}}\mathfrak{g}), \partial)$ is acyclic with homotopy h and homology $C^\infty(C)$ in degree 0: we have*

$$h_{k-1}\,\partial_k + \partial_{k+1}\,h_k = \text{id}_{C^\infty(M_{\text{nice}}, \Lambda^k_{\mathbb{C}}\mathfrak{g})} \quad \textit{for } k \geq 1 \tag{2.57}$$

$$\text{prol}\,\iota^* + \partial_1\,h_0 = \text{id}_{C^\infty(M_{\text{nice}})} \tag{2.58}$$

as well as $\iota^*\,\partial_1 = 0$. Thus the Koszul complex is a free resolution of $C^\infty(C)$ as $C^\infty(M_{\text{nice}})$-module. We have

$$h_0\,\text{prol} = 0, \tag{2.59}$$

and all the homotopies h_k are G-equivariant.

Here resolution means that the homology at $k = 0$ is isomorphic to $C^\infty(C)$ as a $C^\infty(M_{\text{nice}})$-module.

Exercise 2.25 Show that the image of ∂_1 is just

$$\ker \iota^* \cap C^\infty(M_{\text{nice}}) = \mathscr{J}_C \cap C^\infty(M_{\text{nice}})$$

using formula (2.58). This gives immediately

$$\begin{aligned} \ker \partial_0 / \text{Im}\,\partial_1 &= \ker \partial_0 / (\mathscr{J}_C \cap C^\infty(M_{\text{nice}})) \\ &= C^\infty(M_{\text{nice}}) / (\mathscr{J}_C \cap C^\infty(M_{\text{nice}})) \cong C^\infty(C), \end{aligned} \tag{2.60}$$

Use the Koszul complex to prove (2.52): f is in $\mathscr{B}_C$ iff $0 = \iota^*\{J_X, f\} = \iota^*(\mathscr{L}_{X^{*M}} f) = \mathscr{L}_{X^{*C}}(\iota^* f)\, \forall X \in \mathfrak{g}$ iff $\iota^* f \in \pi^* C^\infty(M_{\text{red}})$. For $u \in C^\infty(M_{\text{red}})$ show that $\text{prol}(\pi^* u) \in \mathscr{B}_C$ whence (2.52) is surjective. The injectivity of (2.52) is clear by definition.

The Poisson bracket on M_{red} can then be defined through (2.52) and gives explicitly

$$\pi^*\{u, v\}_{\text{red}} = \iota^*\{\text{prol}(\pi^* u), \text{prol}(\pi^* v)\} \tag{2.61}$$

for $u, v \in C^\infty(M_{\text{red}})$, since the left hand side of (2.52) is canonically a Poisson algebra.

Since for the phase space reduction in deformation quantization we will only need a very small neighbourhood of C, the neighbourhood M_{nice} is sufficient; the geometry of M far away from C plays no role and we may assume without restriction $M_{\text{nice}} = M$ in the following.

2.7.2 The Quantized Koszul Complex

Before defining the deformed Koszul operator we make some further assumptions on the star product $\star$ on M: we assume it to be strongly invariant, i.e. $\mathfrak{g}$-*covariant*,

$$J_X \star J_Y - J_Y \star J_X = \nu J_{[X,Y]} \qquad \forall X, Y \in \mathfrak{g} \tag{2.62}$$

and G-*invariant*

$$\rho(g)^*(f \star h) = (\rho(g)^* f) \star (\rho(g)^* h) \tag{2.63}$$

for all $g \in G$ and $f, h \in C^\infty(M)[[\lambda]]$.

Using the exterior ($\wedge$) product for $\Lambda^\bullet_{\mathbb{C}}\mathfrak{g}$ we extend $\star$ to $C^\infty(M, \Lambda^\bullet_{\mathbb{C}}\mathfrak{g})$ in the canonical way.

Definition 2.22 (*Quantized Koszul operator*) Let $\kappa \in \mathbb{C}[[\nu]]$. The quantized Koszul operator $\partial^{(\kappa)} : C^\infty(M, \Lambda^\bullet_{\mathbb{C}}\mathfrak{g})[[\nu]] \longrightarrow C^\infty(M, \Lambda^{\bullet+1}_{\mathbb{C}}\mathfrak{g})[[\nu]]$ is defined to be

$$\partial^{(\kappa)} x = \iota(e^a) x \star J_a + \frac{\nu}{2} C^c_{ab} e_c \wedge \iota(e^a)\iota(e^b) x + \nu\kappa\, \iota(\Delta) x, \tag{2.64}$$

(with summation over repeated indices) where $C^c_{ab} = e^c([e_a, e_b])$ are the structure constants of $\mathfrak{g}$ and $\Delta(X) = \text{Tr ad}(X)$ for $X \in \mathfrak{g}$ is the modular one-form, $\Delta \in \mathfrak{g}^*$, of $\mathfrak{g}$ (with respect to the chosen basis we have $\Delta = C^b_{ab} e^a$).

Lemma 2.2 ([106, Lemma 3.4]) *Let $\star$ be a strongly invariant $\star$-product and $\kappa \in \mathbb{C}[[\nu]]$. Then $\partial^{(0)}\iota(\Delta) + \iota(\Delta)\partial^{(0)} = 0$, $\partial^{(\kappa)}$ is left $\star$-linear, the classical limit of $\partial^{(\kappa)}$ is ∂, $\partial^{(\kappa)}$ is G-equivariant, and $\partial^{(\kappa)} \circ \partial^{(\kappa)} = \mathbf{0}$.*

The element κ can be arbitrary; in particular, $\kappa = 0$ gives a very simple choice; however, we set $\partial = \partial^{(\kappa = \frac{1}{2})}$. The following constructions will depend on κ; if we omit the reference to κ in our notation, we always mean the particular value $\kappa = \frac{1}{2}$.

Following [40] we define a deformation of the restriction map ι^* and the homotopy:

$$\iota_\kappa^* = \iota^* \left(\mathrm{id} + \left(\partial_1^{(\kappa)} - \partial_1\right) h_0\right)^{-1} : C^\infty(M)[[\nu]] \longrightarrow C^\infty(C)[[\nu]] \tag{2.65}$$

and

$$\mathbf{h}_\mathbf{0}^{(\kappa)} = h_0 \left(\mathrm{id} + \left(\partial_\mathbf{1}^{(\kappa)} - \partial_1\right) h_0\right)^{-1} : C^\infty(M)[[\nu]] \longrightarrow C^\infty(M, \mathfrak{g})[[\nu]], \tag{2.66}$$

which are both well-defined since $\partial^{(\kappa)}$ is a deformation of ∂. From [40, Proposition 25]

$$\mathbf{h}_0^{(\kappa)} \operatorname{prol} = 0, \quad \iota_\kappa^* \partial_1^{(\kappa)} = 0, \quad \text{and} \quad \iota_\kappa^* \operatorname{prol} = \mathrm{id}_{C^\infty(C)[[\nu]]} . \tag{2.67}$$

The homotopy equation becomes

$$\operatorname{prol} \iota_\kappa^* + \partial_\mathbf{1}^{(\kappa)} \mathbf{h}_\mathbf{0}^{(\kappa)} = \mathrm{id}_{\mathbf{C}^\infty(\mathbf{M})[[\nu]]} . \tag{2.68}$$

2.7.3 The Reduced Star Product

We now give an explicit description of the quotient $\mathscr{B}_\mathrm{C} / \mathscr{J}_\mathrm{C}$ where

$$\mathscr{B}_C = \left\{ f \in C^\infty(M)[[\nu]] \mid [f, \mathscr{J}_\mathrm{C}]_\star \subseteq \mathscr{J}_\mathrm{C} \right\}. \tag{2.69}$$

Proposition 2.11 [40, Theorems 29 and 32] *Let $f \in C^\infty(M)[[\nu]]$ and $u, v \in C^\infty(M_{\mathrm{red}})[[\nu]]$.*

- *We have $f \in \mathscr{B}_\mathrm{C}$ iff $\mathscr{L}_{X^{*C}} \iota_\kappa^* f = 0$ for all $X \in \mathfrak{g}$ iff $\iota_\kappa^* f \in \pi^* C^\infty(M_{\mathrm{red}})[[\nu]]$.*
- *The quotient algebra $\mathscr{B}_\mathrm{C} / \mathscr{J}_\mathrm{C}$ is isomorphic to $C^\infty(M_{\mathrm{red}})[[\nu]]$ via the mutually inverse maps*

$$\mathscr{B}_\mathrm{C} / \mathscr{J}_\mathrm{C} \ni [f] \mapsto \iota_\kappa^* f \in \pi^* C^\infty(M_{\mathrm{red}})[[\nu]] \tag{2.70}$$

and

$$C^\infty(M_{\mathrm{red}})[[\nu]] \ni u \mapsto [\operatorname{prol}(\pi^* u)] \in \mathscr{B}_\mathrm{C} / \mathscr{J}_\mathrm{C}. \tag{2.71}$$

- *The induced associative product $\star_{\mathrm{red}\kappa}$ on $C^\infty(M_{\mathrm{red}})[[\nu]]$ from $\mathscr{B}_\mathrm{C} / \mathscr{J}_\mathrm{C}$ is explicitly given by*

$$\pi^*(u \star_{\mathrm{red}\kappa} v) = \iota_\kappa^* \left(\operatorname{prol}(\pi^* \mathbf{u}) \star \operatorname{prol}(\pi^* \mathbf{v})\right). \tag{2.72}$$

This is a bidifferential star product quantizing the Poisson bracket (2.61).

Proof We give a sketch of the proof. For the first part note that $\mathscr{J}_\mathrm{C} = \ker \iota_\kappa^*$ according to (2.68). Now let $g = g^a \star J_a + \nu\kappa C_{ba}^a g^b$ with $g^a \in C^\infty(M)[[\nu]]$ be in the image of $\partial_\mathbf{1}^{(\kappa)}$. For $f \in C^\infty(M)[[\nu]]$ we have by a straightforward computation

$$[f, g]_\star = \partial_1^{(\kappa)} h + \nu g^a \star \mathscr{L}_{(e_a)^{*M}} f$$

with some $h \in C^\infty(M, \mathfrak{g})$ using the strong invariance of $\star$. Thus $[f, g]_\star$ is in $\mathscr{I}_C$ iff $g^a \star \mathscr{L}_{(e_a)^{*M}} f$ is in the image of $\partial_1^{(\kappa)}$ for all g^a. This shows that $f \in \mathscr{B}_C$ iff $\mathscr{L}_{X^{*M}} f \in \operatorname{Im}\partial_1^{(\kappa)} = \ker \iota_\kappa^*$. Since ι_κ^* is G-invariant the first part follows. The second part is then clear from the first part, and (2.72) is a straightforward translation using the isomorphisms (2.70) and (2.71). One can show that $\star_{\mathrm{red}\kappa}$ is bidifferential and that it is indeed a star product on M_{red}.

Remark 2.16 The algebra of quantum observables is not only an associative algebra but is has a *-involution (see Definition 2.4); in the usual picture, where observables are represented by operators, this *-involution corresponds to the passage to the adjoint operator. In the framework of deformation quantization, complex conjugaison is a *-involution on $\mathscr{A} = (\mathbf{C}^\infty(\mathbf{M})[[\lambda]], \star)$ if the star product is Hermitian. We study in [106] the existence of natural *-involutions on the reduced quantum algebra assuming that $\star$ is Hermitian: the choice of a formal series of smooth densities on the embedded coisotropic submanifold $C = J^{-1}(0)$, with some equivariance property, defines a *-involution for $\star_{\mathrm{red}}$ on the reduced space. Whether the corresponding *-involution is the complex conjugation (which is a *-involution in the Marsden–Weinstein context) yields to define a new notion of quantized unimodular class. We study representations (in the sense of [37]) of the reduced algebra with the *-involution given by complex conjugation, relating the categories of modules of the big and of the reduced algebras.

2.8 Some Remarks About Convergence

A formal deformation is not enough for physics; $\hbar$ is a constant of nature and not a formal parameter. Although a nice representation theory has been introduced for $*$-algebras [37], there is no reasonable general notion of spectra for formal star product algebras (except for a few examples with convergence as in [16]); thus formal deformation quantization can not predict in general values of measurements, and hence is not a complete answer to the quantization problem.

Many examples of star products, like the global symbolic calculus on cotangent bundles or like Berezin or Toeplitz quantization of Kňahler manifolds, are obtained as asymptotic expansions for $\hbar \to 0$ of some convergent counterpart in usual quantization (see for instance [38, 51]). Whether the asymptotics can be used to recover the convergent quantization is still unknown. Some partial convergence results in this context were obtained, for instance in [51] for the product of two given functions and in [33] for subalgebras were the product converges.

The framework of C^*-algebras provides the background for a good notion of spectra (the spectrum of an element a in a unital C^*-algebra is the set of $\lambda \in \mathbb{C}$ such that $a - \lambda 1$ is not invertible); an ideal situation would be to construct a C^*-algebra with a physical interpretation of some elements. It is not enough to know a

C^*-algebra of observables of a system, one still needs a rule stating which algebra element corresponds to which physical observable. A problem is that, except for some simple situations, it is hard to write down a C^*-algebra corresponding to a quantum system of which one knows the classical counterpart, although that is the aim of quantization (build a quantum description, given a classical physical system). Formal deformation quantization is not a solution but might be a first step: one can try to use the powerful results for that theory (in particular concerning existence, classification, invariance and constructions) to build, in a second step, a C^*-algebraic framework.

Rieffel introduced the notion of strict deformation quantization (see [150–152]): A **strict deformation quantization** [150] of a dense $*$-subalgebra $\mathbb{A}'$ of a C^*-algebra, in the direction of a Poisson bracket $\{.,.\}$ defined on $\mathbb{A}'$, is an open interval $I \subset \mathbb{R}$ containing 0, and the assignment, for each $\hbar \in I$, of an associative product $\times_\hbar$, an involution $*_\hbar$ and a C^*-norm $\| \ \|_\hbar$ (for $\times_\hbar$ and $*_\hbar$) on $\mathbb{A}'$, which coincide for $\hbar = 0$ to the original product, involution and C^*-norm on $\mathbb{A}'$, such that the corresponding field of C^*-algebras, with continuity structure given by the elements of $\mathbb{A}'$ as constant fields, is a continuous field of C^*-algebras, and such that for all $a, b \in \mathbb{A}'$, $\|\frac{(a \times_\hbar b - ab)}{i\hbar} - \{a, b\}\|_\hbar \to 0$ as $\hbar \to 0$.

Group actions appear here in an essential way: Rieffel introduced a general way to construct such C^*-algebraic deformations based on a strongly continuous isometrical action of $\mathbb{R}^d$ on a C^*-algebra $\mathbb{A}$

$$\alpha : \mathbb{R}^d \times \mathbb{A} \to \mathbb{A} : (x, a) \mapsto \alpha_x a.$$

The product formula for the smooth vectors $\mathbb{A}^\infty$ with respect to this action is defined, using an oscillatory integral, choosing a fixed element θ in the orthogonal Lie algebra $so(d)$, by

$$a \times_\hbar b := a *_\theta^\alpha b := \left(\frac{1}{\pi\hbar}\right)^d \int_{\mathbb{R}^d \times \mathbb{R}^d} \alpha_x(a)\alpha_y(b) \exp\left(\frac{2i}{\hbar} x \cdot \theta y\right) dxdy$$

and it gives a pre C^* associative algebra structure on $\mathbb{A}^\infty$. This generalizes the Weyl quantization of $\mathbb{R}^{2n}$. Indeed formula (2.14) can be rewritten

$$F \times_\hbar G = \left(\frac{1}{\pi\hbar}\right)^{2n} \int_{\mathbb{R}^{2n} \times \mathbb{R}^{2n}} \tau_v(F)\tau_w(G) e^{\frac{2i}{\hbar}\Omega(v,w)} dvdw$$

where τ denotes the action of $\mathbb{R}^{2n}$ on functions on $\mathbb{R}^{2n}$ by translation.

Bieliavsky et al. generalize the construction to actions of Lie groups that admit negatively curved left-invariant Kähler structure. An important observation due to Weinstein is the relevance in the phase appearing in the product kernel (see Eq. (2.15)) of the symplectic flux $S(x, y, z) = \Omega(x, y) + \Omega(y, z) + \Omega(z, x)$ through a geodesic triangle that admits the points x, y and z as mid-points of its geodesic edges. This lead to the study of symmetric symplectic spaces, and, more precisely here to symplectic groups which have a structure of symmetric symplectic spaces. Bieliavsky et al. build analogues of Weyl's quantization which give universal deformation formulas for those groups and obtain new examples of strict deformation quantization [24–26, 28].

A possible drawback of considering "convergent star products" given by integral formulas (like the convergent star product defined on the space of Schwartz functions on $\mathbb{R}^{2n}$ given by formula (2.14)) is the difficulty to extend the construction to infinite dimensional cases, which are unavoidable when dealing with quantum field theory.

Another approach to the convergence problem is the following. Taking the formal power series defining the star product, one can ask for convergence in a mathematically meaningful way. This has been achieved by Waldmann et al. in a growing number of examples [17, 39, 80, 169]. They build seminorms which garantee the convergence of the deformed multiplication. In this way, they construct topological non-commutative algebras, essentially of Fréchet type. It does not yet reach the C^*-framework but it already gives an algebra over $\mathbb{C}$ and not just over $C[[\nu]]$. One can then study Hilbert space representations of this algebra by (still a priori unbounded) operators. Convergence of the Moyal star product on a Fréchet algebra has also been studied by Omori et al. in [138].

Acknowledgements I thank the organisers of the summer school in Villa de Leyva for their very remarkable organization and their most kind hospitality. I aknowledge support from an ARC grant of the Communauté française de Belgique and from the PAI grant "Dygest" of the Belgian science policy.

References

1. A. Alekseev, E. Meinrenken, On the Kashiwara-Vergne conjecture. Invent. Math. **164**(3), 615–634 (2006)
2. A. Alekseev, C. Torossian, deformation quantization and flat connections. Commun. Math. Phys. **300**, 47–64 (2010)
3. A. Alekseev, C. Torossian, The Kashiwara-Vergne conjecture and Drinfeld's associators. Ann. Math. **175**(2), 415–463 (2012)
4. M. Andler, A. Dvorsky, S. Sahi, Kontsevich Quantization and invariant distributions on Lie groups. Annales de l'ENS **35**(3), 371–390 (2002)
5. D. Arnal, $*$ products and representations of nilpotent Lie groups. Pacific J. Math. **114**, 285–308 (1984) and D. Arnal, J.-C. Cortet, $*$ products in the method of orbits for nilpotent Lie groups and J. Geom. Phys. 2, 83–116 (1985)
6. D. Arnal, J.-C. Cortet, Nilpotent Fourier-transform and applications. Lett. Math. Phys. **9**, 25–34 (1985) and D. Arnal, S. Gutt, Décomposition de $L^2(G)$ et transformation de Fourier adaptée pour un groupe G nilpotent. C. R. Acad. Sci. Paris Sér. I Math. **306**, 25–28 (1988)
7. D. Arnal, Le produit star de Kontsevich sur le dual d'une algèbre de Lie nilpotente. C. R. Acad. Sci. Paris Sér. I Math. **237**, 823–826 (1998)
8. D. Arnal, J.C. Cortet, P. Molin, G. Pinczon, Covariance and geometrical invariance in * quantization. J. Geom. Phys. **24**, 276–283 (1983)
9. D. Arnal, M. Cahen, S. Gutt, Deformations on coadjoint orbits. J. Geom. Phys. **3**, 327–351 (1986)
10. D. Arnal, J. Ludwig, M. Masmoudi, Déformations covariantes sur les orbites polarisées d'un groupe de Lie. J. Geom. Phys. **14**, 309–331 (1994)

11. D. Arnal, N. Ben Amar, M. Masmoudi, Cohomology of good graphs and Kontsevich linear star products. Lett. Math. Phys. **48**, 291–306 (1999)
12. D. Arnal, D. Manchon, M. Masmoudi, Choix des signes pour la formalité de M. Kontsevich. Pac. J. Math. **203**(1), 23–66 (2002)
13. P. Aschieri, I. Bakovic, B. Jurco, P. Schupp, Noncommutative gerbes and deformation quantization. J. Geom. Phys. **60**, 1754–1761 (2010)
14. H. Basart, M. Flato, A. Lichnerowicz, D. Sternheimer, Deformation theory applied to quantization and statistical mechanics. Lett. Math. Phys. **8**, 483–494 (1984)
15. F. Bayen, M. Flato, C. Fronsdal, A. Lichnerowicz, D. Sternheimer, Quantum mechanics as a deformation of classical mechanics. Lett. Math. Phys. **1**, 521–530 (1977) and Deformation theory and quantization, part I. Ann. Phys. **111**, 61–110 (1978)
16. F. Bayen, M. Flato, C. Fronsdal, A. Lichnerowicz, D. Sternheimer, Deformation theory and quantization, part II. Ann. Phys. **111**, 111–151 (1978)
17. S. Beiser, S. Waldmann, Fréchet algebraic deformation quantization of the Poincaré disk. Crelle's J. reine angew. Math. **688**, 147–207 (2014)
18. S. Beiser, H. Römer, S. Waldmann, Convergence of the Wick Star Product. Commun. Math. Phys. **272**, 25–52 (2007)
19. F.A. Berezin, General concept of quantization. Commun. Math. Phys. **40**, 153–174 (1975)
20. F.A. Berezin, Quantization. Math. USSR Izvestija **8**(5), 1109–1165 (1975)
21. M. Bertelson, Equivalence de produits star. Mémoire de Licence, U.L.B. (1995) and M. Bertelson, M. Cahen, S. Gutt, Equivalence of star products. Class. Quan. Grav. **14** A 93–107 (1997)
22. M. Bertelson, P. Bieliavsky, S. Gutt, Parametrizing equivalence classes of invariant star products. Lett. Math. Phys. **46**, 339–345 (1998)
23. F. Bidegain, G. Pinczon, Quantization of Poisson-Lie groups and applications. Commun. Math. Phys. **179**, 295–332 (1996) and $*$-product approach to non-compact quantum groups. Lett. Math. Phys. **33**, 231–240 (1995)
24. P. Bieliavsky, Non-formal deformation quantizations of solvable Ricci-type symplectic symmetric spaces. J. Phys. Conf. Ser. **103** (2008)
25. P. Bieliavsky, V. Gayral, Deformation quantization for actions of Kählerian Lie groups. Mem. AMS, **236**(1115) (2015)
26. P. Bieliavsky, Strict quantization of solvable symmetric spaces. J. Sympl. Geom. **1**, 269–320 (2002)
27. P. Bieliavsky, M. Massar, Oscillatory integral formulae for left- invariant star products on a class of lie groups. Lett. Math. Phys. **58**, 115–128 (2001)
28. P. Bieliavsky, S. Detournay, P. Spindel, The deformation quantizations of the hyperbolic plane. Commun. Math. Phys. **289**(2), 529–559 (2009)
29. P. Bonneau, Fedosov star products and one-differentiable deformations. Lett. Math. Phys. **45**(4), 363–376 (1998)
30. P. Bonneau, M. Flato, M. Gerstenhaber, G. Pinczon, The hidden group structure of quantum groups: strong duality, rigidity and preferred deformations. Commun. Math. Phys. **161**, 125–156 (1994)
31. P. Bonneau, M. Gerstenhaber, A. Giaquinto, D. Sternheimer, Quantum groups and deformations quantization: explicit approaches and implicit aspects. J. Math. Phys. **45**, 3704–3741 (2004)
32. M. Bordemann, Deformation quantization: a mini lecture, in Geometric and Topological methods for Quantum Field Theory, summer school, ed. by S. Paycha, B. Uribe, Contemporary Maths. 434 (2005)
33. M. Bordemann, M. Brischle, C. Emmrich, S. Waldmann, Phase space reduction, for star products: an explicit construction for $\mathbb{C}P^n$. Lett. Math. Phys. 36, 357–371 (1996) and Subalgebras with converging star products in deformation quantization: an algebraic construction for for $\mathbb{C}P^n$. J. Math. Phys. **37**, 6311–6323 (1996)
34. M. Bordemann, H.-C. Herbig, M.J. Pflaum, A homological approach to singular reduction in deformation quantization, in *Singularity Theory: Proceedings of the 2005 Marseille Singularity School and Conference*, ed. by D. Chéniot, et al. (World Scientific Publication, 2007)

35. M. Bordemann, N. Neumaier, S. Waldmann, Homogeneous fedosov, star products on cotangent bundles I, Commun. Math. Phys. **198**, 363–396 (1998) and Homogeneous fedosov star products on cotangent bundles II. J. Geom. Phys. **29**, 199–234 (1999)
36. M. Bordemann, (Bi)Modules, morphismes et réduction des star-produits. Travaux Mathématiques, Université du Luxembourg **16**, 9–40 (2005)
37. M. Bordemann, S. Waldmann, Formal GNS construction and states in deformation quantization. Commun. Math. Phys. **195**, 549–583 (1998)
38. M. Bordemann, E. Meinrenken, M. Schlichenmaier, Toeplitz quantization of Kähler manifolds and $gl(N)N \to \infty$ limit. Commun. Math. Phys. **165**, 281–296 (1994)
39. M. Bordemann, H. Römer, S. Waldmann, A remark on formal KMS states in deformation quantization. Lett. Math. Phys. **45**, 49–61 (1998)
40. M. Bordemann, H.-C. Herbig, S. Waldmann, BRST cohomology and phase space reduction in deformation quantization. Commun. Math. Phys. **210**, 107–144 (2000)
41. M. Bordemann, N. Neumaier, S. Waldmann, S. Weiss, Deformation quantization of surjective submersions and principal fibre bundles. Crelle's J. reine angew. Math. **639**, 1–38 (2010)
42. P. Bressler, A. Gorokhovsky, R. Nest, B. Tsygan, Deformation quantization of gerbes. Adv. Math. **214**, 230–266 (2007)
43. P. Bressler, A. Gorokhovsky, R. Nest, B. Tsygan, Algebraic index theorem for symplectic deformations of gerbes. Contemp. Math. **546**, 23–37 (2011)
44. A. Bruguières, A. Cattaneo, B. Keller, C. Torossian, Déformation, quantification, théorie de Lie, Panoramas et synthèse 20 (Soc. Mah. de France, Paris, 1995)
45. H. Bursztyn, S. Waldmann, On positive deformations of $*$-algebras, in *conference Moshe Flato 1999*, ed. by Dito, Sternheimer, Math phys studies 22 (Kluwer, 2000), pp. 69–80
46. H. Bursztyn, S. Waldmann, Deformation quantization of hermitian vector bundles. Lett. Math. Phys. **53**, 349–365 (2000)
47. H. Bursztyn, S. Waldmann, The characteristic classes of Morita equivalent star products on symplectic manifolds. Commun. Math. Phys. **228**, 103–121 (2002)
48. H. Bursztyn, S. Waldmann, Bimodule deformations, Picard groups and contravariant connections. K-Theory **31**, 1–37 (2004)
49. H. Burzstyn, V. Dolgushev, S. Waldmann, Morita equivalence and characteristic classes of star products. Crelle's J. reine angew. Math. **662**, 95–163 (2012)
50. M. Cahen, S. Gutt, Produits $*$ sur les orbites des groupes semi-simples de rang 1, C. R. Acad. Sci. Paris **296**, 821–823 (1983) and An algebraic construction of $*$ product on the regular orbits of semisimple Lie groups, Bibliopolis Ed. Naples, volume in honour of I. Robinson (1987), pp. 71–82
51. M. Cahen, S. Gutt, J. Rawnsley, Quantization of Kähler manifolds I, II, III and IV, J. Geom. Phys. **7**, 45–62 (1990); Trans. Amer. Math. Soc. **337**, 73–98 (1993); Lett. Math. Phys. **30**, 291–305 (1994); Lett. Math. Phys. **34**, 159–168 (1995)
52. M. Cahen, S. Gutt, Regular $*$ representations of lie algebras. Lett. Math. Phys. **6**, 395–404 (1982)
53. M. Cahen, S. Gutt, Produits $*$ sur les espaces affins symplectiques localement symétriques. C. R. Acad. Sci. Paris **297**, 417–420 (1983)
54. M. Cahen, M. De Wilde, S. Gutt, Local cohomology of the algebra of smooth functions on a connected manifold. Lett. Math. Phys. **4**, 157–167 (1980)
55. M. Cahen, M. Flato, S. Gutt, D. Sternheimer, Do different deformations lead to the same spectrum? J. Geom. Phys. **2**, 35–48 (1985)
56. M. Cahen, S. Gutt, J. Rawnsley, On tangential star products for the coadjoint Poisson structure. Commun. Math. Phys. **180**, 99–108 (1996)
57. A.S. Cattaneo, From topological field theory to deformation quantization and reduction, in *Proceedings of the International Congress of Mathematicians, Madrid, Spain, 2006*, ed. by M. Sanz-Solé, J. Soria, J. L. Varona, J. Verdera, vol. III (European Mathematical Society, 2006), pp. 338–365
58. A. Cattaneo, G. Felder, A path integral approach Poisson sigma models and symplectic groupoids, in *Quantization of Singular Symplectic Quotients*, ed. by N.P. Landsman, M. Pflaum, M. Schlichenmeier, Progress in Mathematics, vol. 198 (Birkhäuser, 2001), pp. 61-93

59. A. Cattaneo, D. Indelicato, Formality and star products, in *Poisson Geometry, Deformation Quantisation and Group Representations*, ed. by S. Gutt, J. Rawnsley, D. Sternheimer, LMS Lecture Note Series, vol. 323 (2005)
60. A. Cattaneo, G. Felder, A path integral approach to the Kontsevich quantization formula. Commun. Math. Phys. **212**, 591–611 (2000)
61. A. Cattaneo, G. Felder, On the globalization of Kontsevich's star product and the perturbative sigma model. Prog. Theor. Phys. Suppl. **144**, 38–53 (2001). arXiv:hep-th/0111028
62. A. Cattaneo, G. Felder, Coisotropic submanifolds in Poisson geometry and branes in the Poisson sigma model. Lett. Math. Phys. **69**, 157–175 (2004)
63. A. Cattaneo, G. Felder, Relative formality theorem and quantisation of coisotropic submanifolds. Adv. Math. **208**, 521–548 (2007)
64. A. Cattaneo, G. Felder, L. Tomassini, From local to global deformation quantization of Poisson manifolds. Duke Math. J. **115**, 329–352 (2002)
65. V. Chloup, Star products on the algebra of polynomials on the dual of a semi-simple Lie algebra. Acad. R. Belg. Bull. Cl. Sci. **8**, 263–269 (1997)
66. A. Connes, Non commutative differential geometry. IHES Publ. Math. **62**, 257–360 (1985)
67. A. Connes, M. Flato, D. Sternheimer, Closed star products and cyclic cohomology. Lett. Math. Phys. **24**, 1–12 (1992)
68. M. De Wilde, P. Lecomte, Existence of star-products and of formal deformations of the Poisson Lie algebra of arbitrary symplectic manifolds. Lett. Math. Phys. **7**, 487–496 (1983)
69. M. De Wilde, Deformations of the algebra of functions on a symplectic manifold: a simple cohomological approach. Publication no. 96.005, Institut de Mathématique, Université de Liège (1996)
70. M. De Wilde, S. Gutt, P.B.A. Lecomte, À propos des deuxième et troisième espaces de cohomologie de l'algèbre de Lie de Poisson d'une variété symplectique. Ann. Inst. H. Poincaré Sect. A (N.S.) **40**, 77–83 (1984)
71. M. De Wilde, P. Lecomte, Formal deformations of the Poisson Lie algebra of a symplectic manifold and star products: existence, equivalence, derivations, in *Deformation Theory of Algebras and Structures and Applications*, ed. by Hazewinkel, Gerstenhaber (Kluwer, 1988), pp. 897–960
72. P. Deligne, Déformations de l'Algèbre des Fonctions d'une Variété Symplectique: Comparaison entre Fedosov et De Wilde Lecomte. Selecta Math. (New Ser.) **1**, 667–697 (1995)
73. G. Dito, Star product approach to quantum field theory: the free scalar field. Lett. Math. Phys. **20**, 125–134 (1990)
74. G. Dito, An example of cancellation of infinities in the star-quantization of fields. Lett. Math. Phys. **27**, 73–80 (1993)
75. G. Dito, Kontsevich star product on the dual of a Lie algebra. Lett. Math. Phys. **48**, 307–322 (1999)
76. G. Dito, D. Sternheimer, Deformation quantization: genesis, developments and metamorphoses, in *Deformation quantization. IRMA Lectures in Mathematics and Theoretical Physics*, vol. 1, ed. by G. Halbout (Walter de Gruyter, Berlin, 2002), pp. 9–54
77. V. Dolgushev, Covariant and equivariant formality theorems. Adv. Math. **191**, 147–177 (2005)
78. V. Dolgushev, V.N. Rubtsov, An algebraic index theorem for Poisson manifolds. J. Reine Angew. Math. **633**, 77–113 (2009)
79. V.G. Drinfeld, Quantum groups. Proc. ICM86, Berkeley, Am. Math. Soc. **1**, 101–110 (1987)
80. C. Esposito, P. Stapor, S. Waldmann, Convergence of the Gutt Star Product, arXiv:1509.09160
81. P. Etingof, C. Walton, Finite Dimensional Hopf Actions on Deformation Quantizations, arXiv:1602.00532
82. B.V. Fedosov, *Deformation Quantization and Index Theory*. Mathematical Topics, vol. 9 (Akademie Verlag, Berlin, 1996)
83. B.V. Fedosov, A simple geometrical construction of deformation quantization. J. Differ. Geom. **40**, 213–238 (1994)
84. B.V. Fedosov, The index theorem for deformation quantization, in *Boundary value problems, Schrödinger operators, deformation quantization, Mathematical Topics*, vol. 8, ed. by M. Demuth, et al. (Akademie Verlag, Berlin, 1996), pp. 206–318

85. B.V. Fedosov, Non abelian reduction in deformation quantization. Lett. Math. Phys. **43**, 137–154 (1998)
86. B.V. Fedosov, On G-Trace and G-Index in deformation quantization. Lett. Math. Phys. **52**, 29–49 (2000)
87. G. Felder, B. Shoikhet, Deformation quantization with traces. Lett. Math. Phys. **53**, 75–86 (2000)
88. R. Fioresi, M.A. Lledo, On the deformation quantization of coadjoint orbits of semisimple groups. Pac. J. Math. **198**(2), 411–436 (2001)
89. R. Fioresi, A. Levrero, M.A. Lledo, Algebraic and differential star products on regular orbits of compact Lie groups coadjoint orbits of semisimple groups. Pac. J. Math. **206**, 321–337 (2002)
90. R. Fioresi, M.A. Lledo, V.S. Varadarajan, On the deformation quantization of affine algebraic varieties. Int. J. Math. **16**, 419–436 (2005)
91. M. Flato, A. Lichnerowicz, D. Sternheimer, Déformations 1-différentiables d'algèbres de Lie attachées à une variété symplectique ou de contact, C. R. Acad. Sci. Paris Sér. A **279**, 877–881 (1974) and Compositio Math. **31**, 47–82 (1975)
92. M. Flato, Deformation view of physical theories. Czec. J. Phys. **B32**, 472–475 (1982)
93. M. Flato, A. Lichnerowicz, D. Sternheimer, Crochet de Moyal-Vey et quantification, C. R. Acad. Sci. Paris I Math. **283**, 19–24 (1976)
94. C. Fronsdal, C. Roger, F. Butin, (eds.) Deformation quantization on the closure of minimal coadjoint orbits. Lett. Math. Phys. **88**, 271–320 (2009)
95. C. Fronsdal, Some ideas about quantization. Rep. Math. Phys. **15**, 111–145 (1978)
96. C. Fronsdal, M. Kontsevich, Quantization on curves. Lett. Math. Phys. **79**, 109–129 (2007)
97. M. Gerstenhaber, A Selection Principle in Deformation Quantization, arXiv:math/0512624
98. M. Gerstenhaber, On the deformation of rings and algebras. Ann. Math. **79**, 59–103 (1964)
99. H.J. Groenewold, On the Principles of elementary quantum mechanics. Physica **12**, 405–460 (1946)
100. S. Gutt, Second et troisième espaces de cohomologie différentiable de l'algèbre de Lie de Poisson d'une variété symplectique, Ann. Inst. H. Poincaré Sect. A (N.S.) **33**, 1–31 (1980)
101. S. Gutt, Equivalence of deformations and associated $*$ products. Lett. Math. Phys. **3**, 297–309 (1979)
102. S. Gutt, An explicit $*$ product on the cotangent bundle of a Lie group. Lett. Math. Phys. **7**, 249–258 (1983)
103. S. Gutt, On some second Hochschild cohomology spaces for algebras of functions on a manifold. Lett. Math. Phys. **39**, 157–162 (1997)
104. S. Gutt, J. Rawnsley, Equivalence of star products on a symplectic manifold; an introduction to Deligne's Čech cohomology classes. J. Geom. Phys. **29**, 347–392 (1999)
105. S. Gutt, J. Rawnsley, Natural star products on symplectic manifolds and quantum moment maps. Lett. Math. Phys. **66**, 123–139 (2003)
106. S. Gutt, S. Waldmann, Involutions and representations for reduced quantum algebras. Adv. Math. **224**, 2583–2644 (2010)
107. G. Halbout (ed.), *Deformation Quantization*. IRMA Lectures in Mathematics and Theoretical Physics. (Walter de Gruyter, 2002)
108. G. Halbout, Globalization of Tamarkin's formality theorem. Lett. Math. Phys. **71**(1), 39–48 (2005)
109. K. Hamachi, A new invariant for G-invariant star products. Lett. Math. Phys. **50**, 145–155 (1999)
110. K. Hamachi, Quantum moment maps and invariants for G-invariant star products. Rev. Math. Phys. **14**, 601–621 (2002)
111. G. Hochschild, On the cohomology groups of an associative algebra. Ann. Math. **46**, 58–67 (1945)
112. S. Jansen, N. Neumaier, G. Schaumann, S. Waldmann, Classification of invariant star products up to equivariant morita equivalence on symplectic manifolds. Lett. Math. Phys. **100**, 203–236 (2012)

113. A.V. Karabegov, On the phase form of a deformation quantization with separation of variables, arXiv:1504.03546
114. A. Karabegov, Berezin's quantization on flag manifolds and spherical modules. Trans. Am. Math. Soc. **359**, 1467–1479 (1998)
115. A. Karabegov, Cohomological classification of deformation quantisations with separation of variables. Lett. Math. Phys. **43**, 347–357 (1998)
116. A. Karabegov, On the canonical normalisation of a trace density of deformation quantisation. Lett. Math. Phys. **45**, 217–228 (1999)
117. A. Karabegov, M. Schlichenmaier, Identification of Berezin-Toeplitz deformation quantisation. J. Reine Angew. Math. **540**, 49–76 (2001)
118. M. Kashiwara, P. Shapira, Modules over deformation quantization algebroids: an overview. Lett. Math. Phys. **88**, 79–99 (2009)
119. V. Kathotia, Kontsevich universal formula for deformation quantisation and the CBH formula, preprint arXIv:math/9811174
120. M. Kontsevich, Operads and motives in deformation quantisation. Lett. Math. Phys. **48**, 35–72 (1999)
121. M. Kontsevich, Deformation quantisation of algebraic varieties. Lett. Math. Phys. **56**, 271–294 (2001)
122. M. Kontsevich, Deformation quantisation of Poisson manifolds, I. IHES preprint q-alg/9709040. Lett. Math. Phys. **66**, 157–216 (2003)
123. O. Kravchenko, Deformation quantisation of symplectic fibrations. Compositio Math. **123**, 131–165 (2000)
124. P.B.A. Lecomte, Application of the cohomology of graded Lie algebras to formal deformations of Lie algebras. Lett. Math. Phys. **13**, 157–166 (1987)
125. E. Leichtnam, X. Tang, A. Weinstein, Poisson geometry and deformation quantisation near a pseudoconvex boundary. J. Eur. Mat. Soc. **9**, 681–704 (2007)
126. A. Lichnerowicz, Déformations d'algèbres associées à une variété symplectique (les $*_\nu$-produits), Ann. Inst. Fourier, Grenoble **32**, 157–209 (1982)
127. A. Lichnerowicz, Cohomologie 1-différentiable des algèbres de Lie attachées à une variété symplectique ou de contact. J. Math. pures et appl. **53**, 459–484 (1974)
128. A. Lichnerowicz, Existence and equivalence of twisted products on a symplectic manifold. Lett. Math. Phys. **3**, 495–502 (1979)
129. M. Masmoudi, Tangential formal deformations of the Poisson bracket and tangential star products on a regular Poisson manifold. J. Geom. Phys. **9**, 155–171 (1992)
130. C. Moreno, Star-products on some Kähler-manifolds. Lett. Math. Phys. **11**, 361–372 (1986)
131. C. Moreno, P. Ortega-Navarro, $*$-products on $D^1(C)$, S^2 and related spectral analysis. Lett. Math. Phys. **7**, 181–193 (1983)
132. M. Müller-Bahns, N. Neumaier, Some remarks on $\mathfrak{g}$-invariant fedosov star products and quantum momentum mappings. J. Geom. Phys. **50**, 257–272 (2004)
133. F. Nadaud, On continuous and differential Hochschild cohomology. Lett. Math. Phys. **47**, 85–95 (1999)
134. O.M. Neroslavsky, A.T. Vlassov, Sur les déformations de l'algèbre des fonctions d'une variété symplectique. C. R. Acad. Sci. Paris Sér. I Math. **292**, 71–76 (1981)
135. R. Nest, B. Tsygan, Algebraic index theorem for families. Adv. Math. **113**, 151–205 (1995)
136. R. Nest, B. Tsygan, Algebraic index theorem. Commun. Math. Phys. **172**, 223–262 (1995)
137. N. Neumaier, Local ν-euler derivations and deligne's characteristic class of fedosov star products and star products of special type. Commun. Math. Phys. **230**, 271–288 (2002)
138. H. Omori, Y. Maeda, N. Miyazaki, A. Yoshioka, Deformation quantization of Fréchet-Poisson algebras: convergence of the Moyal product, in *Conférence Moshe Flato 1999. Quantization, Deformations, and Symmetries*, ed. by G. Dito, D. Sternheimer, pp. 233–245 [19]
139. H. Omori, Y. Maeda, A. Yoshioka, Deformation quantisations of Poisson algebras, in *Symplectic Geometry and Quantisation (Sanda and Yokohama, 1993)*, ed. by Y. Maeda et al., vol. 179 (Contemporary Mathematics, 1994), pp. 213–240

140. H. Omori, Y. Maeda, A. Yoshioka, Weyl manifolds and deformation quantisation. Adv. Math. **85**, 224–255 (1991)
141. H. Omori, Y. Maeda, A. Yoshioka, Existence of a closed star product. Lett. Math. Phys. **26**, 285–294 (1992)
142. H. Omori, Y. Maeda, A. Yoshioka, The uniqueness of star-products on $P_n(C)$, in *Differential Geometry (Shanghai 1991)*, ed. by C.H. Gu, et al. (World Science Publishing, River Edge, 1993), pp. 170–176
143. H. Omori, Y. Maeda, N. Miyazaki, A. Yoshioka, Convergent star product on Fréchet-Poisson algebras of Heisenberg type. Contemp. Math. **434**, 99–123 (2007)
144. H. Omori, Y. Maeda, N. Miyazaki, A. Yoshioka, Orderings and non-formal deformation quantization. Lett. Math. Phys. **82**, 153–175 (2007)
145. G. Pinczon, On the equivalence between continuous and differential deformation theories. Lett. Math. Phys. **39**, 143–156 (1997)
146. D. Rauch, Equivalence de produits star et classes de Deligne, Mémoire de Licence U.L.B. (1998)
147. J. Rawnsley, M. Cahen, S. Gutt, Quantisation of Kähler manifolds I. J. Geom. Phys. **7**, 45–62 (1990)
148. T. Reichert, S. Waldmann, Classification of equivariant star products on symplectic manifolds, arXiv:1507:00146
149. N. Reshetikhin, L. Takhtajan, Deformation quantisation of Kähler manifolds, L.D. Fadeev's Seminar on Mathematical Physics, AMS Translation Series 2, vol. 201 (2000), pp. 257–276
150. M. Rieffel, Deformation quantization for actions of $\mathbb{R}^d$, Mem. Am. Math. Soc. **106** (1993)
151. M. Rieffel, Questions on quantisation, in *Operator Algebras and Operator Theory (Shanghai, 1997)*, ed. by L. Ge et al., vol. 228, Contemporary Mathematics, pp. 315–328
152. M. Rieffel, Deformation quantization of heisenberg manifolds. Commun. Math. Phys. **122**, 531–562 (1989)
153. M. Rieffel, Quantization and C^*-algebras. Contemp. Math. **167**, 67–97 (1994)
154. P. Severa, On deformation quantisation of Dirac structures, in Geometry and Topology of Manifolds: The Mathematical Legacy of Charles Ehresmann; on the Occasion of the Hundredth Anniversary of His Birthday (Bedlewo, Poland, 8-15 May 2005)
155. B. Shoikhet, On the Kontsevich and the Campbell-Baker-Hausdorff Deformation Quantisations Of A Linear Poisson Structure, arXiv:math/9903036
156. B. Shoikhet, Vanishing of the Kontsevich integrals of the wheels. Lett. Math. Phys. **56**, 141–149 (2001)
157. B. Shoikhet, Koszul duality in deformation quantization and Tamarkin's approach to Kontsevich formality. Adv. Math. **224**, 731–771 (2010)
158. J.-M. Souriau, Structure des systèmes dynamiques, édition Dunod (1970)
159. D. Sternheimer, Deformation quantization twenty years after, in *Particles, Fields and Gravitation (Lodz, 1998)*. AIP Conference Proceedings, ed. by J. Rembieliński, vol. 453 (1998), pp. 107–145
160. D. Sternheimer, Phase-space representations, in *Applications of Group Theory in Physics and Mathematical Physics (Chicago, 1982)*. Lectures in Applied Mathematics, vol. 21 (American Mathematical Society, Providence RI, 1985), pp. 255–267
161. D. Tamarkin, Quantisation of Poisson structures on $\mathbb{R}^2$. Elec. Res. Ann. AMS **3**, 119–120 (1997)
162. D. Tamarkin, Another proof of M. Kontsevich formality theorem, preprint math/9803025, and formality of chain operad of small squares. Lett. Math. Phys. **66**, 65–72 (2003)
163. A. Unterberger, The calculus of pseudodifferential operators of Fuchs type. Commun. Partial Differ. Equ. **9**, 1179–1236 (1984)
164. M. Van den Bergh, On global deformation quantisation in the algebraic case. J. Algebra **315**, 326–395 (2007)
165. L. van Hove, Sur le problème des relations entre les transformations unitaires de la mécanique quantique et les transformations canoniques de la mécanique classique. Acad. R. Belgique Bull. Cl. Sci. **37**(5), 610–620 (1951)

166. J. Vey, Déformation du crochet de Poisson sur une variété symplectique. Comment. Math. Helvet. **50**, 421–454 (1975)
167. S. Waldmann, *Poisson-Geometrie und Deformationsquantisierung*. Eine Einführung. (Springer, Heidelberg, 2007)
168. S. Waldmann, States and representations in deformation quantization. Rev. Math. Phys. **17**, 15–75 (2005)
169. S. Waldmann, A nuclear Weyl algebra. J. Geom. Phys. **81**, 10–46 (2014)
170. A. Weinstein, P. Xu, Hochschild cohomology and characteristic classes for star-products, in *Geometry of Differential Equations*, AMS Translations Series 2, ed. by A. Khovanskii, et al. (1998), pp. 177–194
171. A. Weinstein, Deformation quantisation, Séminaire Bourbaki 1994, exposé 789. Astérisque **227**, 389–409 (1995)
172. P. Xu, Fedosov $*$-products and quantum momentum maps. Commun. Math. Phys. **197**, 167–197 (1998)

Chapter 3
Principal Fiber Bundles in Non-commutative Geometry

Christian Kassel

Al álgebra le dediqué mis mejores ánimos,
no sólo por respeto a su estirpe clásica
sino por mi cariño y mi terror al maestro.

Gabriel García Márquez,
Vivir para contarla [21]

Abstract These are the expanded notes of a course given at the Summer school "Geometric, topological, and algebraic methods for quantum field theory" held at Villa de Leyva, Colombia, in July 2015. We first give an introduction to non-commutative geometry and to the language of Hopf algebras. We next build up a theory of non-commutative principal fiber bundles and consider various aspects of such objects. Finally, we illustrate the theory using the quantum enveloping algebra $U_q\,\mathfrak{sl}(2)$ and related Hopf algebras.

3.1 Introduction

These are the expanded notes of a course given at the Summer school "Geometric, topological, and algebraic methods for quantum field theory" held at Villa de Leyva, Colombia, in July 2015. The main objective of this course was twofold:

1. to give an introduction to non-commutative geometry and to the language of Hopf algebras;
2. to build up a theory of non-commutative principal fiber bundles, consider various aspects of these non-commutative objects, highlight the similarities and the differences with their classical counterparts, and illustrate the theory with significant examples.

C. Kassel (✉)
Institut de Recherche Mathématique Avancée, CNRS & Université de Strasbourg,
7 rue René Descartes, 67084 Strasbourg, France
e-mail: kassel@math.unistra.fr

A. Cardona et al. (eds.), *Quantization, Geometry and Noncommutative Structures in Mathematics and Physics*, Mathematical Physics Studies,
DOI 10.1007/978-3-319-65427-0_3

Non-commutative geometry is based on the idea that instead of working with the points of a topological space X (or a C^∞-manifold, or an algebraic variety), we may just as well work with the algebra $\mathcal{O}(X)$ of continuous (or C^∞, or regular) functions on X. Many geometrical constructions on X can be expressed by algebraic constructions on the commutative algebra $\mathcal{O}(X)$, which in turn can be extended to non-necessarily commutative algebras. The necessity of passing from commutative algebra to non-commutative ones originates from physics; according to [9],

> [it] arises from the general indication that the small-scale structure of space-time is not well-modeled by usual continuous geometry. At the Planck scale one may reasonably expect that our notion of geometry has to be modified to include quantum effects as well. Non-commutative geometry has the potential to do this.

Keeping in mind the geometric origin of such non-commutative constructions, it is natural to use the phrase "non-commutative spaces" for non-commutative algebras. In mathematics, such generalized spaces have appeared in the 1980s not only in the work of Connes on group actions and on foliations (see [13]), but also in the theory of quantum groups, which originated in the work of Faddeev's school, of Drinfeld, of Jimbo, and of Woronowicz (see [17, 18, 30, 51, 61]).

Quantum groups are non-commutative algebras depending on a parameter q. When q takes the value 1, then quantum groups specialize to classical objects such as groups of symmetries. The construction of quantum groups was inspired by the "quantum inverse scattering method," a method devised for constructing integrable quantum systems and mostly developed by L. D. Faddeev and his collaborators. The discovery of quantum groups was a major event with spectacular applications not only in quantum physics, but also in domains of pure mathematics such as representation theory and low-dimensional topology. Let us quote Drinfeld on quantization from the introduction of [18]:

> ... both in classical and quantum mechanics there are two basic concepts: state and observable. In classical mechanics [...] observables are functions on [a manifold] M. In the quantum case [...] observables are operators in [a Hilbert space] H [...] [O]bservables form an associative algebra which is commutative in the classical case and noncommutative in the quantum case. So quantization is something like replacing commutative algebras by noncommutative ones.

Technically speaking, quantum groups are what algebraists and topologists call Hopf algebras. Therefore, the first aim of this course was to introduce the concept of a Hopf algebra and to illustrate it with significant examples, such as the ones related to the special linear group $SL_2(\mathbb{C})$.

Our second aim was to define non-commutative analogues of principal fiber bundles. Principal fiber bundles are ubiquitous geometrical objects in mathematics and gauge theory. For instance, given a Lie (or algebraic) group G and a closed subgroup G', the projection $G \to G/G'$ onto the homogeneous space G/G' is a principal fiber bundle. To quantize homogeneous spaces, we need an adequate notion of quotient of Hopf algebras and more precisely the concepts of comodule algebras and Hopf Galois extensions. There are numerous meaningful examples of non-commutative principal fiber bundles; see [9, 14, 24, 25, 40, 41, 49, 50].

Let us give an overview of these notes. In Sect. 3.2, we review the definition of classical principal fiber bundles and state their main properties. In Sect. 3.3, we undertake the crucial passage from commutative to non-commutative algebras; we concentrate on two simple situations in which a space X can easily be replaced by its function algebra $\mathscr{O}(X)$, namely when X is a finite set or when it is an affine algebraic variety. To make things even simpler, all objects and algebras considered in these notes are defined over the field $\mathbb{C}$ of complex numbers. We also give in Sect. 3.3 our first example of a non-commutative space, namely the "quantum plane," a one-parameter deformation of the ordinary complex plane, and we extend certain basic operations from ordinary spaces to non-commutative ones.

In Sect. 3.4, we consider the case when a space has an additional group structure. This naturally leads to the notion of a Hopf algebra. In Sect. 3.4.4, we present two mutually dual Hopf algebras constructed from a finite group.

In Sect. 3.5, we introduce two quantum groups associated with the Lie group $SL_2(\mathbb{C})$; one is its quantum coordinate algebra $\mathrm{SL}_q(2)$, the other one is the quantum enveloping algebra $U_q\,\mathfrak{sl}(2)$ of the Lie algebra of $SL_2(\mathbb{C})$. We also construct a duality map between them and consider two interesting quotients.

In Sect. 3.6, we extend the notion of a group action to the non-commutative world. This leads us to the concept of a comodule algebra over a Hopf algebra. We give various examples of comodule algebras, thus showing that this concept covers much more than just group actions. In particular, any group-graded algebra is a comodule algebra over a suitable Hopf algebra. We also show how to equip the quantum plane with the structure of a comodule algebra over the quantum coordinate algebra of $SL_2(\mathbb{C})$.

Section 3.7 is entirely devoted to Hopf Galois extensions, which are non-commutative analogues of principal fiber bundles. We pose the problem of classifying them and show that, contrary to the classical case, there may exist (infinitely many) non-isomorphic non-commutative principal fiber bundles over a point. We also define the non-commutative version of the pull-back of a bundle.

In the final section (Sect. 3.8), for any Hopf algebra H, we construct a non-commutative principal fiber bundle in the form of a deformation $\mathscr{A}_H$ of H over a parameter space $\mathscr{B}_H$ which is the coordinate algebra of a smooth affine algebraic variety of the same dimension as H. We give explicit formulas for this non-commutative principal fiber bundle when H is the quantum enveloping algebra $U_q\,\mathfrak{sl}(2)$ or some of its finite-dimensional quotients.

We will not give the proofs of all statements in these notes. For some of them, we will refer to the relevant publications or to exercises if they turn out to be rather simple. Except for Theorems 3.7 and 3.8 in Sect. 3.8.3, the material presented in these notes already exists in the literature.

3.2 Review of Principal Fiber Bundles

La geometría fue más compasiva tal vez por obra y gracia de su prestigio literario. [21]

We start by recalling the definition and the basic properties of fiber bundles and of principal fiber bundles. In Sect. 3.7, we will define non-commutative analogues of such bundles.

3.2.1 *Fiber Bundles*

Let F be a topological space. Recall that a *fiber bundle* with fiber F is a locally trivial continuous map $\pi : P \to X$ from a topological space P, called the *total space* of the bundle, to a topological space X, called the *base space*, such that each *fiber* $\pi^{-1}(\{x\})$ is homeomorphic to F. *Locally trivial* means that for each $x \in X$, there is a neighborhood $U \subset X$ of x and a homeomorphism $\psi : \pi^{-1}(U) \cong U \times F$ such that $\pi = p_1 \circ \psi$, where $p_1 : U \times F \to U$ is the first projection onto U.

In the sequel, we assume that the topological spaces we consider are Hausdorff and paracompact (the latter means that every open cover has a locally finite open refinement). These conditions are satisfied by most spaces generally considered.

A *fiber bundle map* from a fiber bundle $\pi' : P' \to X'$ to another fiber bundle $\pi : P \to X$ with the same fiber F is a pair $(\widetilde{\varphi} : P' \to P, \varphi : X' \to X)$ of continuous maps such that $\pi \circ \widetilde{\varphi} = \varphi \circ \pi'$. The composition of two such maps is again a fiber bundle map. A fiber bundle map is said to be a *homeomorphism of fiber bundles* if both $\widetilde{\varphi} : P' \to P$ and $\varphi : X' \to X$ are homeomorphisms.

The simplest example of a fiber bundle with fiber F and base space X is given by the first projection $p_1 : X \times F \to X$. Any fiber bundle homeomorphic to such a fiber bundle is called a *trivial fiber bundle*.

3.2.2 *Pull-Backs*

We now deal with an important functoriality property. Any fiber bundle $\pi : P \to X$ with fiber F and base space X together with any continuous map $\varphi : X' \to X$ induces a fiber bundle $\pi' : \varphi^*(P) \to X'$ with the same fiber F and with base space X'. The space $\varphi^*(P)$ is defined by

$$\varphi^*(P) = \{(x', p) \in X' \times P \mid \varphi(x') = \pi(p)\}$$

and the map $\pi' : \varphi^*(P) \to X'$ is equal to the composite map $\varphi^*(P) \subset X' \times P \xrightarrow{p_1} X'$. The fiber bundle $\pi' : \varphi^*(P) \to X'$ is called the *pull-back* of the bundle $\pi : P \to X$ along the map $\varphi : X' \to X$.

Clearly, if $\varphi' : X'' \to X'$ is another continuous map, then

$$\varphi'^*(\varphi^*(P)) \cong (\varphi \circ \varphi')^*(P).$$

If $\mathrm{id} : X \to X$ is the identity map of X, then $\mathrm{id}^*(P) = P$. It follows that any homeomorphism $\varphi : X' \to X$ induces a homeomorphism $\varphi^*(P) \cong P$.

Exercise 3.1 (a) Let $\pi : P \to X$ be a fiber bundle. Prove that if $i : \{x\} \to X$ is the inclusion of a point x in X, then $i^*(P) = \pi^{-1}(\{x\})$ is the fiber of the bundle at x.
(b) Show that any fiber bundle with base space equal to a point is trivial.
(c) Prove that the pull-back of a trivial fiber bundle is trivial.
(d) Let X be a *contractible* space, such that there is an element $x_0 \in X$ and a continuous map $\eta : X \times [0, 1] \to X$ such that $\eta(x, 0) = x$ and $\eta(x, 1) = x_0$ for all $x \in X$. Show that any fiber bundle with base space X is trivial.

For more on fiber bundles, see the classical references [26, 57].

3.2.3 Principal Fiber Bundles

We fix now a topological group G.

Definition 3.1 A *principal G-bundle* is a fiber bundle $\pi : P \to X$ with a continuous left action $G \times P \to P$ satisfying the following two conditions:

(i) we have $\pi(gp) = \pi(p)$ for all $g \in G$ and $p \in P$,
(ii) for all $p, p' \in P$ with $\pi(p) = \pi(p')$ there is a unique element $g \in G$ such that $gp = p'$.

In other words, in a principal G-bundle, the group action preserves each fiber $\pi^{-1}(x)$ and the action of G on each fiber is free and transitive. It follows that each fiber is in bijection with G and that the space of orbits $G\backslash P$ is homeomorphic to the base space X.

An equivalent way to express Conditions (i) and (ii) above is to require that the map

$$\gamma : G \times P \to P \times P; \quad (g, p) \mapsto (gp, p) \tag{3.1}$$

is a bijection from $G \times P$ onto the subspace

$$P \times_X P = \left\{(p, p') \in P \times P \mid \pi(p) = \pi(p')\right\}.$$

Given principal G-bundles $\pi : P' \to X'$ and $\pi : P \to X$, a *map of principal G-bundles* from the first one to the second one is a fiber bundle map $(\widetilde{\varphi}, \varphi)$ compatible with the G-action, such that $\widetilde{\varphi}(gp') = g\widetilde{\varphi}(p')$ for all $g \in G$ and $p' \in P'$.

Example 3.1 Given a topological space X, let G act on $P = G \times X$ by $g'(g, x) = (g'g, x)$ $(g, g' \in G, x \in X)$. This is a principal G-bundle. Any principal G-bundle homeomorphic to such a bundle is called a *trivial principal G-bundle*.

Example 3.2 Consider the group S^1 of complex numbers of modulus one. Given an integer $n \geq 1$, the map $\pi_n : S^1 \to S^1$ defined by $\pi_n(z) = z^n$ is a principal G-bundle, where G is the cyclic group $\mathbb{Z}/n$ of order n.

Exercise 3.2 Prove that the principal $\mathbb{Z}/n$-bundle $\pi_n : S^1 \to S^1$ of Example 3.2 is trivial if and only if $n = 1$.

3.2.4 *Functoriality and Classification*

We now record important properties of principal G-bundles. For the proofs we refer to [26, Chap. 4] or to [57].

Theorem 3.1 *(a) If $\pi : P \to X$ is a principal G-bundle and $\varphi : X' \to X$ is a continuous map, then the pull-back $\pi' : \varphi^*(P) \to X'$ is a principal G-bundle.*

(b) If $\pi : P \to X$ is a principal G-bundle and $\varphi_0, \varphi_1 : X' \to X$ are homotopic[1] *continuous maps, then the principal G-bundles $\varphi_0^*(P)$ and $\varphi_1^*(P)$ are homeomorphic.*

(c) There exists a principal G-bundle $\pi_G : EG \to BG$ such that for any principal G-bundle $\pi : P \to X$ there is a continuous map $\varphi : X \to BG$ such that $\varphi^*(EG)$ is homeomorphic to $\pi : P \to X$; the map φ is unique up to homotopy.

The base space of the principal G-bundle $\pi_G : EG \to BG$ is called the *classifying space* of the group G. The terminology is justified by the following immediate consequence of the theorem.

Corollary 3.1 *The map $\varphi \mapsto \varphi^*(EG)$ induces a bijection between the set $[X, BG]$ of homotopy classes of continuous maps from X to BG and the set $\mathrm{Iso}_G(X)$ of homeomorphism classes of principal G-bundles with base space X:*

$$[X, BG] \cong \mathrm{Iso}_G(X).$$

Starting from the next section, we shall build up the algebraic language necessary to define non-commutative analogues of principal fiber bundles.

3.3 Basic Ideas of Non-commutative Geometry

As we stated in the introduction, non-commutative geometry is based on the idea of (a) replacing a space X by its (commutative) function algebra $\mathcal{O}(X)$, (b) passing from commutative algebras to non-commutative algebras. In this section, we start with two simple geometric situations, namely when X is a finite set and when it is an affine algebraic variety. In Sect. 3.3.2, we present our first elementary example of

---

[1]That is, there exists a continuous map $\Phi : X' \times [0, 1] \to X$ such that $\Phi(x, 0) = \varphi_0(x)$ and $\Phi(x, 1) = \varphi_1(x)$ for all $x \in X'$.

a non-commutative space, namely the quantum plane, and in Sect. 3.3.3 we extend certain basic operations from spaces to non-commutative ones.

For *deformation quantization*, which is another way, inspired by quantum mechanics, to pass from commutative algebras to non-commutative algebras see the lectures [23] by Simone Gutt.

3.3.1 Two Classical Dualities Between Spaces and Algebras

Let us now present two well-known correspondences between spaces and algebras. All algebras we consider in these notes are $\mathbb{C}$-algebras (i.e., defined over the field $\mathbb{C}$ of complex numbers). We furthermore assume that all algebras are associative and unital. We denote the unit of an algebra A by 1, or by 1_A to avoid any confusion.

3.3.1.1 Finite Sets

In the first example, the spaces which we consider are merely sets, or if one prefers, discrete topological spaces. To any set X, we associate its *function algebra* $\mathcal{O}(X)$, which consists of all complex-valued functions on X. Given two such functions $u_1, u_2 : X \to \mathbb{C}$, we may consider any linear combination $\lambda_1 u_1 + \lambda_2 u_2$, where λ_1 and λ_2 are complex numbers; the function $\lambda_1 u_1 + \lambda_2 u_2$ is defined by

$$(\lambda_1 u_1 + \lambda_2 u_2)(x) = \lambda_1 u_1(x) + \lambda_2 u_2(x)$$

for all $x \in X$. Similarly, the product $u_1 u_2$ of two functions $u_1, u_2 \in \mathcal{O}(X)$ is defined by $(u_1 u_2)(x) = u_1(x) u_2(x)$ for all $x \in X$. These operations provide $\mathcal{O}(X)$ with the structure of a commutative associative and unital $\mathbb{C}$-algebra. The unit is the constant function whose values are all equal to 1.

For any $x \in X$, consider the δ-function δ_x defined for all $y \in X$ by $\delta_x(y) = \delta_{x,y}$, where $\delta_{x,y}$ is the Kronecker symbol.[2] The product of two δ-functions is clearly given by

$$\delta_x \, \delta_y = \delta_{x,y} \, \delta_x .$$

This means that each δ-function is an idempotent, i.e., $\delta_x^2 = \delta_x$, and that the product of two distinct δ-functions is zero.

If the set X is *finite*, then the set $\{\delta_x\}_{x \in X}$ of δ-functions forms a basis of $\mathcal{O}(X)$ considered as a vector space over the complex numbers. Indeed, we can expand any function $u : X \to \mathbb{C}$ in the following unique way:

$$u = \sum_{x \in X} u(x) \, \delta_x .$$

Note that the unit of $\mathcal{O}(X)$ is the sum of the δ-functions: $1 = \sum_{x \in X} \delta_x$.

[2] Recall that $\delta_{x,y} = 1$ if $x = y$ and $\delta_{x,y} = 0$ otherwise.

If the set X is of cardinality N, we can order the elements of X and assume that $X = \{x_1, \ldots, x_N\}$. Consider the linear map

$$u \in \mathcal{O}(X) \mapsto (u(x_1), \ldots u(x_N)) \in \mathbb{C}^N.$$

This map is clearly an isomorphism from $\mathcal{O}(X)$ onto the N-dimensional vector space $\mathbb{C}^N$. It is also an algebra isomorphism if we endow $\mathbb{C}^N$ with the product

$$(x_1, \ldots x_N)(y_1, \ldots y_N) = (x_1 y_1, \ldots x_N y_N).$$

In particular, the dimension of $\mathcal{O}(X)$ is equal to the cardinality of X. Since a finite set is determined up to bijection by its cardinality, it follows that a finite set X can be recovered (up to bijection) from its function algebra $\mathcal{O}(X)$.

3.3.1.2 Algebraic Varieties

The next correspondence is more substantial, namely the one between algebraic varieties and commutative algebras. Recall that a *complex algebraic variety* is the set of solutions of a system of polynomial equations over the complex numbers: more precisely, let Σ be a set of polynomials in $\mathbb{C}[X_1, \ldots, X_n]$; then the corresponding algebraic variety is given by

$$V = \left\{ (x_1, \ldots, x_n) \in \mathbb{C}^n \mid P(x_1, \ldots, x_n) = 0 \ \text{ for all } P \in \Sigma \right\}.$$

To V we associate the quotient-algebra

$$\mathcal{O}(V) = \mathbb{C}[X_1, \ldots, X_n]/I_\Sigma,$$

where I_Σ is the ideal of $\mathbb{C}[X_1, \ldots, X_n]$ generated by Σ. We say that $\mathcal{O}(V)$ is the *coordinate algebra* of the algebraic variety V. The algebra $\mathcal{O}(V)$ is a finitely generated commutative $\mathbb{C}$-algebra.

Conversely, let us start from a finitely generated commutative $\mathbb{C}$-algebra A. It can be written as the quotient of a polynomial algebras with finitely many variables, i.e., it is of the form

$$A = \mathbb{C}[X_1, \ldots, X_n]/I$$

for some ideal $I \subset \mathbb{C}[X_1, \ldots, X_n]$. Then $A = \mathcal{O}(V)$, where V is the set of points $(x_1, \ldots, x_n) \in \mathbb{C}^n$ satisfying the system of polynomial equations $P(x_1, \ldots, x_n) = 0$ for all $P \in I$.

There is another way to find V such that $A = \mathcal{O}(V)$ for a given finitely generated commutative $\mathbb{C}$-algebra A. Namely consider the set $\mathrm{Alg}(A, \mathbb{C})$ of characters of A. A *character* of A is an algebra homomorphism χ from A to $\mathbb{C}$, i.e., a linear form satisfying the conditions

$$\chi(ab) = \chi(a)\chi(b) \quad \text{and} \quad \chi(1) = 1.$$

Now, if $A = \mathbb{C}[X_1, \dots, X_n]/I$, then a character $\chi : A \to \mathbb{C}$ is determined by its values $\chi(X_i) = x_i \in \mathbb{C}$ on the generators $X_1, \dots, X_n$. Since χ must be zero on the ideal I, this means that the n-tuple $(x_1, \dots, x_n) \in \mathbb{C}^n$ of values must be a solution of the equations $P(x_1, \dots, x_n) = 0$ for all $P \in I$. Such solutions form an algebraic variety V, and we have $A = \mathscr{O}(V)$.

Let us also observe that the characters of a finitely generated commutative $\mathbb{C}$-algebra A are in bijection with its maximal ideals. Indeed, start from a character $\chi : A \to \mathbb{C}$; its kernel $\mathfrak{m}$ is an ideal of A. Since χ is surjective, we have $A/\mathfrak{m} \cong \mathbb{C}$ by Noether's first isomorphism theorem. Therefore, $\mathfrak{m}$ is a maximal ideal. Conversely, let $\mathfrak{m}$ be a maximal ideal of A. Then $A/\mathfrak{m}$ is a field which is isomorphic to $\mathbb{C}$ by Zarisky's lemma or by Hilbert's Nullstellensatz. The composed algebra map $\chi : A \to A/\mathfrak{m} \cong \mathbb{C}$ is a character of A.

Let us now give some elementary examples of commutative algebras corresponding to algebraic varieties.

Example 3.3 The coordinate algebra of a point is $\mathbb{C}$ since $\mathrm{Alg}(\mathbb{C}, \mathbb{C})$ consists only of one element, namely the identity map. This follows also from the description of the function algebra of a finite set given in Sect. 3.3.1.1.

Example 3.4 The one-variable polynomial algebra $\mathbb{C}[X]$ is the coordinate algebra of the *complex line* $\mathbb{C}$ since any algebra homomorphism $\mathbb{C}[X] \to \mathbb{C}$ is determined by its value on the variable X; equivalently, $\mathrm{Alg}(\mathbb{C}[X], \mathbb{C}) \cong \mathbb{C}$.

Similarly, the two-variable polynomial algebra $\mathbb{C}[X, Y]$ is the coordinate algebra of the *complex plane* $\mathbb{C}^2$: any algebra homomorphism $\mathbb{C}[X, Y] \to \mathbb{C}$ is determined by its values on X and Y. We have $\mathrm{Alg}(\mathbb{C}[X, Y], \mathbb{C}) \cong \mathbb{C}^2$.

Example 3.5 Let us now consider the algebra $A = \mathbb{C}[X, X^{-1}]$ of Laurent polynomials in the variable X. Since $XX^{-1} = 1$, this algebra can also be seen as the quotient-algebra $\mathbb{C}[X, Y]/(XY - 1)$. Here also any algebra homomorphism $\chi : A \to \mathbb{C}$ is determined by its value $\chi(X) = x \in \mathbb{C}$ on the variable X, but contrary to the case of $\mathbb{C}[X]$, the fact that X is invertible in A puts the following restriction on x, namely

$$x\,\chi(X^{-1}) = \chi(X)\,\chi(X^{-1}) = \chi(XX^{-1}) = \chi(1) = 1.$$

Therefore, x is invertible in the field $\mathbb{C}$, which is equivalent to $x \neq 0$. We deduce $\mathrm{Alg}(A, \mathbb{C}) \cong \mathbb{C}^\times$, where $\mathbb{C}^\times = \mathbb{C} \setminus \{0\}$. In other words, the algebra $\mathbb{C}[X, X^{-1}]$ of Laurent polynomials is the coordinate algebra of the *once-punctured complex line*.

Example 3.6 The algebra $\mathbb{C}[X, Y]/(Y^2 - X^3 + X - 1)$ is the coordinate algebra of the *elliptic curve* consisting of the points $(x, y) \in \mathbb{C}^2$ satisfying the equation

$$y^2 = x^3 - x + 1.$$

Example 3.7 Let $x_1, \ldots, x_N$ be distinct points in the complex line $\mathbb{C}$. Consider the quotient-algebra $A = \mathbb{C}[X]/(X - x_1, \ldots, X - x_n)$. Since the polynomials $X - x_i$ are coprime, we also have $A = \mathbb{C}[X]/(P)$, where P is the degree N polynomial

$$P = (X - x_1) \cdots (X - x_n).$$

The assignment $Q \in \mathbb{C}[X] \mapsto (Q(x_1), \ldots, Q(x_N)) \in \mathbb{C}^N$ induces an algebra isomorphism $A \cong \mathbb{C}^N$. This example shows that a finite set can be seen as a special case of an algebraic variety.

3.3.2 Non-commutative Algebras

From now on we deal with non-necessarily commutative algebras. We recall that all algebras we consider are associative unital $\mathbb{C}$-algebras.

3.3.2.1 Non-commutative Polynomials

The prototype of a finitely generated complex commutative algebra is the algebra of polynomials $\mathbb{C}[X_1, \ldots, X_n]$ in finitely many variables. In an analogous way, the prototype of a finitely generated not necessarily commutative complex algebra is the algebra $\mathbb{C}\langle X_1, \ldots, X_n \rangle$ of polynomials in *n non-commuting variables* $X_1, \ldots, X_n$. Any element of $\mathbb{C}\langle X_1, \ldots, X_n \rangle$ is a finite linear combination (with complex coefficients) of finite words in the letters $X_1, \ldots, X_n$. Such a linear combination is unique because such words form a basis of $\mathbb{C}\langle X_1, \ldots, X_n \rangle$ considered as a vector space over the complex numbers.

Mind the difference between these two kinds of polynomial algebras: the element $XY - YX$ is nonzero in $\mathbb{C}\langle X, Y \rangle$ whereas it vanishes in $\mathbb{C}[X, Y]$.

Any finitely generated complex algebra A is a quotient-algebra of $\mathbb{C}\langle X_1, \ldots, X_n \rangle$ for some n, which means that A can be expressed as

$$A = \mathbb{C}\langle X_1, \ldots, X_n \rangle / I$$

for some two-sided ideal I of $\mathbb{C}\langle X_1, \ldots, X_n \rangle$. For instance, for the algebra of ordinary polynomials in n variables, we have

$$\mathbb{C}[X_1, \ldots, X_n] = \mathbb{C}\langle X_1, \ldots, X_n \rangle / I,$$

where I is the two-sided ideal generated by all elements of the form $X_i X_j - X_j X_i$ $(i, j \in \{1, \ldots, n\}^2)$.

3.3.2.2 The Quantum Plane

Let q be a nonzero complex number. Consider the algebra $\mathbb{C}\langle X, Y\rangle$ of polynomials in two non-commuting variables X, Y and the two-sided ideal I_q of $\mathbb{C}\langle X, Y\rangle$ generated by $YX - qXY$. The quotient-algebra

$$\mathbb{C}_q[X, Y] = \mathbb{C}\langle X, Y\rangle / I_q$$

is *not commutative* unless $q = 1$.

When $q = 1$, then the algebra $\mathbb{C}_q[X, Y]$ is isomorphic to $\mathbb{C}[X, Y]$, which is the coordinate algebra of the plane. Thus, $\mathbb{C}_q[X, Y]$ is a one-parameter non-commutative deformation (or a quantization) of the coordinate algebra of the plane. For this reason and by extension, $\mathbb{C}_q[X, Y]$ can be considered as the coordinate algebra of a "space" in an extended sense, of a so-called *non-commutative space*. In this particular instance, this non-commutative space is known in the literature under the name *quantum plane*.

The set $\{X^i Y^j\}_{i,j\geq 0}$ forms a basis of $\mathbb{C}_q[X, Y]$, independently of q (see Exercise 3.3 below). Notice that the defining relation $YX = qXY$ implies the following product formula for two monomials in $\mathbb{C}_q[X, Y]$:

$$(X^i Y^j)(X^k Y^\ell) = q^{jk}\, X^{i+k} Y^{j+\ell}. \qquad (i, j, k, \ell \geq 0)$$

In Sect. 3.3.1.2, we showed how to recover an algebraic variety V from its coordinate algebra, using its characters. Let us look at the set $\mathrm{Alg}(\mathbb{C}_q[X, Y], \mathbb{C})$ of characters of $\mathbb{C}_q[X, Y]$. As with the usual polynomial algebra $\mathbb{C}[X, Y]$, a character $\chi : \mathbb{C}_q[X, Y] \to \mathbb{C}$ is determined by its values $\chi(X) = x$ and $\chi(Y) = y$ on the generators X and Y. Now the set $\mathrm{Alg}(\mathbb{C}_q[X, Y], \mathbb{C})$ is in bijection with the set of points $(x, y) \in \mathbb{C}^2$ such that $yx = qxy$. In $\mathbb{C}$, the values x and y commute, so that $yx = qxy$ is equivalent to $(q-1)xy = 0$. When $q \neq 1$, then $\mathrm{Alg}(\mathbb{C}_q[X, Y], \mathbb{C})$ can be identified with the subset of $\mathbb{C}^2$ defined by $xy = 0$; this subset is the union of the lines $L_1 = \{0\} \times \mathbb{C}$ and $L_2 = \mathbb{C} \times \{0\} \subset \mathbb{C}^2$. The coordinate algebra of $L_1 \cup L_2$ is the commutative algebra $\mathbb{C}[X, Y]/(XY)$. We thus have bijections

$$\mathrm{Alg}(\mathbb{C}_q[X, Y], \mathbb{C}) = \begin{cases} \mathrm{Alg}(\mathbb{C}[X, Y], \mathbb{C}) = \mathbb{C}^2 & \text{if } q = 1, \\ \mathrm{Alg}(\mathbb{C}[X, Y]/(XY), \mathbb{C}) = L_1 \cup L_2 & \text{if } q \neq 1. \end{cases}$$

This shows that from the point of view of characters, there is a jump when we pass from $q = 1$ to an arbitrary complex number q. Observe also that as a vector space, $\mathbb{C}[X, Y]/(XY)$ has a basis given by $\{X^i\}_{i\geq 0} \cup \{Y^j\}_{j\geq 1}$; this basis is clearly very different from the basis $\{X^i Y^j\}_{i,j\geq 0}$ of $\mathbb{C}_q[X, Y]$.

Exercise 3.3 (*A basis of the quantum plane*)

(a) Let τ and υ be the endomorphisms of the polynomial algebra $\mathbb{C}[t]$ defined on any polynomial $P(t)$ by $\tau(P(t)) = tP(t)$ and $\upsilon(P(t)) = P(qt)$. Show that there is a unique algebra morphism $\rho : \mathbb{C}_q[X, Y] \to \mathrm{End}(\mathbb{C}[t])$ such that $\rho(X) = \tau$ and $\rho(Y) = \upsilon$.

(b) Deduce that $\{X^i Y^j\}_{i,j \in \mathbb{N}}$ is a basis of $\mathbb{C}_q[X, Y]$. Hint: use the morphism ρ to prove linear independence.

3.3.2.3 Non-commutative Spaces

In view of the previous examples, non-commutative algebras will henceforth often be called *non-commutative spaces*. The special case of the quantum plane shows that characters are not sufficient to characterize non-commutative spaces. As written in the introduction of [49],

> ... in noncommutative geometry there are no points.

This is a significant difference with ordinary spaces. Such a difference is also well explained in [55, Sect. 2].

3.3.3 *Extending Basic Operations to Non-commutative Spaces*

We now show how to extend certain basic operations on spaces to the world of non-commutative spaces, i.e., of non-necessarily commutative algebras.

3.3.3.1 From Maps to Algebra Homomorphisms

Let $\varphi : X \to Y$ be a map between algebraic varieties. Then we can define a map $\varphi^* : \mathcal{O}(Y) \to \mathcal{O}(X)$ by

$$\varphi^*(u) = u \circ \varphi \tag{3.2}$$

for all $u \in \mathcal{O}(Y)$. It is easy to check that φ^* is a morphism of algebras.

If $\psi : Y \to Z$ is another map between algebraic varieties and $\psi^* : \mathcal{O}(Z) \to \mathcal{O}(Y)$ is the corresponding morphism of algebras, then we have the following equality of morphisms from $\mathcal{O}(Z)$ to $\mathcal{O}(X)$:

$$(\psi \circ \varphi)^* = \varphi^* \circ \psi^*.$$

3.3.3.2 From Products to Tensor Products

Given algebraic varieties X, Y, we can consider their product $X \times Y$. We denote by $\pi_X : X \times Y \to X$ and $\pi_Y : X \times Y \to Y$ the canonical projections. The product $X \times Y$

satisfies the following universal property: for all maps $\varphi_X : Z \to X$ and $\varphi_Y : Z \to Y$ from another algebraic variety Z, there exists a unique map $\varphi : Z \to X \times Y$ such that $\pi_X \circ \varphi = \varphi_X$ and $\pi_Y \circ \varphi = \varphi_Y$.

Applying the contravariant functor $\varphi \mapsto \varphi^*$ defined by (3.2), we see that the coordinate algebra $\mathscr{O}(X \times Y)$ comes with two algebra morphisms

$$\varphi_X^* : \mathscr{O}(X) \to \mathscr{O}(X \times Y) \quad \text{and} \quad \varphi_Y^* : \mathscr{O}(Y) \to \mathscr{O}(X \times Y)$$

satisfying a universal property that is easily deduced from the universal property of the product $X \times Y$. It follows that we have a canonical algebra isomorphism

$$\mathscr{O}(X \times Y) \cong \mathscr{O}(X) \otimes \mathscr{O}(Y), \tag{3.3}$$

where $\mathscr{O}(X) \otimes \mathscr{O}(Y)$ is the tensor product of the algebras $\mathscr{O}(X)$ and $\mathscr{O}(Y)$.

Let us recall that the *tensor product* $U \otimes V$ of two complex vector spaces U and V consists of $\mathbb{C}$-linear combinations of symbols of the form $u \otimes v$, where $u \in U$ and $v \in V$. By definition, the map $U \times V \to U \otimes V$ sending each couple $(u, v) \in U \times V$ to $u \otimes v$ is $\mathbb{C}$-bilinear, i.e., $\mathbb{C}$-linear both in u and in v. It satisfies the following universal property: For any $\mathbb{C}$-bilinear map $f : U \times V \to W$ to another vector space W, there is a unique $\mathbb{C}$-linear map $\widetilde{f} : U\text{Ł}\otimes V \to W$ such that $f(u, v) = \widetilde{f}(u \otimes v)$ for all $(u, v) \in U \times V$. Moreover, if $\{u_i\}_{i \in I}$ is a basis of U and $\{v_j\}_{j \in J}$ is a basis of V, then

$$\{u_i \otimes v_j\}_{(i,j) \in I \times J}$$

is a basis of $U \otimes V$. As a consequence, $\dim(U \otimes V) = \dim(U) \dim(V)$.

If A, B are (not necessarily commutative) algebras, then their tensor product $A \otimes B$ carries a structure of algebra with multiplication determined by

$$(a_1 \otimes b_1)(a_2 \otimes b_2) = a_1 a_2 \otimes b_1 b_2$$

for all $a_1, a_2 \in A$ and $b_1, b_2 \in B$. The algebra $A \otimes B$ has a unit given by

$$1_{A \otimes B} = 1_A \otimes 1_B.$$

The tensor product of algebras satisfies the following universal property.

Proposition 3.1 *Let $f : A \to C$ and $g : B \to C$ be morphisms of algebras such that $f(a)g(b) = g(b)f(a)$ in C for all $a \in A$ and $b \in B$. Then there exists a unique morphism of algebras $f \otimes g : A \otimes B \to C$ such that $(f \otimes g)(a \otimes b) = f(a)g(b)$ for all $a \in A$ and $b \in B$.*

Using the notation $\mathrm{Alg}(A_1, A_2)$ for the set of morphisms of algebras from A_1 to A_2, we can paraphrase the previous proposition by saying that $\mathrm{Alg}(A \otimes B, C)$ is isomorphic to the subset of $\mathrm{Alg}(A, C) \times \mathrm{Alg}(B, C)$ consisting of all pairs (f, g) of morphisms whose images commute in C. In particular, if C is commutative, then

$$\mathrm{Alg}(A \otimes B, C) \cong \mathrm{Alg}(A, C) \times \mathrm{Alg}(B, C).$$

For this reason, we may consider the tensor product of algebras as the non-commutative analogue of the product of spaces.

Exercise 3.4 Prove Proposition 3.1.

3.4 From Groups to Hopf Algebras

In this section, we introduce the concept of a Hopf algebra and illustrate it with several examples which will show up repeatedly in these notes. For general references on Hopf algebras, see [1, 31, 46, 58].

3.4.1 Algebraic Groups

Let G be an *algebraic group*, i.e., an algebraic variety equipped with the structure of a group such that the product map $\mu : G \times G \to G$ is a map of algebraic varieties.

The basic example of an algebraic group is the *general linear group* $GL_N(\mathbb{C})$, which consists of all invertible $N \times N$-matrices with complex entries, equipped with the usual matrix product. This product is given by polynomial formulas in the entries. The coordinate algebra of $GL_N(\mathbb{C})$ is the algebra

$$\mathcal{O}(GL_N(\mathbb{C})) = \mathbb{C}[t, (a_{i,j})_{1 \leq i,j \leq N}]/(t \det(a_{i,j}) - 1). \tag{3.4}$$

Any subgroup of $GL_N(\mathbb{C})$ defined by the vanishing of polynomials is also an algebraic group. For instance, the *special linear group* $SL_N(\mathbb{C})$, which consists of all $N \times N$-matrices whose determinant is 1, is an algebraic group. Its coordinate algebra is the algebra

$$\mathcal{O}(SL_N(\mathbb{C})) = \mathbb{C}[(a_{i,j})_{1 \leq i,j \leq N}]/(\det(a_{i,j}) - 1).$$

It is obtained from $\mathcal{O}(GL_N(\mathbb{C}))$ by setting $t = 1$.

By (3.2) the product map $\mu : G \times G \to G$ of an algebraic group induces a morphism of algebras $\mu^* : \mathcal{O}(G) \to \mathcal{O}(G \times G)$. We can compose μ^* with the canonical isomorphism $\mathcal{O}(G \times G) \cong \mathcal{O}(G) \otimes \mathcal{O}(G)$ (see (3.3)), which yields a morphism of algebras

$$\Delta : \mathcal{O}(G) \to \mathcal{O}(G) \otimes \mathcal{O}(G),$$

which we call the *coproduct* of $\mathcal{O}(G)$.

The product μ of G is *associative*, which means that we have

$$\mu\,(\mu(g_1, g_2), g_3) = \mu\,(g_1, \mu(g_2, g_3))$$

for all $g_1, g_2, g_3 \in G$. This identity, which reads $\mu \circ (\mu \otimes \mathrm{id}) = \mu \circ (\mathrm{id} \otimes \mu)$, transposes to the following *coassociativity* identity for the coproduct:

$$(\Delta \otimes \mathrm{id}) \circ \Delta = (\mathrm{id} \otimes \Delta) \circ \Delta. \tag{3.5}$$

Similarly, the *unit* e of the group G, which can be seen as a homomorphism $\bar{e} : \{1\} \to G$ (sending 1 to e), induces the morphism of algebras

$$\varepsilon = \bar{e}^* : \mathcal{O}(G) \to \mathcal{O}(\{1\}) = \mathbb{C},$$

which we call the *counit* of $\mathcal{O}(G)$. The identities $\mu(e, g) = g = \mu(g, e)$ $(g \in G)$ read

$$\mu \circ (\bar{e} \otimes \mathrm{id}) = \mathrm{id} = \mu \circ (\mathrm{id} \otimes \bar{e}),$$

where we have identified $\{1\} \times G$ and $G \times \{1\}$ with G. They transpose to the *counitality* identities

$$(\varepsilon \otimes \mathrm{id}) \circ \Delta = \mathrm{id} = (\mathrm{id} \otimes \varepsilon) \circ \Delta : \mathcal{O}(G) \to \mathcal{O}(G), \tag{3.6}$$

where we use the natural identifications $\mathbb{C} \otimes \mathcal{O}(G) \cong \mathcal{O}(G)$ and $\mathcal{O}(G) \otimes \mathbb{C} \cong \mathcal{O}(G)$.

In a group G, any element g possesses an *inverse*, i.e., an element g^{-1} such that

$$\mu(g, g^{-1}) = e = \mu(g^{-1}, g). \tag{3.7}$$

The map $\mathrm{inv} : g \mapsto g^{-1}$ induces a map $S = \mathrm{inv}^* : \mathcal{O}(G) \to \mathcal{O}(G)$, which we call the *antipode* of $\mathcal{O}(G)$. The identities (3.7) imply identities for the antipode, which we shall display in Sect. 3.4.3.

When $G = GL_N(\mathbb{C})$ is the general linear group, the coproduct of the coordinate algebra $\mathcal{O}(GL_N(\mathbb{C}))$ is defined on the generators $t, a_{i,j}$ of $\mathcal{O}(GL_N(\mathbb{C}))$ by

$$\Delta(t) = t \otimes t \quad \text{and} \quad \Delta(a_{i,j}) = \sum_{k=1}^{N} a_{i,k} \otimes a_{k,j} \tag{3.8}$$

and the counit by

$$\varepsilon(t) = 1 \quad \text{and} \quad \varepsilon(a_{i,j}) = \delta_{i,j} \tag{3.9}$$

for all $i, j \in \{1, \ldots, N\}$. For the antipode, let A be the $N \times N$-matrix $A = (a_{i,j})_{1 \leq i,j \leq N}$. Denote by $A_{i,j}$ the determinant of the $(N-1) \times (N-1)$ matrix obtained from deleting Row i and Column j of A. Then for each generator $a_{i,j}$ $(i, j \in \{1, \ldots, N\})$ we have

$$S(a_{i,j}) = (-1)^{i+j} \frac{A_{j,i}}{\det(A)}. \tag{3.10}$$

By the definition (3.4), the generator t is invertible with inverse $t^{-1} = \det(A)$ and its antipode is given by $S(t) = t^{-1} = \det(A)$.

The values of $\Delta(a_{i,j})$, $\varepsilon(a_{i,j})$, and $S(a_{i,j})$ given in Formulas (3.8)–(3.10) above also determine the coproduct, counit, and antipode of $\mathcal{O}(SL_N(\mathbb{C}))$, where $SL_N(\mathbb{C})$ is the special linear group.

Exercise 3.5 Prove the claims of this section.

3.4.2 Bialgebras

Before defining Hopf algebras, we present the concept of a bialgebra.

Definition 3.2 A *bialgebra* is an associative unital algebra equipped with two linear maps $\Delta : H \to H \otimes H$ and $\varepsilon : H \to \mathbb{C}$ satisfying the following conditions:

(i) The maps Δ and ε are morphisms of algebras.
(ii) We have the following equalities:

$$(\Delta \otimes \mathrm{id}) \circ \Delta = (\mathrm{id} \otimes \Delta) \circ \Delta. \tag{3.11}$$

and, identifying $\mathbb{C} \otimes H$ and $H \otimes \mathbb{C}$ with H,

$$(\varepsilon \otimes \mathrm{id}) \circ \Delta = \mathrm{id} = (\mathrm{id} \otimes \varepsilon) \circ \Delta. \tag{3.12}$$

The map Δ is called the *coproduct* of H, and ε is its *counit*. It is sometimes convenient to denote the product of the bialgebra H by $\mu : H \otimes H \to H$ and to introduce the unique morphism of algebras $\eta : \mathbb{C} \to H$, which we call the *unit* of H; we have $\eta(1) = 1_H$.

Given a bialgebra H with coproduct Δ, we define the *opposite coproduct*

$$\Delta^{\mathrm{op}} : H \to H \otimes H$$

by $\Delta^{\mathrm{op}} = \tau \circ \Delta$, where $\tau : H \otimes H \to H \otimes H$ is the *flip* defined by $\tau(x \otimes y) = y \otimes x$ for all $x, y \in H$. We say that H is *cocommutative* if $\Delta^{\mathrm{op}} = \Delta$.

Exercise 3.6 Let $\mathbb{C}[t]$ be the polynomial algebra in one variable t. Show that $\mathbb{C}[t]$ is a bialgebra with coproduct Δ and counit ε determined by $\Delta(t) = t \otimes t$ and $\varepsilon(t) = 1$. Check that this bialgebra is cocommutative.

Exercise 3.7 (a) Let H be a bialgebra with coproduct Δ and counit ε. Consider the *linear dual* $H^{\vee} = \mathrm{Hom}(H, \mathbb{C})$ of H. Define a product $\mu^{\vee} : H^{\vee} \otimes H^{\vee} \to H^{\vee}$ for all $x \in H$ and $\alpha, \beta \in H^{\vee}$ by

$$\mu^{\vee}(\alpha \otimes \beta)(x) = (\alpha \otimes \beta)(\Delta(x)) = \sum_i \alpha(x_i') \, \beta(x_i''), \tag{3.13}$$

when $\Delta(x) = \sum_i x_i' \otimes x_i''$. Show that

(i) $\mu^\vee$ is an associative product with unit equal to $\varepsilon \in H^\vee$,
(ii) $H^\vee$ is cocommutative if H is a commutative algebra.

(b) Now assume that H is finite-dimensional as a vector space over $\mathbb{C}$.

(i) Show that $H^\vee$ is a bialgebra with coproduct $\Delta^\vee : H^\vee \to H^\vee \otimes H^\vee$ and counit $\varepsilon^\vee : H^\vee \to \mathbb{C}$ defined by

$$\Delta^\vee(\alpha)(x \otimes y) = \alpha(xy)$$

and $\varepsilon^\vee(\alpha) = \alpha(1_H)$ for all $\alpha \in H^\vee$.
(ii) Prove that $H^\vee$ is commutative if H is cocommutative.

Remark 3.1 It follows from Exercise 3.7 that the dual of a finite-dimensional bialgebra is another (finite-dimensional) bialgebra. To extend such a duality to the case when H is an infinite-dimensional bialgebra, we have to replace the linear dual $H^\vee$ by the *restricted dual* H° defined by

$$H^\circ = \left\{\alpha \in H^\vee \,|\, \alpha(I) = 0 \text{ for some ideal } I \text{ such that } \dim\, H/I < \infty\right\}.$$

See [46, Sect. 1.2] or [58]. We have $H^\circ = H^\vee$ if $\dim\, H < \infty$.

3.4.3 *Hopf Algebras*

Let H be a bialgebra with product μ, unit η, coproduct Δ, and counit ε. Given two linear endomorphisms f, g of H we define a new linear endomorphism $f * g$ of H by

$$f * g = \mu \circ (f \otimes g) \circ \Delta \in \mathrm{End}(H). \tag{3.14}$$

We now define the concept of a Hopf algebra.

Definition 3.3 Let H be a bialgebra.

(a) An *antipode* of H is a linear endomorphism S of H such that

$$S * \mathrm{id}_H = \eta \circ \varepsilon = \mathrm{id}_H * S. \tag{3.15}$$

(b) A *Hopf algebra* is a bialgebra together with an antipode.

(c) A *morphism of Hopf algebras* $f : H \to H'$ between Hopf algebras is a morphim of bialgebras such that

$$\Delta' \circ f = (f \otimes f) \circ \Delta, \quad \varepsilon' \circ f = \varepsilon, \quad S' \circ f = f \circ S,$$

where Δ (resp. Δ') is the coproduct, ε (resp. ε') is the counit and S (resp. S') is the antipode of H (resp. of H').

Example 3.8 If G is an *algebraic group*, then its coordinate algebra $\mathcal{O}(G)$ equipped with the maps Δ, ε, and S defined in Sect. 3.4.1 is a Hopf algebra. Actually, the axioms of a Hopf algebra are derived from this example.

Hopf algebras have two important features which are worth emphasizing:

- The concept of Hopf algebras is *self-dual*: the restricted dual H° of a Hopf algebra H is again a Hopf algebra (see Exercises 3.7 (b) and 3.10 for finite-dimensional Hopf algebras). This duality allows also to extend the Pontryagin duality of abelian groups to non-abelian ones (see Exercise 3.13).
- The category of left H-modules, where H is a Hopf algebra, is a *tensor category*. Recall that a left H-module V is a vector space together with a bilinear map $H \times V \to V; (x, v) \mapsto xv$ $(x, \in H, v$Ł$\in V)$ such that

$$(xy)v = x(y(v)) \quad \text{and} \quad 1_H v = v \tag{3.16}$$

for all $x, y \in H$ and $v \in V$. The map $(x, v) \mapsto xv$ is called the action.
If V_1 and V_2 are left H-modules, then so is the tensor product $V_1 \otimes V_2$. Indeed one defines an action of H on $V_1 \otimes V_2$ by

$$x(v_1 \otimes v_2) = \Delta(x)(v_1 \otimes v_2) = \sum_i x_i' v_1 \otimes x_i'' v_2 \tag{3.17}$$

if $\Delta(x) = \sum_i x_i' \otimes x_i''$.

Exercise 3.8 Check that the action (3.17) of H on $V_1 \otimes V_2$ satisfies (3.16).

Remark 3.2 In many cases, for instance when H is a quantum group as in Sect. 3.5, $V_1 \otimes V_2$ is naturally isomorphic as an H-module to $V_2 \otimes V_1$. It is this feature that leads to braid group representations and knot invariants. We will not say more about this; see [31, Part Three] for details on this vast subject.

Exercise 3.9 Show that the product $*$ on the algebra $\mathrm{End}(H)$ of linear endomorphisms of a Hopf algebra H given by (3.14) is associative with unit equal to $\eta \circ \varepsilon$. Prove that an antipode is unique if it exists.

Exercise 3.10 Show that the dual $H^\vee$ of a finite-dimensional Hopf algebra H is a Hopf algebra.

Exercise 3.11 (*A bialgebra without antipode*) Let $\mathbb{C}[t]$ be the bialgebra considered in Exercise 3.6. Prove that it has no antipode [hint: apply (3.15) to the element t].

The following properties of the antipode of a Hopf algebra are worth mentioning (see [31, III.3] or [58]).

Proposition 3.2 *Let H be a Hopf algebra with coproduct Δ, counit ε, and antipode S.*
(a) The antipode S is an anti-morphism of algebras, i.e., for all $x, y \in H$,

$$S(xy) = S(y)S(x) \quad \textit{and} \quad S(1) = 1,$$

and we have

$$(S \otimes S) \circ \Delta = \Delta^{\mathrm{op}} \circ S \quad \textit{and} \quad \varepsilon \circ S = \varepsilon.$$

(b) If H is commutative or cocommutative, then the antipode S is an involution, i.e., $S^2 = \mathrm{id}_H$.

Another useful concept is the following. An element x of a Hopf algebra H is called *group-like* if

$$\Delta(x) = x \otimes x \quad \text{and} \quad \varepsilon(x) = 1. \tag{3.18}$$

Let $\mathrm{Gr}(H)$ be the set of group-like elements of H.

Proposition 3.3 *The set $\mathrm{Gr}(H)$ of group-like elements of H is a group under the product in H. The inverse of an element x in $\mathrm{Gr}(H)$ is $S(x)$.*

Proof Let $x, y \in H$ be group-like elements. Since Δ and ε are morphisms of algebras, we have

$$\Delta(xy) = \Delta(x)\Delta(y) = (x \otimes x)(y \otimes y) = xy \otimes xy$$

and $\varepsilon(xy) = \varepsilon(x)\varepsilon(y) = 1$. This shows that $\mathrm{Gr}(H)$ is preserved under the product. Clearly, the unit 1 of H is group-like and is a unit for the product in $\mathrm{Gr}(H)$.

Applying (3.15) to a group-like element x, we obtain $S(x)x = 1 = xS(x)$, which shows that $S(x)$ is the inverse of x. To conclude that $\mathrm{Gr}(H)$ is a group, it remains to check that $S(x)$ is group-like. Indeed, by Proposition 3.2 (a),

$$\Delta^{\mathrm{op}}(S(x)) = (S \otimes S)(\Delta(x)) = S(x) \otimes S(x),$$

which implies $\Delta(S(x)) = S(x) \otimes S(x)$. We also have $\varepsilon(S(x)) = \varepsilon(x) = 1$. Thus, $S(x)$ is group-like. □

Examples of group-like elements and computations of $\mathrm{Gr}(H)$ will be given in Exercise 3.16 below.

3.4.4 Examples of Hopf Algebras from Finite Groups

To familiarize the reader with the concept of a Hopf algebra, we now present the following two basic examples, both constructed from a group.

3.4.4.1 The Function Algebra of a Finite Group

Let G be a finite group with unit e and $\mathcal{O}(G)$ be its function algebra, as defined in Sect. 3.3.1.1. It is a Hopf algebra with coproduct Δ, counit ε, and antipode S given by

$$\Delta(u)(g,h) = u(gh), \qquad \varepsilon(u) = u(e), \qquad S(u)(g) = u(g^{-1}) \tag{3.19}$$

for all $g, h \in G$ and $u \in \mathcal{O}(G)$. Here we have identified $\mathcal{O}(G) \otimes \mathcal{O}(G)$ with the function algebra $\mathcal{O}(G \times G)$ of the product group $G \times G$.

We can also express Δ, ε, and S in terms of the δ-functions introduced in *loc. cit.* Namely we have

$$\Delta(\delta_g) = \sum_{h \in G} \delta_h \otimes \delta_{h^{-1}g}, \quad S(\delta_g) = \delta_{g^{-1}}, \quad \varepsilon(\delta_g) = \begin{cases} 1 & \text{if } g = e, \\ 0 & \text{otherwise.} \end{cases}$$

Since the inverse map $g \mapsto g^{-1}$ in a group is an involution, it follows from (3.19) that the antipode S is an involution as well, which is in agreement with Proposition 3.2 (b) applied to the *commutative* Hopf algebra $\mathcal{O}(G)$.

3.4.4.2 The Convolution Algebra of a Group

Let G now be an arbitrary group, not necessarily finite. We define $\mathbb{C}[G]$ to be the vector space spanned by the elements of G. This means that any element of $\mathbb{C}[G]$ is a linear combination of the form

$$\sum_{g \in G} \lambda_g \, g,$$

where the coefficients λ_g are complex numbers, all of which are zero except for a finite number. We also assume that the set $\{g\}_{g \in G}$ is a basis of $\mathbb{C}[G]$, which is equivalent to the implication

$$\left(\sum_{g \in G} \lambda_g \, g = 0 \right) \Rightarrow \left(\lambda_g = 0 \text{ for all } g \in G \right).$$

The vector space $\mathbb{C}[G]$ is equipped with a product, often called the *convolution product*, defined by the formula

$$\left(\sum_{g \in G} \lambda_g \, g \right) \left(\sum_{g \in G} \mu_g \, g \right) = \sum_{g \in G} \left(\sum_{h \in G} \lambda_h \, \mu_{h^{-1}g} \right) g.$$

The convolution product possesses a unit, which is $1_{\mathbb{C}[G]} = e$, where e is the unit of the group G. The algebra $\mathbb{C}[G]$ is called the *convolution algebra* of G, or simply the *group algebra* of G.

We now claim that $\mathbb{C}[G]$ is a Hopf algebra. Its coproduct, counit, and antipode are given by

$$\Delta\left(\sum_{g\in G} \lambda_g\, g\right) = \sum_{g\in G} \lambda_g\, g \otimes g, \qquad \varepsilon\left(\sum_{g\in G} \lambda_g\, g\right) = \sum_{g\in G} \lambda_g, \tag{3.20}$$

$$S\left(\sum_{g\in G} \lambda_g\, g\right) = \sum_{g\in G} \lambda_g\, g^{-1} = \sum_{g\in G} \lambda_{g^{-1}}\, g. \tag{3.21}$$

We can see on Formula (3.20) for the coproduct that $\Delta^{\mathrm{op}} = \Delta$, which means that the Hopf algebra $\mathbb{C}[G]$ is *cocommutative*. By Proposition 3.2 (b), this implies that the antipode S is an involution, which can easily be seen on (3.21).

Exercise 3.12 Prove the claims in Sect. 3.4.4.2.

Exercise 3.13 (*Duality between the function algebra and the group algebra*) Let G be a *finite* group. Define a bilinear form $\mathcal{O}(G) \times \mathbb{C}[G] \to \mathbb{C}$ by

$$\left\langle u, \sum_{g\in G} \lambda_g\, g\right\rangle = \sum_{g\in G} \lambda_g\, u(g)$$

for all $u \in \mathcal{O}(G)$, $g \in G$, and $\lambda_g \in \mathbb{C}$. It induces a linear map $\omega : \mathcal{O}(G) \to \mathbb{C}[G]^{\vee}$ by $\omega(u) = \langle u, -\rangle$ ($u \in \mathcal{O}(G)$). Recall that $\mathbb{C}[G]^{\vee}$ is the dual Hopf algebra of $\mathbb{C}[G]$, as defined in Exercise 3.7. Prove the following:

(i) The linear map $\omega : \mathcal{O}(G) \to \mathbb{C}[G]^{\vee}$ is bijective.
(ii) For all $u, v \in \mathcal{O}(G)$, $g, h \in G$ we have

$$\begin{aligned} \langle uv, g\rangle &= \langle u, g\rangle\, \langle v, g\rangle, \\ \langle \Delta(u), g \otimes h\rangle &= \langle u, gh\rangle, \\ \varepsilon(u) &= \langle u, e\rangle, \\ \langle S(u), g\rangle &= \langle u, g^{-1}\rangle. \end{aligned}$$

(iii) Deduce that $\omega : \mathcal{O}(G) \to \mathbb{C}[G]^{\vee}$ is an isomorphism of Hopf algebras.

Exercise 3.14 (*Duality for finite abelian groups*) Let G be a finite *abelian* group and $\widehat{G} = \mathrm{Hom}(G, \mathbb{C}^{\times})$ be its group of characters. We recall that a *character* of G is a group homomorphism from G to the multiplicative group $\mathbb{C}^\times$ of nonzero complex numbers. Since any element of G is of finite order, the values of a character of G are roots of unity, which are complex numbers of modulus 1.

The set $\widehat{G}$ is a group under pointwise multiplication; it is also called the *Pontryagin dual* of G.

(i) Show that $\widehat{G_1 \times G_2} \cong \widehat{G_1} \times \widehat{G_2}$ whenever G_1 and G_2 are finite abelian groups.

(ii) Determine all characters of a cyclic group of order n and conclude that there is a (non-unique) isomorphism $\widehat{\mathbb{Z}/n} \cong \mathbb{Z}/n$.

(iii) Deduce from (i) and (ii) that $\widehat{G} \cong G$ for any finite abelian group G.

Exercise 3.15 (*The Hopf algebras* $\mathbb{C}[G]$ and $\mathcal{O}(\widehat{G})$) Let G be a finite abelian group and $\widehat{G}$ be its group of characters, as defined in the previous exercise. Consider the function algebra $\mathcal{O}(\widehat{G})$, which is a Hopf algebra by Sect. 3.4.4.1. Observe that this Hopf algebra is not only commutative, but also cocommutative since $\widehat{G}$ is abelian (see Formula (3.19) for the coproduct). On the other hand, we have the cocommutative Hopf algebra $\mathbb{C}[G]$, which is commutative because G is abelian. Prove that the linear map $\mathbb{C}[G] \to \mathcal{O}(\widehat{G})$ defined by $g \in G \mapsto (\chi \mapsto \chi(g))_{\chi \in \widehat{G}}$ is an isomorphism of Hopf algebras.

Exercise 3.16 (*Group-like elements*)

(a) Show that the only group-like elements of a group algebra $\mathbb{C}[G]$ are of the form $\sum_{g\in G} \lambda_g\, g$, where all coefficients λ_g are zero, except for exactly one, which is equal to 1. Deduce a group isomorphism $\mathrm{Gr}(\mathbb{C}[G]) \cong G$.

(b) Given a finite group G, show that an element $u \in \mathcal{O}(G)$ is group-like if and only if $u(e) = 1$ and $u(gh) = u(g)u(h)$ for all $g, h \in G$, i.e., if and only if u is a character of G. Deduce a group isomorphism $\mathrm{Gr}(\mathcal{O}(G)) \cong \widehat{G} = \mathrm{Hom}(G, \mathbb{C}^\times)$.

3.4.5 The Heyneman–Sweedler Sigma Notation

Let H be a Hopf algebra with coproduct Δ, counit ε, and antipode S. It is often convenient to use the following notation (due to Heyneman and Sweedler) for the image of an element $x \in H$ under the coproduct:

$$\Delta(x) = \sum_{(x)} x_{(1)} \otimes x_{(2)}.$$

The coassociativity identity (3.11) expressed in this notation becomes

$$\sum_{(x)} (x_{(1)})_{(1)} \otimes (x_{(1)})_{(2)} \otimes x_{(2)} = \sum_{(x)} x_{(1)} \otimes (x_{(2)})_{(1)} \otimes (x_{(2)})_{(2)}.$$

To simplify, we will express both sides of the previous equality by

$$\sum_{(x)} x_{(1)} \otimes x_{(2)} \otimes x_{(3)}.$$

In this notation the counitality identity (3.12) becomes

$$\sum_{(x)} \varepsilon(x_{(1)})\, x_{(2)} = x = \sum_{(x)} x_{(1)}\, \varepsilon(x_{(2)}). \tag{3.22}$$

The defining Eq. (3.15) for the antipode becomes

$$\sum_{(x)} S(x_{(1)})\, x_{(2)} = \varepsilon(x)1 = \sum_{(x)} x_{(1)}\, S(x_{(2)}). \tag{3.23}$$

The fact that Δ is a morphism of algebras can be expressed in this notation by

$$\sum_{(xy)} (xy)_{(1)} \otimes (xy)_{(2)} = \left(\sum_{(x)} x_{(1)} \otimes x_{(2)}\right)\left(\sum_{(y)} y_{(1)} \otimes y_{(2)}\right).$$

It is convenient to write the previous right-hand side simply as

$$\sum_{(x)(y)} x_{(1)}y_{(1)} \otimes x_{(2)}y_{(2)}.$$

3.5 Quantum Groups Associated with $SL_2(\mathbb{C})$

In this section, we will present two Hopf algebras which were discovered in the 1980s and are quantizations of the special linear group $SL_2(\mathbb{C})$ and of its Lie algebra $\mathfrak{sl}(2)$, the latter consisting of all 2×2-matrices of trace 0. These Hopf algebras depend on a parameter q. They have the particularity of being neither commutative, nor cocommutative. They are instances of so-called *quantum groups*.

The term "quantum group" was introduced by Drinfeld in his Berkeley 1986 ICM address [18].[3] As we mentioned in the introduction, the discovery of quantum groups was a major event with spectacular applications in representation theory, low-dimensional topology, and theoretical physics. The reader may learn more on quantum groups in the monographies [11, 29, 31, 38, 42].

[3]Drinfeld along with other invited mathematicians from the Soviet Union was prevented by the Soviet authorities to attend the conference; in Drinfeld's absence his contribution was read by Cartier.

3.5.1 The Quantum Coordinate Algebra of $SL_2(\mathbb{C})$

In Sect. 3.4.1, we considered the special linear group $SL_N(\mathbb{C})$ and its coordinate algebra

$$\mathcal{O}(SL_N(\mathbb{C})) = \mathbb{C}[(a_{i,j})_{1\leq i,j\leq N}]/(\det(a_{i,j}) - 1).$$

Let us now restrict to the case $N = 2$. For simplicity, set $\mathrm{SL}(2) = \mathcal{O}(SL_2(\mathbb{C}))$. We have

$$\mathrm{SL}(2) = \mathbb{C}[a, b, c, d]/(ad - bc - 1),$$

where $a = a_{1,1}$, $b = a_{1,2}$, $c = a_{2,1}$ and $d = a_{2,2}$. We can rewrite Formulas (3.8)–(3.10) for the coproduct Δ, the counit ε and the antipode S of the Hopf algebra SL(2) in the following compact matrix form:

$$\Delta \begin{pmatrix} a & b \\ c & d \end{pmatrix} = \begin{pmatrix} a & b \\ c & d \end{pmatrix} \otimes \begin{pmatrix} a & b \\ c & d \end{pmatrix}, \tag{3.24}$$

$$\varepsilon \begin{pmatrix} a & b \\ c & d \end{pmatrix} = \begin{pmatrix} 1 & 0 \\ 0 & 1 \end{pmatrix}, \tag{3.25}$$

$$S \begin{pmatrix} a & b \\ c & d \end{pmatrix} = \begin{pmatrix} d & -b \\ -c & a \end{pmatrix}. \tag{3.26}$$

This is a compact version for the formulas

$$\Delta(a) = a \otimes a + b \otimes c, \quad \Delta(b) = a \otimes b + b \otimes d,$$
$$\Delta(c) = c \otimes a + d \otimes c, \quad \Delta(d) = c \otimes b + d \otimes d,$$

$$\varepsilon(a) = \varepsilon(d) = 1, \quad \varepsilon(b) = \varepsilon(d) = 0,$$

$$S(a) = d, \quad S(b) = -b, \quad S(c) = -c, \quad S(d) = a.$$

The Hopf algebra SL(2) is commutative, but not cocommutative, which can be seen for instance on the formula for $\Delta(a)$. Its antipode is clearly an involution, which follows of course from the fact that the map inv : $g \mapsto g^{-1}$ is involutive.

Now we introduce a *non-commutative deformation* of the Hopf algebra SL(2). The deformation depends on a parameter q which we take to be a nonzero complex number. Define $\mathrm{SL}_q(2)$ to be the algebra generated by four generators a, b, c, d subject to the relations

$$ba = qab, \quad ca = qac,$$
$$db = qbd, \quad dc = qcd,$$

$$bc = cb, \quad ad - da = (q^{-1} - q)\, bc,$$
$$ad - q^{-1}bc = 1.$$

If $q = 1$, the previous relations reduce to the fact that the generators a, b, c, d commute and satisfy the additional relation $ad - bc = 1$. Thus in this case, we have $\mathrm{SL}_q(2) = \mathrm{SL}(2)$. If $q \neq 1$, then clearly $\mathrm{SL}_q(2)$ is not commutative, so it cannot be isomorphic to $\mathrm{SL}(2)$.

The algebra $\mathrm{SL}_q(2)$ is a Hopf algebra. Its coproduct Δ and counit ε are given by the same formulas as for $\mathrm{SL}(2)$, namely by (3.24) and (3.25). However the antipode S of $\mathrm{SL}_q(2)$ is given, not by (3.26), but by another formula (depending on q), namely in compact matrix form by

$$S\begin{pmatrix} a & b \\ c & d \end{pmatrix} = \begin{pmatrix} d & -qb \\ -q^{-1}c & a \end{pmatrix}. \tag{3.27}$$

The Hopf algebra $\mathrm{SL}_q(2)$ provides our first example of a Hopf algebra that is (for general q) *neither commutative, nor cocommutative*, and with non-involutive antipode (for the latter, see Exercise 3.18 below). The Hopf algebra $\mathrm{SL}_q(2)$ is a *quantization* of the coordinate algebra $\mathrm{SL}(2)$; this is another way of saying that $\mathrm{SL}_q(2)$ is a deformation of $\mathrm{SL}(2)$ as a Hopf algebra.

The Hopf algebra $\mathrm{SL}_q(2)$ is an example of a *quantum group*. The Hopf algebras $\mathscr{O}(GL_N(\mathbb{C}))$ and $\mathscr{O}(SL_N(\mathbb{C}))$ can be quantized in a similar fashion.

Exercise 3.17 (a) Compute the following expressions in $\mathrm{SL}_q(2) \otimes \mathrm{SL}_q(2)$ involving the coproduct Δ defined by (3.24):

$$\begin{aligned}
&\Delta(b)\Delta(a) - q\Delta(a)\Delta(b), \quad && \Delta(c)\Delta(a) - q\Delta(a)\Delta(c),\\
&\Delta(d)\Delta(b) - q\Delta(b)\Delta(d), && \Delta(d)\Delta(c) - q\Delta(c)\Delta(d),\\
&\Delta(b)\Delta(c) - \Delta(c)\Delta(b), && \Delta(a)\Delta(d) - q^{-1}\Delta(b)\Delta(c) - 1 \otimes 1,\\
&\Delta(a)\Delta(d) - \Delta(d)\Delta(a) - (q - q^{-1})\,\Delta(b)\Delta(c).
\end{aligned}$$

Deduce that $\Delta : \mathrm{SL}_q(2) \to \mathrm{SL}_q(2) \otimes \mathrm{SL}_q(2)$ is a morphism of algebras.

(b) Check that $\mathrm{SL}_q(2)$ satisfies all axioms of a Hopf algebra.

Exercise 3.18 (*The square of the antipode*)

(a) Use (3.27) to compute the square S^2 of the antipode of $\mathrm{SL}_q(2)$ on the generators a, b, c, d.

(b) Show that S^2 has infinite order if q is not a root of unity.

(c) If $q = \exp(\pi\sqrt{-1}/N)$ for some integer $N > 1$, prove that S^2 is a Hopf algebra automorphism of $\mathrm{SL}_q(2)$ of order N.

Exercise 3.19 For $\varepsilon = \pm 1$ define $\mathrm{SL}_{(\varepsilon)}(2)$ to be the algebra generated by X, Y, Z, T and the relations $XY = YX$, $XZ = ZX$, $XT = TX$, $YZ = \varepsilon ZY$, $YT = \varepsilon TY$, $ZT = \varepsilon TZ$, and $X^2 - \varepsilon Y^2 - \varepsilon Z^2 + \varepsilon T^2 = 1$.

(a) Let $\varepsilon = 1$. Show that there is an algebra isomorphism $\varphi : \mathrm{SL}_{(\varepsilon)}(2) \to \mathrm{SL}(2)$ such that $\varphi(X) = (a+d)/2, \varphi(Y) = (a-d)/2, \varphi(Z) = (b+c)/2, \varphi(T) = (b-c)/2$. Deduce $\mathrm{Alg}(\mathrm{SL}_{(\varepsilon)}(2), \mathbb{C}) \cong SL_2(\mathbb{C})$.

(b) Let $\varepsilon = -1$. Show that $\mathrm{Alg}(\mathrm{SL}_{(\varepsilon)}(2), \mathbb{C})$ is the union of three quadrics lying in three distinct planes (for further details, see [22, Sect. 4.2]).

3.5.2 A Quotient of $\mathrm{SL}_q(2)$

Let q be again a nonzero scalar. Consider the algebra $\mathbb{C}_q[X, X^{-1}, Y]$ generated by three generators X, X^{-1}, Y subject to the relations

$$XX^{-1} = X^{-1}X = 1, \qquad YX = qXY.$$

This algebra is non-commutative when $q \neq 1$. Proceeding as in Exercise 3.3, the reader may check that the set $\{X^iY^j\}$ where i runs over $\mathbb{Z}$ and j over $\mathbb{N}$ is a basis of $\mathbb{C}_q[X, X^{-1}, Y]$. The algebra $\mathbb{C}_q[X, X^{-1}, Y]$ contains the quantum plane $\mathbb{C}_q[X, Y]$ of Sect. 3.3.2.2 as a subalgebra.

The algebra $\mathbb{C}_q[X, X^{-1}, Y]$ has the structure of a Hopf algebra with coproduct Δ, counit ε and antipode S given on the generators X, Y by

$$\Delta(X) = X \otimes X, \qquad \Delta(Y) = X \otimes Y + Y \otimes X^{-1}, \tag{3.28}$$

$$\varepsilon(X) = 1, \quad \varepsilon(Y) = 0, \quad S(X) = X^{-1}, \quad S(Y) = -qY. \tag{3.29}$$

The formula for $\Delta(Y)$ shows that $\mathbb{C}_q[X, X^{-1}, Y]$ is a non-cocommutative Hopf algebra.

Moreover, $\mathbb{C}_q[X, X^{-1}, Y]$ is a quotient of the Hopf algebra $\mathrm{SL}_q(2)$ introduced in Sect. 3.5.1; we have the following precise statement, whose proof we leave to the reader.

Lemma 3.1 *There is a surjective morphism of Hopf algebras*

$$\pi : \mathrm{SL}_q(2) \to \mathbb{C}_q[X, X^{-1}, Y]$$

such that $\pi(a) = X$, $\pi(b) = Y$, $\pi(c) = 0$, *and* $\pi(d) = X^{-1}$.

Since the morphism π kills the generator c of $\mathrm{SL}_q(2)$, we can see $\mathbb{C}_q[X, X^{-1}, Y]$ as a quantization of the coordinate algebra of the subgroup B of upper triangular matrices in $SL_2(\mathbb{C})$.

3.5.3 The Quantum Enveloping Algebra of $\mathfrak{sl}(2)$

We now describe another important quantum group, which is dual to the quantum group $\mathrm{SL}_q(2)$ in a sense which will be made precise in Lemma 3.2 below.

This new algebra, denoted $U_q\,\mathfrak{sl}(2)$, also depends on a nonzero complex parameter q; we furthermore assume $q \neq \pm 1$, so that $q - q^{-1} \neq 0$.

We define $U_q\,\mathfrak{sl}(2)$ to be the algebra generated by four elements E, F, K, K^{-1} subject to the relations

$$KK^{-1} = K^{-1}K = 1,$$

$$KE = q^2 EK, \qquad KF = q^{-2} FK,$$

$$EF - FE = \frac{K - K^{-1}}{q - q^{-1}}.$$

The algebra $U_q\,\mathfrak{sl}(2)$ is called the *quantum enveloping algebra*[4] of the Lie algebra $\mathfrak{sl}(2)$. The set $\{E^i F^j K^\ell\}_{i,j\in\mathbb{N};\,\ell\in\mathbb{Z}}$ is a basis of $U_q\,\mathfrak{sl}(2)$ considered as a complex vector space (for a proof, see [31, Proposition VI.1.4]).

The algebra $U_q\,\mathfrak{sl}(2)$ is a Hopf algebra with coproduct Δ, counit ε, and antipode S given on the generators by

$$\Delta(K^{\pm 1}) = K^{\pm 1} \otimes K^{\pm 1}, \quad \varepsilon(K^{\pm 1}) = 1, \quad S(K^{\pm 1}) = K^{\mp 1},$$

$$\Delta(E) = 1 \otimes E + E \otimes K, \quad \varepsilon(E) = 0, \quad S(E) = -EK^{-1},$$

$$\Delta(F) = K^{-1} \otimes F + F \otimes 1, \quad \varepsilon(F) = 0, \quad S(F) = -q^{-1} FK.$$

The algebra $U_q\,\mathfrak{sl}(2)$ first appeared in a paper by Kulish and Reshetikhin; its Hopf algebra structure is due to Sklyanin (*cf.* [39, 56]).

Consider the morphism of algebras $\rho : U_q\,\mathfrak{sl}(2) \to M_2(\mathbb{C})$ given by

$$\rho(K^{\pm 1}) = \begin{pmatrix} q^{\pm 1} & 0 \\ 0 & q^{\mp 1} \end{pmatrix}, \quad \rho(E) = \begin{pmatrix} 0 & 1 \\ 0 & 0 \end{pmatrix}, \quad \rho(F) = \begin{pmatrix} 0 & 0 \\ 1 & 0 \end{pmatrix}.$$

It is a two-dimensional representation of $U_q\,\mathfrak{sl}(2)$. For any $u \in U_q\,\mathfrak{sl}(2)$, the matrix $\rho(u)$ is of the form

$$\rho(u) = \begin{pmatrix} A(u) & B(u) \\ C(u) & D(u) \end{pmatrix}.$$

This equality defines four linear forms A, B, C, D on $U_q\,\mathfrak{sl}(2)$, hence four elements A, B, C, D on the dual algebra $U_q\,\mathfrak{sl}(2)^\vee$ whose product is given by (3.13).

---

[4]The concept of enveloping algebra of a Lie algebra is a classical concept of the theory of Lie algebras; see for instance [15, 28, 31, 54]. The relationship between the quantum enveloping algebra $U_q\,\mathfrak{sl}(2)$ and the enveloping algebra of the Lie algebra $\mathfrak{sl}(2)$ is explained in [31, VI.2].

Lemma 3.2 *There is a unique morphism of algebras* $\psi : \mathrm{SL}_q(2) \to U_q\,\mathfrak{sl}(2)^\vee$ *such that*

$$\psi(a) = A, \quad \psi(b) = B, \quad \psi(c) = C, \quad \psi(d) = D.$$

For a proof, we refer to [31, Sect. VII.4]. Takeuchi [60] showed that ψ is injective; thus, $\mathrm{SL}_q(2)$ embeds into the dual of the quantum enveloping algebra $U_q\,\mathfrak{sl}(2)$. Actually, the image of the morphism ψ lies inside the restricted dual Hopf algebra $U_q\,\mathfrak{sl}(2)^\circ$, as defined in Remark 3.1.

Exercise 3.20 Prove that the map $\rho : U_q\,\mathfrak{sl}(2) \to M_2(\mathbb{C})$ defined above is a morphism of algebras. Give a proof of Lemma 3.2.

Exercise 3.21 Check that the group-like elements of $U_q\,\mathfrak{sl}(2)$ consist of the powers K^k of K $(k \in \mathbb{Z})$.

Exercise 3.22 Show that the following element of $U_q\,\mathfrak{sl}(2)$ belongs to its center:

$$EF + \frac{q^{-1}K + qK^{-1}}{(q - q^{-1})^2}.$$

Remark 3.3 Drinfeld [17, 18] and Jimbo [30] generalized the construction of $U_q\,\mathfrak{sl}(2)$ to any symmetrizable Kac–Moody Lie algebra $\mathfrak{g}$. The resulting Hopf algebra $U_q\,\mathfrak{g}$ is a quantization of the universal enveloping algebra of $\mathfrak{g}$.

3.5.4 A Finite-Dimensional Quotient of $U_q\,\mathfrak{sl}(2)$

The quantum enveloping algebra $U_q\,\mathfrak{sl}(2)$ has an interesting quotient when q is a root of unity of order d ($d \geq 3$ since $q \neq \pm 1$). Assume q is such a root of unity. Set $e = d$ if d is odd, and $e = d/2$ if d is even; we have $e \geq 2$.

Let I be the two-sided ideal of $U_q\,\mathfrak{sl}(2)$ generated by E^e, F^e, and $K^e - 1$. Define the quotient-algebra

$$\mathfrak{u}_d = U_q\,\mathfrak{sl}(2)/I.$$

It can be shown that the set $\{E^i F^j K^\ell\}_{1 \leq i,j,\ell \leq e-1}$ of elements of $U_q\,\mathfrak{sl}(2)$ maps to a basis of $\mathfrak{u}_d$ (for a proof, see [31, Proposition VI.5.8]). Therefore, $\mathfrak{u}_d$ is finite-dimensional of dimension equal to e^3.

Moreover, there is a unique Hopf algebra structure on $\mathfrak{u}_d$ such that the natural projection $U_q\,\mathfrak{sl}(2) \to \mathfrak{u}_d$ is a morphism of Hopf algebras (see [31, Proposition IX.6.1]).

Exercise 3.23 Let q be a root of unity of order $d \geq 3$ and e as above. Show that the elements E^e, F^e, K^e lie in the center of $U_q\,\mathfrak{sl}(2)$.

We will come back to $U_q\,\mathfrak{sl}(2)$ and $\mathfrak{u}_d$ in Sect. 3.8.3.

3.6 Group Actions in Non-commutative Geometry

Our next step is to extend the concept of a group action to the world of non-commutative spaces. We need to introduce the concept of a comodule algebra over a Hopf algebra. As we shall see, such a concept covers various situations.

3.6.1 Comodule-Algebras

Fix a Hopf algebra H with coproduct Δ and counit ε.

Definition 3.4 A *(right) H-comodule algebra* is an (associative unital) algebra A equipped with a morphism of algebras $\delta = A \to A \otimes H$, called the *coaction*, satisfying the following properties:

(a) (Coassociativity)

$$(\delta \otimes \mathrm{id}_H) \circ \delta = (\mathrm{id}_A \otimes \Delta) \circ \delta, \tag{3.30}$$

(b) (Counitarity)

$$(\mathrm{id}_A \otimes \varepsilon) \circ \delta = \mathrm{id}_A, \tag{3.31}$$

where we have identified $A \otimes \mathbb{C}$ with A.

Any H-comodule algebra A contains a subalgebra, which will turn out to be of importance to us, namely the subalgebra of A on which the coaction δ is trivial:

$$A^{\mathrm{co}-H} = \{a \in A \mid \delta(a) = a \otimes 1\}.$$

The elements of $A^{\mathrm{co}-H}$ are called *coinvariant*.

Exercise 3.24 Show that $A^{\mathrm{co}-H}$ is a subalgebra of A and that the unit 1_A of A belongs to $A^{\mathrm{co}-H}$.

The following example of a comodule algebra shows that this concept extends group actions to non-commutative algebra.

Example 3.9 Let G be a finite group acting on the right on a finite set X. Then the action, which is a map $X \times G \to X$, induces a morphism of algebras δ between the corresponding function algebras

$$\delta : \mathscr{O}(X) \to \mathscr{O}(X \times G) = \mathscr{O}(X) \otimes \mathscr{O}(G).$$

Equipped with δ, the algebra $\mathscr{O}(X)$ becomes an H-comodule algebra for the Hopf algebra $H = \mathscr{O}(G)$.

Let $Y = X/G$ be the set of orbits of the action of G on X. Then the projection $X \to Y$ sending each element $x \in X$ to its orbit xG induces an injective morphism of algebras $\mathcal{O}(Y) \to \mathcal{O}(X)$. It can be checked that $\mathcal{O}(Y)$ coincides with the subalgebra $\mathcal{O}(X)^{\mathrm{co}-\mathcal{O}(G)}$ of coinvariant elements of $\mathcal{O}(X)$.

Example 3.10 In Definition 3.4 set A to be equal to the Hopf algebra H and the coaction δ to be equal to the coproduct Δ of H. Then H becomes an H-comodule algebra. We claim that any coinvariant element $x \in H$ is a scalar multiple of the unit 1 of H. Indeed, applying $\varepsilon \otimes \mathrm{id}$ to both sides of the equality $\Delta(x) = x \otimes 1$ and using (3.12), we obtain $x = \varepsilon(x)\, 1$, which yields the desired conclusion.

We now give more examples of comodule algebras.

3.6.2 Group-Graded Algebras

Let G be a group.

Definition 3.5 A *G-graded algebra* is an algebra A together with a vector space decomposition

$$A = \bigoplus_{g \in G} A_g ,$$

where each A_g is a linear subspace of A such that

(a) $A_g A_h \subset A_{gh}$ for all $g, h \in G$, which means that the product ab belongs to A_{gh} whenever $a \in A_g$ and $b \in A_h$;

(b) the unit 1_A of A is in A_e, where e is the unit of the group G.

It follows from the definition that A_e is a subalgebra of A and that each A_g is an A_e-bimodule under the product of A.

When $G = \mathbb{Z}/2$ is the cyclic group of order 2, then a G-graded algebra is often called a *superalgebra*.

We next show that a G-graded algebra is the same as a $\mathbb{C}[G]$-comodule algebra, where $\mathbb{C}[G]$ is the convolution algebra of the group G with its Hopf algebra structure defined in Sect. 3.4.4.2 (see also [7, Lemma 4.8]).

Proposition 3.4 *(a) Any G-graded algebra A is a $\mathbb{C}[G]$-comodule algebra. Moreover, $A^{\mathrm{co}-\mathbb{C}[G]} = A_e$.*

(b) Conversely, any $\mathbb{C}[G]$-comodule algebra is a G-graded algebra.

Proof (a) We define a linear map $\delta : A \to A \otimes \mathbb{C}[G]$ by

$$\delta(a) = a \otimes g \quad \text{for all } a \in A_g .$$

The map δ is a morphism of algebras in view of Conditions (a) and (b) of Definition 3.5. Let us check the coassociativity and counitarity conditions of Definition 3.4 for δ. Firstly, for any $a \in A_g$,

$$(\delta \otimes \mathrm{id}_H) \circ \delta(a) = (\delta \otimes \mathrm{id}_H)(a \otimes g) = a \otimes g \otimes g.$$

Similarly,

$$(\mathrm{id}_A \otimes \Delta) \circ \delta(a) = (\mathrm{id}_A \otimes \Delta)(a \otimes g) = a \otimes g \otimes g$$

in view of (3.20). Therefore, $(\delta \otimes \mathrm{id}_H) \circ \delta = (\mathrm{id}_A \otimes \Delta) \circ \delta$ holds on each subspace A_g, hence on A. Secondly, for any $a \in A_g$,

$$(\mathrm{id}_A \otimes \varepsilon) \circ \delta(a) = (\mathrm{id}_A \otimes \varepsilon)(a \otimes g) = a\, \varepsilon(g) = a$$

again in view of (3.20).

The inclusion $A_e \subset A^{\mathrm{co}-\mathbb{C}[G]}$ follows from the definition of δ and from the fact that e is the unit of $\mathbb{C}[G]$. Let us prove the converse inclusion. For a general element $a = \sum_{g \in G} a_g \in A$ with each $a_g \in A_g$, we have

$$\delta(a) = \sum_{g \in G} a_g \otimes g.$$

Since the elements $g \in G$ are linearly independent in $\mathbb{C}[G]$, we see that, if a is coinvariant, i.e., $\delta(a) = a \otimes e$, then $a_g = 0$ for all $g \neq e$. Thus any coinvariant element belongs to A_e.

(b) Assume now that A is a $\mathbb{C}[G]$-comodule algebra with coaction δ. Using the natural basis $\{g\}_{g \in G}$ of $\mathbb{C}[G]$, we can expand $\delta(a) \in A \otimes \mathbb{C}[G]$ for any $a \in A$ uniquely as

$$\delta(a) = \sum_{g \in G} p_g(a) \otimes g$$

where each $p_g(a)$ belongs to A. It is clear that $a \mapsto p_g(a)$ defines a linear endomorphism p_g of A.

Let us now express the coassociativity of the coaction δ. On one hand, we have

$$(\delta \otimes \mathrm{id}_H) \circ \delta(a) = (\delta \otimes \mathrm{id}_H) \left(\sum_{g \in G} p_g(a) \otimes g \right) = \sum_{g \in G} \sum_{h \in G} p_h(p_g(a)) \otimes h \otimes g.$$

On the other hand,

$$(\mathrm{id}_A \otimes \Delta) \circ \delta(a) = (\mathrm{id}_A \otimes \Delta) \left(\sum_{g \in G} p_g(a) \otimes g \right) = \sum_{g \in G} p_g(a) \otimes g \otimes g.$$

Identifying both right-hand sides in view of (3.30), we obtain

$$p_h \circ p_g = \begin{cases} p_g & \text{if } g = h, \\ 0 & \text{otherwise.} \end{cases} \tag{3.32}$$

Next, the counitarity condition (3.31) implies that

$$a = (\mathrm{id}_A \otimes \varepsilon) \circ \delta(a) = (\mathrm{id}_A \otimes \varepsilon) \left(\sum_{g \in G} p_g(a) \otimes g \right)$$
$$= \sum_{g \in G} p_g(a)\, \varepsilon(g) = \sum_{g \in G} p_g(a).$$

In other words,

$$\sum_{g \in G} p_g = \mathrm{id}_A\,. \tag{3.33}$$

Define the linear subspace $A_g = p_g(A)$ of A for all $g \in G$. The equality (3.33) implies $\sum_{g \in G} A_g = A$. Let us check that this sum is a direct sum. Indeed, let us assume that $\sum_{g \in G} p_g(a_g) = 0$ in A for a family (a_g) of elements of A and apply p_h to it for a fixed element $h \in G$. By (3.32), we obtain

$$0 = p_h \left(\sum_{g \in G} p_g(a_g) \right) = \sum_{g \in G} p_h(p_g(a_g)) = p_h(a_h).$$

Since this holds for any $h \in G$, we see that each summand in the sum $\sum_{g \in G} p_g(a_g)$ vanishes.

We claim that $\delta(a) = a \otimes g$ for any $a \in A_g$. Indeed, an element of A_g is of the form $a = p_g(a')$ for some $a' \in A$. Using (3.32), we obtain

$$\delta(a) = \sum_{h \in G} p_h(a) \otimes h = \sum_{h \in G} p_h(p_g(a')) \otimes h = p_g(a') \otimes g = a \otimes g.$$

It remains to check that ab belongs to A_{gh} for all $a \in A_g$ and $b \in A_h$, and that 1_A belongs to A_e. For the first requirement, we have $\delta(a) = a \otimes g$ and $\delta(b) = b \otimes h$. Since δ is a morphism of algebras, we have

$$\delta(ab) = \delta(a)\delta(b) = (a \otimes g)(b \otimes h) = ab \otimes gh,$$

which proves that the product ab belongs to A_{gh}.

For the second requirement, we have $\delta(1_A) = 1_A \otimes e$; thus, the unit of the algebra belongs to the component A_e indexed by the unit e of the group. □

Let us give a few examples of group-graded algebras.

Example 3.11 By Example 3.10 we know that the Hopf algebra $\mathbb{C}[G]$ is itself a $\mathbb{C}[G]$-comodule algebra with coaction equal to the coproduct Δ of $\mathbb{C}[G]$. Since $\Delta(g) = g \otimes g$ by (3.20), we deduce from Proposition 3.4 and its proof that $\mathbb{C}[G]$ is a G-graded algebra $\mathbb{C}[G] = \bigoplus_{g \in G} A_g$, where each g-component A_g is one-dimensional and consists of all scalar multiples of the element g.

Example 3.12 (*Gradings on matrix algebras*)

(a) Consider the algebra $M_N(\mathbb{C})$ of $N \times N$-matrices. Let $E_{i,j} \in M_N(\mathbb{C})$ be the matrix whose entries are all zero, except for the (i, j)-entry which is equal to 1. The N^2 matrices $E_{i,j}$ $(1 \leq i, j \leq N)$ form a basis of $M_N(\mathbb{C})$.

The algebra $M_N(\mathbb{C})$ can be given many group gradings. Indeed, let G be a group and $(g_1, \ldots, g_N)$ be an N-tuple of elements of G. For any $g \in G$, let A_g be the vector space spanned by all matrices $E_{i,j}$ such that $g_i g_j^{-1} = g$; we set $A_g = 0$ is there is no couple (i, j) such that $g_i g_j^{-1} = g$. Then the decomposition $M_N(\mathbb{C}) = \bigoplus_{g \in G} A_g$ yields the structure of a G-graded algebra on $M_N(\mathbb{C})$ (check this claim!).

(b) As a special case of the previous gradings, take $G = \mathbb{Z}/N$ to be the cyclic group generated by an element t of order N and

$$(g_1, \ldots, g_N) = (e, t, t^2, \ldots, t^{N-1}).$$

Then $M_N(\mathbb{C})$ has a grading $M_N(\mathbb{C}) = \bigoplus_{k=0}^{N-1} A_{t^k}$ for which A_{t^k} consists of all matrices $(a_{i,j})_{1 \leq i,j \leq N}$ such that $a_{i,j} = 0$ if $i - j \not\equiv k \pmod N$. In particular, A_e is the subalgebra of diagonal matrices. Each A_{t^k} is N-dimensional.

Example 3.13 Let $\mathbb{H}$ be the four-dimensional algebra of *complex quaternions*. Recall that it has a basis $\{1, i, j, k\}$ such that the multiplication of $\mathbb{H}$ is given by the following rules: 1 is the unit and

$$i^2 = j^2 = k^2 = -1, \quad ij = -ji = k, \quad jk = -kj = i, \quad ki = -ik = j.$$

The algebra $\mathbb{H}$ is G-graded, where G is the group $(\mathbb{Z}/2)^2$ of order 4: We have

$$A_{(0,0)} = \mathbb{C}\,1, \quad A_{(1,0)} = \mathbb{C}\,i, \quad A_{(0,1)} = \mathbb{C}\,j, \quad A_{(1,1)} = \mathbb{C}\,k.$$

There is an isomorphism of algebras $\psi : \mathbb{H} \to M_2(\mathbb{C})$ given by

$$\psi(1) = \begin{pmatrix} 1 & 0 \\ 0 & 1 \end{pmatrix}, \quad \psi(i) = \begin{pmatrix} 0 & \sqrt{-1} \\ \sqrt{-1} & 0 \end{pmatrix},$$
$$\psi(j) = \begin{pmatrix} 0 & -1 \\ 1 & 0 \end{pmatrix}, \quad \psi(k) = \begin{pmatrix} \sqrt{-1} & 0 \\ 0 & -\sqrt{-1} \end{pmatrix}.$$

This isomorphism induces a $(\mathbb{Z}/2)^2$-grading on $M_2(\mathbb{C})$. Such a grading is not of the form presented in Example 3.12 (b) above.

3.6.3 Algebras with Group Actions

Let G be a group.

Definition 3.6 A *G-algebra* is an algebra A together with a group homomorphism $\rho : G \to \mathrm{Aut}(A)$ such that each $\rho(g)$ is an algebra automorphism of A.

The subspace A^G consisting of all elements $a \in A$ such that $\rho(g)(a) = a$ for all $g \in A$ forms a subalgebra of G. The elements of A^G are called *G-invariants*.

Any algebra has the structure of a G-algebra with G taken to be (a subgroup of) the group of algebra automorphisms of A. Let us give a few more examples of G-algebras.

Example 3.14 If K is a finite *Galois extension* of a number field k with Galois group G, then G acts by automorphisms on K and we have $K^G = k$.

Example 3.15 The general linear group $GL_N(\mathbb{C})$ acts by conjugation on the matrix algebra $M_N(\mathbb{C})$. The $GL_N(\mathbb{C})$-invariants are the scalar multiples of the identity matrix.

Assume now that the group G is finite. Consider the Hopf algebra $\mathcal{O}(G)$ (introduced in Sect. 3.4.4.1) and its basis $\{\delta_g\}_{g\in G}$ of δ-functions.

Proposition 3.5 *(a) Any G-algebra A is an $\mathcal{O}(G)$-comodule algebra with coaction $\delta : A \to A \otimes \mathcal{O}(G)$ given for all $a \in A$ by*

$$\delta(a) = \sum_{g\in G} \rho(g)(a) \otimes \delta_g .$$

Moreover, the subalgebra $A^{\mathrm{co}-\mathcal{O}(G)}$ of coinvariant elements coincides with the subalgebra A^G of G-invariant elements of A:

$$A^{\mathrm{co}-\mathcal{O}(G)} = A^G .$$

(b) Conversely, any $\mathcal{O}(G)$-comodule algebra is a G-algebra.

The proof is left to the reader, who is invited to take inspiration from the proof of Proposition 3.4.

3.6.4 The Quantum Plane and Its $\mathrm{SL}_q(2)$*-coaction*

The special linear group $SL_2(\mathbb{C})$ acts on the two-dimensional vector space $\mathbb{C}^2$ by matrix multiplication. As a special case of Example 3.9, the coordinate algebra $\mathbb{C}[X, Y]$ of $\mathbb{C}^2$ becomes a SL(2)-comodule algebra. Recall from Sect. 3.5.1 that

$$\mathrm{SL}(2) = \mathbb{C}[a, b, c, d]/(ad - bc - 1)$$

is the coordinate algebra of $SL_2(\mathbb{C})$. It is easy to check that the corresponding coaction $\delta : \mathbb{C}[X, Y] \to \mathbb{C}[X, Y] \otimes SL_2(\mathbb{C})$ is given by

$$\delta(X, Y) = (X, Y) \otimes \begin{pmatrix} a & b \\ c & d \end{pmatrix}, \tag{3.34}$$

which is short for

$$\delta(X) = X \otimes a + Y \otimes c \quad \text{and} \quad \delta(Y) = X \otimes b + Y \otimes d.$$

In Sect. 3.5.1, we quantized SL(2) using a complex parameter $q \neq 0$. We now proceed to quantize the previous coaction. To this end, we replace $\mathbb{C}[X, Y]$ by the quantum plane $\mathbb{C}_q[X, Y] = \mathbb{C}\langle X, Y\rangle/(YX - qXY)$ introduced in Sect. 3.3.2.2.

Theorem 3.2 *The map δ given by* (3.34) *equips the quantum plane $\mathbb{C}_q[X, Y]$ with the structure of a* $\mathrm{SL}_q(2)$*-comodule algebra. Moreover, the subalgebra of coinvariants of $\mathbb{C}_q[X, Y]$ is* $\mathbb{C}1$.

The second assertion is the non-commutative analogue of the fact that the only point of the plane which is invariant under the action of $\mathrm{SL}_2(\mathbb{C})$ is the origin.

Proof (a) We first have to establish that δ is a morphism of algebras. It suffices to check that $\delta(Y)\delta(X) = q\,\delta(X)\delta(Y)$. Using (3.34), we have

$$\begin{aligned} \delta(Y)\delta(X) &= (X \otimes b + Y \otimes d)(X \otimes a + Y \otimes c) \\ &= X^2 \otimes ba + YX \otimes da + XY \otimes bc + Y^2 \otimes dc. \end{aligned}$$

Similarly,

$$\begin{aligned} \delta(X)\delta(Y) &= (X \otimes a + Y \otimes c)(X \otimes b + Y \otimes d) \\ &= X^2 \otimes ab + YX \otimes cb + XY \otimes ad + Y^2 \otimes cd. \end{aligned}$$

Now using the defining relations of $\mathrm{SL}_q(2)$ and the relation $YX = qXY$, we obtain

$$\begin{aligned} \delta(Y)\delta(X) - q\,\delta(X)\delta(Y) &= X^2 \otimes (ba - qab) + YX \otimes (da - qcb) \\ &\quad + XY \otimes (bc - qad) + Y^2 \otimes (dc - qcd) \\ &= XY \otimes q(da - qcb + q^{-1}bc - ad) \\ &= -XY \otimes q(ad - da - (q^{-1} - q)bc) = 0. \end{aligned}$$

The map δ being a morphism of algebras, it is enough to check its coassociativity and its counitarity on the generators X, Y, which is easy to do.

(b) Let $\omega \in \mathbb{C}_q[X, Y]$ be a coinvariant element, i.e., $\delta(\omega) = \omega \otimes 1$. Recall the morphism of Hopf algebras $\pi : \mathrm{SL}_q(2) \to \mathbb{C}_q[X, X^{-1}, Y]$ of Lemma 3.1. The composed map

$$\delta' = (\mathrm{id} \otimes \pi) \circ \delta : \mathbb{C}_q[X, Y] \to \mathbb{C}_q[X, Y] \otimes \mathbb{C}_q[X, X^{-1}, Y]$$

turns the quantum plane $\mathbb{C}_q[X, Y]$ into a $\mathbb{C}_q[X, X^{-1}, Y]$-comodule algebra. We have $\delta'(\omega) = (\mathrm{id} \otimes \pi)(\omega \otimes 1) = \omega \otimes \pi(1) = \omega \otimes 1$. Thus ω is coinvariant for the $\mathbb{C}_q[X, X^{-1}, Y]$-coaction. Now, it follows from (3.34) and the formula for π that

$$\delta'(X) = X \otimes \pi(a) + Y \otimes \pi(c) = X \otimes X$$

and

$$\delta'(Y) = X \otimes \pi(b) + Y \otimes \pi(d) = X \otimes Y + Y \otimes X^{-1}.$$

Comparing with Formula (3.28) for the coproduct Δ of the Hopf algebra $\mathbb{C}_q[X, X^{-1}, Y]$, we see that δ' is the restriction of Δ to the subalgebra $\mathbb{C}_q[X, Y]$. It follows from this remark and from Example 3.10 that ω is a scalar multiple of the unit of $\mathbb{C}_q[X, X^{-1}, Y]$, which is also the unit of $\mathbb{C}_q[X, Y]$. □

Exercise 3.25 Let q be a nonzero complex number. For any integer $r > 0$ define the *q-integer* $[r]$ by

$$[r] = 1 + q + \cdots + q^{r-1} = \frac{q^r - 1}{q - 1}.$$

and the *q-factorial* $[r]!$ by

$$[r]! = \prod_{k=1}^{r} [k] = \frac{(q-1)(q^2-1)\cdots(q^r-1)}{(q-1)^r}.$$

We agree that $[0]! = 1$. For $0 \le r \le n$ we define the *q-binomial coefficient*

$$\begin{bmatrix} n \\ r \end{bmatrix} = \frac{[n]!}{[r]!\,[n-r]!}.$$

(a) For $0 < r < n$ show the following q-analogue of the *Pascal identity*

$$\begin{bmatrix} n \\ r \end{bmatrix} = \begin{bmatrix} n-1 \\ r-1 \end{bmatrix} + q^r \begin{bmatrix} n-1 \\ r \end{bmatrix}.$$

(b) Let X, Y be variables subject to the relation $YX = qXY$. Prove the *q-binomial formula*

$$(X+Y)^n = \sum_{r=0}^{n} \begin{bmatrix} n \\ r \end{bmatrix} X^r Y^{n-r}.$$

Exercise 3.26 Recall the basis $\{X^i Y^j\}_{i,j \in \mathbb{N}}$ of the quantum plane $\mathbb{C}_q[X, Y]$. Compute $\delta(X^i Y^j)$ for the coaction (3.34).

3.6.5 Quantum Homogeneous Spaces

Let G be an algebraic group and G' be an algebraic subgroup. To this data, we associate the *homogeneous space* G/G', whose elements are the left cosets gG' of G' in G with respect to $g \in G$; in other words, two elements $g_1, g_2 \in G$ represent the same element of G/G' if and only if there exists $g' \in G'$ such that $g_2 = g_1 g'$.

To the inclusion $i : G' \hookrightarrow G$ corresponds the morphism of Hopf algebras $\pi = i^* : \mathcal{O}(G) \to \mathcal{O}(G')$, which sends a function $u \in \mathcal{O}(G)$ to its restriction to G'. The map π is surjective. The composition

$$\delta = (\mathrm{id} \otimes \pi) \circ \Delta : \mathcal{O}(G) \to \mathcal{O}(G) \otimes \mathcal{O}(G')$$

turns $\mathcal{O}(G)$ into an $\mathcal{O}(G')$-comodule algebra. Let us consider the subalgebra

$$\mathcal{O}(G)^{\mathrm{co}-\mathcal{O}(G')} \subset \mathcal{O}(G)$$

of coinvariant elements.

Lemma 3.3 *An element $u \in \mathcal{O}(G)$ belongs to the subalgebra $\mathcal{O}(G)^{\mathrm{co}-\mathcal{O}(G')}$ if and only if $u(gg') = u(g)$ for all $g \in G$ and $g' \in G'$.*

Proof Identifying $\mathcal{O}(G) \otimes \mathcal{O}(G')$ with $\mathcal{O}(G \times G')$ and using Formula (3.19) for the coproduct of $\mathcal{O}(G)$, we see that the above coaction δ sends an element $u \in \mathcal{O}(G)$ to the function $\delta(u) \in \mathcal{O}(G \times G')$ given by

$$\delta(u)(g, g') = u(gg')$$

for all $g \in G$ and $g' \in G'$. Such an element u is coinvariant if and only if $\delta(u) = u \otimes 1$, which is equivalent to $\delta(u)(g, g') = u(g)1$ for all $g \in G$ and $g' \in G'$. □

It follows from the lemma and the above description of G/G' that the subalgebra $\mathcal{O}(G)^{\mathrm{co}-\mathcal{O}(G')}$ of coinvariant elements can be identified with the coordinate algebra $\mathcal{O}(G/G')$ of the homogeneous space G/G'.

The non-commutative analogue of a homogeneous space is the following. Let $\pi : H \to \bar{H}$ be a surjective morphism of Hopf algebras. The map

$$\delta = (\mathrm{id} \otimes \pi) \circ \Delta : H \to H \otimes \bar{H}$$

turns H into an $\bar{H}$-comodule algebra. Let us consider the subalgebra $H^{\mathrm{co}-\bar{H}}$ of coinvariant elements; by analogy with the previous classical case we call $H^{\mathrm{co}-\bar{H}}$ a *quantum homogeneous space*.

This general construction provides many examples of quantum homogeneous spaces; see [9, 14, 24, 25, 40, 41, 50, 53]. We have already encountered such a situation with the surjective morphism of Hopf algebras $\pi : \mathrm{SL}_q(2) \to \mathbb{C}_q[X, X^{-1}, Y]$ in Sect. 3.5.2, where $\mathbb{C}_q[X, X^{-1}, Y]$ has been hinted at as a quantization of the coordinate algebra of the subgroup B of upper triangular matrices in $SL_2(\mathbb{C})$. It is well known that the homogeneous space $SL_2(\mathbb{C})/B$ is in bijection with the *projective line* $\mathbb{CP}^1$. Therefore the subalgebra $\mathrm{SL}_q(2)^{\mathrm{co}-\mathbb{C}_q[X,X^{-1},Y]}$ can be seen as a quantization of $\mathbb{CP}^1$.

3.7 Hopf Galois Extensions

It was noticed in the 1990s (see [9, 19, 53]) that the right non-commutative version of a principal fiber bundle is the concept of a Hopf Galois extension, a notion which had been introduced in the 1960s by algebraists in order to extend the classical Galois theory of field extensions to a more general framework.

Let us now define Hopf Galois extensions. The use of the word "Galois" in this expression will be justified by Example 3.16 below.

3.7.1 Definition and Examples

Definition 3.7 Let H be a Hopf algebra and B an (associative unital) algebra. An *H-Galois extension* of B is an H-comodule algebra A with coaction $\delta : A \to A \otimes H$ such that the following three conditions hold:

(i) A contains B as a subalgebra;
(ii) $B = A^{\mathrm{co}-H} = \{a \in A \mid \delta(a) = a \otimes 1\}$;
(iii) the linear map

$$\beta : A \otimes A \to A \otimes H\,; \quad a \otimes a' \mapsto (a \otimes 1)\,\delta(a') \tag{3.35}$$

induces a linear isomorphism $A \otimes_B A \xrightarrow{\cong} A \otimes H$.

Let us comment on Condition (iii). Firstly, the vector space $A \otimes_B A$ is by definition the quotient of $A \otimes A$ by the subspace U spanned by all tensors of the form

$$ab \otimes a' - a \otimes ba'. \qquad (a, a' \in A,\ b \in B)$$

Condition (iii) implies that the map β factors through the quotient space $A \otimes_B A$. Let us check this: It is enough to verify that β vanishes on the generators of the subspace U. Indeed,

$$\begin{aligned}\beta(ab \otimes a' - a \otimes ba') &= (ab \otimes 1)\,\delta(a') - (a \otimes 1)\,\delta(ba')\\ &= (a \otimes 1)(b \otimes 1)\,\delta(a') - (a \otimes 1)\,\delta(b)\,\delta(a') = 0\end{aligned}$$

in view of the fact that b is coinvariant, hence satisfies $\delta(b) = b \otimes 1$.

The map β in Condition (iii) is the non-commutative analogue of the map $\gamma : G \times P \to P \times P$ defined by (3.1), and the isomorphism $A \otimes_B A \xrightarrow{\cong} A \otimes H$ is the non-commutative analogue of the bijection $\gamma : G \times P \to P \times_X P$. For this reason, a Hopf Galois extension can be seen as a *non-commutative principal fiber bundle.*

Remark 3.4 Let A be an H-Galois extension of B. Observe that, if dim A is finite, then so are dim $A \otimes A$ and dim $A \otimes_B A$. In view of the isomorphism $A \otimes_B A \cong A \otimes H$, we deduce that the Hopf algebra H is finite-dimensional and that dim $H \leq$ dim A. If in addition $B = \mathbb{C}$ is the ground field, then $A \otimes_B A = A \otimes A$ and dim $H =$ dim A.

Remark 3.5 Sometimes in the definition of an H-Galois extension A of B one also requires A to be *faithfully flat* as a left B-module. This means that taking the tensor product $\otimes_B M$ with a sequence of right B-modules produces an exact sequence if and only if the original sequence is exact. Finite-rank free or projective modules are examples of faithfully flat modules. The Hopf Galois extensions we will consider in Sect. 3.8 satisfy this extra condition.

According to [12, Sect. 7], Definition 3.7 was introduced to give a generalization of Galois theory to arbitrary commutative rings, the finite group of automorphisms in the classical theory being replaced by a Hopf algebra.

Let us now present the prototypical example of a Hopf Galois extension, which justifies the terminology used.

Example 3.16 If K is a finite *Galois extension* of a number field k with Galois group G, then by Proposition 3.5 (a) the field K is an $\mathcal{O}(G)$-comodule k-algebra with coaction δ given for all $a \in K$ by

$$\delta(a) = \sum_{g \in G} ga \otimes \delta_g .$$

We know that the subalgebra of coinvariant elements of K is the subalgebra of G-invariant elements, therefore coinciding with the field k. The map

$$\beta : K \otimes_k K \to K \otimes_k \mathcal{O}(G)$$

defined by (3.35) is an isomorphism (see e.g., [46, Sect. 8.1.2]). Therefore, K is an $\mathcal{O}(G)$-Galois extension of k.

Here are more examples of Hopf Galois extensions.

Example 3.17 If $P \to X$ is a *principal G-bundle*, then $\mathcal{O}(P)$ is an $\mathcal{O}(G)$-Galois extension of $\mathcal{O}(X)$.

Example 3.18 Let $A = \mathbb{C}[x, x^{-1}]$ be the algebra of Laurent polynomials in one variable and let $n \geq 1$ be an integer. We can give A a $\mathbb{Z}/n$-grading by setting $\deg(x^i) \equiv i \pmod{n}$. This is a strong grading in the sense defined above. The algebra A becomes a $\mathbb{C}[\mathbb{Z}/n]$-Galois extension of the subalgebra $B = \mathbb{C}[x^n, x^{-n}]$. This is the algebraic version of the principal $\mathbb{Z}/n$-bundle $\pi_n : S^1 \to S^1$ of Example 3.2.

Example 3.19 (*Strongly graded algebras*) Let G be a group. We know (see Proposition 3.4) that any G-graded algebra A is a $\mathbb{C}[G]$-comodule algebra. Recall that the subalgebra of coinvariants is the e-component A_e. Such a comodule algebra is a $\mathbb{C}[G]$-Galois extension of A_e if and only if A is a *strongly G-graded algebra*, i.e., a G-graded algebra such that $A_g A_h = A_{gh}$ for all $g, h \in G$ (see [46, Theorem 8.1.7]).

The matrix algebra $M_N(\mathbb{C})$ with the $\mathbb{Z}/N$-grading given in Example 3.12 (b) and the algebra of quaternions with the $(\mathbb{Z}/2)^2$-grading of Example 3.13 are strongly graded algebras.

Remark 3.6 In classical differential geometry once one has a principal G-bundle, one can construct a vector bundle associated with it and with an additional representation of G, equip this vector bundle with a connection, and derive various characteristic classes. Nowadays these classical constructions have non-commutative counterparts; for details, see [9, 14, 24, 25, 49, 62].

3.7.2 The Classification Problem

We say that two H-Galois extensions A, A' of B are *isomorphic* if there is an isomorphism of H-comodule algebras $A \to A'$.

In Sect. 3.2.4 (see Corollary 3.1), we showed how to classify principal G-bundles: there exists a bijection

$$[X, BG] \xrightarrow{\cong} \mathrm{Iso}_G(X)$$

which is functorial in X. Recall that $\mathrm{Iso}_G(X)$ is the set of homeomorphism classes of principal G-bundles with base space X and $[X, BG]$ is the set of homotopy classes of continuous maps from X to BG.

We wish likewise to classify all H-Galois extensions of B up to isomorphism for a given Hopf algebra H and a given algebra B. In other words, we would like to compute the set $\mathrm{Gal}_H(B)$ of isomorphism classes of H-Galois extensions of B.

So far not many general results on $\mathrm{Gal}_H(B)$ are available. Here is one.

Theorem 3.3 *The set* $\mathrm{Gal}_H(B)$ *is non-empty.*

This is a consequence of the following result.

Proposition 3.6 *The tensor product algebra $A = B \otimes H$ is an H-Galois extension of $B = B \otimes 1$ with coaction $\delta = \mathrm{id}_B \otimes \Delta : A = B \otimes H \to A \otimes H = B \otimes H \otimes H$, where Δ is the coproduct of H.*

This Hopf Galois extension is called the *trivial Hopf Galois extension*. Its isomorphism class is thus a special point of $\mathrm{Gal}_H(B)$, just as the trivial principal G-bundle is a special element of the set $\mathrm{Iso}_G(X)$ of homeomorphism classes of principal G-bundles with given base space X.

Proof The map δ turns A into an H-comodule algebra. Proceeding as in Example 3.10, we prove that the subalgebra of coinvariant elements coincides with $B \otimes 1 = B$.

Finally we have to establish that the map $\beta : A \otimes_B A \to A \otimes H$ of (3.35) is an isomorphism. Now

$$A \otimes_B A = (B \otimes H) \otimes_B (B \otimes H) = B \otimes H \otimes H$$

and $A \otimes H = B \otimes H \otimes H$. It suffices to check that the map $\beta_1 : H \otimes H \to H \otimes H$ defined for all $x, y \in H$ by

$$\beta_1(x \otimes y) = (x \otimes 1)\, \Delta(y) = \sum_{(y)} xy_{(1)} \otimes y_{(2)}$$

is a linear isomorphism (here again we use the Heyneman–Sweedler sigma notation of Sect. 3.4.5). Define a map β_2 in the other direction by

$$\beta_2(x \otimes y) = (x \otimes 1)(S \otimes \mathrm{id})(\Delta(y)) = \sum_{(y)} xS(y_{(1)}) \otimes y_{(2)}.$$

On the one hand, by (3.23) and (3.22) we have

$$\begin{aligned}(\beta_1 \circ \beta_2)(x \otimes y) &= \sum_{(y)} xS(y_{(1)})y_{(2)} \otimes y_{(3)} = \sum_{(y)} x\varepsilon(y_{(1)}) \otimes y_{(2)} \\ &= x \otimes \sum_{(y)} \varepsilon(y_{(1)})y_{(2)} = x \otimes y,\end{aligned}$$

which proves $\beta_1 \circ \beta_2 = \mathrm{id}_{H \otimes H}$. On the other,

$$\begin{aligned}(\beta_2 \circ \beta_1)(x \otimes y) &= \sum_{(y)} xy_{(1)}S(y_{(2)}) \otimes y_{(3)} = \sum_{(y)} x\varepsilon(y_{(1)}) \otimes y_{(2)} \\ &= x \otimes \sum_{(y)} \varepsilon(y_{(1)})y_{(2)} = x \otimes y.\end{aligned}$$

This completes the proof of the bijectivity of β_1, hence of β. □

3.7.3 *The Set* $\mathrm{Gal}_H(\mathbb{C})$ *May Be Non-trivial*

We observed in Sect. 3.2.1 that any fiber bundle over a point is trivial. The corresponding result for H-Galois extensions of the ground field $\mathbb{C}$ may not hold. To show this let us present examples of Hopf algebras H for which $\mathrm{card}\,\mathrm{Gal}_H(\mathbb{C}) > 1$.

It is convenient to introduce the following definition.

Definition 3.8 Let H be a Hopf algebra. An H-*Galois object* is an H-Galois extension of $\mathbb{C}$.

3.7.3.1 The Case of a Group Algebra

Let us consider $H = \mathbb{C}[G]$ for some group G. We now describe $\mathrm{Gal}_H(\mathbb{C})$ for this Hopf algebra.

By Example 3.19, we know that any $\mathbb{C}[G]$-Galois extension A of $\mathbb{C}$ is a strongly G-graded algebra $A = \bigoplus_{g\in G} A_g$ such that $A_e = \mathbb{C}$. Since it is strongly graded, it follows that each component A_g is one-dimensional. Let us pick a nonzero element u_g in each A_g. Then the product structure of the algebra A is determined by the products $u_g u_h$ for each pair (g, h) of elements of G. We have

$$u_g u_h = \lambda(g, h)\, u_{gh} \in A_{gh} \tag{3.36}$$

for some scalar $\lambda(g, h)$ depending on g and h. Such a scalar is nonzero since by definition the multiplication map $A_g \times A_h \to A_{gh}$ is surjective. Thus, the family of scalars $\lambda(g, h)$ defines a map $\lambda : G \times G \to \mathbb{C}^\times$, where $\mathbb{C}^\times = \mathbb{C} \setminus \{0\}$.

The map λ satisfies an additional relation called *cocyclicity*, originating from the fact that the product of A is associative. Indeed, we have $(u_g u_h)u_k = u_g(u_h u_k)$ for all $g, h, k \in G$. Using (3.36), we obtain the following equality

$$\lambda(g, h)\, \lambda(gh, k) = \lambda(h, k)\, \lambda(g, hk) \tag{3.37}$$

for all $g, h, k \in G$. A map $\lambda : G \times G \to \mathbb{C}^\times$ satisfying the identity (3.37) is called a 2-*cocycle* for the group G.

It can be checked (see any textbook on group cohomology, for instance [8]) that the pointwise multiplication of maps from $G \times G$ to $\mathbb{C}^\times$ induce an abelian group structure on the set $Z^2(G, \mathbb{C}^\times)$ of 2-cocycles for G.

Let us choose another nonzero element v_g in each A_g. Then we have $v_g = \mu(g)\, u_g$ for some nonzero scalar $\mu(g)$. Combining this with (3.36), we obtain $v_g v_h = \lambda'(g, h)\, v_{gh}$, where

$$\lambda'(g, h) = \frac{\mu(g)\mu(h)}{\mu(gh)}\, \lambda(g, h) \tag{3.38}$$

for all $g, h \in G$. We say that two 2-cocycles λ, λ' are *cohomologous* if they are related by an equation of the form (3.38). It is easy to check that for any map $\mu : G \to \mathbb{C}^\times$ the assignment $(g, h) \mapsto \mu(g)\mu(h)/\mu(gh)$ is a 2-cocycle, which we call a *coboundary*. Moreover, the set $B^2(G, \mathbb{C}^\times)$ of coboundaries is a subgroup of $Z^2(G, \mathbb{C}^\times)$.

We define the *second cohomology group* of G as the quotient

$$H^2(G, \mathbb{C}^\times) = Z^2(G, \mathbb{C}^\times)/B^2(G, \mathbb{C}^\times).$$

It follows from the previous arguments that we have a bijection

$$\mathrm{Gal}_{\mathbb{C}[G]}(\mathbb{C}) \cong H^2(G, \mathbb{C}^\times). \tag{3.39}$$

Example 3.20 It is well known (see [8, V.6]) that for a cyclic group G (infinite or not) we have $H^2(G, \mathbb{C}^\times) = 0$; for such a group $\mathrm{Gal}_{\mathbb{C}[G]}(\mathbb{C})$ is then trivial by (3.39), i.e., any $\mathbb{C}[G]$-Galois object is trivial.

Example 3.21 Let $G = (\mathbb{Z}/N)^r$ for some integer $r \geq 2$. Then

$$H^2(G, \mathbb{C}^\times) \cong (\mathbb{Z}/N)^{r(r-1)/2},$$

which implies that $\mathrm{Gal}_{\mathbb{C}[G]}(\mathbb{C}) > 1$ for such a group. This is of course a rather surprising result, which again shows that non-commutative geometry has features which classical geometry does not have.

Example 3.22 Even more surprising, if $G = \mathbb{Z}^r$ is the free abelian group of rank $r \geq 2$, then

$$H^2(G, \mathbb{C}^\times) \cong (\mathbb{C}^\times)^{r(r-1)/2}.$$

Hence, for $r \geq 2$ there are *infinitely many* isomorphism classes of $\mathbb{C}[\mathbb{Z}^r]$-Galois objects.

Remark 3.7 In contrast with Example 3.20, the cohomology group $H^2(\mathbb{Z}/2, \mathbb{R}^\times)$ of the cyclic group of order 2, now with coefficients in $\mathbb{R}^\times = \mathbb{R} \setminus \{0\}$, is not trivial:

$$H^2(G, \mathbb{R}^\times) = \mathbb{R}^\times/(\mathbb{R}^\times)^2 \cong \mathbb{Z}/2.$$

Proceeding as above, we deduce that, up to isomorphism, there are two real $\mathbb{Z}/2$-Galois extensions of $\mathbb{R}$. The trivial one is $\mathbb{R}[\mathbb{Z}/2] = \mathbb{R}[x]/(x^2 - 1) \cong \mathbb{R} \times \mathbb{R}$, which has zero divisors. The second one is the field $\mathbb{C} = \mathbb{R}[x]/(x^2 + 1)$ of complex numbers. Both are two-dimensional superalgebras, with the even part spanned by the unit 1 and the odd part by the image of x.

Remark 3.8 Group algebras are cocommutative Hopf algebras and by (3.39) the group $\mathrm{Gal}_H(\mathbb{C})$ is abelian in this case. More generally, for any *cocommutative* Hopf algebra H, the set $\mathrm{Gal}_H(\mathbb{C})$ has the structure of an abelian group; its product is induced by the cotensor product[5] of comodule algebras (see for example [10, 10.5.3]).

3.7.3.2 Taft Algebras

Let N be an integer ≥ 2 and q a root of unity of order N. The *Taft algebra* of dimension N^2 is the algebra H_{N^2} generated by two generators g, x subject to the relations

[5] The concept of the cotensor product of comodules was first introduced in [20]. See also [46, 58].

$$g^N = 1, \quad x^N = 0, \quad xg = q\, gx.$$

It is a Hopf algebra with

$$\Delta(g) = g \otimes g, \quad \Delta(x) = 1 \otimes x + x \otimes g, \quad \varepsilon(g) = 1, \quad \varepsilon(x) = 0.$$

This Hopf algebra is neither commutative, nor cocommutative. When $N = 2$, the four-dimensional Hopf algebra H_4 is known under the name of *Sweedler algebra*.

For any $s \in \mathbb{C}$ consider the algebra

$$A_s = \mathbb{C}\langle G, X \rangle / \big(G^N - 1,\ X^N - s,\ XG - q\, GX\big).$$

It is a right H_{N^2}-Galois object with coaction given by

$$\Delta(G) = G \otimes g, \quad \Delta(X) = 1 \otimes x + X \otimes g.$$

By [44, Propositions 2.17 and 2.22] (see also [16]), any H_{N^2}-Galois object is isomorphic to A_s for some scalar s, and any two such Galois objects A_s and A_t are isomorphic if and only if $s = t$. Therefore,

$$\mathrm{Gal}_{H_{N^2}}(\mathbb{C}) \cong \mathbb{C},$$

which is an abelian group although the Hopf algebra H_{N^2} is not cocommutative.

See also [5, 6, 47, 48] for the determination of $\mathrm{Gal}_H(\mathbb{C})$ for other finite-dimensional Hopf algebras H generalizing the Sweedler algebra.

3.7.3.3 The Quantum Enveloping Algebra $U_q\,\mathfrak{g}$

Masuoka [45] determined $\mathrm{Gal}_H(\mathbb{C})$ when $H = U_q\,\mathfrak{g}$ is Drinfeld–Jimbo's quantum enveloping algebra mentioned in Sect. 3.5.3, Remark 3.3. A partial result had been given in [37, Theorem 4.5] under the form of a surjection

$$\mathrm{Gal}_H(\mathbb{C}) \twoheadrightarrow H^2(\mathbb{Z}^r, \mathbb{C}^\times) \cong (\mathbb{C}^\times)^{r(r-1)/2},$$

where r is the size of the corresponding Cartan matrix (see also [4]).

3.7.4 Push-Forward of Central Hopf Galois Extensions

In Sect. 3.2.4, we saw that given a continuous map $\varphi : X' \to X$, there is a functorial map

$$\varphi^* : \mathrm{Iso}_G(X) \to \mathrm{Iso}_G(X')$$

induced by $P \mapsto \varphi^*(P)$.

In our algebraic setting we may wonder whether, given a Hopf algebra H and a morphism of algebras $f : B \to B'$, there exists a functorial map

$$f_* : \mathrm{Gal}_H(B) \to \mathrm{Gal}_H(B')$$

which would be the algebraic analogue of the pull-back of bundles. The most natural way to construct such a *push-forward* map f_* is the following. Let A be an H-Galois extension of B. Since B is a subalgebra of A, we can consider A as a left B-module. Given a morphism of algebras $f : B \to B'$, we can then define the left B'-module $f_*(A)$ as

$$f_*(A) = B' \otimes_B A.$$

Here we have used the fact that B' is a right B-module via the morphism of algebras f. It is clear that if $g : B' \to B''$ is another morphism of algebras, then we have a natural isomorphism $(g \circ f)_*(A) \cong g_*(f_*(A))$ of B''-modules.

There is however a serious problem with this construction: In general, $f_*(A) = B' \otimes_B A$ is not an algebra! To circumvent this difficulty, we will restrict to *central H-Galois extensions*, namely to those for which B is contained in the center of A; this implies of course that B is a commutative algebra (central Hopf Galois extensions were first discussed in [52]). The algebra $\mathscr{A}_H$ defined in Sect. 3.8.2.2 below is an (important) example of a central H-Galois extension.

We denote by $\mathrm{Zgal}_H(B)$ the set of isomorphism classes of central H-Galois extensions of B. Then a morphism of *commutative* algebras $f : B \to B'$ induces a push-forward map $f_* : \mathrm{Zgal}_H(B) \to \mathrm{Zgal}_H(B')$ given by $A \mapsto f_*(A)$ and satisfying the desired functorial properties[6] (see [32, 37]).

In particular, let $\chi : B \to \mathbb{C}$ be a character of B. Then $A \mapsto \chi_*(A)$ induces a map $\chi_* : \mathrm{Zgal}_H(B) \to \mathrm{Zgal}_H(\mathbb{C})$. Observe that $\mathrm{Zgal}_H(\mathbb{C}) = \mathrm{Gal}_H(\mathbb{C})$ when $B = \mathbb{C}$ is the ground field, as the latter is always central. In analogy with the case of a fiber bundle (see Exercise 3.1 (a)), we call $\chi_*(A) = \mathbb{C} \otimes_B A$ the *fiber* of the H-Galois extension A at χ. Note that $\chi_*(A) = A/\mathfrak{m}A$, where $\mathfrak{m}$ is the kernel of χ.

3.7.5 Universal Central Hopf Galois Extensions

A non-commutative analogue of the classifying space BG mentioned in Sect. 3.2.4 would be a central H-Galois extension $\mathscr{A}_H$ of some commutative algebra $\mathscr{B}_H$ such that for any commutative algebra B and any central H-Galois extension A of B there exists a morphism of algebras $f : \mathscr{B}_H \to B$ such that $f_*(\mathscr{A}_H) \cong A$. In other words, we would have a functorial surjection

$$\mathrm{Alg}(\mathscr{B}_H, B) \twoheadrightarrow \mathrm{Zgal}_H(B)$$

[6]For this to hold we need the extra faithful flatness condition mentioned in Sect. 3.7.1, Remark 3.5.

induced by $f \mapsto f_*(\mathscr{A}_H)$. Here $\mathrm{Alg}(\mathscr{B}_H, B)$ is the set of morphisms of algebras from $\mathscr{B}_H$ to B.

Does such a central H-Galois extension $\mathscr{A}_H$ exist for an arbitrary Hopf algebra H? It is an open question. We do not even know whether in general there exists a central H-Galois extension $\mathscr{B}_H \subset \mathscr{A}_H$ with a natural surjection

$$\mathrm{Alg}(\mathscr{B}_H, \mathbb{C}) \twoheadrightarrow \mathrm{Zgal}_H(\mathbb{C}) = \mathrm{Gal}_H(\mathbb{C})$$

from the set of characters of $\mathscr{B}_H$ to the set of isomorphism classes of H-Galois objects. If such a surjection existed and was even bijective, then the H-Galois objects would be classified up to isomorphism by the characters of $\mathscr{B}_H$.

Example 3.23 Let us give an example for which H-Galois objects can be classified by the characters of a commutative algebra $\mathscr{B}$. Take the Taft algebra H_{N^2} introduced in Sect. 3.7.3.2. Let $\mathscr{B}$ be the polynomial algebra $\mathbb{C}[s]$ and $\mathscr{A} = A_s$ considered as a $\mathbb{C}[s]$-module, where A_s is the Galois object defined in *loc. cit.* Each complex number s gives rise to a unique character χ_s of $\mathbb{C}[s]$; it is tautologically defined by $\chi(s) = s$. The map $s \mapsto \chi_s$ induces a bijection $\mathbb{C} \to \mathrm{Alg}(\mathbb{C}[s], \mathbb{C}) = \mathrm{Alg}(\mathscr{B}, \mathbb{C})$. Now the assignment $\chi_s \mapsto (\chi_s)_*(\mathscr{A})$ induces a bijection

$$\mathrm{Alg}(\mathscr{B}, \mathbb{C}) \xrightarrow{\cong} \mathrm{Gal}_{H_{N^2}}(\mathbb{C}).$$

When in 2005 I lectured on Hopf Galois extensions at the *XVIo Coloquio Latinoamericano de Álgebra* in Colonia del Sacramento, Uruguay, I raised the question of the existence of a universal central Hopf Galois extension. Eli Aljadeff immediately suggested the use of an appropriate theory of polynomial identities, based on his joint work [3] with Haile and Natapov on group-graded algebras. In [2], we implemented Aljadeff's idea, using a theory of polynomial identities for comodule algebras. Given a Hopf algebra H and an H-comodule algebra A, we constructed a "universal H-comodule algebra" $\mathscr{U}_H(A)$ out of these identities. Localizing $\mathscr{U}_H(A)$, we obtained a central H-Galois extension $\mathscr{A}_H$ of some commutative algebra $\mathscr{B}_H$, the latter being a nice domain. The Hopf Galois extension $\mathscr{B}_H \subset \mathscr{A}_H$ comes with a map of the form

$$\mathrm{Alg}(\mathscr{B}_H, \mathbb{C}) \to \mathrm{Gal}_H(\mathbb{C})\,; \quad \chi \mapsto \chi_*(\mathscr{A}_H).$$

In the next section, we will construct this central H-Galois extension directly, without passing through polynomial identities. Nevertheless the reader interested in polynomial identities, the universal H-comodule algebra $\mathscr{U}_H(A)$, and the precise connection with the central H-Galois extension constructed in Sect. 3.8.2 may learn the details from [2, 33].

3.8 Flat Deformations of Hopf Algebras

De pronto me sentí poseído por un aura
de inspiración que me permitió improvisar
respuestas creíbles y chiripas milagrosas.
Salvo en las matemáticas, que no se me
rindieron ni en lo que Dios quiso. [21]

Let H be a Hopf algebra. The aim of this final section is to construct the commutative algebra $\mathscr{B}_H$ and the central H-Galois extension $\mathscr{A}_H$ of $\mathscr{B}_H$ we have just mentioned. When H is finite-dimensional, the algebra $\mathscr{B}_H$ is the coordinate algebra of a smooth algebraic variety whose dimension is equal to dim H. The algebra $\mathscr{A}_H$ is a deformation of H as an H-comodule algebra; this deformation is parametrized by the characters of $\mathscr{B}_H$.

We conclude these notes by showing how to apply these constructions to the quantum enveloping algebra $U_q\,\mathfrak{sl}(2)$ and to its finite-dimensional quotients $\mathfrak{u}_d$.

3.8.1 A Universal Construction by Takeuchi

Let C be a *coalgebra*, that is a vector space equipped with two linear maps $\Delta : C \to C \otimes C$ (called the *coproduct*) and $\varepsilon : C \to \mathbb{C}$ (called the *counit*) satisfying the coassociativity identity (3.5) and the counitality identity (3.6). There is a coalgebra underlying any bialgebra or any Hopf algebra.

Takeuchi [59, Chap. IV] proved the following result.

Theorem 3.4 *Given a coalgebra C, there exist a commutative Hopf algebra $\mathscr{S}_C$ and a morphism of coalgebras $t : C \to \mathscr{S}_C$ such that for any morphism of coalgebras $f : C \to H'$ to a commutative Hopf algebra H' there is a unique morphism of Hopf algebras*

$$\tilde{f} : \mathscr{S}_C \to H'$$

satisfying $f = \tilde{f} \circ t$. The Hopf algebra $\mathscr{S}_C$ is unique up to unique isomorphism.

We say that $\mathscr{S}_C$ is the *free commutative Hopf algebra* over the coalgebra C. It can be constructed as follows.

3.8.1.1 Construction of $\mathscr{S}_C$

Pick a copy t_C of the underlying vector space of C, that is to say we assign a symbol t_x to each element $x \in C$ so that the map $x \mapsto t_x$ is linear and defines a linear isomorphism $t : C \to t_C$. Let $\mathrm{Sym}(t_C)$ be the *symmetric algebra* over the vector space t_C. It means concretely the following: if $\{x_i\}_{i \in I}$ is a basis of C, then $\mathrm{Sym}(t_C)$ is the algebra $\mathbb{C}[t_{x_i}]_{i\in I}$ of polynomials in the variables t_{x_i}.

The commutative algebra $\mathrm{Sym}(t_C)$ is a bialgebra with coproduct and counit given on the generators t_x (in terms of the Heyneman–Sweedler notation) by

$$\Delta(t_x) = \sum_{(x)} t_{x_{(1)}} \otimes t_{x_{(2)}} \quad \text{and} \quad \varepsilon(t_x) = \varepsilon(x). \qquad (x \in C) \tag{3.40}$$

In general, the bialgebra $\mathrm{Sym}(t_C)$ does not have an antipode: Indeed, if $x \in C$ is a group-like element, then by (3.18) we have $\Delta(t_x) = t_x \otimes t_x$ and $\varepsilon(t_x) = 1$. If there existed an antipode S, then it would follow from the previous equalities and from (3.15) that $S(t_x)t_x = 1$, hence $S(t_x) = 1/t_x$, which is not a polynomial. But this computation gives us hope that we may turn the bialgebra $\mathrm{Sym}(t_C)$ into a Hopf algebra by using rational algebraic fractions instead of mere polynomials. This can indeed be done thanks to the following fact.

Let us denote by $\mathrm{Frac\,Sym}(t_C)$ the field of fractions of $\mathrm{Sym}(t_C)$: If $\{x_i\}_{i Ł \in I}$ is a basis of C, then $\mathrm{Frac\,Sym}(t_C)$ is the algebra of rational algebraic fractions in the variables t_{x_i} $(i \in I)$. There exists a unique linear map $t^{-1} : C \to \mathrm{Frac\,Sym}(t_C)$ such that

$$\sum_{(x)} t^{-1}_{x_{(1)}} t_{x_{(2)}} = \varepsilon(x)1 = \sum_{(x)} t_{x_{(1)}} t^{-1}_{x_{(2)}}$$

for all $x \in C$ (for a proof, see [2, Lemma A.1]). Then the subalgebra of $\mathrm{Frac\,Sym}(t_C)$ generated by all elements t_x and t_x^{-1} $(x \in C)$ satisfies the requirements of Theorem 3.4 to be the free commutative Hopf algebra $\mathscr{S}_C$. This subalgebra is a Hopf algebra with coproduct and counit given by (3.40) and the additional formulas

$$\Delta(t_x^{-1}) = \sum_{(x)} t^{-1}_{x_{(2)}} \otimes t^{-1}_{x_{(1)}} \quad \text{and} \quad \varepsilon(t_x^{-1}) = \varepsilon(x). \qquad (x \in C)$$

The antipode is given on the generators t_x and t_x^{-1} by

$$S(t_x) = t_x^{-1} \quad \text{and} \quad S(t_x^{-1}) = t_x.$$

To check the universal property in Theorem 3.4, define the morphism $\widetilde{f} : \mathscr{S}_C \to H'$ by $\widetilde{f}(t_x) = f(x)$ and $\widetilde{f}(t_x^{-1}) = S'(f(x))$, where S' is the antipode of H'.

It follows by construction that $\mathscr{S}_C$, being a subalgebra of some field of rational functions, is a *domain*, i.e., an algebra without zero divisors.

In the sequel, we will apply Takeuchi's construction to the underlying coalgebra of an arbitrary Hopf algebra H, thus leading to the commutative algebra $\mathscr{S}_H$.

3.8.1.2 Pointed Hopf Algebras

A Hopf algebra is *pointed* if any simple subcoalgebra is one-dimensional. Group algebras, Taft algebras, enveloping algebras of Lie algebras, Drinfeld–Jimbo quantum enveloping algebras $U_q\,\mathfrak{g}$ and their quotients are examples of pointed Hopf algebras.

When H is a pointed Hopf algebra, then the free commutative Hopf algebra $\mathscr{S}_H$ over the coalgebra underlying H has a simple description in terms of the group $\mathrm{Gr}(H)$ of group-like elements introduced in Sect. 3.4.3, namely

$$\mathscr{S}_H = \mathrm{Sym}(t_H)\left[\frac{1}{t_g}\right]_{g\in \mathrm{Gr}(H)}. \tag{3.41}$$

Example 3.24 If $H = \mathbb{C}[G]$ is a group algebra, then $\mathrm{Sym}(t_H)$ is the polynomial algebra

$$\mathrm{Sym}(t_H) = \mathbb{C}[t_g]_{g\in G}.$$

Since H is pointed and $\mathrm{Gr}(H) = G \subset \mathbb{C}[G]$, then by (3.41) the free commutative Hopf algebra $\mathscr{S}_H$ is the algebra of *Laurent polynomials* on the symbols t_g ($g \in G$), or equivalently the algebra of the free abelian group $\mathbb{Z}^{(G)}$ generated by the symbols t_g:

$$\mathscr{S}_H = \mathbb{C}[t_g, t_g^{-1}]_{g\in G} = \mathbb{C}[\mathbb{Z}^{(G)}].$$

Example 3.25 Let G be a finite group and H be the function algebra $\mathscr{O}(G)$ (this Hopf algebra is not pointed when G is not abelian). Then $\mathrm{Sym}(t_H) = \mathbb{C}[t_g \mid g \in G]$ and

$$\mathscr{S}_H = \mathbb{C}[t_g]_{g\in G}\left[\frac{1}{\Theta_G}\right],$$

where $\Theta_G = \det(t_{gh^{-1}})_{g,h\in G}$ is *Dedekind's group determinant* (see [2, Appendix B]).

3.8.2 *The Generic Hopf Galois Extension Associated with a Hopf Algebra*

In this section, we associate with any Hopf algebra H a central H-Galois extension $\mathscr{B}_H \subset \mathscr{A}_H$, where the "base space" $\mathscr{B}_H$ is a nice commutative algebra whose size is related to the dimension of H. We can see $\mathscr{A}_H$ as a deformation of H over the parameter space $\mathscr{B}_H$.

3.8.2.1 The Algebra $\mathscr{B}_H$

Let H be a Hopf algebra. In order to construct the "base space" $\mathscr{B}_H$, we apply Takeuchi's theorem to the situation where C is the coalgebra underlying H and $H' = H_{\mathrm{ab}}$ is the largest commutative Hopf algebra quotient of H: it is the quotient of H by the ideal generated by all commutators $xy - yx$ ($x, y \in H$).

Let $\pi : H \to H_{\mathrm{ab}}$ be the canonical Hopf algebra surjection. Then by Theorem 3.4, for the free commutative Hopf algebra $\mathscr{S}_H$, there exists a unique morphism of Hopf algebras $\tilde{\pi} : \mathscr{S}_H \to H_{\mathrm{ab}}$ such that $\pi = \tilde{\pi} \circ t$. The Hopf algebra $\mathscr{S}_H$ becomes an H_{ab}-comodule algebra with coaction

$$\delta = (\mathrm{id} \otimes \tilde{\pi}) \circ \Delta. \tag{3.42}$$

On the generators of $\mathscr{S}_H$, the coaction is given by

$$\delta(t_x) = \sum_{(x)} t_{x_{(1)}} \otimes \tilde{\pi}(x_{(2)}) \quad \text{and} \quad \delta(t_x^{-1}) = \sum_{(x)} t_{x_{(2)}}^{-1} \otimes \tilde{\pi}\left(S(x_{(1)})\right).$$

Definition 3.9 The *algebra* $\mathscr{B}_H$ associated with a Hopf algebra H is the subalgebra of coinvariants of $\mathscr{S}_H$ for this coaction:

$$\mathscr{B}_H = \mathscr{S}_H^{\mathrm{co}-H_{\mathrm{ab}}} = \{a \in \mathscr{S}_H \mid \delta(a) = a \otimes 1\}.$$

We call $\mathscr{B}_H$ the *generic base algebra* of the Hopf algebra H. It has the following nice properties (see [35, Theorem 3.6 and Corollary 3.7] and [36, Proposition 3.4]).

Theorem 3.5 *Let H be a finite-dimensional Hopf algebra.*

(a) The algebra $\mathscr{B}_H$ is a finitely generated smooth Noetherian domain; its Krull dimension[7] *is equal to* $\dim H$.

(b) $\mathscr{S}_H$ is a finitely generated projective $\mathscr{B}_H$-module.

(c) If in addition H is pointed, then

$$\mathscr{B}_H = \mathbb{C}[u_1^{\pm 1}, \ldots, u_\ell^{\pm 1}, u_{\ell+1}, \ldots, u_n],$$

where $n = \dim H$ and $\ell = \mathrm{cardGr}(H)$ and where $u_1, \ldots, u_n$ are monomials in the generators t_x of $\mathrm{Sym}(t_H)$.

Example 3.26 If $H = \mathbb{C}[G]$ be a group algebra, then $H_{\mathrm{ab}} = \mathbb{C}[\Gamma]$, where $\Gamma = G/[G, G]$ is the maximal abelian quotient of G, i.e., the quotient by the normal subgroup generated by all elements of the form $ghg^{-1}h^{-1}$. Let $p : \mathbb{Z}^{(G)} \to \Gamma$ be the homomorphism sending each generator t_g to the image of g in Γ. Let Y_G be the kernel of p. Then by [3, Propositions 9 and 14],

$$\mathscr{B}_H = \mathbb{C}[Y_G].$$

When G is a finite group, then Y_G is a free abelian subgroup of $\mathbb{Z}^{(G)}$ of finite index (equal to the order of Γ). A basis of Y_G is given in [36, Lemma 4.7] (see also [27, Appendix A]).

Example 3.27 For a Hopf algebra H it may happen that $H_{\mathrm{ab}} = \mathbb{C}[\Gamma]$ is the algebra of an abelian group Γ, for instance when the commutative Hopf algebra H_{ab} is finite-dimensional and pointed (see [36, Lemma 2.1]). Then by Proposition 3.4 the algebra $\mathscr{S}_H$ is Γ-graded with $\mathscr{S}_H = \bigoplus_{\gamma \in \Gamma} \mathscr{S}_H(\gamma)$, where

$$\mathscr{S}_H(\gamma) = \{a \in \mathscr{S}_H \mid \delta(a) = a \otimes \gamma\},$$

and $\mathscr{B}_H = \mathscr{S}_H(0)$ is the component of $\mathscr{S}_H$ corresponding to the unit element $0 \in \Gamma$.

---

[7]The Krull dimension of $\mathscr{B}_H$ is the dimension of the algebraic variety V such that $\mathscr{B}_H = \mathscr{O}(V)$.

Example 3.28 Let G be a finite group and $H = \mathscr{O}(G)$. Since this Hopf algebra is commutative, we have $H_{\mathrm{ab}} = H$. Therefore the morphism of Hopf algebras $\widetilde{\pi} : \mathscr{S}_H \to H$ is split by the morphism of coalgebras $t : H \to \mathscr{S}_H$, i.e., $\widetilde{\pi} \circ t = \mathrm{id}_H$. The coaction (3.42) turns $\mathscr{S}_H$ into an $\mathscr{O}(G)$-comodule algebra. Thus by Proposition 3.5, $\mathscr{S}_H$ is a G-algebra. One checks that G acts on $\mathscr{S}_H = \mathbb{C}[t_g]_{g\in G}[1/\Theta_G]$ by $g \cdot t_h = t_{gh}$ $(g, h \in G)$ and that the square Θ_G^2 of the Dedekind group determinant is G-invariant. Therefore,

$$\mathscr{B}_H = \mathbb{C}[t_g]_{g\in G}^G \left[\frac{1}{\Theta_G^2}\right],$$

where $\mathbb{C}[t_g]_{g\in G}^G$ is the subalgebra of G-invariant polynomials.

The algebra $\mathscr{B}_H$ has also been completely described for the Sweedler algebra in [2] (see also [34]), for the Taft algebras and other natural generalizations of the Sweedler algebra in [27].

3.8.2.2 The Algebra $\mathscr{A}_H$

To construct what we call the *generic H-Galois extension* $\mathscr{A}_H$ we need the bilinear form $\sigma : H \times H \to \mathscr{S}_H$ with values in $\mathscr{S}_H$ defined by

$$\sigma(x, y) = \sum_{(x)(y)} t_{x_{(1)}} t_{y_{(1)}} t^{-1}_{x_{(2)}y_{(2)}}. \qquad (x, y \in H) \tag{3.43}$$

By [36, Proposition 3.4], the bilinear map σ actually takes values in the subalgebra $\mathscr{B}_H$ of $\mathscr{S}_H$. We can then equip the vector space $\mathscr{A}_H = \mathscr{B}_H \otimes H$ with the following product:

$$(b \otimes x) * (c \otimes y) = \sum_{(x)(y)} bc\, \sigma(x_{(1)}, y_{(1)})\, x_{(2)} y_{(2)} \tag{3.44}$$

$(b, c \in \mathscr{B}_H$ and $x, y \in H)$.

The following properties of $\mathscr{A}_H$ were established in [2, 35] (see also [34]).

Theorem 3.6 *Let H be a finite-dimensional Hopf algebra.*

(a) The product $*$ turns $\mathscr{A}_H$ into an associative unital algebra.

(b) The algebra $\mathscr{A}_H$ is a central H-Galois extension of $\mathscr{B}_H = \mathscr{B}_H \otimes 1$ with coaction $\delta = \mathrm{id}_{\mathscr{B}_H} \otimes \Delta$, where Δ is the coproduct of H. Moreover, $\mathscr{A}_H$ is free as a $\mathscr{B}_H$-module.

(c) Let $\chi_0 : \mathscr{B}_H \to \mathbb{C}$ be the character defined as the restriction to $\mathscr{B}_H$ of the counit of $\mathscr{S}_H$. Then there is an isomorphism of H-comodule algebras

$$\mathbb{C} \otimes_{\mathscr{B}_H} \mathscr{A}_H = \mathscr{A}_H / \ker(\chi_0)\, \mathscr{A}_H \cong H.$$

(d) For any character $\chi : \mathscr{B}_H \to \mathbb{C}$ of $\mathscr{B}_H$, the fiber of $\mathscr{A}_H$ at χ

$$\mathbb{C} \otimes_{\mathscr{B}_H} \mathscr{A}_H = \mathscr{A}_H / \ker(\chi)\, \mathscr{A}_H$$

is an H-Galois object.

This means that $\mathscr{B}_H \subset \mathscr{A}_H$ is a "non-commutative principal fiber bundle" with "fiber" H. We can also see $\mathscr{A}_H$ as a deformation of H over the parameter space $\mathscr{B}_H$ or, if one prefers, over the set $\mathrm{Alg}(\mathscr{B}_H, \mathbb{C})$ of characters of $\mathscr{B}_H$. By the last statement of the theorem, $\chi \mapsto \chi_*(\mathscr{A}_H)$ induces a map $\mathrm{Alg}(\mathscr{B}_H, \mathbb{C}) \to \mathrm{Gal}_H(\mathbb{C})$.

Exercise 3.27 Check that the product (3.44) is associative with unit $t_1^{-1} \otimes 1_H$.

3.8.3 *Multiparametric Deformations of $U_q\,\mathfrak{sl}(2)$ and of $\mathfrak{u}_d$*

We now illustrate the previous constructions on the cases where H is the quantum enveloping algebra $U_q = U_q\,\mathfrak{sl}(2)$ (defined in Sect. 3.5.3) and its finite-dimensional quotients $\mathfrak{u}_d$ (defined in Sect. 3.5.4). Both U_q and $\mathfrak{u}_d$ are pointed Hopf algebras. Theorems 3.7 and 3.8 below are new.

3.8.3.1 The Generic Base Algebra of U_q

The Hopf algebra U_q is infinite-dimensional with basis $\{E^i F^j K^\ell\}_{i,j \in \mathbb{N};\, \ell \in \mathbb{Z}}$. Its group $\mathrm{Gr}(U_q)$ of group-like elements consists of all powers (positive and negative) of K. Therefore, by (3.41) the free commutative Hopf algebra $\mathscr{S}_{U_q}$ is described by

$$\mathscr{S}_{U_q} = \mathbb{C}\,[t_{E^i F^j K^\ell}]_{i,j \in \mathbb{N};\, \ell \in \mathbb{Z}} \left[\frac{1}{t_{K^m}} \right]_{m \in \mathbb{Z}} .$$

The maximal commutative quotient Hopf algebra $(U_q)_{\mathrm{ab}}$ is generated by four generators $\overline{E}$, $\overline{F}$, $\overline{K}$, $\overline{K}^{-1}$ subject to the same relations as the corresponding generators in U_q in Sect. 3.5.3 plus the additional relations expressing that $(U_q)_{\mathrm{ab}}$ is commutative. We thus have

$$\overline{E}\,\overline{K} = \overline{K}\,\overline{E} = q^2\, \overline{E}\,\overline{K},$$

which implies $\overline{E} = 0$ in $(U_q)_{\mathrm{ab}}$ since $q^2 \neq 1$ and $\overline{K}$ is invertible. Similarly, $\overline{F} = 0$. Finally the relation

$$\overline{K} - \overline{K}^{-1} = (q - q^{-1})\, \left(\overline{E}\,\overline{F} - \overline{F}\,\overline{E}\right) = 0$$

shows that $\overline{K} = \overline{K}^{-1}$, hence $\overline{K}^2 = 1$ in $(U_q)_{\mathrm{ab}}$. Therefore

$$(U_q)_{\mathrm{ab}} = \mathbb{C}[\overline{K}]/(\overline{K}^2 - 1) \cong \mathbb{C}[\mathbb{Z}/2],$$

which is the algebra of the group $\mathbb{Z}/2$.

As noted in Example 3.27, the isomorphism $(U_q)_{\mathrm{ab}} \cong \mathbb{C}[\mathbb{Z}/2]$ implies that $\mathscr{S}_{U_q}$ is a superalgebra: $\mathscr{S}_{U_q} = \mathscr{S}_{U_q}(0) \bigoplus \mathscr{S}_{U_q}(1)$, and that the generic base algebra $\mathscr{B}_{U_q}$ coincides with the 0-degree component:

$$\mathscr{B}_{U_q} = \mathscr{S}_{U_q}(0).$$

On the generators t_E, t_F, t_K, the coproduct of $\mathscr{S}_{U_q}$ is given by

$$\Delta(t_E) = t_1 \otimes t_E + t_E \otimes t_K, \quad \Delta(t_F) = t_{K^{-1}} \otimes t_F + t_F \otimes t_1, \quad \Delta(t_K) = t_K \otimes t_K.$$

Since $\widetilde{\pi}(t_E) = \overline{E} = 0$, $\widetilde{\pi}(t_F) = \overline{F} = 0$, and $\widetilde{\pi}(t_K) = \overline{K}$, the coaction δ of $(U_q)_{\mathrm{ab}}$ on $\mathscr{S}_{U_q}$ satisfies

$$\delta(t_E) = t_E \otimes \overline{K}, \quad \Delta(t_F) = t_F \otimes 1, \quad \Delta(t_K) = t_K \otimes \overline{K}.$$

Therefore, t_F is an even element, i.e., it belongs to $\mathscr{S}_{U_q}(0) = \mathscr{B}_{U_q}$ while t_E and t_K are both odd, that is belong to $\mathscr{S}_{U_q}(1)$. It can be proved more generally that $t_{E^i F^j K^\ell}$ belongs to $\mathscr{B}_{U_q}$ if and only if $i + \ell$ is even, and that $t_{K^m}^{-1}$ belongs to $\mathscr{B}_{U_q}$ if and only m is even.

Exercise 3.28 Set $u_{E^i F^j K^\ell} = t_{E^i F^j K^\ell}$ if $i + \ell$ is even, and $u_{E^i F^j K^\ell} = t_{E^i F^j K^\ell} t_K^{-1}$ if $i + \ell$ is odd. Show that

$$\mathscr{B}_{U_q} = \mathbb{C}\,[u_{E^i F^j K^\ell}]_{i,j \in \mathbb{N};\, \ell \in \mathbb{Z}} \left[\frac{1}{u_{K^m}}\right]_{m \in \mathbb{Z}}.$$

3.8.3.2 The Algebra $\mathscr{A}_{U_q}$

We have the following result.

Theorem 3.7 *The generic U_q-Galois extension $\mathscr{A}_{U_q}$ is the $\mathscr{B}_H$-algebra generated by E, F, K, K^{-1} subject to the relations*

$$K * K^{-1} = K^{-1} * K = \frac{t_K t_{K^{-1}}}{t_1},$$

$$K * E = q^2 E * K + (1 - q^2) \frac{t_E}{t_K} K * K,$$

$$K * F = q^{-2} F * K + (1 - q^{-2})\, t_F\, K,$$

$$E * F - F * E = t_1 \frac{(t_{K^{-1}}/t_K)\, K - K^{-1}}{q - q^{-1}} + (q^{-2} - 1)\left(\frac{t_E}{t_K} F * K - \frac{t_E t_F}{t_K} K\right).$$

The algebra $\mathscr{A}_{U_q}$ is an U_q-comodule algebra with coaction given by the same formulas as for the coproduct of U_q. The algebra depends continuously on the parameters t_E, t_F which can take any complex values and on the parameters t_1, t_K, $t_{K^{-1}}$ which can take any *nonzero* complex values. Note that all monomials in the t-variables occurring in the previous relations belong to $\mathscr{B}_{U_q}$ (they are all of degree 0 in the superalgebra $\mathscr{S}_{U_q}$).

If we specialize the parameters t_1, t_K, $t_{K^{-1}}$ to 1 and the parameters t_E, t_F to 0, we recover the defining relations of U_q and $\mathscr{A}_{U_q}$ becomes U_q. In other words, $\mathscr{A}_{U_q}$ is a 5-*parameter deformation* of U_q as a non-commutative principal bundle.

Proof We use an observation made in [2, Sect. 6]: in order to find relations between elements $1 \otimes x$ in $\mathscr{A}_H$, where x is an arbitrary element of a Hopf algebra H, it is enough to find the relations between the following elements of the tensor product algebra $\mathscr{B}_H \otimes H$:

$$X_x = \sum_{(x)} t_{x_{(1)}} \otimes x_{(2)}.$$

It follows from the formula for the coproduct of U_q (see Sect. 3.5.3) that we have

$$X_1 = t_1\, 1, \qquad X_K = t_K\, K, \qquad X_{K^{-1}} = t_{K^{-1}}\, K^{-1},$$

$$X_E = t_1\, E + t_E\, K, \qquad X_F = t_{K^{-1}}\, F + t_F\, 1.$$

(Here we dropped the tensor product signs since we may consider the commutative algebra $\mathscr{B}_H$ as an extended algebra of scalars.)

To prove the relations between K and K^{-1}, it suffices to compute $X_K X_{K^{-1}}$ and $X_{K^{-1}} X_K$. We have

$$X_K X_{K^{-1}} = t_K t_{K^{-1}}\, K K^{-1} = t_K t_{K^{-1}} = \frac{t_K t_{K^{-1}}}{t_1} X_1,$$

which is also equal to $X_{K^{-1}} X_K$; this implies the desired formulas for $K * K^{-1}$ and $K^{-1} * K$.

For the relation between K and E in $\mathscr{A}_H$, it is enough to compute the following:

$$\begin{aligned} X_K X_E - q^2 X_E X_K &= t_K t_1\, KE + t_K t_E\, K^2 - q^2\, t_1 t_K\, EK - q^2\, t_E t_K\, K^2 \\ &= t_1 t_K\, (KE - q^2\, EK) + (1 - q^2)\, t_E t_K\, K^2 \\ &= (1 - q^2)\, t_E t_K\, K^2. \end{aligned}$$

Now, $(X_K)^2 = t_K^2\, K^2$. Therefore,

$$X_K X_E - q^2\, X_E X_K = (1-q^2)\, t_E t_K / t_K^2\, (X_K)^2 = (1-q^2)\, t_E / t_K\, (X_K)^2 .$$

We leave the computation of the relation between K and F in $\mathscr{A}_H$ as an exercise to the reader. For the commutator of E and F in $\mathscr{A}_H$, we have

$$\begin{aligned} X_E X_F - X_F X_E &= (t_1\, E + t_E\, K)(t_{K^{-1}}\, F + t_F\, 1) - (t_{K^{-1}}\, F + t_F\, 1)(t_1\, E + t_E\, K) \\ &= t_1 t_{K^{-1}}\, (EF - FE) + (q^{-2} - 1)\, t_E t_{K^{-1}} FK \\ &= \frac{1}{q - q^{-1}}\, t_1 t_{K^{-1}}\, (K - K^{-1}) + (q^{-2} - 1)\, t_E t_{K^{-1}} FK \\ &= \frac{1}{q - q^{-1}}\, t_1 \left(\frac{t_{K^{-1}}}{t_K}\, X_K - X_{K^{-1}} \right) + (q^{-2} - 1)\, t_E t_{K^{-1}} FK . \end{aligned}$$

It remains to compute FK in terms of the X-variables. We have

$$X_F X_K = t_K t_{K^{-1}}\, FK + t_F t_K\, K = t_K t_{K^{-1}}\, FK + t_F\, X_K ,$$

so that

$$t_E t_{K^{-1}} FK = \frac{t_E}{t_K}\, X_F X_K - \frac{t_E t_F}{t_K}\, X_K .$$

Combining these equalities, we obtain a formula for $X_E X_F - X_F X_E$ in terms of the X-variables, hence the desired formula for $E * F - F * E$. □

3.8.3.3 A Deformation of $\mathfrak{u}_d$

Let q be a root of unity of order $d \geq 3$. Consider the finite-dimensional Hopf algebra $\mathfrak{u}_d$ defined in Sect. 3.5.4. We know that it has a basis consisting of the e^3 elements $E^i F^j K^\ell$, where $1 \leq i, j, \ell \leq e - 1$. Recall that $e = d/2$ if d is even and $e = d$ if d is odd. The group $\mathrm{Gr}(\mathfrak{u}_d)$ consists of the e elements $1, K, K^2, \ldots, K^{e-1}$; it is a cyclic group of order e.

By (3.41), the free commutative Hopf algebra $\mathscr{S}_{\mathfrak{u}_d}$ is given by

$$\mathscr{S}_{\mathfrak{u}_d} = \mathbb{C}\, [t_{E^i F^j K^\ell}]_{0 \leq i,j,\ell \leq e-1} \left[\frac{1}{t_{K^m}} \right]_{0 \leq m \leq e-1} .$$

The maximal commutative quotient Hopf algebra $(\mathfrak{u}_d)_{\mathrm{ab}}$ is the quotient of $(U_q)_{\mathrm{ab}}$ by the additional relation $\overline{K}^e = 1$. Since $\overline{K}^2 = 1$, we conclude that

$$(\mathfrak{u}_d)_{\mathrm{ab}} = \begin{cases} \mathbb{C} & \text{if } e \text{ is odd,} \\ (U_q)_{\mathrm{ab}} \cong \mathbb{C}[\mathbb{Z}/2] & \text{if } e \text{ is even.} \end{cases}$$

Therefore, if e is odd, then $\mathscr{S}_{\mathfrak{u}_d}$ is trivially graded, which implies $\mathscr{B}_{\mathfrak{u}_d} = \mathscr{S}_{\mathfrak{u}_d}$. If e is even, then $\mathscr{S}_{\mathfrak{u}_d}$ is a superalgebra and the generic base algebra is $\mathscr{B}_{\mathfrak{u}_d}$ is its even part (see Exercise 3.29 below for a complete description).

Theorem 3.8 *The algebra $\mathscr{A}_{\mathfrak{u}_d}$ is the quotient of $\mathscr{A}_{U_q}$ by the two-sided ideal generated by the relations*

$$K^{*e} - \frac{t_K^e}{t_1} = 0, \qquad \left(E - \frac{t_E}{t_K} K\right)^{*e} = 0, \qquad \left(F - \frac{t_F}{t_1}\right)^{*e} = 0.$$

If we set $t_1 = t_K = t_{K^{-1}} = 1$ and $t_E = t_F = 0$ in the defining relations of $\mathscr{A}_{\mathfrak{u}_d}$ (see Theorems 3.7 and 3.8), we recover those of $\mathfrak{u}_d$.

Proof We proceed as in the proof of Theorem 3.7 by checking the relations between the corresponding X-variables in $\mathscr{B}_{\mathfrak{u}_d} \otimes \mathfrak{u}_d$. We have

$$(X_K)^e - \frac{t_K^e}{t_1} X_1 = t_K^e\, K^e - t_K^e = 0$$

since $K^e = 1$ in $\mathfrak{u}_d$. Next, in view of $E^e = F^e = 0$ in $\mathfrak{u}_d$, we have

$$\left(X_E - \frac{t_E}{t_K}\, X_K\right)^{*e} = t_1^e\, E^e = 0 \quad \text{and} \quad \left(X_F - \frac{t_F}{t_1}\, X_1\right)^{*e} = t_{K^{-1}}^e\, F^e = 0.$$

This completes the proof. □

Let us determine the "parameter space" $\mathrm{Alg}(\mathscr{B}_{\mathfrak{u}_d}, \mathbb{C})$ when e is odd. In this case, $\mathscr{B}_{\mathfrak{u}_d} = \mathscr{S}_{\mathfrak{u}_d}$. Since $\mathscr{S}_{\mathfrak{u}_d} = \mathbb{C}\,[t_{E^iF^jK^\ell}]_{0\le i,j,\ell\le e-1}\,[1/t_{K^m}]_{0\le m\le e-1}$, a character of $\mathscr{B}_{\mathfrak{u}_d}$ is completely determined by its values on the generators $t_{E^iF^jK^\ell}$; each of these generators can take any complex value, except in the case $(i, j) = (0, 0)$, where the corresponding value has to be nonzero. It follows that

$$\mathrm{Alg}(\mathscr{B}_{\mathfrak{u}_d}, \mathbb{C}) \cong \mathbb{C}^{e(e^2-1)} \times (\mathbb{C}^\times)^e,$$

which is an open Zarisky subset of the affine space of dimension e^3.

Exercise 3.29 Assume e is even (equivalently, d is divisible by 4). Define $u_{E^iF^jK^\ell}$ as in Exercise 3.28. Show that

$$\mathscr{B}_{\mathfrak{u}_d} = \mathbb{C}\,[u_{E^iF^jK^\ell}]_{0\le i,j,\ell\le e-1}\,[1/u_{K^m}]_{0\le m\le e-1}.$$

Hence, $\mathrm{Alg}(\mathscr{B}_{\mathfrak{u}_d}, \mathbb{C}) \cong \mathbb{C}^{e(e^2-1)} \times (\mathbb{C}^\times)^e$ holds in this case too.

Acknowledgements I thank the organizers of the Summer school "Geometric, topological and algebraic methods for quantum field theory" held at Villa de Leyva, Colombia, in July 2015 for inviting me to give the course that led to these notes. I am also grateful to the students for their feedback and to Sylvie Paycha for her careful reading of these notes and her suggestions. Finally let me mention the travel support I received from the Institut de Recherche Mathématique Avancée in Strasbourg.

References

1. E. Abe, *Hopf Algebras*, Cambridge Tracts in Mathematics, 74 (Cambridge University Press, Cambridge, 1980)
2. E. Aljadeff, C. Kassel, Polynomial identities and noncommutative versal torsors. Adv. Math. **218**, 1453–1495 (2008)
3. E. Aljadeff, D. Haile, M. Natapov, Graded identities of matrix algebras and the universal graded algebra. Trans. Am. Math. Soc. **362**, 3125–3147 (2010)
4. T. Aubriot, Classification des objets galoisiens de $U_q(\mathfrak{g})$ à homotopie près. Commun. Algebra **35**, 3919–3936 (2007)
5. J. Bichon, Galois and bigalois objects over monomial non-semisimple Hopf algebras. J. Algebra Appl. **5**, 653–680 (2006)
6. J. Bichon, Hopf-Galois objects and cogroupoids. Rev. Un. Mat. Argentina **5**, 11–69 (2014)
7. R.J. Blattner, S. Montgomery, A duality theorem for Hopf module algebras. J. Algebra **95**, 153–172 (1985)
8. K.S. Brown, *Cohomology of Groups*, Graduate Texts in Mathematics 87 (Springer, New York, 1982)
9. T. Brzeziński, S. Majid, Quantum group gauge theory on quantum spaces. Commun. Math. Phys. **157**, 591–638 (1993). Erratum: **167**, 235 (1995)
10. S. Caenepeel, *Brauer Groups, Hopf Algebras and Galois Theory* (Kluwer Academic Publishers, Dordrecht, 1998)
11. V. Chari, A. Pressley, *A Guide to Quantum Groups* (Cambridge University Press, Cambridge, 1994)
12. S.U. Chase, M.E. Sweedler, *Hopf Algebras and Galois Theory*, vol. 97, Lecture Notes in Mathematics (Springer, Berlin, 1969)
13. A. Connes, *Noncommutative Geometry* (Academic Press Inc, San Diego, 1994)
14. L. Dąbrowski, H. Grosse, P.M. Hajac, Strong connections and Chern-Connes pairing in the Hopf-Galois theory. Commun. Math. Phys. **220**, 301–331 (2001)
15. J. Dixmier, *Enveloping Algebras* (North-Holland Publ. Co, Amsterdam, 1977)
16. Y. Doi, M. Takeuchi, Quaternion algebras and Hopf crossed products. Commun. Algebra **23**, 3291–3325 (1995)
17. V.G. Drinfeld, Hopf algebras and the quantum Yang-Baxter equation. Dokl. Akad. Nauk SSSR **283**, 1060–1064 (1985). English translation: Soviet Math. Dokl. **32**, 254–258 (1985)
18. V.G. Drinfeld, Quantum groups, in *Proceedings of the International Congress of Mathematicians*, Vol. 1, 2 (Berkeley, Calif., 1986), pp. 798–820. Amer. Math. Soc., Providence, RI (1987)
19. M. Đurđević, Quantum principal bundles as Hopf-Galois extensions (1995), arXiv:q-alg/9507022
20. S. Eilenberg, J.C. Moore, Homology and fibrations. I. Coalgebras, cotensor product and its derived functors. Comment. Math. Helv. **40**, 199–236 (1966)
21. G. García Márquez, *Vivir para contarla* (Editorial Diana, México, 2002)
22. P. Guillot, C. Kassel, A. Masuoka, Twisting algebras using non-commutative torsors: explicit computations. Math. Z. **271**, 789–818 (2012)
23. S. Gutt, Deformation quantization and group actions, this volume

24. P.M. Hajac, Strong connections on quantum principal bundles. Commun. Math. Phys. **182**, 579–617 (1996)
25. P.M. Hajac, S. Majid, Projective module description of the q-monopole. Commun. Math. Phys. **206**, 247–264 (1999)
26. D. Husemoller, *Fibre Bundles*, 3rd edn., Graduate Texts in Mathematics 20 (Springer, New York, 1994)
27. U.N. Iyer, C. Kassel, Generic base algebras and universal comodule algebras for some finite-dimensional Hopf algebras Trans. Am. Math. Soc. **367**, 8465–8486 (2015)
28. N. Jacobson, *Lie Algebras* (Dover Publications, New York, 1979)
29. J.C. Jantzen, *Lectures on Quantum Groups*, Graduate Studies in Mathematics 6 (American Mathematical Society, Providence, 1996)
30. M. Jimbo, A q-difference analogue of $U(\mathfrak{g})$ and the Yang-Baxter equation. Lett. Math. Phys. **10**, 63–69 (1985)
31. C. Kassel, *Quantum Groups*, Graduate Texts in Mathematics 155 (Springer, New York, 1995)
32. C. Kassel, Quantum principal bundles up to homotopy equivalence, in *The Legacy of Niels Henrik Abel, The Abel Bicentennial, Oslo, 2002*, ed. by O.A. Laudal, R. Piene (Springer, New York, 2004), pp. 737–748, arXiv:math.QA/0507221
33. C. Kassel, Hopf algebras and polynomial identities, in *Quantum Groups and Quantum Topology*, RIMS Kôkyûroku 1714 (2010), pp. 49–62, arXiv:1009.3180
34. C. Kassel, Generic Hopf Galois extensions, in *Quantum Groups and Noncommutative Spaces*, Aspects Math. E41 (Vieweg + Teubner, Wiesbaden, 2011), pp. 104–120, arXiv:0809.0638
35. C. Kassel, A. Masuoka, Flatness and freeness properties of the generic Hopf Galois extensions. Rev. Un. Mat. Argentina **51**, 79–94 (2010), arXiv:0911.3719
36. C. Kassel, A. Masuoka, The Noether problem for Hopf algebras. J. Noncommut. Geom. (2016), arXiv:1404.4941
37. C. Kassel, H.-J. Schneider, Homotopy theory of Hopf Galois extensions. Ann. Inst. Fourier (Grenoble) **55**, 2521–2550 (2005)
38. C. Kassel, M. Rosso, V. Turaev, *Quantum Groups and Knot Invariants*, Panoramas et Synthèses 5 (Société Mathématique de France, Paris, 1997)
39. P.P. Kulish, N.Y. Reshetikhin, Quantum linear problem for the sine-Gordon equation and higher representations. Zap. Nauchn. Sem. Leningrad. Otdel. Mat. Inst. Steklov. (LOMI) **101**, 101–110 (1981). English translation: J. Soviet Math. **23**, 2435–2441 (1983)
40. G. Landi, W. van Suijlekom, Principal fibrations from noncommutative spheres. Commun. Math. Phys. **260**, 203–225 (2005)
41. G. Landi, C. Pagani, C. Reina, A Hopf bundle over a quantum four-sphere from the symplectic group. Commun. Math. Phys. **263**, 65–88 (2006)
42. G. Lusztig, *Introduction to Quantum Groups*, Progress in Mathematics 110 (Birkhäuser Boston, Inc., Boston, 1993)
43. S. Majid, *Foundations of Quantum Group Theory* (Cambridge University Press, Cambridge, 1995)
44. A. Masuoka, Cleft extensions for a Hopf algebra generated by a nearly primitive element. Commun. Algebra **22**, 4537–4559 (1994)
45. A. Masuoka, Abelian and non-abelian second cohomologies of quantized enveloping algebras. J. Algebra **320**, 1–47 (2008)
46. S. Montgomery, *Hopf Algebras and Their Actions on Rings*, vol. 82, CBMS Regional Conference Series in Mathematics (American Mathematical Society, Providence, 1993)
47. A. Nenciu, Cleft extensions for a class of pointed Hopf algebras constructed by Ore extensions. Commun. Algebra **29**, 1959–1981 (2001)
48. F. Panaite, F. Van Oystaeyen, Clifford-type algebras as cleft extensions for some pointed Hopf algebras. Commun. Algebra **28**, 585–600 (2000)
49. M.J. Pflaum, Quantum groups on fibre bundles. Commun. Math. Phys. **166**, 279–315 (1994)
50. P. Podleś, Quantum spheres. Lett. Math. Phys. **14**, 193–202 (1987)
51. N.Y. Reshetikhin, L.A. Takhtadzhyan, L.D. Faddeev, Quantization of Lie groups and Lie algebras. Algebra i Analiz **1**, 178–206 (1989). English translation: Leningrad Math. J. **1**, 193–225 (1990)

52. D. Rumynin, Hopf-Galois extensions with central invariants and their geometric properties. Algebr. Represent. Theory **1**, 353–381 (1998)
53. H.-J. Schneider, Principal homogeneous spaces for arbitrary Hopf algebras. Israel J. Math. **72**, 167–195 (1990)
54. J.-P. Serre, *Lie Algebras and Lie Groups* (W. A. Benjamin Inc., New York, 1965)
55. A. Sitarz, Pointless geometry, in *Mathematical Structures of the Universe* (Copernicus Center Press, 2014)
56. E.K. Sklyanin, On an algebra generated by quadratic relations. Uspekhi Mat. Nauk **40**, 214 (1985). (in Russian)
57. N. Steenrod, *The Topology of Fibre Bundles*, vol. 14, Princeton Mathematical Series (Princeton University Press, Princeton, 1951)
58. M.E. Sweedler, *Hopf Algebras* (W. A. Benjamin Inc., New York, 1969)
59. M. Takeuchi, Free Hopf algebras generated by coalgebras. J. Math. Soc. Japan **23**, 561–582 (1971)
60. M. Takeuchi, Some topics on $GL_q(n)$. J. Algebra **147**, 379–410 (1992)
61. S.L. Woronowicz, Compact matrix pseudogroups. Commun. Math. Phys. **111**, 613–665 (1987)
62. S.L. Woronowicz, Differential calculus on compact matrix pseudogroups (quantum groups). Commun. Math. Phys. **122**, 125–170 (1989)

Chapter 4
An Introduction to Nichols Algebras

Nicolás Andruskiewitsch

Abstract Nichols algebras, Hopf algebras in braided categories with distinguished properties, were discovered several times. They appeared for the first time in the thesis of W. Nichols [72], aimed to construct new examples of Hopf algebras. In this same paper, the small quantum group $u_q(sl_3)$, with q a primitive cubic root of one, was introduced. Independently they arose in the paper [84] by Woronowicz as the invariant part of his non-commutative differential calculus. Later there were two unrelated attempts to characterize abstractly the positive part $U_q^+(\mathfrak{g})$ of the quantized enveloping algebra of a simple finite-dimensional Lie algebra $\mathfrak{g}$ at a generic parameter q. First, Lusztig showed in [64] that $U_q^+(\mathfrak{g})$ can be defined through the radical of a suitable invariant bilinear form. Second, Rosso interpreted $U_q^+(\mathfrak{g})$ in [74, 75] via quantum shuffles. These two viewpoints were conciliated later, as alternative definitions of the same notion of Nichols algebra. Other early appearances of Nichols algebras are in [65, 77]. As observed in [17, 18], Nichols algebras are basic invariants of pointed Hopf algebras, their study being crucial in the classification program of Hopf algebras; see also [10]. More recently, they are the subject of an intriguing proposal in Conformal Field Theory [79]. This is an introduction from scratch to the notion of Nichols algebra. I was invited to give a mini-course of two lessons, 90 min each, at the Geometric, Algebraic and Topological Methods for Quantum Field Theory, Villa de Leyva, Colombia, in July 2015. The theme was Nichols algebras that requires several preliminaries and some experience to be appreciated; a selection of the ideas to be presented was necessary. These notes intend to preserve the spirit of the course, discussing some motivational background material in Sect. 4.1, then dealing with braided vector spaces and braided tensor categories in Sect. 4.2, arriving at last to the definition and main calculation tools of Nichols algebras in Sect. 4.3. I hope that the various examples and exercises scattered through the text would serve the reader to absorb the beautiful concept of Nichols algebra and its many facets. Section 4.4 is a survey of the main examples of, and results on, Nichols algebras that I am aware of; here the pace is faster and the precise formulation of some statements

N. Andruskiewitsch (✉)
FaMAF-CIEM (CONICET), Universidad Nacional de Córdoba, Medina Allende s/n, Ciudad Universitaria, 5000 Córdoba, República Argentina
e-mail: andrus@famaf.unc.edu.ar

A. Cardona et al. (eds.), *Quantization, Geometry and Noncommutative Structures in Mathematics and Physics*, Mathematical Physics Studies,
DOI 10.1007/978-3-319-65427-0_4

is referred to the literature. I apologize in advance for any possible omission. This section has intersection with, and is an update of, the surveys [1, 2, 19], to which I refer for further information.

4.1 Preliminaries

4.1.1 Conventions

We assume the conventions $\mathbb{N} = \{1, 2, 3, \dots\}$, $\mathbb{N}_0 = \mathbb{N} \cup \{0\}$, $\mathbb{N}_{\geq 2} = \mathbb{N} - \{1\}$, etc. If $k < \theta \in \mathbb{N}_0$, then we denote $\mathbb{I}_{k,\theta} = \{n \in \mathbb{N}_0 : k \leq n \leq \theta\}$ and $\mathbb{I}_\theta = \mathbb{I}_{1,\theta}$.

If $N \in \mathbb{N}$, then $\mathbb{G}_N$ denotes the group of N-roots of unity in $\Bbbk$, while $\mathbb{G}'_N$ is the subset of primitive roots of order N. Also $\mathbb{G}_\infty = \bigcup_{N\in\mathbb{N}} \mathbb{G}_N$, $\mathbb{G}'_\infty = \mathbb{G}_\infty - \{1\}$.

If V is a vector space, then $V^* := \hom_{\Bbbk}(V, \Bbbk)$ and $\langle\, ,\, \rangle : V^* \times V \to \Bbbk$ is the evaluation.

The finite field with q elements is denoted $\mathbb{F}_q$.

We abbreviate $W \leq V$ for W *is a subobject of* V, where subobject means submodule, subgroup, subspace, subrack, according to the context.

4.1.2 Groups

We fix a field $\Bbbk$; later we shall assume that $\Bbbk$ is algebraically closed and has characteristic 0. We expect that the reader is familiar with the notions of group, module and representation; we use indistinctly the languages of modules and representations. As customary, we denote by $GL(V)$ the group of bijective linear transformations of a vector space V onto itself. We remind some basic definitions:

- A module is *simple* if it has exactly two submodules, 0 and itself (thus, it is different from 0). In the representation-theoretic language, one says *irreducible* instead of simple.
- A module is *semisimple* if it is a direct sum of simple submodules. In the representation theory, *completely reducible* is the translation of semisimple.

Let G be a group. We denote by $\Bbbk G$ the group algebra of G, with the canonical basis $(e_g)_{g\in G}$. Thus, there is a bijective correspondence between representations of G and of $\Bbbk G$. We observe that $\Bbbk G$ can be identified with (a subspace of) the linear dual of the vector space of functions from G to $\Bbbk$, where $e_g(f) = f(g)$, for $f : G \to \Bbbk$ and $g \in G$.

We denote by $\operatorname{Irr} G$ the set of classes of simple G-modules, up to isomorphism. For instance, $\varepsilon \in \operatorname{Irr} G$ is the class of the trivial representation, the one-dimensional vector space where every $g \in G$ acts by 1. If $\xi \in \operatorname{Irr} G$ and V is a G-module, then

$$V_\xi := \sum_{W \le V : W \in \xi} W$$

is the isotypical component of V of type ξ. Particularly, $V^G := V_\varepsilon$, the isotypical component of trivial type, is the submodule of G-invariants of V.

Example 4.1 Let U and V be G-modules. Then $\operatorname{Hom}(V, W)$ is a G-module with the action $g \cdot T = gTg^{-1}$ and $\operatorname{Hom}_G(V, W) = \operatorname{Hom}(V, W)^G$.

Theorem 4.1 (Maschke) *Let G be a finite group. Then the following are equivalent:*

(1) The characteristic of $\Bbbk$ does not divide $|G|$.
(2) Every finite-dimensional representation of G is completely reducible.

Assume that (1) holds. Let V be a finite-dimensional G-module. The action of

$$\int_G = \frac{1}{|G|} \sum_{g \in G} e_g \in \Bbbk G \tag{4.1}$$

on V is a G-morphism and a projector $V \to V^G$. (If $\Bbbk = \mathbb{C}$, then $\int_G$ is a normalized Haar measure on the discrete group G). To prove (2), it is enough to show, arguing recursively, that any $W \le V$ admits a complement U that is also a G-submodule. So, consider $p \in \operatorname{Hom}(V, U)$ a projector onto U; then $q := \int_G \cdot p \in \operatorname{Hom}_G(V, U)$ is a projector onto U and $\ker q$ is the desired complement.

To prove (1), it is enough to assume that the representation of G on $\Bbbk G$ by left multiplication is completely reducible. Then the kernel of the projection $p : \Bbbk G \to G$, $e_g \mapsto 1$ for all $g \in G$, admits a complement U that is also a G-submodule. It turns out that U has to be the span of $x = \sum_{g \in G} e_g$; since $p(x) = |G|$, this could not be 0.

Remark 4.1 There is a natural notion of integral in finite-dimensional Hopf algebras that permits a generalization of the classical Maschke Theorem. This can be extended further to Hopf algebras with arbitrary dimension, but the complete reducibility in question is of comodules. See, e.g., [78] for details.

Let X be a set. We denote by $\mathbb{S}_X$ the group of bijections from X onto itself, with multiplication being the composition. In particular, $\mathbb{S}_n$ is the symmetric group on n letters, i.e., $\mathbb{S}_n = \mathbb{S}_{\mathbb{I}_n}$, where $\mathbb{I}_n := \{1, \dots, n\}$. Let τ_i be the the transposition $(i\ i+1)$. Then $\mathbb{S}_n$ is generated by the τ_i, with $i \in \mathbb{I}_{n-1}$, subject to the defining relations

$$\tau_i^2 = e, \qquad i \in \mathbb{I}_{n-1}, \tag{4.2}$$

$$\tau_i\tau_j = \tau_j\tau_i, \qquad |i-j| \ge 2, \tag{4.3}$$

$$\tau_i\tau_j\tau_i = \tau_j\tau_i\tau_j, \qquad |i-j| = 1. \tag{4.4}$$

The group $\mathbb{S}_n$ together with $S = \{\tau_i,\ i \in \mathbb{I}_{n-1}\}$ is a Coxeter group. In particular, there is a length function $\ell : \mathbb{S}_n \to \mathbb{N}_0$, measuring the minimum of the possible expressions of an element as product of τ_i's. Thus, we have the sign representation $\operatorname{sgn}: \mathbb{S}_n \to \Bbbk^\times$, $w \mapsto (-1)^{\ell(w)}$, $w \in \mathbb{S}_n$.

E. Artin introduced in 1926 the braid group $\mathbb{B}_n$ that has important applications in various areas and plays a central role in our story. Concretely, $\mathbb{B}_n$ is the group generated by σ_i, $i \in \mathbb{I}_{n-1}$, with defining relations (4.3) and (4.4) (with σ instead of τ). By definition, there is a surjective group homomorphism $\pi : \mathbb{B}_n \to \mathbb{S}_n$, $\sigma_i \mapsto \tau_i$; it admits a set-theoretical section $M : \mathbb{S}_n \to \mathbb{B}_n$ (i. e., not a group homomorphism), sometimes called the Matsumoto section, determined by

$$\begin{aligned} M(\tau_i) &= \sigma_i, && i \in \mathbb{I}_{n-1}, \\ M(uw) &= M(u)M(w), && \text{if } \ell(uw) = \ell(u) + \ell(w). \end{aligned} \tag{4.5}$$

4.1.3 The Tensor Algebra

We denote by $\tau : V \otimes W \to W \otimes V$ the usual flip $v \otimes w \mapsto w \otimes v$ between the tensor products of vector spaces V and W.

We expect that the reader is familiar with the notions of associative, commutative and Lie algebra. The ideal, respectively the subalgebra, of an algebra A generated by a subset S is denoted by $\langle S \rangle$, respectively by $\Bbbk\langle S \rangle$. A graded vector space is a vector space with a fixed grading $V = \oplus_{n \in \mathbb{N}_0} V^n$; it is locally finite if $\dim V^n < \infty$ for all $n \in \mathbb{N}_0$. In such case, its Hilbert–Poincaré series is

$$\mathscr{H}_V = \sum_{n \in \mathbb{N}_0} \dim V^n t^n \in \mathbb{Z}[[t]].$$

The *graded dual* of a locally finite graded vector space $V = \oplus_{n \in \mathbb{N}_0} V^n$ is

$$V^\star = \oplus_{n \in \mathbb{N}_0} V^{\star n}, \qquad V^{\star n} = \hom_\Bbbk(V^n, \Bbbk). \tag{4.6}$$

A graded algebra is a graded vector space $A = \oplus_{n \in \mathbb{N}_0} A^n$ with an algebra structure such that $A^n A^m \subset A^{n+m}$.

We also assume that the reader knows the basics of the theory of categories. Let $\mathrm{Vec}_\Bbbk$, $\mathrm{Assoc}_\Bbbk$, $\mathrm{Comm}_\Bbbk$, $\mathrm{Lie}_\Bbbk$, be the categories of vector spaces, associative algebras, associative and commutative algebras, Lie algebras, over $\Bbbk$, respectively.

Let V be a vector space. As customary, we set $T^0(V) = \Bbbk$, $T^{n+1}(V) = V \otimes T^n(V)$, $n \geq 0$, and $T(V) = \oplus_{n \geq 0} T^n(V)$. We abridge

$$v_1 v_2 \dots v_n := v_1 \otimes v_2 \otimes \dots \otimes v_n, \qquad v_1, v_2, \dots, v_n \in V.$$

The natural identifications

$$\mu_{m,n} : T^m(V) \otimes T^n(V) \simeq T^{m+n}(V)$$

patch together to an associative product $\mu : T(V) \otimes T(V) \to T(V)$, giving rise to the tensor algebra $T(V)$. This is also the free algebra on V, meaning that it satisfies the universal property:

(i) There is a linear map $\iota : V \to T(V)$, the inclusion $V = T^1(V) \hookrightarrow T(V)$.
(ii) Every linear map $\varphi : V \to A$, where A is an associative algebra, extends to a morphism of algebras $\Phi : T(V) \to A$ such that $\Phi \circ \iota = \varphi$.

In categorical terms, this means that we have a functor T: $\mathrm{Vec}_{\Bbbk} \to \mathrm{Assoc}_{\Bbbk}$ that is left adjoint to the forgetful functor $\mathrm{Assoc}_{\Bbbk} \to \mathrm{Vec}_{\Bbbk}$.

Among the plentiful applications of the tensor algebra, let us single out the construction of the enveloping algebra of a Lie algebra $\mathfrak{g}$, as the quotient

$$U(\mathfrak{g}) := T(\mathfrak{g})/\langle xy - yx - [x, y] : x, y \in \mathfrak{g}\rangle.$$

Again, this is a functor U: $\mathrm{Lie}_{\Bbbk} \to \mathrm{Assoc}_{\Bbbk}$ left adjoint to the forgetful functor $\mathrm{Assoc}_{\Bbbk} \to \mathrm{Lie}_{\Bbbk}$; indeed, every associative algebra became a Lie algebra with the commutator $[a, b] = ab - ba$.

Remark 4.2 Let V be a vector space. By the universal property, the linear map $\delta : V \to T(V) \otimes T(V), \delta(v) = v \otimes 1 + 1 \otimes v, v \in V$, extends to $\Delta : T(V) \to T(V) \otimes T(V)$; then $T(V)$ becomes a Hopf algebra. It is cocommutative, i.e. $\Delta = \tau\Delta$.

Remark 4.3 Let $\mathfrak{g}$ be a Lie algebra. The linear map $\delta : \mathfrak{g} \to U(\mathfrak{g}) \otimes U(\mathfrak{g})$, $\delta(v) = v \otimes 1 + 1 \otimes v, v \in \mathfrak{g}$, extends to $\Delta : U(\mathfrak{g}) \to U(\mathfrak{g}) \otimes U(\mathfrak{g})$, so that $U(\mathfrak{g})$ is a cocommutative Hopf algebra.

Exercise 4.1 Let V be a vector space and
$$L(V) := \mathrm{Prim}\, T(V) = \{x \in T(V) : \Delta(x) = x \otimes 1 + 1 \otimes x\}$$

1. $\mathrm{Prim}\, T(V)$ is a Lie subalgebra of $T(V)$ (this is valid for every Hopf algebra).
2. $T(V) \simeq U(L(V))$.
3. $L(V)$ is the free Lie algebra on V. This provides the left adjoint to the forgetful functor $\mathrm{Lie}_{\Bbbk} \to \mathrm{Vec}_{\Bbbk}$.

4.1.4 The Symmetric Algebra

Let V be a vector space. The symmetric algebra $S(V)$ is the free *commutative* algebra on V, meaning that it satisfies the analogous universal property as above but with respect to linear maps from V to commutative algebras. Categorically, it gives a functor S: $\mathrm{Vec}_{\Bbbk} \to \mathrm{Comm}_{\Bbbk}$ left adjoint to the forgetful functor $\mathrm{Comm}_{\Bbbk} \to \mathrm{Vec}_{\Bbbk}$. Concretely,

$$S(V) := T(V)/\langle xy - yx : x, y \in V\rangle = \oplus_{n\geq 0} S^n(V),$$

so that $S(V)$ is the enveloping algebra of V with the zero bracket. In passing, we mention also the exterior algebra

$$\Lambda(V) := T(V)/\langle xy + yx : x, y \in V\rangle = \oplus_{n\geq 0}\Lambda^n(V).$$

A *quadratic* algebra is one of the form $T(W)/\langle J\rangle$, where W is a vector space and $J \leq T^2(W)$ (recall our convention in Sect. 4.1.1, this means that J is a suspace of $T^2(W)$).

Both $S(V)$ and $\Lambda(V)$ are quadratic algebras, of the form $T(V)/\langle J_{\mp}\rangle$, respectively, where $J_{\pm} = \{xy \pm yx : x, y \in V\}$.

The symmetric group $\mathbb{S}_n$ acts on $T^n(V)$ by $w \cdot v_1 \dots v_n = v_{w(1)} \dots v_{w(n)}$, for $n \in \mathbb{N}_0$ (where $\mathbb{S}_0 = \mathbb{S}_1$ are trivial). In particular, the isotypic components of $T^2(V)$ with respect to the action of $\mathbb{S}_2 \simeq \mathbb{Z}/2$ are J_+ for the trivial, respectively J_- for the sign, representation. It turns out that the nth homogeneous components of the ideals $\langle J_{\pm}\rangle$ are $\mathbb{S}_n$-submodules of $T^n(V)$. Hence, $S^n(V)$ and $\Lambda^n(V)$ are $\mathbb{S}_n$-modules, and it is not difficult to see that the former is a trivial module.

Assume now that char $\Bbbk = 0$. Then the various $\mathbb{S}_n$-modules $T^n(V)$ are all completely reducible and we may consider $\widetilde{S}(V) = \oplus_{n\geq 0}T^n(V)^{\mathbb{S}_n}$.

Proposition 4.1 *The natural projection* $T(V) \to S(V)$ *induces a linear isomorphism* $\widetilde{S}(V) \simeq S(V)$. *Consequently, cf.* (4.1),

$$\ker(T(V) \to S(V)) = \langle J_-\rangle = \oplus_{n\geq 2}\ker \int_{\mathbb{S}_n}. \tag{4.7}$$

Similarly, the polynomial algebra $\Bbbk[X_1, \dots, X_d]$ is the free commutative algebra on the set $\mathbb{I}_d$. Thus, if $\dim V = d$, then every choice of a basis in V induces an isomorphism of algebras $S(V) \simeq \Bbbk[X_1, \dots, X_d]$.

4.1.5 Coalgebras and Hopf Algebras

We expect that the reader has acquaintance with the notions of coalgebra, bialgebra and Hopf algebra. There are several books and monographs to be initiated on these topics; some of them are [25, 66, 71, 73, 78, 80]. The reader willing to learn these matters is advised to acquire first some experience with groups and Lie algebras.

As usual, the comultiplication of a coalgebra C is denoted by Δ, for which the Sweedler notation is $\Delta(c) = c_{(1)} \otimes c_{(2)}$, and the counit by ε. If D, E are subspaces of the coalgebra C, then

$$D \wedge E := \{c \in C : \Delta(c) \in D \otimes C + C \otimes E\}.$$

Coalgebras and comodules have a distinguished feature: they are locally finite, i.e., they are union of their finite-dimensional subcoalgebras, respectively subcomodules.

A coalgebra without proper subcoalgebras (remember that 0 is not a coalgebra) is called *simple*; thus a simple coalgebra is finite-dimensional. If $\Bbbk$ is algebraically closed, then every simple coalgebra is the dual of a matrix algebra.

The *coradical* of a coalgebra C is the sum of all its simple subcoalgebras, denoted by C_0; it is analogous to the socle of a module (in fact it is the socle of a coalgebra as a comodule over itself). By a standard argument, the coradical is a direct sum of simple coalgebras. A coalgebra is cosemisimple if it coincides with its coradical, i.e., if it is a (direct) sum of simple subcoalgebras. A one-dimensional coalgebra is of course simple; a coalgebra is *pointed* if its coradical is a (direct) sum of one-dimensional coalgebras. Basic examples are:

- The group algebra $H = \Bbbk G$ of a group G, with $\Delta(g) = g \otimes g, g \in G$. Here $H_0 = H$.
- The enveloping algebra $U(\mathfrak{g})$ of a Lie algebra $\mathfrak{g}$, with $\Delta(x) = x \otimes 1 + 1 \otimes x$, $x \in \mathfrak{g}$. Here $H_0 = \Bbbk$.

The study of pointed Hopf algebras started in the 1970s by Taft, Wilson, Radford, Nichols and others, being those with the simplest possible coradical. Some examples beyond group algebras and enveloping algebras were discovered. In the early 1980s, Reshetikhin, Kulish and Sklyanin introduced the Hopf algebra nowadays known as $U_q(sl_2)$ and soon after that, Drinfeld and Jimbo defined the quantized enveloping algebras $U_q(\mathfrak{g})$ for every finite-dimensional simple Lie algebra $\mathfrak{g}$; these are pointed Hopf algebras. Finite-dimensional pointed Hopf algebras related to $U_q(\mathfrak{g})$ appeared in the work of Lusztig [62–64]. The ICM report [28] made a deep impact in the area of Hopf algebras–and in many others. After some time, the classification program of finite-dimensional pointed Hopf algebras was launched [17, 18], see the survey [19], and the classification under some hypothesis in [20]. For more references and details, see [2].

The notions of filtration and grading are ubiquitous in algebra. For instance, it is useful for many purposes to filter an algebra by powers of an ideal. A coalgebra filtration of a coalgebra C is a family of subspaces $(D_n)_{n\in\mathbb{N}_0}$ such that

$$D_n \subseteq D_{n+1}, \qquad C = \bigcup_{N\in\mathbb{N}_0} D_n, \qquad \Delta(D_n) \subseteq \sum_{0\leq i\leq n} D_i \otimes D_{n-i}.$$

Here the first condition says that the filtration is ascending and the second that it is exhaustive. The coradical filtration is defined recursively by

$$C_0 = \text{the coradical}, \qquad C_{n+1} = C_n \wedge C_0.$$

Exercise 4.2 1. Let $\mathscr{G} = \oplus_{n\in\mathbb{N}_0}\mathscr{G}^n$ be a graded coalgebra, i.e.,

$$\Delta(\mathscr{G}^n) \subseteq \oplus_{0\leq i\leq n}\mathscr{G}^i \otimes \mathscr{G}^{n-i}.$$

Let $\mathfrak{D}_n := \oplus_{0 \le i \le n} \mathscr{G}^i$. Prove that $(\mathfrak{D}_n)_{n \in \mathbb{N}_0}$ is a coalgebra filtration. We say that $\mathscr{G} = \oplus_{n \in \mathbb{N}_0} \mathscr{G}^n$ is *coradically graded* if $\mathfrak{D}_n = \mathscr{G}_n$ (in words, the coradical filtration coincides with the filtration associated with the grading).

Let A be a finite-dimensional algebra and $C = A^*$ the dual coalgebra (with the transpose of the multiplication and the unit). If $I \subseteq A$, then we set

$$I^{\perp} := \{c \in C : \langle c, x \rangle = 0 \text{ for all } x \in I\}.$$

2. $I \subseteq A$ is a two-sided ideal if and only if $I^{\perp} \subseteq C$ is a subcoalgebra.
3. If $I, J \subseteq A$, then $(IJ)^{\perp} = I^{\perp} \wedge J^{\perp}$.
4. Let $(I_n)_{n \in \mathbb{N}}$ be a family of subspaces of A and $D_n := I_{n+1}^{\perp}$. Then $(I_n)_{n \in \mathbb{N}}$ is a descending algebra filtration if and only if $(D_n)_{n \in \mathbb{N}_0}$ is a coalgebra filtration. Prove that gr $C = \oplus_{n \in \mathbb{N}_0} D_n / D_{n-1}$ is a graded coalgebra (where $D_{-1} = 0$).
5. Let J be the Jacobson radical of A. Then $C_0 = J^{\perp}$. Conclude that $C_{n+1} = (J^n)^{\perp}$ and that the coradical filtration is a coalgebra filtration. Show that gr C, with respect to the coradical filtration, is coradically graded.
6. Let H be a Hopf algebra with bijective antipode $\mathscr{S}$. Assume that the coradical H_0 is a subalgebra. Prove that H_0 is a Hopf subalgebra and that the coradical filtration is an ascending filtration of algebras, each term being stable under the antipode. Conclude that gr H is a graded Hopf algebra.
7. If C is coalgebra, then $G(C) = \{x \in C - 0 : \varDelta(x) = x \otimes x\}$ is linearly independent. If H is a Hopf algebra, then $G(H)$ is a group with the multiplication of H and inverse $x^{-1} = \mathscr{S}(x)$, $x \in G(H)$.
8. The coradical of a pointed Hopf algebra H is a Hopf subalgebra: $H_0 \simeq \Bbbk G(H)$.

4.1.6 The Tensor Coalgebra

Let V be a vector space. We shall need later the tensor coalgebra $T^c(V)$; this is the vector space $T(V)$ with the comultiplication Δ given by

$$\Delta(v_1 v_2 \dots v_n) := \sum_{j \in \mathbb{I}_n} v_1 \dots v_j \otimes v_{j+1} \dots v_n, \qquad v_1, \dots, v_n \in V. \tag{4.8}$$

Clearly $\Delta(v) = v \otimes 1 + 1 \otimes v$ for $v \in V$, but $\Delta(v_1 v_2) = v_1 v_2 \otimes 1 + v_1 \otimes v_2 + 1 \otimes v_1 v_2 \neq \Delta(v_1)\Delta(v_2)$, thus $\Delta \neq \varDelta$ from Remark 4.2.

Remark 4.4 The coalgebra $T^c(V)$ is dual to the tensor algebra $T(V^*)$, but it is not the cofree coalgebra on V (cofree means universal with respect to maps $C \to V$, C a coalgebra). The construction of the cofree coalgebra is more delicate [80].

4.1.7 Gelfand–Kirillov Dimension

The notion of dimension pervades all mathematics. In the dictionary affine algebraic geometry–commutative algebra, the Krull dimension is the translation of the topological dimension. A guiding principle in non-commutative algebra is to adapt ideas and tools from geometry; in this sense, there are different attempts to generalize the Krull dimension. Perhaps the best adapted is the Gelfand–Kirillov dimension, GK-dim for short; a comprehensive account is [59].

Let A be a finitely generated $\Bbbk$-algebra. Let V be a finite-dimensional subspace of A such that $A = \Bbbk\langle V\rangle$. Set

$$V^j = \underbrace{V \cdot V \cdots V}_{j \text{ times}}, \qquad A_n = \sum_{0\le j\le n} V^j$$

The Gelfand–Kirillov dimension is defined as

$$\text{GK-dim}\, A := \overline{\lim}_{n\to\infty} \log_n \dim A_n. \tag{4.9}$$

It can be shown that GK-dim A does not depend on the choice of V [59, 1.1]. When A is not finitely generated, the definition is extended as follows:

$$\text{GK-dim}\, A := \sup\{\text{GK-dim}\, B | B \text{ finitely generated subalgebra of } A\}. \tag{4.10}$$

Example 4.2 Let V be a vector space of dimension $1 < d \in \mathbb{N}$ and $A = T(V)$. Then

$$\dim A_n = \sum_{0\le j\le n} \dim T^j(V) = \sum_{0\le j\le n} d^j = \frac{d^{n+1}-1}{d-1} \implies \log_n \dim A_n \sim \frac{n}{\log n},$$

hence GK-dim $T(V) = \infty$.

Exercise 4.3 Let A be a finitely generated $\Bbbk$-algebra and V a finite-dimensional subspace such that $A = \Bbbk\langle V\rangle$. Show that

$$\begin{aligned}\text{GK-dim} A &= \inf\{r \in \mathbb{R} : \dim V^j \le cj^r \text{ for some } c \in \mathbb{R}, \forall j \in \mathbb{N}\}\\ &= \inf\{r \in \mathbb{R} : \dim V^j \le j^r \text{ for large } j\}.\end{aligned}$$

Exercise 4.4 Let V be a vector space of dimension $d \in \mathbb{N}$ and $A = S(V)$. Let $\Bbbk[X_1, \dots, X_d]^j \simeq S^j(V)$ be the subspace of homogeneous polynomials of degree j.

1. Prove that $\dim S^j(V) = \binom{d+j-1}{j}$ (e.g., argue recursively and use that $S^j(V) \simeq \Bbbk[X_1, \dots, X_d]^{j-1} \cdot X_d \oplus \Bbbk[X_1, \dots, X_{d-1}]^j$).
2. Prove that $\dim A_n = \sum_{0\le j\le n} \dim S^j(V) = \binom{d+n}{n}$ (e.g., use the linear isomorphism $\Bbbk[X_1, \dots, X_{d+1}]^n \to \oplus_{0\le j\le n} \Bbbk[X_1, \dots, X_d]^j$, $f(X_1, \dots, X_{d+1}) \mapsto f(X_1, \dots, X_d, 1)$).

3. Since $\binom{d+n}{n}$ is a polynomial of degree d in n, conclude that GK-dim $S(V) = d$.
4. If dim $V = \infty$, then GK-dim $S(V) = \infty$.

Exercise 4.5 Let A be a finitely generated algebra. Then GK-dim $A = 0$ if and only if A is finite-dimensional.

If A is arbitrary, then GK-dim $A = 0$ if and only if every finitely generated subalgebra is finite-dimensional. For example, if dim $V = \infty$, then GK-dim $\Lambda(V) = 0$.

Example 4.3 If A is a finitely generated commutative algebra, then

$$\text{GK-dim} A = \text{Krull dim } A = \dim \text{Spec } A.$$

Here Spec A is the Zariski spectrum of A; it could be replaced by its subset of closed points, that is the affine variety defined by A. In other words, the Gelfand–Kirillov dimension coincides with the usual dimension in the commutative case. Therefore, if A is a commutative algebra, then GK-dim $A \in \mathbb{N}_0 \cup \infty$. However there are examples of non-commutative algebras A with GK-dim $A = r$ for any $r \in [2, \infty)$. But there is no algebra A with GK-dim $A = r$ for any $r \in (1, 2)$. See [59].

Example 4.4 A finitely generated group G is *virtually nilpotent* or *nilpotent-by-finite* if it has a normal nilpotent subgroup N such that G/N is finite.

- J.A. Wolf, J. Milnor and others showed that the group algebra of a virtually nilpotent group has finite Gelfand–Kirillov dimension (in an equivalent formulation).
- A celebrated Theorem of Gromov establishes the converse: if G is a finitely generated group and GK-dim $\Bbbk G < \infty$, then G is virtually nilpotent.

Example 4.5 Let A be an algebra with an ascending algebra filtration. Then

$$\text{GK-dim} A \geq \text{GK-dim gr } A;$$

also, the equality holds if gr A is finitely generated. Let $\mathfrak{g}$ be a Lie algebra; we conclude that GK-dim $U(\mathfrak{g}) = \dim \mathfrak{g}$.

4.2 Braided Tensor Categories

We first discuss the notion of braided vector space, the input of the definition of Nichols algebra, and illustrate it through various examples. Then we review braided tensor categories and the example of our main interest, Yetter–Drinfeld modules.

4.2.1 Braided Vector Spaces

The Yang–Baxter equation, introduced independently by C.N. Yang in 1968, and R.J. Baxter in 1971 in statistical mechanics, has important applications in various areas of mathematics. Here we consider the equivalent braid equation

$$(c \otimes \mathrm{id})(\mathrm{id} \otimes c)(c \otimes \mathrm{id}) = (\mathrm{id} \otimes c)(c \otimes \mathrm{id})(\mathrm{id} \otimes c), \quad c \in GL(V \otimes V), \quad (4.11)$$

where V is a vector space. Solutions of the braid equation (4.11) are the input for the definition of Nichols algebras. Following the common usage, we say that a pair (V, c), with c satisfying (4.11), is a *braided vector space*.

We first justify the adjective braided: we claim that the assignment

$$\sigma_j \mapsto \mathrm{id}_{T^{(j-1)}(V)} \otimes c \otimes \mathrm{id}_{T^{(n-j-1)}(V)} \quad (4.12)$$

gives rise to a representation $\rho_n : \mathbb{B}_n \to GL(T^n(V))$, for every $n \geq 2$. Indeed, (4.11) insures that (4.4) holds, while (4.3) is free from the definition. The applications of the Yang–Baxter equation mostly arise from these representations. For us, they will useful to present Nichols algebras. But let us discuss before some classes of examples of braided vector spaces.

4.2.1.1 Symmetries

Here char $\Bbbk \neq 2$. A *symmetry* is a solution c of (4.11) such that $c^2 = \mathrm{id}$. The name alludes to the fact that ρ_n factorizes through the representation $\widetilde{\rho}_n : \mathbb{S}_n \to GL(T^n(V))$ given by

$$\tau_j \mapsto \mathrm{id}_{T^{(j-1)}(V)} \otimes c \otimes \mathrm{id}_{T^{(n-j-1)}(V)}, \qquad j \in \mathbb{I}_{n-1}.$$

Prominent examples of symmetries are:

- The transposition, i.e., the usual flip $\tau : V \otimes V \to V \otimes V$, $v \otimes w \mapsto w \otimes v$.
- The super transposition of a super vector space $V = V_0 \oplus V_1$; i.e., the linear map $s\tau : V \otimes V \to V \otimes V$, determined by $v \otimes w \mapsto (-1)^{ij} w \otimes v$ for $v \in V_i$, $w \in V_j$.

Clearly, we have the decomposition

$$T^2(V) = \ker(\mathrm{id} + c) \oplus \ker(\mathrm{id} - c).$$

4.2.1.2 Hecke Type

Here char $\Bbbk = 0$. Let $q \in \Bbbk^\times$, $q \neq -1$. The Hecke algebra of parameter q is the associative algebra $\mathrm{H}_n(q)$ generated by $(T_i)_{i\in\mathbb{I}_{n-1}}$ with relations (4.4) (with T instead of τ) and

$$(T_i - q\mathrm{id})(T_i + \mathrm{id}) = 0, \qquad i \in \mathbb{I}_{n-1}. \tag{4.13}$$

A braided vector space (V, c) is of *Hecke type* with label q if

$$(c - q\mathrm{id})(c + \mathrm{id}) = 0. \tag{4.14}$$

The name refers to the fact that in this case, ρ_n factorizes through the representation $\widetilde{\rho}_n : \mathrm{H}_n(q) \to GL(T^n(V))$ given by $T_j \mapsto \mathrm{id}_{T^{(j-1)}(V)} \otimes c \otimes \mathrm{id}_{T^{(n-j-1)}(V)}$, $j \in \mathbb{I}_{n-1}$.

4.2.1.3 Diagonal Type

We fix $\theta \in \mathbb{N}$ and abbreviate $\mathbb{I} = \mathbb{I}_\theta$. Let $\mathfrak{q} = (q_{ij}) \in (\Bbbk^\times)^{\mathbb{I}\times\mathbb{I}}$ and let V a vector space with a basis $(x_i)_{i\in\mathbb{I}}$. We define $c^{\mathfrak{q}} \in GL(T^2(V))$ by

$$c^{\mathfrak{q}}(x_i \otimes x_j) = q_{ij}\, x_j \otimes x_i, \qquad i, j \in \mathbb{I}, \tag{4.15}$$

Then $c^{\mathfrak{q}}$ satisfies (4.11). When $q_{ij} = 1$, we recover the transposition τ, and the super transposition also has this shape. By technical reasons, we say that a braided vector space (V, c) with $c = c^{\mathfrak{q}}$ as in (4.15) is of *diagonal type* if in addition

$$q_{ii} \neq 1, \qquad i \in \mathbb{I}. \tag{4.16}$$

Instead of the matrix $\mathfrak{q}$, we also give the associated Dynkin diagram,[1] that has

- set of vertices $\mathbb{I}$, the i-th vertex being labeled with q_{ii};
- an edge between the vertices i and j only if $\widetilde{q}_{ij} := q_{ij}q_{ji} \neq 1$, in which case the edge is decorated by $\widetilde{q}_{ij}$.

Notice that we loose some information, but this is justified by Example 4.29.

We introduce the important subclass of Cartan type. Let $A = (a_{ij}) \in \mathbb{Z}^{\mathbb{I}\times\mathbb{I}}$ be a *generalized Cartan matrix*, that is, it satisfies

$$a_{ii} = 2, \qquad i \in \mathbb{I}; \tag{4.17}$$

$$a_{ij} \leq 0, \qquad i \neq j \in \mathbb{I}; \tag{4.18}$$

$$a_{ij} = 0 \iff a_{ji} = 0, \qquad i \neq j \in \mathbb{I};. \tag{4.19}$$

[1] Actually this is called a generalized Dynkin diagram but we omit generalized.

These are the input for the definition of Kac-Moody algebras [57]; among them, there are the celebrated Cartan matrices classifying finite-dimensional Lie algebras. Let (V, c) be a braided vector space of diagonal type with respect to a matrix $\mathfrak{q} = (q_{ij}) \in (\Bbbk^{\times})^{\mathbb{I}\times\mathbb{I}}$. We say that (V, c) is of *Cartan type* (with matrix A) if

$$q_{ij}q_{ji} = q_{ii}^{a_{ij}}, \qquad i \neq j \in \mathbb{I}. \tag{4.20}$$

Suppose that q_{ii} is a root of 1 of order N_i, for all i. Observe that if the matrix $\mathfrak{q}$ satisfies (4.20) for some integers a_{ij}, then we get a generalized Cartan matrix by taking $a_{ii} = 2$ for all i and normalizing the a_{ij}'s by

$$-N_i < a_{ij} \leq 0, \qquad i \neq j \in \mathbb{I}.$$

4.2.1.4 Triangular Type

Let (V, c) be a braided vector space with a basis $(x_i)_{i\in\mathbb{I}}$. Let V_j be the subspace generated by $(x_i)_{i\in\mathbb{I}_j}$. We say that (V, c) is of *triangular type* if there exists $\mathfrak{q} = (q_{ij}) \in (\Bbbk^{\times})^{\mathbb{I}\times\mathbb{I}}$ such that

$$c(x_i \otimes x_j) \in q_{ij}\, x_j \otimes x_i + V_{j-1} \otimes V, \qquad i, j \in \mathbb{I}. \tag{4.21}$$

Example 4.6 Let $\varepsilon \in \Bbbk^{\times}$ and $\ell \in \mathbb{N}_{\geq 2}$. The *block* $\mathcal{V}(\varepsilon, \ell)$ is the braided vector space with a basis $(x_i)_{i\in\mathbb{I}_\ell}$ such that for $i, j \in \mathbb{I}_\ell = \{1, 2, \dots, \ell\}$, $1 < j$:

$$c(x_i \otimes x_1) = \varepsilon x_1 \otimes x_i, \qquad c(x_i \otimes x_j) = (\varepsilon x_j + x_{j-1}) \otimes x_i. \tag{4.22}$$

Later on, we call $\mathcal{V}(\varepsilon, 2)$ and ε-block; this is justified by Theorem 4.7.

4.2.1.5 Rack Type

To define this class of braided vector spaces, we need to discuss the notion of rack, that is an abstract version of the conjugation in a group; see [15] for more information. We start with the general notion of braided set; we leave to the reader to fill in the details of the proofs.

The braid Eq. (4.11) makes sense in any monoidal category, a basic example being the category of sets with Cartesian product as the tensor one. So, a *braided set* is a pair (X, c), where $X \neq \emptyset$ is a set and $\mathrm{c} : X \times X \to X \times X$ is a bijection such that

$$(\mathrm{c} \times \mathrm{id})(\mathrm{id} \times \mathrm{c})(\mathrm{c} \times \mathrm{id}) = (\mathrm{id} \times \mathrm{c})(\mathrm{c} \times \mathrm{id})(\mathrm{id} \times \mathrm{c}). \tag{4.23}$$

Also, c is called a set-theoretical solution of the quantum Yang–Baxter equation; it has been studied in many papers [30, 32, 61], etc.

Notice that any braided set $(X, \mathfrak{c})$ gives rise to a braided vector space $(\Bbbk X, c)$ by linearization, namely $\Bbbk X$ is the vector space with basis $(e_x)_{x\in X}$ and c extends linearly the map defined on the basis by $\mathfrak{c}$.

Let X be a non-empty set and $\mathfrak{c} : X \times X \to X \times X$ be a bijection. If $p_1, p_2 : X \times X \to X$ are the standard projections, then we write

$$x \triangleright y = p_1\mathfrak{c}(x, y), \quad x \triangleleft y = p_2\mathfrak{c}(x, y), \quad \text{so that} \quad \mathfrak{c}(x, y) = (x \triangleright y, x \triangleleft y), x, y \in X.$$

Clearly, to give $\triangleright$ and $\triangleleft$ is equivalent to give $\mathfrak{c}$.

Exercise 4.6 1. Find necessary and sufficient conditions on the pair $(\triangleright, \triangleleft)$ so that $\mathfrak{c}$ satisfies (4.23).
2. Let $\mathfrak{q} : X \times X \to \Bbbk^\times$ be a function denoted $(x, y) \mapsto \mathfrak{q}_{x,y}$ and let

$$c^{\mathfrak{q}} : \Bbbk X \otimes \Bbbk X \to \Bbbk X \otimes \Bbbk X, \quad c^{\mathfrak{q}}(e_x \otimes e_y) = \mathfrak{q}_{x,y} e_{x\triangleright y} \otimes e_{x\triangleleft y}, \quad x, y \in X. \tag{4.24}$$

Prove that if $c^{\mathfrak{q}}$ satisfies (4.11), then $(X, \mathfrak{c})$ is a braided set.
3. Let $\mathfrak{q}$ be as in the previous item. Assume that $(X, \mathfrak{c})$ is a braided set. Find necessary and sufficient conditions on $\mathfrak{q}$ so that $c^{\mathfrak{q}}$ satisfies (4.11).

The definition of rack arises by considering the trivial $\triangleleft$, i.e., $x \triangleleft y = x$ for all $x, y \in X$. That is, consider $\triangleright : X \times X \to X$ and correspondingly $\mathfrak{c} : X \times X \to X \times X$ given by $\mathfrak{c}(x, y) = (x \triangleright y, x)$, $x, y \in X$. Then $\mathfrak{c}$ is bijective if and only if

$$\text{the map } \phi_x = x \triangleright \_ \text{ is bijective for any } x \in X \tag{4.25}$$

while $\mathfrak{c}$ satisfies (4.23) if and only if

$$x \triangleright (y \triangleright z) = (x \triangleright y) \triangleright (x \triangleright z) \text{ for all } x, y, z \in X. \tag{4.26}$$

We say that $(X, \triangleright)$ is a *rack* if (4.25) and (4.26) hold. Morphisms of racks and subracks are defined as usual; $\operatorname{Aut} X$ denotes the group of rack automorphisms of X.

Exercise 4.7 Let X be a non-empty set and $\triangleright : X \times X \to X$ a function.

1. Let $\mathfrak{q} : X \times X \to \Bbbk^\times$ be a function, $(x, y) \mapsto \mathfrak{q}_{x,y}$, and let $c^{\mathfrak{q}} : \Bbbk X \otimes \Bbbk X \to \Bbbk X \otimes \Bbbk X$ be given by

$$c^{\mathfrak{q}}(e_x \otimes e_y) = \mathfrak{q}_{x,y} e_{x\triangleright y} \otimes e_x, \qquad x, y \in X.$$

Prove that $(\Bbbk X, c^{\mathfrak{q}})$ is a braided vector space if and only if $(X, \triangleright)$ is a rack and

$$q_{x,y\triangleright z} q_{y,z} = q_{x\triangleright y, x\triangleright z} q_{x,z}, \qquad \forall x, y, z \in X. \tag{4.27}$$

2. Here is a generalization. Let W be a vector space and let $\mathfrak{q} : X \times X \to GL(W)$ be a function. Set $V = \Bbbk X \otimes W$, $e_x v := e_x \otimes v$. Let $c^{\mathfrak{q}} : V \otimes V \to V \otimes V$ be given by

$$c^{\mathfrak{q}}(e_x v \otimes e_y w) = e_{x \triangleright y} \mathfrak{q}_{x,y}(w) \otimes e_x v, \qquad x, y \in X, \qquad v, w \in W. \tag{4.28}$$

Prove that $(V, c^{\mathfrak{q}})$ is a braided vector space if and only if $(X, \triangleright)$ is a rack and (4.27) holds.
3. Let $\mathfrak{q}, \mathfrak{p} : X \times X \to \Bbbk^{\times}$ be two functions satisfying (4.27) and let $\mathfrak{b} : X \to \Bbbk^{\times}$ be a function. Let $T : \Bbbk X \to \Bbbk X$ be given by $T(e_x) = \mathfrak{b}_x e_x$, $x \in X$. Find necessary and sufficient conditions such that $T : (\Bbbk X, c^{\mathfrak{q}}) \to (\Bbbk X, c^{\mathfrak{p}})$ is a morphism of braided vector spaces.

The condition (4.27) says that $\mathfrak{q}$ is a 2-cocycle; when $\dim W = 1$, it is part of a cohomology theory, while for $n > 1$ it is a non-abelian cocycle. Observe that any constant function $\mathfrak{q}$ is a 2-cocycle.

Braided vector spaces as in the previous Exercise are called of *rack type* and play an important role in the classification of finite-dimensional pointed Hopf algebras.

4.2.1.6 Racks

We discuss now examples of racks; once again the reader is encouraged to work out the details.

Example 4.7 Let $(X, \mathfrak{c})$ be a braided set with associated $\triangleright$ and $\triangleleft$ as above. Then we say that $\mathfrak{c}$ is non-degenerate if for all $x, y \in X$ the maps

$$x \triangleright \_ : X \to X, \qquad\qquad \_ \triangleleft y : X \to X$$

are both bijective. Assume that this is the case. Write $\_ \triangleleft y^{-1}$ for the inverse of $\_ \triangleleft y$. Define $\blacktriangleright : X \times X \to X$ by

$$x \blacktriangleright y = \big((x \triangleleft y^{-1}) \triangleright y\big) \triangleleft x. \tag{4.29}$$

Then $(X, \blacktriangleright)$ is a rack, called the derived rack of $(X, \mathfrak{c})$.

Example 4.8 Let X be a non-empty set. Given $\sigma \in \mathbb{S}_X$, the associated permutation rack $(X, \triangleright)$ is defined by $x \triangleright y = \sigma(y)$ for all $x, y \in X$.

Example 4.9 A group G is a rack with $x \triangleright y = xyx^{-1}$, $x, y \in G$. If $X \subset G$ is stable under conjugation by G (e.g., if X is a conjugacy class), then it is a subrack of G.

Example 4.10 A set $X \neq \emptyset$ with $\triangleright$ defined by $x \triangleright y = y$ for all $x, y \in X$ is a rack; such racks are called *abelian*. The abelian rack with elements $\{1, \dots, n\}$ is denoted $\mathbb{I}_n$.

Exercise 4.8 1. If X is a rack, then $\phi : X \to \mathbb{S}_X$, $x \mapsto \phi_x$, is a morphism of racks. The subgroup of $\mathbb{S}_X$ generated by the image of ϕ is denoted by $\operatorname{Inn} X$. Thus $\operatorname{Inn} X$ acts on X. Show that $\operatorname{Inn} X$ is a normal subgroup of $\operatorname{Aut} X$ (the group of rack automorphisms).
2. When a subrack of a group is a permutation rack?
3. Let X be a subrack of a group. Then, for all $x, y \in X$, we have

$$x \triangleright x = x, \tag{4.30}$$

$$x \triangleright y = y \implies y \triangleright x = x, \tag{4.31}$$

A rack with these properties is a *crossed set*. Can a crossed set be realized always as a subrack of a group?

The following examples can be identified with subracks of groups, but they deserve a separate consideration.

Example 4.11 Let G be a group and $T \in \operatorname{Aut} G$. Let $\rightharpoonup_T$ be the action of G on itself given by $x \rightharpoonup_T y = x\, y\, T(x^{-1})$, $x, y \in G$. Then the orbit $\mathscr{O}_x^{G,T}$ of $x \in G$ by this action is a rack with operation

$$y \triangleright_T z = y\, T(z\, y^{-1}), \quad y, z \in \mathscr{O}_x^{G,u}. \tag{4.32}$$

The rack $(\mathscr{O}_x^{G,T}, \triangleright_T)$ is called a *twisted conjugacy class* of type (G, T).

Example 4.12 Let A be an abelian group and $T \in \operatorname{Aut} A$. We define the operation $\triangleright$ by

$$x \triangleright y = (1 - T)x + Ty, \qquad x, y \in A.$$

Then $(A, \triangleright)$ is a rack, denoted $\operatorname{Aff}(A, T)$. If T is multiplication by a fixed m, then the rack is denoted by $\operatorname{Aff}(A, m)$. The rack $\operatorname{Aff}(A, T)$ is isomorphic to the subrack $A \times \operatorname{id}$ of $A \rtimes \langle T \rangle$. Racks of this sort are called *affine*. For instance, the dihedral rack $\mathscr{D}_n$, $n \geq 3$, is $\operatorname{Aff}(\mathbb{Z}/n, T)$, where T is multiplication by -1.

Exercise 4.9 Let X be a rack; below $\dot\cup$ means disjoint union.

1. A *decomposition* of X is a pair of subracks (Y, Z) such that $X = Y \dot\cup Z$; X is *decomposable* if it admits a decomposition, *indecomposable* otherwise.

   Let $\emptyset \neq Y \subsetneq X$ and $Z = X - Y$. If X is finite, then

   $$(Y, Z) \text{ is a decomposition of } X \iff Y \triangleright Z \subseteq Z \text{ and } Z \triangleright Y \subseteq Y \iff X \triangleright Y \subseteq Y.$$

   If X is not finite, which of the implications remain true? Find counterexamples for the rest.
2. X is indecomposable $\iff X = \mathscr{O}_x^{\operatorname{Inn} X}$ for any $x \in X$.

3. Let $n \geq 3$. Compute all subracks of $\mathscr{D}_n$. Conclude that $\mathscr{D}_n$ is indecomposable if n is odd. Prove that $\operatorname{Inn} \mathscr{D}_4 \neq \operatorname{Aut}\mathscr{D}_4$, what about $\operatorname{Inn} \mathscr{D}_n$ for $n \neq 4$?

Exercise 4.10 Let Y, Z be two racks and $X = Y \dot{\cup} Z$. The following are equivalent:

1. Structures of rack on X such that (Y, Z) is a decomposition.
2. Pairs (ς, ϖ) of morphisms of racks $\varsigma : Y \to \operatorname{Aut}Z$, $\varpi : Z \to \operatorname{Aut}Y$ such that

$$y \triangleright \varpi_z(u) = \varpi_{\varsigma_y(z)}(y \triangleright u), \quad \forall y, u \in Y,\ z \in Z, \quad \text{i.e., } \phi_y \varpi_z = \varpi_{\varsigma_y(z)}\phi_y; \tag{4.33}$$

$$z \triangleright \varsigma_y(w) = \varsigma_{\varpi_z(y)}(z \triangleright w), \quad \forall y \in Y,\ z, w \in Z, \quad \text{i.e., } \phi_z \varsigma_y = \varsigma_{\varpi_z(y)}\phi_z. \tag{4.34}$$

The rack X is denoted $Y\,{}_{\varsigma}\coprod_{\varpi} Z$, with ς omitted if $\varsigma_y = \mathrm{id}_Z$ for all $y \in Y$, idem for ϖ. Assume that Y and Z are crossed sets and that (4.33) and (4.34) hold. Then X is a crossed set if and only if the following condition holds:

$$\varsigma_y(z) = z \text{ if and only if } \varpi_z(y) = y, \qquad \forall y \in Y,\ z \in Z. \tag{4.35}$$

Exercise 4.11 Assume that $Y = \mathbb{I}_n$. Then the previous setting reduces to a family $(\varsigma_i)_{i \in \mathbb{I}_n}$ of commuting elements in $\operatorname{Aut}Z$ and a morphism of racks $\varpi : Z \to \mathbb{S}_n$ such that

$$\varpi_z = \varpi_{\varsigma_j(z)}, \qquad z \triangleright \varsigma_j(w) = \varsigma_{\varpi_z(j)}(z \triangleright w), \qquad \forall j \in \mathbb{I}_n,\ z, w \in Z.$$

Suppose that $Y = \mathbb{I}_n$ and $Z = \mathbb{I}_m$. Then the previous setting consists of families $(\varsigma_i)_{i \in \mathbb{I}_n}$ and $(\varpi_h)_{h \in \mathbb{I}_m}$ of commuting elements in $\mathbb{S}_m$ and $\mathbb{S}_n$, respectively, such that

$$\varpi_h = \varpi_{\varsigma_j(h)}, \qquad \varsigma_j = \varsigma_{\varpi_h(j)}, \qquad \forall j \in \mathbb{I}_n,\ h \in \mathbb{I}_m. \tag{4.36}$$

In particular, let $\sigma \in \mathbb{S}_m$ and $\pi \in \mathbb{S}_n$ and consider the constant families $\varsigma_i = \sigma, i \in \mathbb{I}_n$, and $\varpi_h = \pi, h \in \mathbb{I}_m$. These families satisfy (4.36), thus we have the rack $\mathbb{I}_n\,{}_{\sigma}\coprod_{\pi} \mathbb{I}_m$.

Here is an important notion for our purposes.

Definition 4.1 A finite rack X is *simple* if

- it has at least 2 elements,
- for any surjective morphism of racks $\pi : X \to Y$, either π is an isomorphism or Y has just one element.

Finite simple racks have been classified in [15, Th. 3.9, Th. 3.12], [56]. Because of its importance in recursive arguments about Nichols algebras, we state this result.

Theorem 4.2 *Let X be a finite simple rack with $|X|$ elements. Then either of the following holds:*

1. *$|X|$ is divisible by at least two primes. In this case, there exist*
   - *a simple non-abelian group L,*
   - *$t \in \mathbb{N}$, and*
   - *$\theta \in \operatorname{Aut} L$,*

   such that X is a twisted conjugacy class of type (G, T), where
   - *$G = L^t$ and*
   - *$T \in \operatorname{Aut}(L^t)$ acts by*

$$T(\ell_1, \dots, \ell_t) = (\theta(\ell_t), \ell_1, \dots, \ell_{t-1}), \qquad \ell_1, \dots, \ell_t \in L.$$

   Furthermore, L and t are unique, and T only depends on its conjugacy class in $\operatorname{Out}(L^t) = \operatorname{Aut}(L^t)/\mathit{Inn}(L^t)$.
2. *$|X| = p^t$ where p is a prime and $t \in \mathbb{N}$. In this case, there are two possibilities:*
   a. *$t = 1$ and $X \simeq \mathbb{I}_p$ is the permutation rack of the cycle $(1, 2, \dots, p)$ (this could not be realized as a conjugacy class in a group).*
   b. *X is the affine rack $({\mathbb{F}_p}^t, T)$, where T is the companion matrix of a monic irreducible polynomial $f \in \mathbb{F}_p[\mathtt{X}]$ of degree t, different from $\mathtt{X}$ and $\mathtt{X} - 1$.*

Particularly, non-trivial conjugacy classes in finite simple groups are simple racks.

4.2.2 *Braided Tensor Categories*

The notion of braided vector space has a counterpart in the notion of braided tensor category, that is both technically convenient and the right formulation for applications. We briefly discuss this notion and refer to [31, 58] for extensive expositions.

4.2.2.1 Tensor Categories

We start by the formal definitions.

A *monoidal category* is a collection $(\mathscr{C}, \otimes, \mathbf{1}, a, l, r)$, where

- $\mathscr{C}$ is a category;
- $\otimes : \mathscr{C} \times \mathscr{C} \to \mathscr{C}$ is a functor, called the tensor product;
- $\mathbf{1} \in \mathscr{C}$ is an object called the unit;
- $a_{X,Y,Z} : (X \otimes Y) \otimes Z \to X \otimes (Y \otimes Z)$ is an invertible natural transformation, called the associativity constraint;
- $l_X : X \to X \otimes \mathbf{1}$, $r_X : X \to \mathbf{1} \otimes X$, are invertible natural transformations, called the left and right unit constraints.

These data are required to satisfy the pentagon and the triangle axioms, expressed by the commutativity of the following diagrams:

$$
\begin{array}{ccccc}
((X \otimes Y) \otimes Z) \otimes U & & \xrightarrow{a_{(X\otimes Y),Z,U}} & & (X \otimes Y) \otimes (Z \otimes U) \\
\downarrow{\scriptstyle a_{X,Y,Z}\otimes \mathrm{id}_U} & & & & \downarrow{\scriptstyle a_{X,Y,Z\otimes U}} \\
(X \otimes (Y \otimes Z)) \otimes U & & & & X \otimes (Y \otimes (Z \otimes U)) \\
& \searrow{\scriptstyle a_{X,Y\otimes Z,U}} & & \nearrow{\scriptstyle \mathrm{id}_X \otimes a_{Y,Z,U}} & \\
& & X \otimes ((Y \otimes Z) \otimes U) & &
\end{array}
$$

and

$$
\begin{array}{ccccc}
(X \otimes \mathbf{1}) \otimes Y & & \xrightarrow{a_{X,\mathbf{1},Y}} & & X \otimes (\mathbf{1} \otimes Y) \\
& \nwarrow{\scriptstyle l_X \otimes \mathrm{id}_Y} & & \nearrow{\scriptstyle \mathrm{id}_X \otimes r_Y} & \\
& & X \otimes Y. & &
\end{array}
$$

The pentagon and triangle axioms guarantee that we can tensor any finite number of objects, the result being independent of the distribution of parentheses up to isomorphism, and that the unit objects can be ignored in such a product. This was shown by S. Mac Lane, who also proved any monoidal category is equivalent to a strict one (one with associative and unit constraints equal to the identity).

Let $\mathscr{C}$ be a monoidal category and $M \in \mathscr{C}$. A *left dual* of M is an object ${}^*M \in \mathscr{C}$ provided with morphisms

$$\mathbf{1} \xrightarrow{\mathrm{coev}_M} {}^*M \otimes M\ , \qquad M \otimes {}^*M \xrightarrow{\mathrm{ev}_M} \mathbf{1},$$

such that the composition

$$M \xrightarrow{l_M} M \otimes \mathbf{1} \xrightarrow{\mathrm{id}\otimes \mathrm{coev}_M} M \otimes {}^*M \otimes M \xrightarrow{\mathrm{ev}_M \otimes \mathrm{id}} \mathbf{1} \otimes M \xrightarrow{r_M^{-1}} M \tag{4.37}$$

equals id_M. Analogously, a *right dual* of M is an object $M^* \in \mathscr{C}$ provided with morphisms

$$\mathbf{1} \xrightarrow{\mathrm{coev}'_M} M \otimes M^*\ , \qquad M^* \otimes M \xrightarrow{\mathrm{ev}'_M} \mathbf{1},$$

such that the following composition equals id_M:

$$M \xrightarrow{r_M} \mathbf{1} \otimes M \xrightarrow{\mathrm{coev}'_M \otimes \mathrm{id}} M \otimes M^* \otimes M \xrightarrow{\mathrm{id}\otimes \mathrm{ev}_M} M \otimes \mathbf{1} \xrightarrow{l_M^{-1}} M \tag{4.38}$$

Clearly, if M has a right dual M^*, then M^* has itself a left dual which is M.

Exercise 4.12 Prove that two left duals of the same object are isomorphic.

A monoidal category $\mathscr{C}$ is *rigid* if every object in $\mathscr{C}$ has right and left duals.

Example 4.13 Assume that $\mathscr{C}$ is a discrete category, i.e., Ob $\mathscr{C}$ is a set X and the only arrows are the identities id_x, $x \in X$. Then a structure of monoidal category on $\mathscr{C}$ is tantamount to a structure of monoid on X. Thus, a structure of rigid monoidal category on $\mathscr{C}$ is tantamount to a structure of group on X. In other words, the notion of rigid monoidal category is a categorification of the notion of group.

Definition 4.2 A *tensor category* (over $\Bbbk$) is a rigid monoidal category such that $\mathscr{C}$ is abelian $\Bbbk$-linear and $\otimes$ is $\Bbbk$-linear in each variable (i.e., $\otimes$ is a bifunctor).

Example 4.14 Let $\mathrm{Vec}_{\Bbbk}$ be the category of vector spaces over $\Bbbk$ and let $\mathrm{vec}_{\Bbbk}$ be the full subcategory of the finite-dimensional ones. Then $\mathrm{Vec}_{\Bbbk}$ is a monoidal (abelian $\Bbbk$-linear) category, with $\otimes = \otimes_{\Bbbk}$ the usual tensor product over $\Bbbk$, $\mathbf{1} \simeq \Bbbk$, and the natural isomorphisms a, l and r from the universal property defining $\otimes_{\Bbbk}$. Also $\mathrm{vec}_{\Bbbk}$ is a tensor category; given $V \in \mathrm{vec}_{\Bbbk}$, we take ${}^*V = V^* = \hom_{\Bbbk}(V, \Bbbk)$ and

$$\Bbbk \xrightarrow{\mathrm{coev}_V} {}^*V \otimes V \qquad\qquad V \otimes {}^*V \xrightarrow{\mathrm{ev}_V} \Bbbk$$

$$1 \longmapsto \textstyle\sum_{i \in \mathbb{I}} \alpha_i \otimes v_i, \qquad\qquad f \otimes v \longmapsto f(v),$$

where $(v_i)_{i \in \mathbb{I}}$ is a basis of V and $(\alpha_i)_{i \in \mathbb{I}}$ is its dual basis.

Exercise 4.13 Prove that $\mathrm{Vec}_{\Bbbk}$ is not rigid.

Example 4.15 Let H be a Hopf algebra with bijective antipode $\mathscr{S}$. Let $\mathrm{Rep}H$ be the category of representations of H and let $\mathrm{rep}H$ be the full subcategory of the finite-dimensional ones. Then $\mathrm{Rep}H$ is a monoidal subcategory of $\mathrm{Vec}_{\Bbbk}$ and $\mathrm{rep}H$ is a tensor subcategory of $\mathrm{vec}_{\Bbbk}$ (but neither is full). Indeed, if $V, W \in \mathrm{Rep}H$, then H acts on $V \otimes W$ via the comultiplication Δ; the unit is $\Bbbk$ with the trivial action given by the counit ε; ${}^*V$, respectively V^*, is $\hom_{\Bbbk}(V, \Bbbk)$ with the action given by the transpose of the antipode, respectively, its inverse.

This class of examples includes the following:

- The category $\mathrm{Rep}\, G$ of representations of a group G over $\Bbbk$ and the subcategory $\mathrm{rep}\, G$.
- The category $\mathrm{Rep}\, \mathfrak{g}$ of representations of a Lie algebra $\mathfrak{g}$ and the subcategory $\mathrm{rep}\, \mathfrak{g}$.
- The category $\mathrm{Vec}_{\Bbbk}^G$ of G-graded vector spaces, where G is a group, and the subcategory $\mathrm{vec}_{\Bbbk}^G$. Here the tensor product of $V = \oplus_{g \in G} V_g$ and $W = \oplus_{g \in G} W_g$ is graded as $V \otimes W = \oplus_{g \in G} (V \otimes W)_g$, where

$$(V \otimes W)_g = \oplus_{h \in G} V_h \otimes W_{h^{-1}g}.$$

The category of super vector spaces is the particular case $\mathrm{Svec}_{\Bbbk} = \mathrm{Vec}_{\Bbbk}^{\mathbb{Z}/2}$; as usual, $\mathrm{svec}_{\Bbbk}$ is the full subcategory of finite-dimensional objects.

Exercise 4.14 Assume that $\Bbbk$ is algebraically closed of characteristic 0. Let G be a finite abelian group and $\widehat{G}$ be its group of characters. Then $\text{Vec}_{\Bbbk}^{G}$ is equivalent to $\text{Rep}\widehat{G}$ as monoidal categories.

Example 4.16 Let H be a Hopf algebra with bijective antipode. Let $\mathscr{M}^H$, respectively ${}^H\mathscr{M}$, be the category of right, respectively left, H-comodules. Then both $\mathscr{M}^H$ and ${}^H\mathscr{M}$ are monoidal subcategories of $\text{Vec}_{\Bbbk}$. Indeed, the tensor product arises via the multiplication and the unit is $\Bbbk$ with the trivial coaction. The subcategories of finite-dimensional comodules are tensor, with duals given by the antipode, respectively its inverse.

4.2.2.2 Braided Tensor Categories

If the notion of monoidal category could be thought as an extension of the notion of monoid (or group), then it is natural to seek for the analog of the notion of abelian group. Such an analogue was already proposed by S. Mac Lane–symmetric monoidal categories. However, the weaker notion of braided category turned out to be much more flexible for applications.

A *braided monoidal category* is a monoidal category $\mathscr{C}$ provided with a natural isomorphism $c_{X,Y} : X \otimes Y \to Y \otimes X$, called the *braiding*, that is required to fulfill the hexagon axioms, meaning that the following diagrams commute:

$$\begin{array}{ccccc} (X \otimes Y) \otimes Z & \xrightarrow{a_{X,Y,Z}} & X \otimes (Y \otimes Z) & \xrightarrow{c_{X,Y\otimes Z}} & (Y \otimes Z) \otimes X \\ \downarrow{\scriptstyle c_{X,Y}\otimes \mathrm{id}} & & & & \downarrow{\scriptstyle a_{Y,Z,X}} \\ (Y \otimes X) \otimes Z & \xrightarrow[a_{Y,X,Z}]{} & Y \otimes (X \otimes Z) & \xrightarrow[\mathrm{id}\otimes c_{X,Z}]{} & Y \otimes (Z \otimes X), \end{array} \tag{4.39}$$

$$\begin{array}{ccccc} X \otimes (Y \otimes Z) & \xrightarrow{a^{-1}_{X,Y,Z}} & (X \otimes Y) \otimes Z & \xrightarrow{c_{X\otimes Y,Z}} & Z \otimes (X \otimes Y) \\ \downarrow{\scriptstyle \mathrm{id}\otimes c_{Y,Z}} & & & & \downarrow{\scriptstyle a^{-1}_{Z,X,Y}} \\ X \otimes (Z \otimes Y) & \xrightarrow[a^{-1}_{X,Y,Z}]{} & (X \otimes Z) \otimes Y & \xrightarrow[c_{X,Z}\otimes \mathrm{id}]{} & (Z \otimes X) \otimes Y, \end{array} \tag{4.40}$$

for all $X, Y, Z \in \mathscr{C}$. In addition, $\mathscr{C}$ is *symmetric* when

$$c_{Y,X}c_{X,Y} = \mathrm{id}_{X\otimes Y}, \qquad \text{for all } X, Y \in \mathscr{C}. \tag{4.41}$$

Loosely, (4.41) is abbreviated as $c^2 = \mathrm{id}$. In this case, c is called a symmetry, instead of a braiding.

Exercise 4.15 Assume that (4.41) holds. Then (4.39) and (4.40) are equivalent.

Needless to say, a braided tensor category is a tensor category that is also braided.

Example 4.17 The super categories $\text{Vec}_{\Bbbk}$ and $\text{vec}_{\Bbbk}$ are symmetric, with symmetry being the transposition τ.

Example 4.18 If G is a group and $\mathfrak{g}$ is a Lie algebra, then the tensor categories Rep$G$ and Rep$\mathfrak{g}$ are symmetric, with symmetry τ.

Example 4.19 The categories $\text{Svec}_{\Bbbk}$ and $\text{svec}_{\Bbbk}$ are symmetric, with symmetry being the super transposition $s\tau$.

Exercise 4.16 1. Classify all possible braidings in the category $\text{Vec}_{\Bbbk}^{\mathbb{Z}/n}$, $1 < n \in \mathbb{N}$.
2. Classify all possible braidings in the category $\text{Vec}_{\Bbbk}^{G}$, where G is an abelian group; determine those that are symmetries.
3. Let G be a group. Prove that the category $\text{Vec}_{\Bbbk}^{G}$ admits a braiding if and only if G is abelian.

Where the adjective braided comes from?

Proposition 4.2 *Let $\mathscr{C}$ be a braided monoidal category. Assume that it is strict, i.e., the associativity and unit constraints are identities. Then for all $X, Y, Z \in \mathscr{C}$,*

$$(c_{Y,Z} \otimes \text{id}_X)(\text{id}_Y \otimes c_{X,Z})(c_{X,Y} \otimes \text{id}_Z) = (\text{id}_Z \otimes c_{X,Y})(c_{X,Z} \otimes \text{id}_Y)(\text{id}_X \otimes c_{Y,Z}), \tag{4.42}$$

equality in $\hom(X \otimes Y \otimes Z, Z \otimes Y \otimes X)$.

Thus, if $X \in \mathscr{C}$, then $c_{X,X}$ is a solution of the braid equation. If $\mathscr{C}$ is not strict, then a version of (4.42) with associators holds.

V.G. Drinfeld found a mechanism to construct solutions of the braid Eq. (4.11). First, he introduced the notion of quasitriangular Hopf algebra as a pair (H, R) where H is a Hopf algebra and $R \in H \otimes H$ is tailored to give Rep$H$ a structure of braided tensor category. Second, he showed how to assign to a Hopf algebra H (say finite-dimensional to avoid technicalities), a quasitriangular Hopf algebra $D(H)$–called nowadays the Drinfeld double of H. For a better understanding of this construction, we give now the categorical version; passing from H to $D(H)$ is a particular instance of the center of a monoidal category.

Exercise 4.17 Let $\mathscr{C}$ be a monoidal category. Prove that $\mathscr{Z}(\mathscr{C})$ (the center of $\mathscr{C}$) defined as follows is a braided monoidal category:

- The objects are pairs (Z, γ) where $Z \in \mathscr{C}$ and γ is a natural isomorphism

$$\gamma_X : X \otimes Z \to Z \otimes X, \qquad X \in \mathscr{C},$$

such that the following diagram commutes:

$$\begin{array}{ccccc} X\otimes(Y\otimes Z) & \xrightarrow{a^{-1}_{X,Y,Z}} & (X\otimes Y)\otimes Z & \xrightarrow{\gamma_{X\otimes Y}} & Z\otimes(X\otimes Y) \\ {\scriptstyle \mathrm{id}\otimes\gamma_Y}\downarrow & & & & \downarrow{\scriptstyle a^{-1}_{Z,X,Y}} \\ X\otimes(Z\otimes Y) & \xrightarrow[a^{-1}_{X,Y,Z}]{} & (X\otimes Z)\otimes Y & \xrightarrow[\gamma_X\otimes\mathrm{id}]{} & (Z\otimes X)\otimes Y. \end{array} \tag{4.43}$$

By the similarity of (4.43) with (4.40), γ is called a half-braiding.

- The morphisms between pairs (Z,γ) and (Z',γ') are maps $f : Z\to Z'$ in $\mathscr{C}$ such that

$$(f\otimes \mathrm{id}_X)\gamma_X = \gamma'_X(\mathrm{id}_X\otimes f) : X\otimes Z'\to Z'\otimes X, \qquad \text{for all } X\in\mathscr{C}.$$

- The tensor product of (Z,γ) and (Z',γ') is $(Z\otimes Z',\widetilde{\gamma})$, where $\widetilde{\gamma}$ is defined by the commutativity of the diagram

$$\begin{array}{ccccc} X\otimes(Z\otimes Z') & \xrightarrow{a^{-1}_{X,Z,Z'}} & (X\otimes Z)\otimes Z' & \xrightarrow{\gamma_X\otimes\mathrm{id}} & (Z\otimes X)\otimes Z' \\ {\scriptstyle \widetilde{\gamma}}\downarrow & & & & \downarrow{\scriptstyle a_{Z,X,Z'}} \\ (Z\otimes Z')\otimes X & \xleftarrow{a^{-1}_{Z,Z',X}} & Z\otimes(Z'\otimes X) & \xleftarrow{\mathrm{id}\otimes\gamma'_X} & Z\otimes(X\otimes Z') \end{array} \tag{4.44}$$

- The unit object is $(\mathbf{1}, r^{-1}l)$.
- The braiding between (Z,γ) and (Z',γ') is

$$c_{(Z,\gamma),(Z',\gamma')} = \gamma'_Z.$$

If $(Z,\gamma)\in\mathscr{Z}(\mathscr{C})$ and Z has a left dual in $\mathscr{C}$, then (Z,γ) has a left dual in $\mathscr{Z}(\mathscr{C})$.

Exercise 4.18 Compute explicitly $\mathscr{Z}(\mathrm{Rep}G)$ and $\mathscr{Z}(\mathrm{Vec}_{\Bbbk}G)$.

4.2.2.3 Yetter–Drinfeld Modules

Let H be a Hopf algebra with bijective antipode $\mathscr{S}$. Let $G(H)$ be the group of group-like elements. This is the point we wanted to reach:

Definition 4.3 A *Yetter Drinfeld module* over H is a vector space V provided with

- a structure of left H-module $\cdot : H\otimes V\to V$ and
- a structure of left H-comodule $\delta : V\to H\otimes V$, such that
- for all $h\in H$ and $v\in V$, the following compatibility condition holds:

$$\delta(h\cdot v) = h_{(1)}v_{(-1)}\mathscr{S}(h_{(3)})\otimes h_{(2)}\cdot v_{(0)}. \tag{4.45}$$

Thus we have the category ${}^H_H\mathcal{YD}$ of Yetter–Drinfeld modules, with morphisms being linear maps that preserve both the action and the coaction.

Exercise 4.19 Prove that ${}^H_H\mathcal{YD}$ is a braided tensor category, with the tensor product of modules and comodules and braiding

$$c_{V,W}(v\otimes w) = v_{(-1)}\cdot w\otimes v_{(0)},\qquad V, W\in {}^H_H\mathcal{YD},\qquad v\in V,\ w\in W. \tag{4.46}$$

Here $c_{V,W}$ is bijective because $\mathcal{S}$ is so; indeed

$$c^{-1}_{W,V}(v\otimes w) = w_{(0)}\otimes \mathcal{S}^{-1}(w_{(-1)})\cdot v,\qquad V, W\in {}^H_H\mathcal{YD},\qquad v\in V,\ w\in W. \tag{4.47}$$

That is, the assignment $H \rightsquigarrow {}^H_H\mathcal{YD}$ is the categorical version of $H \rightsquigarrow D(H)$; indeed, when H is finite-dimensional, ${}^H_H\mathcal{YD}$ is equivalent to $\operatorname{Rep}D(H)$.

Exercise 4.20 Show that ${}^H_H\mathcal{YD}$ is equivalent as tensor category to $\mathcal{Z}({}^H\mathcal{M})$.

Notice that there are four versions of Yetter–Drinfeld categories, the other three being ${}^H\mathcal{YD}_H$ (left comodules and right modules), $\mathcal{YD}^H_H$ and ${}_H\mathcal{YD}^H$.

Summarizing, given H as above, every $V\in {}^H_H\mathcal{YD}$ provides a braided vector space, namely $(V, c_{V,V})$. Two questions come up naturally: Does every braided vector space (V, c) arise as a Yetter–Drinfeld module over some H? (For short, we say that (V, c) is *realized* over H.) If yes, then in so many ways? The answer to the first is affirmative, up to a technical hypothesis:

Definition 4.4 A finite-dimensional braided vector space (V, c) is *rigid* if the map $c^\flat : V^*\otimes V\to V\otimes V^*$ given by

$$f\otimes v \longmapsto \sum_i (\mathrm{ev}\otimes\mathrm{id}\otimes\mathrm{id})(f\otimes c(v\otimes v_i)\otimes\alpha^i)$$

is invertible, where (v_i) is a basis of V and (α^i) its dual basis.

Proposition 4.3 ([33, 43, 77, 81]) *Let (V, c) be a rigid braided vector space. Then there is a Hopf algebra $H(V)$ such that $V\in {}^{H(V)}_{H(V)}\mathcal{YD}$ and $c = c_{V,V}$.*

The construction of $H(V)$ is done in two steps: first, one attaches a bialgebra $A(V)$ such that $V\in {}^{A(V)}_{A(V)}\mathcal{YD}$–this is the celebrated FRT-construction. Second, and here rigidity is needed, one passes from $A(V)$ to $H(V)$.

However, $H(V)$ does not provide, by far, the unique realization and the problem of classifying or even characterizing all of them contains some subtleties.

Example 4.20 A pair $(g,\chi) \in G(H) \times \hom_{\text{alg}}(H,\Bbbk)$ is called a *YD-pair* for H provided that

$$\chi(h)\,g = \chi(h_{(2)})h_{(1)}\,g\,\mathcal{S}(h_{(3)}), \qquad h \in H. \tag{4.48}$$

If (g,χ) is a YD-pair, then $g \in Z(G(K))$.

YD-pairs classify the $V \in {}^H_H\mathcal{YD}$ with $\dim V = 1$. Indeed, if (g,χ) is a YD-pair, then $\Bbbk_g^\chi = \Bbbk$ with action and coaction given by χ and g respectively, is in ${}^H_H\mathcal{YD}$. In fact, (4.48) is just (4.45). Clearly, the braiding of $\Bbbk_g^\chi$ is multiplication by $q = \chi(g)$.

Example 4.21 Let $\mathfrak{q} = (q_{ij}) \in (\Bbbk^\times)^{\mathbb{I}\times\mathbb{I}}$ satisfying (4.16) and let V the corresponding braided vector space of diagonal type with respect to a basis $(x_i)_{i\in\mathbb{I}}$. A principal realization of (V,c) is a collection $(g_i,\chi_i)_{i\in\mathbb{I}}$ of YD-pairs such that $q_{ij} = \chi_j(g_i)$ for all $i, j \in \mathbb{I}$. But there might be realizations different from these.

Example 4.22 Assume that $\Bbbk$ is algebraically closed and $\operatorname{char}\Bbbk = 0$. If $H = \Bbbk\Gamma$, where Γ is a finite abelian group, then ${}^H_H\mathcal{YD}$ is semisimple and its simple objects have dimension 1. Now (4.48) always holds. In conclusion, every $V \in {}^H_H\mathcal{YD}$ of dimension $\theta \in \mathbb{N}$ is determined by families $(g_i)_{i\in\mathbb{I}_\theta}$ and $(\chi_i)_{i\in\mathbb{I}_\theta}$; the braiding of V is of diagonal type with matrix $\mathfrak{q} = (q_{ij})$, $q_{ij} = \chi_j(g_i)$, for all $i, j \in \mathbb{I}_\theta$.

Example 4.23 We now explain how to realize blocks of dimension 2, cf. Example 4.6. A *YD-triple* for H is a collection (g,χ,η) where (g,χ), is a YD-pair for H, $\eta \in \operatorname{Der}_{\chi,\chi}(H,\Bbbk)$, $\eta(g) = 1$ and

$$\eta(h)g = \eta(h_{(2)})h_{(1)}g\mathcal{S}(h_{(3)}), \qquad h \in H. \tag{4.49}$$

Let (g,χ,η) be a YD-triple. Let $\mathcal{V}_g(\chi,\eta)$ be a vector space with a basis $(x_i)_{i\in\mathbb{I}_2}$, where action and coaction are given by

$$h \cdot x_1 = \chi(h)x_1, \qquad h \cdot x_2 = \chi(h)x_2 + \eta(h)x_1, \qquad \delta(x_i) = g \otimes x_i,$$

$h \in H$, $i \in \mathbb{I}_2$. Then $\mathcal{V}_g(\chi,\eta) \in {}^H_H\mathcal{YD}$, the compatibility being granted by (4.48), (4.49). Since $\eta(g) \neq 0$, then $\mathcal{V}_g(\chi,\eta)$ is indecomposable in ${}^H_H\mathcal{YD}$. As braided vector space, $\mathcal{V}_g(\chi,\eta)$ is the block $\mathcal{V}(\varepsilon,2)$, where $\varepsilon := \chi(g)$.

Exercise 4.21 Find a realization of the block $\mathcal{V}(\varepsilon,2)$ over $\Bbbk\mathbb{Z}$.

Exercise 4.22 Let G be a group. Prove that $M \in {}^{\Bbbk G}_{\Bbbk G}\mathcal{YD}$ if and only if M is a G-module with a G-grading $M = \oplus_{\gamma\in G} M_\gamma$ such that $g \cdot M_\gamma = M_{g\gamma g^{-1}}$. Consequently, if $N \leq M$ is a Yetter–Drinfeld submodule, then N inherits the grading; in particular $N \neq 0$ implies $N_\gamma \neq 0$ for some $\gamma \in G$.

Example 4.24 Let G be a finite group. Let $\mathcal{O}$ be a conjugacy class in G, pick $x \in \mathcal{O}$ and (W,ρ) an irreducible representation of $G^x = \{g \in G : gx = xg\}$, i.e., the centralizer (or the isotropy subgroup) of x. Let

$$M(\mathcal{O}, \rho) = \mathrm{Ind}_{G^x}^G \rho = \Bbbk G \otimes_{\Bbbk G^x} W. \tag{4.50}$$

We want to show that $M(\mathcal{O}, \rho) \in {}^{\Bbbk G}_{\Bbbk G}\mathcal{YD}$, for which we need to define the coaction. Let $(x_i)_{i\in\mathbb{I}_m}$ be a numeration of $\mathcal{O}$, $m = |\mathcal{O}|$. Then there are $(z_i)_{i\in\mathbb{I}_m}$ in G such that

$$z_i \triangleright x = z_i x z_i^{-1} = x_i, \qquad i \in \mathbb{I}_m.$$

Thus $G = \coprod_{i\in\mathbb{I}_m} z_i G^x$. We may normalize the choice by $x_1 = x$ and $z_1 = e$. Now

$$M(\mathcal{O}, \rho) = \oplus_{i\in\mathbb{I}_m} \Bbbk z_i \otimes W; \tag{4.51}$$

the action of G is explicitly given by

$$g \cdot (z_i \otimes w) = z_j \otimes \rho(y)(w), \qquad \text{if } gz_i = z_j y, \qquad g \in G,\ i \in \mathbb{I},\ w \in W.$$

We define $\delta : M(\mathcal{O}, \rho) \to \Bbbk G \otimes M(\mathcal{O}, \rho)$ by

$$\delta(z_i \otimes w) = x_i \otimes (z_i \otimes w), \qquad i \in \mathbb{I},\ w \in W.$$

(In the formulation of Exercise 4.22, the grading is (4.51) with $\Bbbk z_i \otimes W$ in degree x_i). We prove the compatibility condition (4.45). Let $g \in G$, $i \in \mathbb{I}$, $w \in W$ and suppose that $gz_i = z_j y$ with $y \in G^x$. Then

$$\delta(g \cdot (z_i \otimes w)) = x_j \otimes (z_j \otimes \rho(y)(w)),$$
$$g(z_i \otimes w)_{(-1)} g^{-1} \otimes g \cdot (z_i \otimes w)_{(0)} = g x_i g^{-1} \otimes z_j(\otimes \rho(y)(w)).$$

and the first line equals the second because

$$g x_i g^{-1} = g z_i x z_i^{-1} g^{-1} = z_j y x y^{-1} z_j^{-1} = z_j x z_j^{-1} = x_j.$$

Using Exercise 4.22, we check that $M(\mathcal{O}, \rho)$ is a simple Yetter–Drinfeld module. Clearly $\dim M(\mathcal{O}, \rho) = |\mathcal{O}| \dim W$. Also it is easy to see that $M(\mathcal{O}, \rho) \simeq M(\mathcal{O}', \rho')$ implies $\mathcal{O} = \mathcal{O}'$ and $\rho = \rho'$ (we have picked one element in each conjugacy class). Since ${}^{\Bbbk G}_{\Bbbk G}\mathcal{YD} \simeq \mathrm{Rep} D(\Bbbk G)$, we conclude that

$$\bigoplus_{\substack{\mathcal{O} \text{ conjugacy class} \\ \rho \in \mathrm{Irr} G^x}} \mathrm{End} M(\mathcal{O}, \rho) \le D(\Bbbk G). \tag{4.52}$$

Assume that $\Bbbk$ is algebraically closed and that char $\Bbbk$ does not divide $|G|$. Then by a counting argument, see [15, p. 63], we see that the equality holds in (4.52); hence

- the category ${}^{\Bbbk G}_{\Bbbk G}\mathcal{YD}$ is semisimple and
- any simple object in ${}^{\Bbbk G}_{\Bbbk G}\mathcal{YD}$ is isomorphic to $M(\mathcal{O}, \rho)$ for a unique $(\mathcal{O}, \rho)$.

Finally, the braiding in $M(\mathscr{O}, \rho)$ is given by

$$c((z_k \otimes v) \cdot (z_i \otimes w)) = x_k \cdot (z_i \otimes w) \otimes (z_k \otimes v) = z_j \otimes \rho(y)(w) \otimes (z_k \otimes v)$$

where $x_k z_i = z_j y$. Now, $x_j = z_j y \triangleright x = x_k z_i \triangleright x = x_k \triangleright x_i$. In other words, $M(\mathscr{O}, \rho)$ is isomorphic to a braided vector space of rack type. Namely consider the rack $\mathscr{O}$; then the map $M(\mathscr{O}, \rho) \to \Bbbk X \otimes W$, $z_i \otimes w \mapsto e_{x_i} \otimes w$ is an isomorphism of braided vector spaces, where $\mathfrak{q} : X \times X \to GL(W)$, $\mathfrak{q}_{x_i, x_j} = \rho(y)$.

4.3 Nichols Algebras

Nichols algebras are a special kind of Hopf algebras in braided tensor categories. Our main interest is in Nichols algebras in the category ${}^H_H\mathscr{YD}$. We start with the definition of Hopf algebra in a braided tensor category; then we discuss the concept of Nichols algebra. Finally, we overview several techniques to compute Nichols algebras. Throughout we refrain from using parentheses and the associativity constraints, as justified by Mac Lane coherence theorem.

4.3.1 *Hopf Algebras in Braided Tensor Categories*

Let $\mathscr{C}$ be a monoidal category. A *monoid* in $\mathscr{C}$ is a triple (M, μ, u), where $M \in \mathscr{C}$, $\mu : M \otimes M \to M$ and $u : \mathbf{1} \to M$ are morphisms in $\mathscr{C}$, such that the following diagrams commute:

$$\begin{array}{ccc} M \otimes M \otimes M & \xrightarrow{\mathrm{id} \otimes \mu} & M \otimes M \\ \downarrow{\scriptstyle \mu \otimes \mathrm{id}} & & \downarrow{\scriptstyle \mu} \\ M \otimes M & \xrightarrow{\mu} & M, \end{array} \qquad \begin{array}{ccc} \mathbf{1} \otimes M & \xrightarrow{u \otimes \mathrm{id}} & M \otimes M \\ \downarrow{\scriptstyle l} & \swarrow{\scriptstyle \mu} & \uparrow{\scriptstyle \mathrm{id} \otimes u} \\ M & \xleftarrow[r]{} & M \otimes \mathbf{1} \end{array} \qquad (4.53)$$

When $\mathscr{C}$ is actually a tensor category, it is also customary to say *algebra* in $\mathscr{C}$ instead of monoid in $\mathscr{C}$. Indeed, a monoid in $\mathrm{Vec}_\Bbbk$ is just an associative algebra over $\Bbbk$.

Example 4.25 Let G be a group. An algebra in $\mathrm{Vec}_\Bbbk^G$ is just a G-graded algebra.

The dual notion of *comonoid* in $\mathscr{C}$ is obtained reversing the arrows. That is, a comonoid is a triple (C, δ, ε), where $C \in \mathscr{C}$, $\delta : C \to C \otimes C$, and $\varepsilon : C \to \mathbf{1}$ are morphisms in $\mathscr{C}$, such that the following diagrams commute:

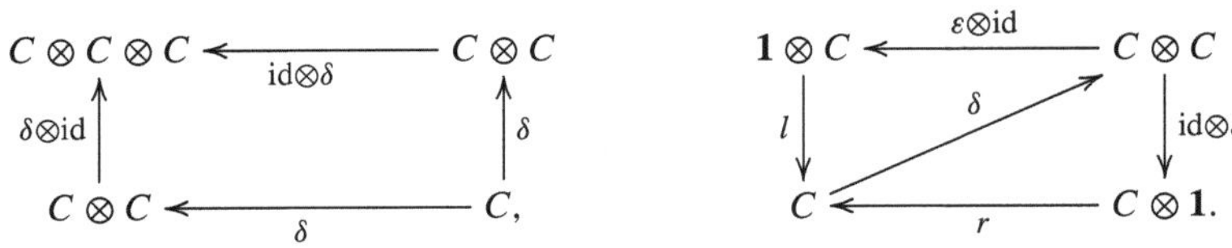
(4.54)

When $\mathscr{C}$ is a tensor category, we say *coalgebra* instead of comonoid.

There are straightforward definitions of morphisms of monoids, and thus of the category of monoids in $\mathscr{C}$, and also of actions of monoids on objects of $\mathscr{C}$, and thus of the category of objects in $\mathscr{C}$ with action of a fixed monoid. However extra structure is needed to define the tensor product of two monoids.

Definition 4.5 Let $\mathscr{C}$ be a *braided* monoidal category. The tensor product of two monoids $M = (M, \mu_M, u_M)$ and $N = (N, \mu_N, u_N)$ in $\mathscr{C}$ is the monoid

$$M \underline{\otimes} N = (M \otimes N, \mu_{M\otimes N}, u_{M\otimes N}),$$

where $\mu_{M\otimes N}$ and $u_{M\otimes N}$ are defined by the following compositions:

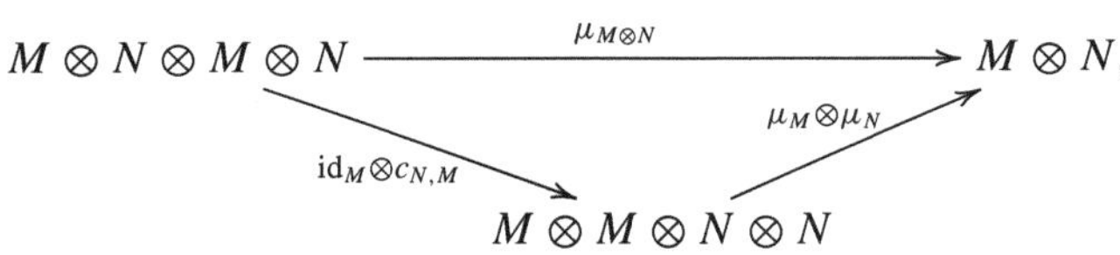

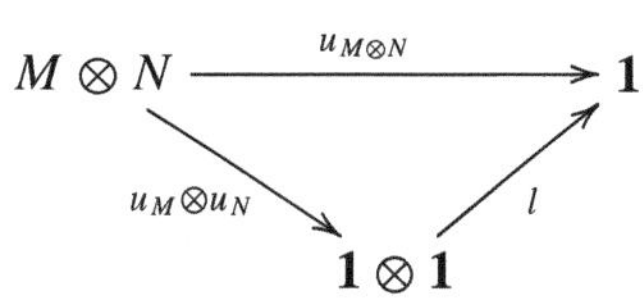

Exercise 4.23 1. Prove that the unit of a monoid is unique. Idem for the counit of a comonoid.
2. Prove that $M\underline{\otimes}N$ is a monoid, i.e., it satisfies (4.53).
3. Define the tensor product comonoid of two comonoids; show that it satisfies (4.54).
4. Let M be a monoid and C a comonoid in $\mathscr{C}$. Define the convolution product

$$* : \operatorname{hom}_{\mathscr{C}}(C, M) \times \operatorname{hom}_{\mathscr{C}}(C, M) \to \operatorname{hom}_{\mathscr{C}}(C, M), \quad f * g = \mu(f \otimes g)\delta.$$

Prove that $*$ is associative and has unit $u\varepsilon$.

Definition 4.6 Let $\mathscr{C}$ be a braided tensor category. A bialgebra in $\mathscr{C}$ is a collection $(B, \mu, u, \Delta, \varepsilon)$ such that

- (B, μ, u) is an algebra (a monoid) in $\mathscr{C}$;
- (B, Δ, ε) is a coalgebra in $\mathscr{C}$;
- $\Delta : B \to B\underline{\otimes}B$ is a morphism of algebras.

A Hopf algebra in $\mathscr{C}$ is a bialgebra B such that the identity $\mathrm{id}_B \in \hom_{\mathscr{C}}(B, B)$ admits an inverse $\mathscr{S}$ for the convolution product $*$; i.e., there exists $\mathscr{S} \in \hom_{\mathscr{C}}(B, B)$ such that

$$\mathscr{S} * \mathrm{id}_B = \mathrm{id}_B * \mathscr{S} = u\varepsilon.$$

Example 4.26 Let H be a Hopf algebra with bijective antipode and $\mathscr{C} = {}^H_H\mathscr{YD}$. Let $V \in {}^H_H\mathscr{YD}$. Then the tensor algebra $T(V)$ is an algebra in ${}^H_H\mathscr{YD}$. Thus we may consider the algebra $T(V)\underline{\otimes}T(V)$, which is not the same as the algebra $T(V) \otimes T(V)$. For instance, if $y, u \in T(V)$, then the product in $T(V)\underline{\otimes}T(V)$ gives

$$(1 \otimes y)(u \otimes 1) = (y_{(-1)} \cdot u) \otimes y_{(0)}.$$

By the universal property, since $T(V)\underline{\otimes}T(V)$ is in particular an associative algebra, there is a unique map $\underline{\Delta} : T(V) \to T(V)\underline{\otimes}T(V)$ such that $\underline{\Delta}(v) = v \otimes 1 + 1 \otimes v$, $v \in V$. Notice that $\underline{\Delta}$ neither coincides with the Δ in Remark 4.2 nor with the Δ in Sect. 4.1.6.

Exercise 4.24 The tensor algebra $T(V)$ with the map $\underline{\Delta} : T(V) \to T(V)\underline{\otimes}T(V)$ defines a Hopf algebra in ${}^H_H\mathscr{YD}$ (define the antipode on $T^n(V)$ recursively on n).

Exercise 4.25 There is a coalgebra map $\mu : T^c(V)\underline{\otimes}T^c(V) \to T^c(V)$ determined by $\mu(v \otimes 1) = v = \mu(1 \otimes v)$, $v \in V$; with this, $T^c(V)$ is a graded Hopf algebra in ${}^H_H\mathscr{YD}$.

We shall see plenty of examples of Hopf algebras in braided tensor categories, *aka* braided Hopf algebras. But before that, let us show how they appear in nature.

4.3.2 Bosonization

A basic result in group theory establishes an equivalence between the following two situations:

(a) $\pi : G \to L$ and $\iota : L \to G$ are morphisms of groups such that $\pi\iota = \mathrm{id}_L$.
(b) L and N are groups with L acting on N by group homomorphisms.

Namely, if (a) holds, then one takes $N = \ker \pi$; while if (b) holds, then $G \simeq N \rtimes L$. The situation is slightly more complicated when we consider the parallel setting for Hopf algebras; as we shall see, braided Hopf algebras appear in a natural way.

We start considering the situation:

(A) $\pi : A \to H$ and $\iota : H \to A$ are morphisms of Hopf algebras such that $\pi\iota = \mathrm{id}_H$.

It turns out that the right analogue of $\ker \pi$ in this setting is

$$R = A^{\mathrm{co}H} = \{a \in A : (\mathrm{id} \otimes \pi_H)\Delta(a) = a \otimes 1\}$$

(Un)fortunately, this is not a Hopf algebra, but, following Radford and Majid, see [66, 73], we claim that R is a braided Hopf algebra in ${}^H_H\mathscr{Y}\mathscr{D}$; explicitly, via

$$\begin{aligned} &h \cdot r = h_{(1)} r \mathscr{S}(h_{(2)}), \\ &r_{(-1)} \otimes r_{(0)} = \pi(r_{(1)}) \otimes r_{(2)}, \\ &R \text{ is a subalgebra of } A, \\ &\Delta_R(r) = r^{(1)} \otimes r^{(2)} = \vartheta_R(r_{(1)}) \otimes r_{(2)}, \quad r \in R,\ h \in H. \end{aligned} \tag{4.55}$$

We leave the proof to the reader, who may find useful the map $\vartheta_R : A \to R$ given by

$$\vartheta_R(a) = a_{(1)} \iota\pi(\mathscr{S}(a_{(2)})), \qquad a \in A; \tag{4.56}$$

it satisfies $$\vartheta_R(rh) = r\varepsilon(h), \qquad \vartheta_R(hr) = h \cdot r, \qquad r \in R,\ h \in H. \tag{4.57}$$

It is tempting to guess that the situation (A) would be equivalent to

(B) H is a Hopf algebra and R is a braided Hopf algebra.

This is indeed the case; it remains to produce a Hopf algebra $R\#H$ from R and H, and this is done by a construction proposed by Radford, and interpreted in terms of braided categories by Majid; see [66, 73]. Concretely, $R\#H = R \otimes H$ as a vector space, so we use the notation $r\#h = r \otimes h$, $r \in R$, $h \in H$. This is a Hopf algebra by

$$\begin{aligned} (r\#h)(s\#f) &= r(h_{(1)} \cdot s)\#h_{(2)} f, \\ \Delta(r\#h) &= r^{(1)}\#(r^{(2)})_{(-1)} h_{(1)} \otimes (r^{(2)})_{(0)}\#h_{(2)}. \end{aligned} \tag{4.58}$$

We call $R\#H$ the bosonization of R (some authors say the Radford biproduct instead). We are back in situation (A) by the maps $\pi : R\#H \to H$ and $\iota : H \to R\#H$,

$$\pi(r\#h) = \varepsilon(r)h, \qquad \iota(h) = 1\#h, \qquad r \in R,\ h \in H.$$

Exercise 4.26 1. Prove that $R\#H$ is a Hopf algebra with the structure (4.55), with antipode $\mathscr{S}_{R\#H}$ determined by

$$\mathscr{S}_{R\#H}(r) = \mathscr{S}(r_{(-1)})\mathscr{S}_R(r_{(0)}), \qquad r \in R. \tag{4.59}$$

Then $p_R : R\#H \to R$, $p_R(r\#h) = r\varepsilon(h)$, $r \in R, h \in H$, is a morphism of coalgebras.

2. Prove that $R = A^{\mathrm{co}H}$ is a Hopf algebra in ${}^H_H\mathscr{YD}$ with the structure (4.58). Along the way, prove that the antipode $\mathscr{S}_R$ of R is given by

$$\mathscr{S}_R(r) = r_{(-1)}\mathscr{S}(r_{(0)}), \qquad r \in R, \tag{4.60}$$

is a morphism in ${}^H_H\mathscr{YD}$ and is anti-multiplicative and anti-comultiplicative in the following sense:

$$\begin{aligned} \mathscr{S}_R\mu &= \mu(\mathscr{S}_R \otimes \mathscr{S}_R)c = \mu c(\mathscr{S}_R \otimes \mathscr{S}_R), \\ \Delta_R\mathscr{S}_R &= (\mathscr{S}_R \otimes \mathscr{S}_R)c\Delta_R = c(\mathscr{S}_R \otimes \mathscr{S}_R)\Delta_R, \end{aligned} \tag{4.61}$$

3. Let R be a Hopf algebra in ${}^H_H\mathscr{YD}$. The adjoint representation of R on itself is the linear map $\mathrm{ad}_c : R \to \mathrm{End}\, R$ given by

$$\mathrm{ad}_c x(y) = \mu(\mu \otimes \mathscr{S})(\mathrm{id} \otimes c)(\Delta \otimes \mathrm{id})(x \otimes y), \qquad x, y \in R.$$

Show that this is indeed an algebra map. Explicitly,

$$\mathrm{ad}_c x(y) = x^{(1)}[(x^{(2)})_{(-1)} \cdot y]\mathscr{S}((x^{(2)})_{(0)}) = \mathrm{ad}\, x(y), \qquad x, y \in R. \tag{4.62}$$

Show the second equality (use (4.55) and the expression of the antipode). Let $\mathscr{P}(R) = \{x \in R : \Delta_R(x) = x \otimes 1 + 1 \otimes x\}$, the space of primitive elements. Then

$$\mathrm{ad}_c x(y) = xy - (x_{(-1)} \cdot y)x_{(0)}, x \in \mathscr{P}(R), y \in R. \tag{4.63}$$

Hence $\mathscr{P}(R)$ is a Yetter–Drinfeld submodule of R. Using (4.61), show that

$$\mathrm{ad}_c x(\mathscr{S}_R(y)) = \mathscr{S}_R(\mathrm{ad}_{c^{-1}} x(y)), \qquad x \in \mathscr{P}(R), y \in R. \tag{4.64}$$

3. Let X be a Yetter–Drinfeld submodule of R. Then $\mathscr{S}_R(\Bbbk\langle X\rangle) = \Bbbk\langle \mathscr{S}_R(X)\rangle$.

4.3.3 Nichols Algebras: Definitions

We are now ready to address the main objective of this paper. Let H be a Hopf algebra with bijective antipode.

Let $V \in {}^H_H\mathscr{YD}$; for simplicity of the exposition, we assume that $\dim V < \infty$, although this is not needed in most places. Recall that the tensor algebra $T(V)$ and the tensor coalgebra $T^c(V)$ are Hopf algebras in ${}^H_H\mathscr{YD}$, see Exercises 4.24 and 4.25. By the universal property of the tensor algebra, there is a morphism of algebras

$$\Omega : T(V) \to T^c(V) \qquad \text{such that} \qquad \Omega(v) = v, \text{ for all } v \in V. \tag{4.65}$$

It is not difficult to see that Ω

- is a morphism in ${}^H_H\mathcal{YD}$,
- it preserves the coalgebra structure,
- it preserves the grading.

Indeed, all properties follow because they hold at the level of V. In short, Ω is a morphism of graded Hopf algebras in ${}^H_H\mathcal{YD}$. We denote

$$\Omega_n = \Omega_{|T^n(V)}; \qquad \text{hence} \qquad \Omega = \sum_n \Omega_n.$$

Definition 4.7 The Nichols algebra $\mathscr{B}(V)$ is the image of the map Ω.

Let $\mathscr{J}(V) := \ker \Omega$; then $\mathscr{J}(V) = \oplus_{n\geq 2}\mathscr{J}^n(V)$, where $\mathscr{J}^n(V) = \ker \Omega^n$. Then

$$\mathscr{B}(V) = \oplus_{n\geq 0}\mathscr{B}^n(V) \simeq T(V)/\mathscr{J}(V), \qquad \mathscr{B}^n(V) \simeq T^n(V)/\mathscr{J}^n(V).$$

We give now a first alternative description of $\mathscr{J}(V)$. Recall the representation $\rho_n : \mathbb{B}_n \to GL(V^{\otimes n})$ of the braid group $\mathbb{B}_n$, cf. (4.12). Recall also the Matsumoto section $M : \mathbb{S}_n \to \mathbb{B}_n$, cf. (4.5).

Proposition 4.4 *If $n \geq 2$, then*

$$\Omega_n = \sum_{\sigma \in \mathbb{S}_n} \rho_n(M(\sigma)) \in End(V^{\otimes n}). \tag{4.66}$$

In particular, the algebra and the coalgebra structures of $\mathscr{B}(V)$ depend on the braided vector space (V, c) but not on the specific realization in ${}^H_H\mathcal{YD}$.

For instance, write $c_1 = c \otimes \mathrm{id}$, $c_2 = \mathrm{id} \otimes c$. Then

$$\Omega_2 = \mathrm{id} + c, \qquad \Omega_3 = \mathrm{id} + c_1 + c_2 + c_1c_2 + c_2c_1 + c_1c_2c_1.$$

Here is an abstract characterization of Nichols algebras.

Proposition 4.5 ([19]) *The ideal $\mathscr{J}(V)$ is maximal in the set*

$$\mathfrak{C} := \{J = \oplus_{n\geq 2}J^n \textit{ is a graded Hopf ideal and Yetter–Drinfeld submodule of } T(V)\}.$$

Let $\mathscr{B} = \oplus_{n\geq 0}\mathscr{B}^n$, $\mathscr{E} = \oplus_{n\geq 0}\mathscr{E}^n$ be graded Hopf algebras in ${}^H_H\mathcal{YD}$ such that

$$\mathscr{B}^1 \simeq \mathscr{E}^1 \simeq V \text{ in } {}^H_H\mathcal{YD}.$$

Assume that $\mathscr{B}$ satisfies $\dim \mathscr{B}^n < \infty$. Then the graded dual of $\mathscr{B}$, denoted by $\mathscr{B}^\star$, is again a graded Hopf algebra in ${}^H_H\mathscr{YD}$, see (4.6).

Definition 4.8 1. If $\mathscr{B} = \Bbbk\langle V\rangle$, then we say that $\mathscr{B}$ is a *pre-Nichols algebra*. By definition, there is a surjective map $T(V) \twoheadrightarrow \mathscr{B}$ of graded Hopf algebras in ${}^H_H\mathscr{YD}$; but the kernel of this map is contained in $\mathscr{J}(V)$ by Proposition 4.5, so that there is also a surjective map $\mathscr{B} \twoheadrightarrow \mathscr{B}(V)$ of graded Hopf algebras in ${}^H_H\mathscr{YD}$.

2. We say $\mathscr{E}$ is a *post-Nichols algebra* if it is coradically graded. Dually, there are injective maps $\mathscr{B}(V) \hookrightarrow \mathscr{E} \hookrightarrow T^c(V)$ of graded Hopf algebras in ${}^H_H\mathscr{YD}$.

Indeed, $\mathscr{B}$ is a pre-Nichols algebra (of V) if and only if $\mathscr{B}^\star$ is a post-Nichols algebra (of V^*); here we need that $\dim V < \infty$. Thus, for $\mathscr{B}$ a pre-Nichols, and $\mathscr{E}$ a post-Nichols, algebra, the situation can be summarized by the following commutative diagram:

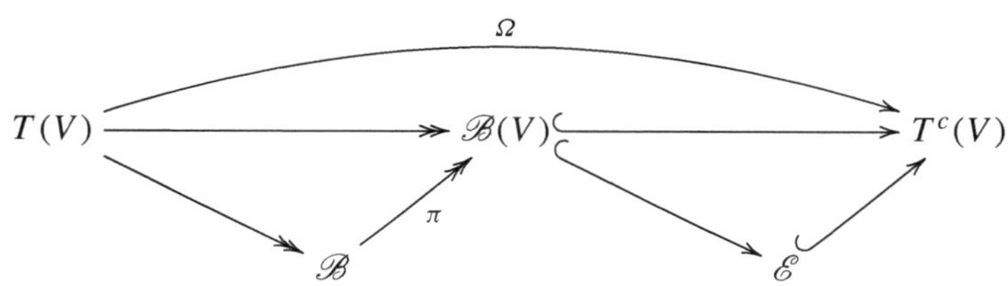

The next characterization is a natural consequence of this discussion, see [19].

Proposition 4.6 *The graded Hopf algebra $\mathscr{B}$ is isomorphic to $\mathscr{B}(V)$ if and only if*

1. *it is generated as an algebra by V, $\mathscr{B} = \Bbbk\langle V\rangle$,*
2. *it is coradically graded.*

We summarize the characterizations, or alternative definitions of the Nichols algebra $\mathscr{B}(V)$, or equivalently the defining ideal $\mathscr{J}(V) = \oplus_{n\geq 2}\mathscr{J}^n(V)$:

- $\mathscr{B}(V) =$ image of $\Omega : T(V) \to T^c(V)$. Thus, $\mathscr{J}(V) = \ker \Omega$.
- $\Omega = \sum_{n\geq 2}\Omega_n$, $\Omega_n = \sum_{\sigma\in\mathbb{S}_n}\rho_n(M(\sigma))$. Thus $\mathscr{J}^n(V) = \ker \sum_{\sigma\in\mathbb{S}_n}\rho_n(M(\sigma))$.
- $\mathscr{J}(V)$ is maximal in the class $\mathfrak{C}$ of graded Hopf ideals as in Proposition 4.5.
- $\mathscr{B}(V)$ is the only graded Hopf algebra both coradically graded and generated in degree 1 (by V). That is the only pre- and simultaneously post-Nichols algebra of V (up to isomorphisms).

There is a useful criterion with skew derivations to find relations of $\mathscr{B}(V)$, $V \in {}^H_H\mathscr{YD}$; see, e.g., [16] for details. Let $f \in V^*$. Let $\partial_f \in \operatorname{End} T(V)$ be given by

$$\partial_f(1) = 0, \qquad \partial_f(v) = f(v), \quad v \in V, \tag{4.67}$$

$$\partial_f(xy) = x\partial_f(y) + \sum_j \partial_{f_j}(x)y_j, \quad \text{where } c^{-1}(y\otimes f) = \sum_j f_j \otimes y_j. \tag{4.68}$$

Here is the criterion:

- Let $x \in T^n(V)$, $n \geq 2$. If $\partial_f(x) = 0$ for all $f \in V^*$, then $x \in \mathscr{J}^n(V)$.

Suppose that there are a basis $(x_i)_{i\in\mathbb{I}}$ of V, with dual basis $(f_i)_{i\in\mathbb{I}}$, and a family $(g_i)_{i\in\mathbb{I}}$ in $G(H)$ such that $\delta(x_i) = g_i \otimes x_i$, for $i \in \mathbb{I}$. Set $\partial_i = \partial_{f_i}$, $i \in \mathbb{I}$. Then (4.68) for all f is equivalent to

$$\partial_i(xy) = x\partial_i(y) + \partial_i(x)\, g_i \cdot y, \qquad x, y \in T(V), \qquad i \in \mathbb{I}. \tag{4.69}$$

The preceding arguments are the gate to the applications of Nichols algebras to the classification of pointed Hopf algebras [4, 19, 20], see also [10]. Indeed, let A be a pointed Hopf algebra and let gr A be the graded coalgebra associated with the coradical filtration. Then

$$\operatorname{gr} A \simeq \mathscr{R}\#\Bbbk G(A), \quad \text{where } \mathscr{R} = \oplus_{n\geq 0}\mathscr{R}^n \text{ is a graded Hopf algebra in } {}^H_H\mathscr{YD}.$$

Set $V = \mathscr{R}^1$. Now $\mathscr{R}$ is coradically graded, as it arises from the coradical filtration, in short it is a post-Nichols algebra of V; while its subalgebra generated by V is isomorphic to $\mathscr{B}(V)$. This leads to following problems:

- When is $\dim \mathscr{B}(V) < \infty$? For such V, classify its finite-dimensional post-Nichols algebras.
- When is the Gelfand–Kirillov dimension of $\mathscr{B}(V)$ finite? For such V, classify its post-Nichols algebras with finite Gelfand–Kirillov dimension.

We do not need only the list of all V positively answering these questions, but also we need to compute the Nichols algebras explicitly. By this we mean:

- Find a basis of $\mathscr{B}(V)$, hence the dimension or the Gelfand–Kirillov dimension, and
- describe the defining relations, i.e., a minimal set of generators of the ideal $\mathscr{J}(V)$.

Analogously for the mentioned post-Nichols algebras.

Needless to say, there is no hope presently to solve in full generality these problems. Toward the first question above, it was proposed in [18]:

Conjecture 4.1 Assume that char $\Bbbk = 0$ and that H is semisimple. Let $V \in {}^H_H\mathscr{YD}$ such that $\dim \mathscr{B}(V) < \infty$. Then there is no finite-dimensional post-Nichols algebra except $\mathscr{B}(V)$ itself.

The conjecture contains the following particular case:

Conjecture 4.2 Assume that char $\Bbbk = 0$. Then every finite-dimensional pointed Hopf algebra is generated by group-like and skew-primitive elements.

The last conjecture is definitely false if either char $\Bbbk > 0$ or else the finite-dimensional requirement is dropped.

Theorem 4.3 ([22]) *Assume that* char $\Bbbk = 0$. *Then every finite-dimensional pointed Hopf algebra* with abelian group of group-likes *is generated by group-like and skew-primitive elements.*

4.3.4 Nichols Algebras: Techniques

Here we discuss approaches to compute Nichols algebras.

4.3.4.1 Direct Computation

First, let $m \in \mathbb{N}_{\geq 2}$. The m-th approximation of $\mathscr{B}(V)$ is

$$\widehat{\mathscr{B}}_m(V) = T(V)/\langle \oplus_{2 \leq n \leq m} \mathscr{J}^n(V) \rangle = T(V)/\Big\langle \oplus_{2 \leq n \leq m} \ker \Big(\sum_{\sigma \in \mathbb{S}_n} \rho_n(M(\sigma)) \Big) \Big\rangle,$$

cf. (4.66). By definition, there is an epimorphism of graded Hopf algebras

$$\pi_m : \widehat{\mathscr{B}}_m(V) \to \mathscr{B}(V).$$

A brutal approach would be to compute $\widehat{\mathscr{B}}_m(V)$ for $m = 2, 3, \ldots$ and at each step try to figure out whether π_m is an isomorphism, using some of the characterizations of $\mathscr{B}(V)$. In principle, $\mathscr{J}^2(V) = \ker(\mathrm{id} + c)$ is effectively computable, but the difficulties mount with m, as Ω_n is the sum of $n!$ terms acting on a vector space of dimension $(\dim V)^n$. Other drawbacks are that the ideal $\mathscr{J}(V)$ need not be finitely generated, nor have quadratic relations at all; even to predict the lowest degree relations is not within reach. A variation of this approach would be:

- Find a set R_1 of relations in $\mathscr{J}(V)$; i.e., $\mathscr{J}^2(V)$ or some relations of small order.
- Compute the pre-Nichols algebra $\mathscr{B}_1 = T(V)/\langle R_1 \rangle$, i.e., find a basis B_1 of $\mathscr{B}_1$.
- Decide whether the image of B_1 in $\mathscr{B}(V)$ is linearly independent (here derivations are the best option). If yes, then $\mathscr{B}(V) \simeq \mathscr{B}_1$.
- If no, then we would have found a new set of relations R_2; set $\mathscr{B}_2 = \mathscr{B}_1/\langle R_1 \rangle$ and start again.

Of course the success of this approach depends on great doses of intuition and luck.

Exercise 4.27 Let $q \in \Bbbk^\times$ and let $\mathscr{A} = \mathbb{Z}[\mathbf{q}]$ be the polynomial algebra.

1. Let $(n)_\mathbf{q} = 1 + \mathbf{q} + \cdots + \mathbf{q}^{n-1}$ and $(n)!_\mathbf{q} = (n)_\mathbf{q} \ldots (2)_\mathbf{q}(1)_\mathbf{q} \in \mathscr{A}$. The $\mathbf{q}$-binomial numbers are

$$\binom{n}{i}_\mathbf{q} := \frac{(n)!_\mathbf{q}}{(n-i)!_\mathbf{q}(i)!_\mathbf{q}}, \qquad n \in \mathbb{N},\ i \in \mathbb{I}_{0,n}.$$

Prove that

$$\mathbf{q}^k \binom{n}{k}_{\mathbf{q}} + \binom{n}{k-1}_{\mathbf{q}} = \binom{n+1}{k}_{\mathbf{q}}, \qquad k \in \mathbb{I}_n. \tag{4.70}$$

Conclude that $\binom{n}{i}_{\mathbf{q}} \in \mathscr{A}$. Let $\binom{n}{i}_q \in \Bbbk$ be the specialization of $\binom{n}{i}_{\mathbf{q}}$ at q.

2. Let A be an associative algebra; let $u, v \in A$ be q-commuting elements, i.e., $uv = qvu$. Then the *quantum binomial formula* holds:

$$(u+v)^n = \sum_{i=0}^{n} \binom{n}{i}_q v^i u^{n-i}, \qquad \text{for every } n \in \mathbb{N}. \tag{4.71}$$

3. Let V be a braided vector space of dimension 1 with braiding $c(x \otimes x) = qx \otimes x$ for all $x \in V$. Fix $x \in V - 0$. Let

$$N := \begin{cases} \operatorname{ord} q, & \text{if } q \in \mathbb{G}'_\infty, \\ 1, & \text{otherwise.} \end{cases}$$

Prove that $x^N \in \mathscr{P}(T(V))$. Conclude that

$$\mathscr{B}(V) \simeq \begin{cases} T(V)/\langle x^N \rangle, & \text{if } q \in \mathbb{G}'_\infty, \\ T(V), & \text{otherwise.} \end{cases}$$

Example 4.27 ([17]) Let $\mathfrak{q} = (q_{ij}) \in (\Bbbk^\times)^{\mathbb{I}\times\mathbb{I}}$. Let V be a braided vector space with basis $(x_i)_{i\in\mathbb{I}}$ and braiding (4.15); (4.16) is not assumed. Let

$$N_k := \begin{cases} \operatorname{ord} q_{kk}, & \text{if } q_{kk} \in \mathbb{G}'_\infty, \\ 1, & \text{otherwise,} \end{cases} \qquad \text{for } k \in \mathbb{I}.$$

Suppose that

$$q_{ij}q_{ji} = 1, \qquad \text{for all } i \neq j \in \mathbb{I}.$$

It is easy to check that

$$x_i x_j - q_{ij} x_j x_i \in \mathscr{J}^2(V), \qquad \text{for all } i \neq j \in \mathbb{I}. \tag{4.72}$$

Then

$$\mathscr{B} = T(V)/\langle x_i x_j - q_{ij} x_j x_i,\ i \neq j \in \mathbb{I};\ \ x_k^{N_k},\ q_{kk} \in \mathbb{G}'_\infty \rangle \tag{4.73}$$

is a pre-Nichols algebra of V. Using linear algebra arguments, one may check that $\mathscr{B} \simeq \mathscr{B}(V)$ and that

$$\{x_1^{a_1} x_2^{a_2} \dots x_\theta^{a_\theta} : 0 \leq a_k \leq N_k, \text{ if } q_{kk} \in \mathbb{G}'_\infty; \quad 0 \leq a_k \text{ otherwise } \} \text{ is a basis of } \mathscr{B}(V).$$

Definition 4.9 The algebra presented by generators and relations as in the right-hand side of (4.73) is called a *quantum linear space*.

Notice that there are examples of quantum planes (quantum linear spaces with $\theta = 2$) that are braided Hopf algebras with respect to braidings not of diagonal type; this was first noticed in [38].

Example 4.28 ([14]) Let (V, c) be a braided vector space. Assume that

- $\dim V = 2$,
- $\mathscr{J}^2(V) \neq 0$,
- c is not of diagonal type.

Then $\mathscr{B}(V)$ is known. The starting point is the classification of braided vector spaces of dimension 2 [55]. The outcome is that, as algebras, the examples arising are variations of quantum planes, variations of the Jordan and super Jordan algebras (see Sect. 4.4.3) and some strange examples.

4.3.4.2 Dual

Let $V \in {}^H_H\mathscr{YD}$ finite-dimensional. As we observed after Definition 4.8, the graded dual of a pre-Nichols of V is a post-Nichols algebra of V^* and vice versa, thus we have:

Remark 4.5 The graded dual of $\mathscr{B}(V)$ is isomorphic to $\mathscr{B}(V^*)$.

This gives some new information without extra effort, as V^* need not be isomorphic to V as braided vector space. For instance, let $X = (X, \triangleright)$ be a rack and $\mathfrak{q}$ a 2-cocycle as in (4.27). Let $X^{-1} = (X, \triangleright^{-1})$, where $x \triangleright^{-1} y = \phi_x^{-1}(y)$, cf. Exercise 4.8. Let also $\mathfrak{q}^* : X^{-1} \times X^{-1} \to \Bbbk^\times$ given by

$$\mathfrak{q}^*_{xy} = \mathfrak{q}_{x, x \triangleright^{-1} y}, \qquad x, y \in X.$$

Then the braided vector space dual to $(\Bbbk X, c^{\mathfrak{q}})$, is $(\Bbbk X^{-1}, c^{\mathfrak{q}^*})$. See [39] for details.

4.3.4.3 Twisting

V.G. Drinfeld introduced in [29] the twisting of quasi-Hopf algebras, meaning conjugation of the comultiplication by a suitable element, to keep account of equivalences of tensor categories. This was specialized to Hopf algebras in [76], with the definition of multiparametric quantum groups as application. The dual version, called twisting of the multiplication, appeared in [27]. We recall this last one. Let H be a Hopf algebra.

Definition 4.10 A linear map $\phi : H \otimes H \to \Bbbk$ is a unitary 2-cocycle if

$$\phi \text{ is invertible with respect to the convolution product } *, \text{ see Exercise 4.23;} \tag{4.74}$$

$$\phi(x_{(1)} \otimes y_{(1)})\, \phi(x_{(2)}y_{(2)} \otimes z) = \phi(y_{(1)} \otimes z_{(1)})\, \phi(x \otimes y_{(2)}z_{(2)}), \tag{4.75}$$

$$\phi(x \otimes 1) = \phi(1 \otimes x) = \varepsilon(x), \tag{4.76}$$

for all $x, y, z \in H$. Let ϕ be a unitary 2-cocycle and define a new multiplication $\cdot_\phi$ in the vector space H by

$$x \cdot_\phi y = \phi(x_{(1)}, y_{(1)}) x_{(2)} y_{(2)} \phi^{-1}(x_{(3)}, y_{(3)}), \qquad x, y \in H.$$

Then $H_\phi = (H, \cdot_\phi, \Delta)$ is a Hopf algebra.

Exercise 4.28 Let G be a group. A unitary 2-cocycle on $\Bbbk G$ is equivalent to a 2-cocycle $\phi \in Z^2(G, \Bbbk^\times)$, i.e., a map $\phi : G \times G \to \Bbbk^\times$ such that

$$\phi(g, h)\phi(gh, t) = \phi(h, t)\phi(g, ht), \quad \phi(g, e) = \phi(e, g) = 1, \quad g, h, t \in G. \tag{4.77}$$

The relation with bosonization was established in [68].

Theorem 4.4 ([68, Theorem 2.7, Corollary 3.4]) *Let $\phi : H \otimes H \to \Bbbk$ be an invertible unitary 2-cocycle.*

(a) There exists an equivalence of braided categories $\mathscr{T}_\phi : {}^H_H\mathscr{YD} \to {}^{H_\phi}_{H_\phi}\mathscr{YD}$, $V \mapsto V_\phi$, which is the identity on the underlying vector spaces, morphisms and coactions, and transforms the action of H on V to $\cdot_\phi : H_\phi \otimes V_\phi \to V_\phi$,

$$h \cdot_\phi v = \phi(h_{(1)}, v_{(-1)})(h_{(2)} \cdot v_{(0)})\_0\, \phi^{-1}((h_{(2)} \cdot v_{(0)})\_{-1}, h_{(3)}),$$

$h \in H_\phi$, $v \in V_\phi$. The monoidal structure on $\mathscr{T}_\phi$ is given by the natural transformation $b_{V,W} : (V \otimes W)_\phi \to V_\phi \otimes W_\phi$

$$b_{V,W}(v \otimes w) = \phi(v_{(-1)}, w_{(-1)}) v\_0 \otimes w\_0, \quad v \in V, w \in W.$$

(b) $\mathscr{T}_\phi$ preserves Nichols algebras: $\mathscr{B}(V)_\phi \simeq \mathscr{B}(V_\phi)$ as objects in ${}^{H_\phi}_{H_\phi}\mathscr{YD}$. In particular, the Hilbert–Poincaré series of $\mathscr{B}(V)$ and $\mathscr{B}(V_\phi)$ are the same.

Example 4.29 Let $\mathfrak{q} = (q_{ij})$, $\mathfrak{q}' = (q'_{ij}) \in (\Bbbk^\times)^{\mathbb{I}\times\mathbb{I}}$ satisfying (4.16). We say that $\mathfrak{q}$ and $\mathfrak{q}'$ are twist-equivalent if

$$q_{ii} = q'_{ii}, \qquad i \in \mathbb{I} \qquad \text{and} \qquad q_{ij}q_{ji} = q'_{ij}q'_{ji}, \qquad i \neq j \in \mathbb{I}.$$

In other words, twist-equivalent means that the matrices $\mathfrak{q}$ and $\mathfrak{q}'$ have the same Dynkin diagram, cf. Sect. 4.2.1.3. Let V and V' be the braided vector spaces of diagonal type associated with $\mathfrak{q}$ and $\mathfrak{q}'$, respectively. If $\mathfrak{q}$ and $\mathfrak{q}'$ are twist-equivalent, then the Hilbert–Poincaré series of $\mathscr{B}(V)$ and $\mathscr{B}(V')$ coincide; this consequence of Theorem 4.4 was observed in [19, Proposition 3.9].

Example 4.30 Let X be a rack (isomorphic to a conjugacy class in a finite group) and let $\mathfrak{q}$ and $\mathfrak{q}'$ be 2-cocycles on X. We say that $\mathfrak{q}$ and $\mathfrak{q}'$ are twist-equivalent if there exists $\phi : X \times X \to \Bbbk^\times$ such that $\mathfrak{q}' = \mathfrak{q}^\phi$, which is

$$q^\phi_{xy} = \phi(x, y)\phi^{-1}(x \triangleright y, x)\, q_{xy}, \quad x, y \in X. \tag{4.78}$$

If $\mathfrak{q}$ and $\mathfrak{q}'$ are twist-equivalent, then the Hilbert–Poincaré series of $\mathscr{B}(X, \mathfrak{q})$ and $\mathscr{B}(X, \mathfrak{q}')$ coincide [13, Sect. 3.4].

Exercise 4.29 Let X be a rack, $\mathfrak{q}$ a 2-cocycle on X and $\phi : X \times X \to \Bbbk^\times$. Show that $\mathfrak{q}^\phi$ defined by (4.78) is a 2-cocycle iff for any $x, y, z \in X$, we have

$$\begin{aligned}\phi(x, z)\phi(x \triangleright y, x \triangleright z)\phi(x \triangleright (y \triangleright z), x)\phi(y \triangleright z, y) \\ = \phi(y, z)\phi(x, y \triangleright z)\phi(x \triangleright (y \triangleright z), x \triangleright y)\phi(x \triangleright z, x)\end{aligned} \tag{4.79}$$

Hence, if X is a subrack of a group G and $\phi \in Z^2(G, \Bbbk^\times)$, then $\phi|_{X \times X}$ satisfies (4.79).

4.3.4.4 Discard

There are techniques to prove that a Nichols algebra has infinite dimension, or GK-dimension. Various of them are related to decompositions, as explained below. Let (V, c) be a braided vector space. We mention in this line of thought:

- If $W \hookrightarrow V$ is a braided subspace, respectively $V \twoheadrightarrow W$ is a quotient braided space, then $\mathscr{B}(W) \hookrightarrow \mathscr{B}(V)$, respectively $\mathscr{B}(V) \twoheadrightarrow \mathscr{B}(W)$. Thus, if V has a braided subspace or a braided quotient whose Nichols algebra has infinite dimension (or GK-dimension), then so has $\mathscr{B}(V)$. There are elaborations of these arguments specific to rack type that are evoked below.
- Assume that V has a filtration of braided subspaces: $0 = V_0 \subsetneq V_1 \cdots \subsetneq V_d = V$. Then this filtration propagates to $\mathscr{B}(V)$ and the associated gr $\mathscr{B}(V)$ turns out to be a pre-Nichols algebra of gr V. Thus, if $\mathscr{B}(\text{gr } V)$ has infinite dimension, (or GK-dimension), then so has $\mathscr{B}(V)$.

4.3.4.5 Decomposition

Let $\theta \geq 2$. Assume that (V, c) satisfies

$$V = V_1 \oplus \cdots \oplus V_\theta, \qquad c(V_i \otimes V_j) = V_j \otimes V_i, \qquad i, j \in \mathbb{I}_\theta. \tag{4.80}$$

Here we suppose that the $\mathscr{B}(V_i)$'s are known and seek to infer $\mathscr{B}(V)$. This idea, mentioned in passing in [1, p. 41], is a roundabout approach, where instead of computing the relations or the basis, one looks for combinatorial invariants reminiscent of the Weyl group. The principal actors are the maps

$$c_{ij} := c_{|V_i \otimes V_j} : V_i \otimes V_j \to V_j \otimes V_i, , \qquad i, j \in \mathbb{I}_\theta.$$

Exercise 4.30 ([41]) If $\theta = 2$ and $c_{21}c_{12} = \mathrm{id}_{V_1 \otimes V_2}$, then $\mathscr{B}(V) \simeq \mathscr{B}(V_1)\underline{\otimes}\mathscr{B}(V_2)$.

Here are some particular instances of this situation:

- Assume that $\dim V_i = 1$ for all i. Then V is of diagonal type, up to (4.16); cf. Sect. 4.4.2.
- Assume that there exists a Hopf algebra H such that $V_i \in {}^H_H\mathscr{YD}$ and is irreducible in this category. This setting was considered in [16, 50].
- See Sect. 4.4.5 for the case $H = \Bbbk G$, G a finite group.
- Assume that either $\dim V_i = 1$ or V_i is a block. Then the classification of all V such that GK-dim $\mathscr{B}(V) < \infty$ was obtained in [4], see Sect. 4.4.4. Here is a crucial remark that should be useful in other settings:

Assume that $\theta = 2$ but that either V_1 or V_2 is not irreducible, or both. The combinatorial invariants from [16, 49] are not available but we may proceed as follows. There are natural morphisms of braided Hopf algebras

$$\pi : \mathscr{B}(V) \to \mathscr{B}(V_1), \qquad \iota : \mathscr{B}(V_1) \to \mathscr{B}(V), \qquad \text{such that } \pi\iota = \mathrm{id}_{\mathscr{B}(V_1)}.$$

As in Sect. 4.3.2, we consider $K = \mathscr{B}(V)^{\mathrm{co}\,\mathscr{B}(V_1)}$, but now in ${}^H_H\mathscr{YD}$. Remarkably,

$$\mathscr{B}(V) \simeq K\#\mathscr{B}(V_1) \qquad \text{and} \qquad K \simeq \mathscr{B}\left(\mathrm{ad}_c\mathscr{B}(V_1)(V_2)\right),$$

[50, Proposition 8.6], cf. also [16, Lemma 3.2]. Ingenuously, one may try to compute the Nichols algebra of $\mathrm{ad}_c\mathscr{B}(V_1)(V_2)$; at a first glance, this appears more complicated, but sometimes this works.

4.4 Classes of Nichols Algebras

We discuss in the last section of this paper several classes of Nichols algebras. From now on, $\Bbbk$ is algebraically closed and char $\Bbbk = 0$.

4.4.1 Symmetries and Hecke Type

Here the situation is quite simple:

Proposition 4.7 *Let (V, c) be a braided vector space such that c is either a symmetry or of Hecke type with label $q \notin \mathbb{G}_\infty$. Then $\mathscr{B}(V) \simeq T(V)/\langle \ker(c + \mathrm{id})\rangle$.*

See [19, Proposition 3.4]; the argument is taken from a paper by Andrés Abella and the author. By [42], it follows that $\mathscr{B}(V)$ is a Koszul algebra, see *loc. cit.*

4.4.2 Diagonal Type

Nichols algebras of diagonal type were studied in depth. In the finite-dimensional setting, there are two main results:

- The classification of all finite-dimensional Nichols algebras of diagonal type appears in [45], using the Weyl groupoid introduced in [44].
- The defining relations of the finite-dimensional Nichols algebras of diagonal type appear in [21, 22].

We refer to the survey [3] for details, since both answers are very long and require a careful preparation. One of the outcomes is that the theory of Nichols algebras of diagonal type embeds into Lie theory. Here is a remarkable instance of this affirmation:

Theorem 4.5 ([18, 44]) *Let V be a braided vector space of Cartan type with Cartan matrix A, see Sect. 4.2.1.3. Then* $\dim \mathscr{B}(V) < \infty$ *if and only if A is a finite Cartan matrix (i.e., corresponds to a finite-dimensional simple Lie algebra).*

This result was proved in [18] under some restrictions on the matrix $\mathfrak{q}$ of the braiding, by reduction to the theory of quantum groups. A proof valid without restrictions appears in [44] based on the beautiful theory of the Weyl groupoid.

As for finite Gelfand–Kirillov dimension, the validity of the following conjecture would say that the classification follows from [45]. Let (V, c) be a braided vector space of diagonal type.

Conjecture 4.3 ([4]) If GK-dim $\mathscr{B}(V) < \infty$, then its Weyl groupoid is finite.

The following partial results support the conjecture.

Theorem 4.6 ([5]) *If either its Weyl groupoid is infinite and* dim $V = 2$, *or else* V *is of affine Cartan type, then GK-dim* $\mathscr{B}(V) = \infty$.

Let us finally discuss an example with many applications.

Example 4.31 Let V be a braided vector space of dimension 2, of diagonal type with braiding matrix $\mathfrak{q} = \begin{pmatrix} q & q \\ q & q \end{pmatrix}$.

- The case $q = 1$ is not of diagonal type, strictly speaking, by our requirement (4.16). Nevertheless, $\mathscr{B}(V) \simeq S(V)$.
- If $q = -1$, then $\mathscr{B}(V) \simeq \Lambda(V)$.
- If $q \in \mathbb{G}'_N$, then V is of Cartan type $\begin{pmatrix} 2 & 2-N \\ 2-N & 2 \end{pmatrix}$. Thus, if $N = 3$, then is of Cartan type A_2 and dim $\mathscr{B}(V) = 27$.
- If $N > 3$, then GK-dim $\mathscr{B}(V) = \infty =$ dim $\mathscr{B}(V)$ by Theorems 4.6, respectively Theorem 4.5.

4.4.3 Triangular Type

Here we give a glimpse to the main results in [4] on Nichols algebras with finite GK-dimension over an abelian group. Succinctly, these results consist of

- The classification of braided vector spaces whose Nichols algebras have finite GK-dim, that admit a decomposition (4.80) whose components are ±1-blocks or points, i.e., $V = V_1 \oplus \cdots \oplus V_t \oplus V_{t+1} \oplus \cdots \oplus V_\theta$ where

$$V_h \quad \varepsilon_h\text{-block}, \quad \varepsilon_h^2 = 1, \ h \in \mathbb{I}_t; \qquad V_i \quad q_{ii}\text{-point}, \quad q_{ii} \in \Bbbk^\times, \ i \in \mathbb{I}_{t+1,\theta},$$

  with $c(V_i \otimes V_j) = V_j \otimes V_i$, $i, j \in \mathbb{I}_\theta$. Set as usual $c_{ij} = c_{|V_i \otimes V_j}$. We assume that

  - V is not of diagonal type, i.e., $t > 0$;
  - the braiding c_{ij} between a block $i \in \mathbb{I}_t$ and a point $j \in \mathbb{I}_{t+1,\theta}$ has the form (4.83).

- The classification of Yetter–Drinfeld modules over abelian groups whose Nichols algebras have finite GK-dim, that admit a decomposition $V = V_1 \oplus V_2$ like (4.80) where V_1 is a ±1-block and V_2 is a point, but c_{12} does not have the form (4.83).

We point out that the first classification mentioned assumes the validity of Conjecture 4.3. The explicit formulation requires some preparation, so we refer to [5] for full details. Instead, we discuss here two relevant steps of the proof–steps that do not require Conjecture 4.3.

To start with, recall the block $\mathscr{V}(\varepsilon, \ell)$, $\varepsilon \in \Bbbk^\times$ and $\ell \in \mathbb{N}_{\geq 2}$, cf. Example 4.6.

Theorem 4.7 ([4, Theorem 1.2, Propositions 3.4, 3.5]) *The Gelfand–Kirillov dimension of the Nichols algebra $\mathscr{B}(\mathscr{V}(\varepsilon, \ell))$ is finite if and only if $\ell = 2$ and $\varepsilon^2 = 1$.*

The algebras $\mathscr{B}(\mathscr{V}(\varepsilon, 2))$ have GK-dim 2 and are presented by generators x_1 and x_2 with defining relations

$$x_2x_1 - x_1x_2 + \frac{1}{2}x_1^2, \qquad \text{if } \varepsilon = 1; \tag{4.81}$$

$$x_2x_{21} - x_{21}x_2 - x_1x_{21}, \qquad x_1^2, \qquad \text{if } \varepsilon = -1, \tag{4.82}$$

where $x_{21} = x_2x_1 + x_1x_2$.

This result explains why we restrict to ± 1-blocks (recall that these means also dimension 2).

Next we turn our attention to the setting *one block plus one point*, i.e., braided vector spaces $V = V_1 \oplus V_2$, where V_1 has a basis $(x_i)_{i \in \mathbb{I}_2}$, V_2 has a basis (x_3). Our key hypothesis is that the braiding has the shape

$$(c(x_i \otimes x_j))_{i,j \in \mathbb{I}_3} = \begin{pmatrix} \varepsilon x_1 \otimes x_1 & (\varepsilon x_2 + x_1) \otimes x_1 & q_{12}x_3 \otimes x_1 \\ \varepsilon x_1 \otimes x_2 & (\varepsilon x_2 + x_1) \otimes x_2 & q_{12}x_3 \otimes x_2 \\ q_{21}x_1 \otimes x_3 & q_{21}(x_2 + ax_1) \otimes x_3 & q_{22}x_3 \otimes x_3 \end{pmatrix}, \tag{4.83}$$

with $\varepsilon^2 = 1$ and $q_{ij} \in \Bbbk^\times$, $i, j \in \mathbb{I}_2$. We do not want to have $c^2_{|V_1 \otimes V_2} \overset{\star}{=} \mathrm{id}$ because we know the answer, see Exercise 4.30; here $\star$ is equivalent to

$$q_{12}q_{21} = 1 \text{ and } a = 0.$$

So $c^2_{|V_1 \otimes V_2}$ is determined by $q_{12}q_{21}$, that we call the *interaction*, and a, of which we consider a normalized version, that we call the *ghost*: $\mathscr{G} = \begin{cases} -2a, & \varepsilon = 1, \\ a, & \varepsilon = -1. \end{cases}$

If $\mathscr{G} \in \mathbb{N}$, then we say that the ghost is *discrete*.

Theorem 4.8 ([4]) *Let V be a braided vector space with braiding* (4.83). *Then GK-dim $\mathscr{B}(V) < \infty$, if and only if the ghost is discrete and V is as in Table* 4.1.

The meanings of the diagrams are as follows:

- ⊞, respectively ⊟, says that V_1 is a 1-block, respectively a -1-block.
- The label over the point is q_{22}.
- The edge $\overset{\mathscr{G}}{\text{———}}$ says that $q_{12}q_{21} = 1$; $\mathscr{G}$ is discrete but arbitrary unless explicitly given.
- The edge $\overset{(-1,1)}{\text{———}}$ says that $q_{12}q_{21} = -1$ and $\mathscr{G} = 1$.

Table 4.1 Nichols algebras of a block and a point with finite GK-dim

V	Diagram	GK-dim	Generators and relations
$\mathfrak{L}(1,\mathscr{G})$	$\boxplus \overset{\mathscr{G}}{\text{———}} \overset{1}{\bullet}$	$\mathscr{G}+3$	$\daleth\langle x_1, x_2, x_3 \vert x_2x_1 - x_1x_2 + \frac{1}{2}x_1^2, x_1x_3 - q_{12}\, x_3x_1, z_{1+\mathscr{G}}, z_tz_{t+1} - q_{21}q_{22}\, z_{t+1}z_t, 0 \le t < \mathscr{G}\rangle$
$\mathfrak{L}(-1,\mathscr{G})$	$\boxplus \overset{\mathscr{G}}{\text{———}} \overset{-1}{\bullet}$	2	$\daleth\langle x_1, x_2, x_3 \vert x_2x_1 - x_1x_2 + \frac{1}{2}x_1^2, x_1x_3 - q_{12}\, x_3x_1, z_{1+\mathscr{G}}, z_t^2, 0 \le t \le \mathscr{G}\rangle$
$\mathfrak{L}_-(1,\mathscr{G})$	$\boxminus \overset{\mathscr{G}}{\text{———}} \overset{1}{\bullet}$	$\mathscr{G}+3$	$\daleth\langle x_1, x_2, x_3 \vert x_1^2, x_2x_{21} - x_{21}x_2 - x_1x_{21}, x_1x_3 - q_{12}\, x_3x_1, x_{21}x_3 - q_{12}^2x_3x_{21}, z_{1+2\mathscr{G}}, z_{2k+1}^2, z_{2k}z_{2k+1} - q_{21}q_{22}\, z_{2k+1}z_{2k}, 0 \le k < \mathscr{G}\rangle$
$\mathfrak{L}_-(-1,\mathscr{G})$	$\boxminus \overset{\mathscr{G}}{\text{———}} \overset{-1}{\bullet}$	$\mathscr{G}+2$	$\daleth\langle x_1, x_2, x_3 \vert x_1^2, x_2x_{21} - x_{21}x_2 - x_1x_{21}, x_3^2, x_1x_3 - q_{12}\, x_3x_1, x_{21}x_3 - q_{12}^2x_3x_{21}, z_{1+2\mathscr{G}}, z_{2k}^2, z_{2k-1}z_{2k} - q_{21}q_{22}\, z_{2k}z_{2k-1}, 0 < k \le \mathscr{G}\rangle$
$\mathfrak{L}(\omega, 1)$	$\boxplus \overset{1}{\text{———}} \overset{\omega}{\bullet}$	2	$\daleth\langle x_1, x_2, x_3 \vert x_2x_1 - x_1x_2 + \frac{1}{2}x_1^2, x_1x_3 - q_{12}\, x_3x_1, z_2, x_3^3, z_1^3, z_{1,0}^3\rangle$
$\mathfrak{C}_1$	$\boxminus \overset{(-1,1)}{\text{———}} \overset{-1}{\bullet}$	2	$\daleth\langle x_1, x_2, x_3 \vert x_1^2, x_2x_{21} - x_{21}x_2 - x_1x_{21}, x_3^2, f_0^2, f_1^2, z_1^2, x_{21}x_3 - q_{12}^2x_3x_{21}, x_2z_1 + q_{12}z_1x_2 - q_{12}\, f_0x_2 - \frac{1}{2}f_1\rangle$

We next deal with the following situation: $V = V_1 \oplus V_2$, where V_1 has dimension 2 and is of diagonal type, V_2 has dimension 1 but the braiding between them is not diagonal. Concretely, V is a braided vector space of dimension 3 with braiding given in the basis $(x_i)_{i\in\mathbb{I}_3}$, for some $\varepsilon, q_{ij} \in \Bbbk^\times$, $i, j \in \mathbb{I}_2$, by

$$(c(x_i \otimes x_j))_{i,j\in\mathbb{I}_3} = \begin{pmatrix} \varepsilon x_1 \otimes x_1 & \varepsilon x_2 \otimes x_1 & q_{12}x_3 \otimes x_1 \\ \varepsilon x_1 \otimes x_2 & \varepsilon x_2 \otimes x_2 & q_{12}x_3 \otimes x_2 \\ q_{21}x_1 \otimes x_3 & q_{21}(x_2 + x_1) \otimes x_3 & q_{22}x_3 \otimes x_3 \end{pmatrix}. \tag{4.84}$$

Theorem 4.9 ([4]) *Let V be as above. Then GK-dim $\mathscr{B}(V) = \infty$ if and only if $\varepsilon = -1$ and either of the following holds:*

1. $q_{12}q_{21} = 1$ *and* $q_{22} = \pm 1$*; in this case GK-dim* $\mathscr{B}(V) = 1$.
2. $q_{22} = -1 = q_{12}q_{21}$*; in this case GK-dim* $\mathscr{B}(V) = 2$.

In conclusion, let G be an abelian group and $V \in {}^{\Bbbk G}_{\Bbbk G}\mathcal{YD}$ of dimension 3 but not of diagonal type. Then GK-dim $\mathscr{B}(V) < \infty$ if and only if as a braided vector space, it has the shape (4.83) or (4.84), and is determined by Theorems 4.8 and 4.9.

4.4.4 Rack Type, Infinite Dimension

From now on, any rack is assumed to be isomorphic to a conjugacy class of a finite group. The problem we deal with is:

For every finite rack X, every finite-dimensional vector space W and a every 2-cocycle $\mathfrak{q} : X \times X \to GL(W)$, see (4.27), compute the Nichols algebra $\mathscr{B}(X, \mathfrak{q}) := \mathscr{B}(V)$, where $V = \Bbbk X \otimes W$ and the braiding is given by (4.28).

Specifically, decide when $\dim \mathscr{B}(X, \mathfrak{q}) < \infty$ or GK-dim $\mathscr{B}(X, \mathfrak{q}) < \infty$.

This is an enormous task and we are far away from a complete answer.[2] Fortunately, there are methods to reduce the problem. Before stating them, we make some comments.

Remark 4.6 Let X, W, and $\mathfrak{q}$ as above. Suppose that Y is an abelian subrack. Then $U = \Bbbk Y \otimes W$ is a braided vector subspace of diagonal type of $(V, c^{\mathfrak{q}})$. Thus, if $\dim \mathscr{B}(Y, \mathfrak{q}_{|Y \times Y}) = \infty$, what can be verified from [45], then $\dim \mathscr{B}(X, \mathfrak{q}) = \infty$.

Remark 4.7 For every finite rack X and every finite-dimensional vector space W, we would need first to compute all 2-cocycles $\mathfrak{q} : X \times X \to GL(W)$, up to some natural equivalence. When $\dim W = 1$ and X is indecomposable, an explicit description of these 2-cocycles was given in [40].

4.4.4.1 Criteria of Types C, D, F

The optimist sees the opportunity in every difficulty, and we proposed:

Definition 4.11 ([11, 2.2]) A finite rack X *collapses* if $\dim \mathscr{B}(X, \mathfrak{q}) = \infty$ for any $\mathfrak{q}$.

Actually, this definition accompanied the discovery of the criterion of type D [11, 3.5]; later we found the criteria of type F [6, 2.4] and C [8, 2.3]. Let us first state concretely these criteria and then discuss their implications

Definition 4.12 We say that a rack X is of type

- **C** when there are a decomposable subrack such that $Y = R \coprod S$, with

$$R = \mathscr{O}_r^{\operatorname{Inn} Y}, \qquad S = \mathscr{O}_s^{\operatorname{Inn} Y}, \qquad \min\{|R|, |S|\} > 2 \text{ or } \max\{|R|, |S|\} > 4;$$

  (see Exercise 4.8 for Inn Y); and elements $r \in R$, $s \in S$ satisfying

$$r \triangleright s \neq s; \tag{4.85}$$

- **D** if there are a decomposable subrack $Y = R \coprod S$, $r \in R$, $s \in S$ such that

$$r \triangleright (s \triangleright (r \triangleright s)) \neq s; \tag{4.86}$$

[2] Technically, it is enough to assume that $\mathfrak{q}$ is faithful, what means that the map $X \to GL(V)$, $x \mapsto (e_y w \mapsto e_{x \triangleright y} \mathfrak{q}_{xy}(w))$ is injective, but we omit this requirement for an easier exposition. As well, for the classification of finite-dimensional pointed Hopf algebras, it is enough to assume that $\mathfrak{q}$ is finite, i.e., that its image is contained in a finite subgroup of $GL(W)$.

- **F** if there are subracks $(R_a)_{a\in\mathbb{I}_4}$ and elements $r_a \in R_a$, $a \in \mathbb{I}_4$, such that
  - $R_a \triangleright R_b = R_b$, $a, b \in \mathbb{I}_4$;
  - $R_a \cap R_b = \emptyset$, $a \neq b \in \mathbb{I}_4$;
  - $r_a \triangleright r_b \neq r_b$ for $a \neq b \in \mathbb{I}_4$.

First of all, these definitions are well adapted to our goal.

Theorem 4.10 ([6, 8, 11]). *A rack X of type D, F or C collapses.*

The proof of this theorem follows is based on results from [26, 48, 53].

Second, these criteria can be phrased in group terms; that is, if we realize X as a conjugacy class in a finite group G, then

- Equation (4.85) means that $rs \neq sr$;
- Equation (4.86) means that $(rs)^2 \neq (sr)^2$;
- the other requirements can be stated in terms of suitable subgroups of G.

In other words, the criteria are really problems in finite group theory. Third, there is another advantage, but to state it succinctly, we introduce more terminology.

Definition 4.13 A rack is *austere* if every subrack generated by two elements is either abelian or indecomposable; *sober* if every subrack is either abelian or indecomposable; *kthulhu* if it is neither of type C, D nor F.

It is easy to see that sober $\Longrightarrow$ austere $\Longrightarrow$ kthulhu. Although the proof of the following result is straightforward, it shows that the criteria are meaningful.

Proposition 4.8 ([6, 8, 11]) *Let $X \to Y$ be a surjective morphism of racks. If Y is not kthulhu, then X is not kthulhu.*

In fact, every finite rack projects onto a simple rack, by an evident recursive argument.

Corollary 4.1 *Let X be a rack that admits a surjective morphism of racks $X \to Y$ with Y simple and not kthulhu. Then X collapses.*

In other words, we do not need to compute cocycles, even less Nichols algebras, for racks as in the Corollary (if we are interested in finite-dimensional Nichols algebras).

Question 4.1 Are the criteria of types C, D, F valid, or adjustable, to finite Gelfand–Kirillov dimension?

In conclusion, we arrive at the next problem.

Determine all simple racks that are not kthulhu.

Remembering now Theorem 4.2, we overview the present status of this problem. The most substantial results are on simple racks associated with triples (L, t, θ). We consider only the case $t = 1$, so that the racks in question are either conjugacy classes in the non-abelian simple group L or in the semidirect product $L \rtimes \langle\theta\rangle$. Indeed, the racks associated with triples (L, t, θ) with $t > 1$ represent an even more serious challenge. Some partial results appear in [24].

Finally, the affine simple racks seem to be insensible to these arguments. For instance, the dihedral rack $\mathscr{D}_n$, see Example 4.12, where $n \geq 3$ is odd, is sober.

4.4.4.2 Alternating and Symmetric Groups

We start by the alternating groups $\mathbb{A}_m$, $m \in \mathbb{N}_{\geq 5}$. Recall that $\mathrm{Aut}\mathbb{A}_m = \mathbb{S}_m$, except for $m = 6$. Thus we need to deal with conjugacy classes in $\mathbb{A}_m$ and $\mathbb{S}_m$. The conjugacy class $\mathscr{O}_\sigma^{\mathbb{S}_m}$ of σ in $\mathbb{S}_m$ is determined by its type $(1^{n_1}, 2^{n_2}, \dots, m^{n_m})$, saying that the action of σ on $\mathbb{I}_m$ has n_1 fixed points, n_2 orbits of 2 elements, etc. If $\sigma \in \mathbb{A}_m$, then $\mathscr{O}_\sigma^{\mathbb{S}_m} \cap \mathbb{A}_m$ is either the conjugacy class $\mathscr{O}_\sigma^{\mathbb{A}_m}$ in $\mathbb{A}_m$, or else the union of two conjugacy classes that are isomorphic as racks. Thus, the type is also an appropriate label for them. We need a name for the set

$$\mathscr{F} = \left\{p \in \mathbb{N} : p \text{ prime}, \ p = \frac{r^k - 1}{r - 1}, \text{ where } r \text{ is a prime power and } k \in \mathbb{N}\right\}.$$

Theorem 4.11 ([11, 34]) *Let $\mathscr{O}$ be either $\mathscr{O}_\sigma^{\mathbb{S}_m}$, if $\sigma \notin \mathbb{A}_m$, or else $\mathscr{O}_\sigma^{\mathbb{A}_m}$ if $\sigma \in \mathbb{A}_m$. If $\mathscr{O}$ is not listed in Table 4.2, then it collapses.*

Table 4.2 Kthulhu classes in a symmetric or alternating group

G	Type	Reference
$\mathbb{S}_m$	$(1^{m-2}, 2)$	Kthulhu, [11, Remark 4.2]
$\mathbb{A}_m$	$(1^{m-3}, 3)$	Austere, idem
$\mathbb{A}_p$, $p = 5, 7$ or $\notin \mathscr{F}$	(p)	Sober, [34, Remark 3.2 (b)]
$\mathbb{A}_{p+1}$, $p = 5$ or $\notin \mathscr{F}$	$(1, p)$	Sober, [34, Remark 3.2 (c)]
$\mathbb{A}_8$	(2^4)	Austere, [11, Remark 4.2]
$\mathbb{A}_7$	$(2^2, 3)$	Austere, idem
$\mathbb{S}_6$	2^3	Kthulhu, isomorphic to the class of type $(1^4, 2)$
$\mathbb{A}_6$	(3^2), $(1^2, 2^2)$	Austere, [11, Remark 4.2]
$\mathbb{S}_5$	$(2, 3)$	Sober, idem
$\mathbb{A}_5$	$(1, 2^2)$	Idem

4.4.4.3 Finite Simple Groups of Lie Type

The first examples of these appeared in the seminal paper of Evariste Galois! We start by observing:

- The finite simple groups of Lie type are (related to) the kernels of the so-called Steinberg endomorphisms of simple algebraic groups in positive characteristic. An exposition of their construction and description, even assuming the classification of the simple algebraic groups, is beyond the limits of this monograph. The interested reader may consult the beautiful account [84] of the classification of the finite simple groups, or the book [69] for a detailed presentation. Steinberg endomorphisms of simple algebraic groups fall into three possible classes [69, 21 & 22.5]; hence, there are three families of finite simple groups of Lie type: Chevalley, Steinberg, and Suzuki-Ree groups. The complete list of the simple groups in each family also appears in [6, p. 38].
- For each finite simple group of Lie type, the classification of the conjugacy classes is a classical problem whose answer, again, is long and difficult. However, there are two special classes, namely unipotent and semisimple, a terminology that correctly suggests a relation with the theory of the Jordan form of a linear transformation.

Here is the main result on these conjugacy classes, summarizing [6–9].

Theorem 4.12 *Let* **G** *be a Chevalley or Steinberg group and let* $\mathcal{O}$ *be a non-trivial unipotent conjugacy class in* **G**. *If* $\mathcal{O}$ *is not listed in Table 4.3, then it collapses.*

Remark 4.8 Let $\mathcal{O}$ be a *non-semisimple* class in a finite simple group of Lie type **G**. Then $\mathcal{O}$ has a subrack that is a unipotent conjugacy class in a smaller group and we may argue inductively, as was effectively performed for $PSL_n(q)$ in [6]. Semisimple classes appear to be more difficult to tackle, see partial results in [8].

Remark 4.9 Since $PSL_3(2) \simeq PSL_2(7)$, the unipotent class of type (3) is really a semisimple class in the former group.

Also, $PSL_2(q) \simeq PSp_2(q)$, so we really have two families of kthulhu unipotent classes in Table 4.3:

Table 4.3 Kthulhu unipotent classes in a finite simple Chevalley or Steinberg group

G	q	Type	Reference
$PSL_2(q)$	Even or not a square	(2)	Sober, [6, Lemma 3.5]
$PSL_3(2)$		(3)	Sober, [6, Lemma 3.7 (b)]
$PSp_{2n}(q), n \geq 2$	Even	$W(1)^a \oplus V(2)$	Austere, [8, Lemma 2.14]
	Odd, 9 or not a square	$(1^{r_1}, 2)$	Idem
$PSp_4(q)$	Even	$W(2)$	Idem
$PSU_n(q)$	Even	$(2, 1, \dots, 1)$	Austere, [9, Lemma 5.16]

- the class $\mathsf{Sp}_{2n,q}$ inside $PSp_{2n}(q)$, $n \geq 1$, and
- the class $\mathsf{SU}_{m,q}$ inside $PSU_m(q)$.

They both correspond to the partition (2, 1, ..., 1) and, up to rack isomorphism, are represented by $x_\beta(1)$, where β is the highest root, see [7].

Indeed, the class $W(2)$ in $PSp_4(q)$ for q even is due to the existence of a non-standard graph automorphism in C_2, in even characteristic, that interchanges short with long roots. Hence, this class is isomorphic, as a rack, to $\mathsf{Sp}_{4,q}$.

These families are related: first, if $q|q'$, then

$$\mathsf{Sp}_{2n,q} \leq \mathsf{Sp}_{2n,q'}, \qquad \mathsf{SU}_{2n,q} \leq \mathsf{SU}_{2n,q'}.$$

Next the morphism of groups $Sp_{2n}(q) \hookrightarrow Sp_{2n+2}(q)$ implies that

$$\mathsf{Sp}_{2n,q} \leq \mathsf{Sp}_{2n+2,q}.$$

When $q = 2^t$ and $m = 2n$ are even, $Sp_{2n}(q) \leq SU_{2n}(q)$, hence

$$\mathsf{Sp}_{2n,2^t} \leq \mathsf{SU}_{2n,2^t}.$$

Finally, there are inclusions between the unitary groups that induce

$$\mathsf{SU}_{n,q} \leq \mathsf{SU}_{n+2,q}, \qquad \mathsf{SU}_{2n,q} \leq \mathsf{SU}_{2n+1,q}.$$

Naturally, we are eager to know:

Are there cocycles for $\mathsf{Sp}_{2n,q}$ or $\mathsf{SU}_{m,q}$ such that the corresponding Nichols algebras are finite-dimensional?

4.4.4.4 Sporadic Groups

The classification of the finite simple groups contains, besides the alternating groups and those of Lie type, 26 more examples that are called the *sporadic* groups; here we discuss also the so-called Tits group. We refer to [84, Chap. 5] for an introduction to these groups.

Theorem 4.13 ([12, 35]) *Let* $\mathbf{G}$ *be a sporadic simple group different from the Moster* M *and let* $\mathcal{O}$ *be a non-trivial conjugacy class in* $\mathbf{G}$ *or* Aut$\mathbf{G}$. *If* $\mathcal{O}$ *is not listed in Table 4.4, then it collapses.*

The proof of this last result was done using the information in the online version of the Atlas, with the computer program GAP.

Table 4.4 Classes in sporadic, or automorphism of sporadic, groups not of type D

Group	Classes	Group	Classes	Group	Classes
T	2A	M_{11}	8A, 8B, 11A, 11B	$\mathrm{Aut}(M_{22})$	2B
M_{12}	11A, 11B	M_{22}	11A, 11B	$\mathrm{Aut}(HS)$	2C
M_{23}	23A, 23B	M_{24}	23A, 23B	$\mathrm{Aut}(Fi_{22})$	2D
Ru	29A, 29B	Suz	3A	$\mathrm{Aut}(J_3)$	34A, 34B
HS	11A, 11B	McL	11A, 11B	$\mathrm{Aut}(ON)$	38A, 38B, 38C
Co_1	3A	Co_2	2A, 23A, 23B	$\mathrm{Aut}(McL)$	22A, 22B
Co_3	23A, 23B	J_1	15A, 15B, 19A, 19B, 19C	$\mathrm{Aut}(Fi'_{24})$	2C
J_2	2A, 3A	J_3	5A, 5B, 19A, 19B	J_4	29A, 43A, 43B, 43C
Ly	37A, 37B, 67A, 67B, 67C	$O'N$	31A, 31B	Fi_{23}	2A
Fi_{22}	2A, 22A, 22B	Fi'_{24}	29A, 29B	B	2A, 46A, 46B, 47A, 47B

Remark 4.10 As for the Monster group M, these conjugacy classes are not known to be of type D: 32A, 32B, 41A, 46A, 46B, 47A, 47B, 59A, 59B, 69A, 69B, 71A, 71B, 87A, 87B, 92A, 92B, 94A, 94B. All the rest are of type D.

The criteria of type C and F were not applied neither to these classes nor to those in Table 4.4.

4.4.5 *Rack Type, Finite Dimension*

Here we discuss finite-dimensional Nichols algebras of rack type. We first present some examples that were computed by ad hoc techniques. Then we summarize the main results on Nichols algebras of decomposable Yetter–Drinfeld modules from [53, 54].

The quadratic approximations of Nichols algebras associated with racks an abelian cocycles are not difficult to describe explicitly by generators and relations, see [37] for a general formulation. Thus the problem is either to see whether the Nichols algebra is quadratic or else to find higher degree relations.

If $\mathscr{B}(V)$ is finite-dimensional, then there exists $N \in \mathbb{N}$ such that $\mathscr{B}^N(V) \neq 0$, $\mathscr{B}^{N+1}(V) = 0$; we call N the *top degree* of $\mathscr{B}(V)$. Notice that $\mathscr{B}^N(V)$ is the space of integral of $\mathscr{B}(V)$, hence $\dim \mathscr{B}^N(V) = 1$ and $\mathscr{B}(V)$ satisfies Poincaré duality $\dim \mathscr{B}^j(V) = \dim \mathscr{B}^{N-j}(V)$ for all $j \in \mathbb{I}_{0,N}$.

4.4.5.1 Fomin-Kirillov Algebras

Let $m \geq 3$. We consider two Nichols algebras associated with the conjugacy class $\mathscr{O}_2^m$ of transpositions in $\mathbb{S}_m$, with respect to the following cocycles:

$$\varepsilon \equiv -1; \qquad \chi(\nu, \zeta) = \begin{cases} 1 & \nu(i) < \nu(j), \\ -1 & \nu(i) > \nu(j), \end{cases} \qquad \text{where } \zeta = (i\ j), i < j.$$

The braided vector spaces $(\Bbbk\mathscr{O}_2^m, c^{\varepsilon})$ and $(\Bbbk\mathscr{O}_2^m, c^{\chi})$ can be realized as Yetter–Drinfeld modules M_1 and M_2 over $\Bbbk\mathbb{S}_m$. Furthermore, if $M \in {}^{\Bbbk\mathbb{S}_m}_{\Bbbk\mathbb{S}_m}\mathscr{YD}$, $M \not\simeq M_1, M_2$, and $m > 6$, then $\dim \mathscr{B}(M) = \infty$ [11, Theorem 1.1].

We start by the quadratic approximations of $\mathscr{B}(\mathscr{O}_2^m, c^{\varepsilon})$ and $\mathscr{B}(\mathscr{O}_2^m, c^{\chi})$.

Definition 4.14 ([36, 70]) Let $\mathtt{FK_m}$ be the algebra presented by generators $(x_{(ij)})_{i<j\in\mathbb{I}_m}$ and relations

$$\begin{aligned} x_{(ij)}^2 &= 0, && i < j \in \mathbb{I}_m, \\ x_{(ij)}x_{(kl)} - x_{(kl)}x_{(ij)} &= 0, && i, j, k, l \in \mathbb{I}_m, \text{ all different}, \\ x_{(jk)}x_{(ik)} - x_{(ij)}x_{(jk)} + x_{(ik)}x_{(ij)} &= 0, && i < j < k \in \mathbb{I}_m, \\ x_{(ik)}x_{(jk)} - x_{(jk)}x_{(ij)} + x_{(ij)}x_{(ik)} &= 0, && i < j < k \in \mathbb{I}_m. \end{aligned}$$

This is the quadratic approximation of $\mathscr{B}(\mathscr{O}_2^m, c^{\chi})$; it is called the m-th Fomin-Kirillov algebra since it appeared first in [36], albeit rediscovered in [70].

Proposition 4.9 *If $m = 3$, 4, or 5, then the dimension, the Hilbert–Poincaré series $\mathscr{H}_{\mathtt{FK_m}}(t)$ and its top degree of the Fomin-Kirillov algebra $\mathtt{FK_m}$ are given in Table 4.5.*

We turn to the quadratic approximation of $\mathscr{B}(\mathscr{O}_2^m, c^{\varepsilon})$.

Definition 4.15 ([70]) Let $\mathtt{B_m}$ be the algebra generated $(x_{(ij)})_{i<j\in\mathbb{I}_m}$ with relations

$$\begin{aligned} x_{(ij)}^2 &= 0, && i < j \in \mathbb{I}_m, \\ x_{(ij)}x_{(kl)} + x_{(kl)}x_{(ij)} &= 0, && i, j, k, l \in \mathbb{I}_m, \text{ all different}, \\ x_{(jk)}x_{(ik)} + x_{(ik)}x_{(ij)} + x_{(ij)}x_{(jk)} &= 0, && i < j < k \in \mathbb{I}_m, \\ x_{(ik)}x_{(jk)} + x_{(jk)}x_{(ij)} + x_{(ij)}x_{(ik)}, &= 0, && i < j < k \in \mathbb{I}_m. \end{aligned}$$

Table 4.5 Fomin-Kirillov algebras $\mathtt{FK_m}$ and their relatives $\mathtt{B_m}$, $m = 3, 4, 5$

m	Dimension	Top degree	Hilbert–Poincaré series
3	12	4	$\mathscr{H}_{\mathtt{FK_3}}(t) = (2)_t^2(3)_t$
4	576	12	$\mathscr{H}_{\mathtt{FK_5}}(t) = (2)_t^2(3)_t^2(4)_t^2$
5	8294400	40	$\mathscr{H}_{\mathtt{FK_5}}(t) = (4)_t^4(5)_t^2(6)_t^4$

These algebras are closely related; cf. Example 4.30.

Proposition 4.10 ([83]) *$\mathscr{B}(\Bbbk\mathscr{O}_2^m, c^{\varepsilon})$ and $\mathscr{B}(\Bbbk\mathscr{O}_2^m, c^{\chi})$ are twist-equivalent.*

Corollary 4.2 *If $m = 3$, 4 or 5, then $\mathtt{B}_\mathtt{m}$ has the same dimension and the same Hilbert–Poincaré series as those of $\mathtt{FK}_\mathtt{m}$ and thus they are also given in Table 4.5.*

Indeed, it is enough to prove Proposition 4.9 or Corollary 4.2, as they are equivalent. If $m = 3$ or 4, then Proposition 4.9 was proved in [36]; if $m = 5$, is due to Jan-Erik Roos, with a computer program. If $m = 3$ or 4, then Corollary 4.2 was proved in [70] using Grobner basis.

Theorem 4.14 *If $m = 3$, 4, or 5, then $\mathtt{FK}_\mathtt{m} \simeq \mathscr{B}(\mathscr{O}_2^\mathtt{m}, \mathtt{c}^{\chi})$.*

The proof appears in [70] for $\mathtt{B}_\mathtt{m}$ and $m \leq 4$, and verified by Graña for $m = 5$ using Deriva–see details in [37]. By Proposition 4.10, it translates at once to $\mathtt{FK}_\mathtt{m}$.

Let $m \geq 6$. The following three assertions are open questions:
- $\mathtt{FK}_\mathtt{m} \simeq \mathscr{B}(\mathscr{O}_2^\mathtt{m}, \mathtt{c}^{\chi})$ is a Nichols algebra (i.e., $\mathscr{B}(\mathscr{O}_2^m, c^{\chi})$ is quadratic).
- The dimension of $\mathtt{FK}_\mathtt{m}$ is finite.
- The dimension of $\mathscr{B}(\mathscr{O}_2^m, c^{\chi})$ is finite.

Some authors suggest that the last two assertions are false, see, e.g., [67].

Needless to say, the analogous question is stated for $\mathtt{B}_\mathtt{m}$ and $\mathscr{B}(\mathscr{O}_2^m, c^{\varepsilon})$; but both questions are equivalent by Proposition 4.10.

The following Example is close to $\mathtt{B}_4$.

Example 4.32 ([15]) The Nichols algebra of the conjugacy class $\mathscr{O}_4^4$ of 4-cycles in $\mathbb{S}_4$ with the constant cocycle -1 is quadratic and has the same dimension and Hilbert–Poincaré series as those of $\mathtt{FK}_4$ and thus are given in Table 4.5.

4.4.5.2 Finite-Dimensional Nichols Algebras of Some Affine Racks

In the examples below, we consider simple affine racks $\operatorname{Aff}(\mathbb{F}_q, T)$ and the constant cocycle $\mathfrak{q} \equiv -1$. We set $\mathscr{B}(\mathbb{F}_q, T) := \mathscr{B}(\operatorname{Aff}(\mathbb{F}_q, T), \mathfrak{q})$. Notice that $\operatorname{Aff}(\mathbb{F}_3, 2) \simeq \mathscr{O}_2^3$; as we have seen, $\dim \mathscr{B}(\mathbb{F}_3, 2) = 12 = 3.2^2$ and the top degree is $4 = 2^2$.

Example 4.33 ([41]) Let $\omega \in \mathbb{F}_4$ such that $\omega^2 + \omega + 1 = 0$. The tetrahedron rack is $\mathscr{T} := \operatorname{Aff}(\mathbb{F}_4, \omega)$. Then $\mathscr{B}(\mathbb{F}_4, \omega)$ is generated by $(x_i)_{i \in \mathbb{F}_4}$ with relations

$$\begin{aligned}
&x_i^2 = 0, && i \in \mathbb{F}_4;\\
&x_i x_j + x_{(\omega+1)i+\omega j} x_i + x_j x_{(\omega+1)i+\omega j} = 0, && i \neq j \in \mathbb{F}_4;\\
&x_\omega x_1 x_0 x_\omega x_1 x_0 + x_1 x_0 x_\omega x_1 x_0 x_\omega + x_0 x_\omega x_1 x_0 x_\omega x_1 = 0.
\end{aligned}$$

Also, $\dim \mathscr{B}(\mathbb{F}_4, \omega) = 72 = 4.2.3^2$ (observe that $2 = \varphi(4)$ where φ is the Euler function) and the top degree is $9 = 3^2$. The Hilbert–Poincaré series is the polynomial

$$(1+t)^2(1+t+t^2)^2(1+t^3) = t^9 + 4t^8 + 8t^7 + 11t^6 + 12t^5 + 12t^4 + 11t^3 + 8t^2 + 4t + 1.$$

Example 4.34 ([46]) There is a cocycle $\mathfrak{q}$ on $\mathscr{T} = \operatorname{Aff}(\mathbb{F}_4, \omega)$ that takes values $\pm\xi$, where $\xi \in \mathbb{G}_3'$, such that $\dim \mathscr{B}(\mathscr{T}, \mathfrak{q}) = 5184$. The Nichols algebra $\mathscr{B}(X_{4,\omega}, \mathfrak{q})$ can be presented by generators $(x_i)_{i\in\mathbb{F}_4}$ with defining relations

$$\begin{aligned} x_0^3 = x_1^3 = x_\omega^3 &= x_{\omega^2}^3 = 0, \\ \xi^2 x_0 x_1 + \xi x_1 x_\omega - x_\omega x_0 = 0, &\quad \xi^2 x_0 x_\omega + \xi x_\omega x_{\omega^2} - x_{\omega^2} x_0 = 0, \\ \xi x_0 x_{\omega^2} - \xi^2 x_1 x_0 + x_{\omega^2} x_1 = 0, &\quad \xi x_1 x_{\omega^2} + \xi^2 x_\omega x_1 + x_{\omega^2} x_\omega = 0, \end{aligned}$$

$$\begin{aligned} x_0^2 x_1 x_\omega x_1^2 + x_0 x_1 x_\omega x_1^2 x_0 + x_1 x_\omega x_1^2 x_0^2 + x_\omega x_1^2 x_0^2 x_1 + x_1^2 x_0^2 x_1 x_\omega + x_1 x_0^2 x_1 x_\omega x_1 \\ + x_1 x_\omega x_1 x_0^2 x_\omega + x_\omega x_1 x_0 x_1 x_0 x_\omega + x_\omega x_1^2 x_0 x_\omega x_0 = 0. \end{aligned}$$

Example 4.35 (Graña, see [15]). We consider the affine racks $\operatorname{Aff}(\mathbb{F}_5, 2)$, $\operatorname{Aff}(\mathbb{F}_5, 3)$. First, $\mathscr{B}(\mathbb{F}_5, 2)$ is generated by $(x_i)_{i\in\mathbb{F}_5}$ with relations

$$\begin{aligned} &x_i^2 = 0, && i \in \mathbb{F}_5; \\ &x_i x_j + x_{-i+2j} x_i + x_{3i-2j} x_{-i+2j} + x_j x_{3i-2j} && i \neq j \in \mathbb{F}_5; \\ &x_1 x_0 x_1 x_0 + x_0 x_1 x_0 x_1. && \end{aligned}$$

Also, $\dim \mathscr{B}(\mathbb{F}_5, 2) = 1280 = 5.4^4$ and the top degree is $16 = 4^2$. The Hilbert–Poincaré series is the polynomial

$$\begin{aligned} &(1+t)^2(1+t+t^2+t^3)(1+t+2t^2+2t^3+2t^4+t^5+t6)(1+t+2t^2+2t^3+t^4+t^5) \\ &= t^{16} + 5t^{15} + 15t^{14} + 35t^{13} + 66t^{12} + 105t^{11} + 145t^{10} + 175t^9 \\ &\quad + 186t^8 + 175t^7 + 145t^6 + 105t^5 + 66t^4 + 35t^3 + 15t^2 + 5t + 1. \end{aligned}$$

Next the braided vector space associated with $\operatorname{Aff}(\mathbb{F}_5, 3)$ with $\mathfrak{q} \equiv -1$ is dual to the preceding; hence, $\dim \mathscr{B}(\mathbb{F}_5, 3) = 1280$ and the Hilbert–Poincaré series is the same.

Example 4.36 (Graña). We consider the affine racks $\operatorname{Aff}(\mathbb{F}_7, 3)$, $\operatorname{Aff}(\mathbb{F}_7, 5)$. First, $\mathscr{B}(\mathbb{F}_7, 3)$ is generated by $(x_i)_{i\in\mathbb{F}_7}$ with relations

$$\begin{aligned} &x_i^2 = 0, && i \in \mathbb{F}_7; \\ &x_i x_j + x_{-2i+3j} x_i + x_j x_{-2i+3j} && i \neq j \in \mathbb{F}_7; \\ &x_2 x_1 x_0 x_2 x_1 x_0 + x_1 x_0 x_2 x_1 x_0 x_2 + x_0 x_2 x_1 x_0 x_2 x_1. && \end{aligned}$$

Also, $\dim \mathscr{B}(\mathbb{F}_7, 3) = 326592 = 7.6^6$ and the top degree is $36 = 6^2$. The Hilbert–Poincaré series is the polynomial

$$\begin{aligned}
&(1+t)^2(1+t+t^2)^2 \\
&\quad\times(1+t+2t^2+3t^3+4t^4+5t^5+4t^6+5t^7+4t^8+3t^9+2t^{10}+t^{11}+t^{12}) \\
&\qquad\times(1+t+t^2+2t^3+2t^4+2t^5+2t^6+t^7+t^8+t^9) \\
&\qquad\times(1+t+2t^2+2t^3+3t^4+3t^5+2t^6+2t^7+t^8+t^9) \\
&= t^{36}+7t^{35}+28t^{34}+84t^{33}+210t^{32}+462t^{31}+918t^{30}+1673t^{29}+2828t^{28} \\
&+4473t^{27}+6664t^{26}+9394t^{25}+12573t^{24}+16023t^{23}+19488t^{22}+22659t^{21} \\
&+25214t^{20}+26873t^{19}+27448t^{18}+26873t^{17}+25214t^{16}+22659t^{15}+19488t^{14} \\
&+16023t^{13}+12573t^{12}+9394t^{11}+6664t^{10}+4473t^9+2828t^8+1673t^7 \\
&+918t^6+462t^5+210t^4+84t^3+28t^2+7t+1.
\end{aligned}$$

Next the braided vector space associated with $\mathrm{Aff}(\mathbb{F}_7, 5)$, $\mathfrak{q} \equiv -1$, is dual to the preceding; hence $\dim \mathscr{B}(\mathbb{F}_7, 5) = 326592$ and the Hilbert–Poincaré series is the same.

It was conjectured that the examples in this §and the preceding exhaust all genuine finite-dimensional Nichols algebras over groups (besides those of diagonal type); see [47] for the precise formulation.

4.4.5.3 Decompositions with 2 Summands

We start by the description of some decomposable braided vector spaces of rack type with finite dimensional Nichols algebra. Then we state the main result of [53]. For simplicity, we assume that $\Bbbk$ is algebraically closed and char $\Bbbk = 0$.

Example 4.37 Let $X = \mathscr{D}_4 = \mathbb{I}_2\, {}_\sigma\coprod_\sigma \mathbb{I}_2$, $\sigma \neq \mathrm{id}$, see Exercise 4.10. Concretely, $X = \{1,2\}_{(34)}\coprod_{(12)}\{3,4\}$. Then $\Bbbk X = V_1 \oplus V_2$, where V_1 is spanned by $(x_i)_{i\in\mathbb{I}_2}$, while V_2 is spanned by $(x_j)_{j\in\mathbb{I}_{3,4}}$. Let $p, q, r, t \in \Bbbk^\times$, $p \neq 1 \neq q$, and $\varepsilon, \varepsilon' \in \mathbb{G}_2$. Define a braiding on $\Bbbk X$ by

$$\begin{aligned}
&c_{|V_1\otimes V_1} \text{ is of diagonal type with matrix } \begin{pmatrix} q & \varepsilon q \\ \varepsilon q & q \end{pmatrix}, \\
&c_{|V_2\otimes V_2} \text{ is of diagonal type with matrix } \begin{pmatrix} p & \varepsilon' p \\ \varepsilon' p & p \end{pmatrix}, \\
&\left(c(x_i\otimes x_j)_{i\in\mathbb{I}_2, j\in\mathbb{I}_{3,4}}\right) = \begin{pmatrix} x_4\otimes x_1 & t^2x_3\otimes x_1 \\ \varepsilon' x_4\otimes x_2 & \varepsilon' t^2 x_3\otimes x_2 \end{pmatrix}, \\
&\left(c(x_j\otimes x_i)_{j\in\mathbb{I}_{3,4}, i\in\mathbb{I}_2}\right) = \begin{pmatrix} x_2\otimes x_3 & r^2x_1\otimes x_3 \\ \varepsilon x_2\otimes x_4 & \varepsilon r^2 x_1\otimes x_4 \end{pmatrix}.
\end{aligned} \tag{4.87}$$

Exercise 4.31 1. Prove that this is indeed a braiding.

2. Assume that $\varepsilon = \varepsilon' = 1$. Consider the basis $(y_h)_{h \in \mathbb{I}_4}$ of $\Bbbk X$ where

$$y_1 = rx_1 + x_2, \quad y_2 = -rx_1 + x_2, \quad y_3 = tx_3 + x_4, \quad y_4 = -tx_3 + x_4.$$

Then c on this basis is of diagonal type, with matrix

$$\begin{pmatrix} q & q & t & -t \\ q & q & t & -t \\ r & -r & p & p \\ r & -r & p & p \end{pmatrix}.$$

If $\dim \mathscr{B}(V) < \infty$, then $p = q = -1$. (Indeed, by Example 4.31, $p, q \in \mathbb{G}'_2 \cup \mathbb{G}'_3$, then inspect the list in [45]). In this case, the Dynkin diagram is

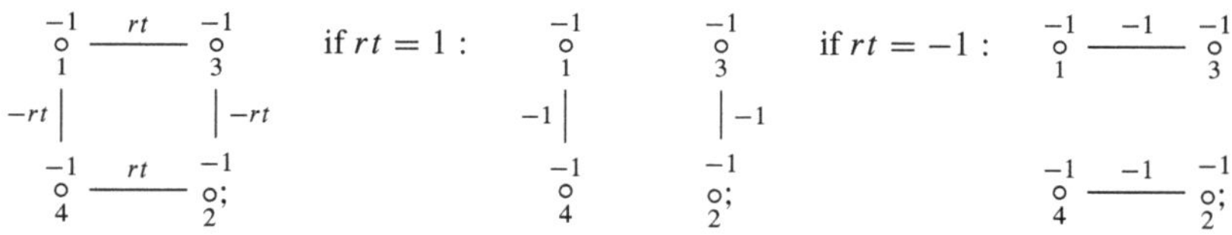

Now $\overset{-1}{\circ}\overset{-1}{\text{———}}\overset{-1}{\circ}$, is a Dynkin diagram of Cartan type A_2 at -1. By elementary arguments, its Nichols algebra has dimension 8. Therefore, if $rt \in \mathbb{G}_2$, then $\dim \mathscr{B}(V) = 64$. If $rt \notin \mathbb{G}_2$, then $\dim \mathscr{B}(V) = \infty$ by inspection of the list in [45].

3. If $\varepsilon = \varepsilon' = -1$, then there is a twist ϕ as in Example 4.30 that reduces to the previous case.

When $\varepsilon, \varepsilon' \in \mathbb{G}_2$ are arbitrary, the same result holds but the proof requires the Weyl groupoid:

Theorem 4.15 ([48, Theorem 4.6]) *Let* $(V, c) = (\Bbbk\mathscr{D}, c)$ *where* c *is given by* (4.87). *Then* $\dim \mathscr{B}(V) = 64$.

Example 4.38 Let $X = \mathscr{D}_3 \coprod \{4\}$. Then $\Bbbk X = V_1 \oplus V_2$, where V_1 is spanned by $(x_i)_{i \in \mathbb{I}_3}$, while V_2 is spanned by x_4. Let $\omega \in \Bbbk^\times$, $\zeta \in \mathbb{G}_3$, $q_1, q_2 \in \Bbbk^\times$. Recall c^ε from Sect. 4.4.5.1 and define a braiding on $\Bbbk X$ by

$$\begin{aligned} c_{|V_1 \otimes V_1} &= c^\varepsilon, & c(x_4 \otimes x_4) &= -\omega x_4 \otimes x_4, & \\ c(x_i \otimes x_4) &= q_1 \zeta^{i-1} x_4 \otimes x_i, & c(x_4 \otimes x_i) &= q_2 x_i \otimes x_4, & i \in \mathbb{I}_3. \end{aligned} \tag{4.88}$$

Exercise 4.32 Check that (4.88) satisfies the braid equation.

Thus $\Bbbk X = V_1 \oplus V_2$ is a decomposition of braided vector spaces where V_1 is $(\Bbbk\mathscr{O}_2^3, c^\varepsilon)$, V_2 is a point with label $-\omega \in \mathbb{G}'_6$ and the braiding between them is prescribed in the second line of (4.88).

Theorem 4.16 ([53, Theorem 8.2]) *Assume that* $\omega \in \mathbb{G}'_3$ *and that* $q_1q_2 = -\omega^2$. *Let* $(V, c) = (\Bbbk(\mathcal{D}_3 \coprod \{4\}), c)$ *where* c *is given by* (4.88). *Then* $\dim \mathscr{B}(V) = 10368 = 3^4 2^7$.

Example 4.39 Let $X = \mathcal{D}_3 \coprod \{4\}$ as in the previous Example. Let $V = V_1 \oplus V_2$, where $V_1 = \Bbbk\mathcal{D}_3$ is spanned by $(x_i)_{i\in\mathbb{I}_3}$, but now V_2 is $\Bbbk x_4 \otimes \Bbbk^2$. Let $y_4 = x_4 \otimes (1, 0)$, $y_5 = x_4 \otimes (0, 1)$. Let $\zeta \in \mathbb{G}_3$, $q_1, q_2 \in \Bbbk^\times$. Define a braiding on $\Bbbk X$ by

$$\begin{aligned} c_{|V_1\otimes V_1} &= c^{\varepsilon}, & c_{|V_2\otimes V_2} &= -\tau, & \\ c(x_i \otimes y_4) &= \zeta^{i-1} y_5 \otimes x_i, & c(x_i \otimes y_5) &= q_1\zeta^{2(i-1)} y_4 \otimes x_i, & \\ c(y_4 \otimes x_i) &= q_2 y_4 \otimes x_i, & c(y_5 \otimes x_i) &= q_2 x_i \otimes y_5, & i \in \mathbb{I}_3. \end{aligned} \tag{4.89}$$

Thus $V = V_1 \oplus V_2$ is a decomposition of braided vector spaces where V_1 is $(\Bbbk\mathcal{O}^3_2, c^{\varepsilon})$ as in Sect. 4.4.5.1, $V_2 = \Bbbk y_4 \oplus \Bbbk y_5$ has dimension 2 and the braiding between them is prescribed in the second and third lines of (4.89).

Exercise 4.33 Check that (4.89) satisfies the braid equation.

Theorem 4.17 ([53, Theorem 8.4]) *Let* (V, c) *be the braided vector space with* c *given by* (4.89). *Assume that* $q_1q_2^2 = 1$. *Then* $\dim \mathscr{B}(V) = 2304 = 3^2 2^8$.

Example 4.40 Let $X = \mathcal{D}_3\, {}_{(45)}\!\coprod_{(132),(123)} \mathbb{I}_{4,5}$; let $\sigma = (132)$. Let $V = \Bbbk X = V_1 \oplus V_2$, where $V_1 = \Bbbk\mathcal{D}_3$ is spanned by $(x_i)_{i\in\mathbb{I}_3}$ and V_2 is spanned by x_4, x_5. Let $\zeta \in \mathbb{G}_3$, $a_1, q_1, q_2 \in \Bbbk^\times$. Define a braiding on V by

$$\begin{aligned} c_{|V_1\otimes V_1} &= c^{\varepsilon}, & c(x_i \otimes x_j) &= a_1\zeta^{2-\delta_{ij}} x_j \otimes x_i, & i, j \in \mathbb{I}_{4,5}; \\ c(x_i \otimes x_4) &= \zeta^{i-1} x_5 \otimes x_i, & c(x_i \otimes x_5) &= q_1\zeta^{2(i-1)} x_4 \otimes x_i, & \\ c(x_4 \otimes x_i) &= q_2 x_{\sigma(i)} \otimes x_4, & c(x_5 \otimes x_i) &= q_2 x_{\sigma^{-1}(i)} \otimes x_5, & i \in \mathbb{I}_3. \end{aligned} \tag{4.90}$$

Exercise 4.34 Check that (4.90) satisfies the braid equation and that $V = V_1 \oplus V_2$ is a decomposition of braided vector spaces.

Theorem 4.18 ([53, Theorem 8.1, 8.3]) *Let* (V, c) *be the braided vector space with* c *given by* (4.90).

1. *Assume that* $\zeta \in \mathbb{G}'_3$, $a_1 = -\zeta^2$ *and* $q_1q_2^2 = \zeta^2$. *Then* $\dim \mathscr{B}(V, c) = 10368$.
2. *Assume that* $\zeta = 1$, $a_1 = -1$ *and* $q_1q_2^2 = 1$. *Then* $\dim \mathscr{B}(V, c) = 2304$.

Example 4.41 Let $X = \mathcal{D}_4\, {}_{(56)}\!\coprod_{\sigma_1,\sigma_2} \mathbb{I}_{5,6}$. Here we number $\mathcal{D}_4$ as follows: $\mathcal{D}_4 = \{1, 3\}\, {}_{\sigma}\!\coprod_{\sigma} \{2, 4\}$, where $\sigma \neq \mathrm{id}$; that is, we change the numeration in Example 4.37 by $2 \leftrightarrow 3$. Also, $\sigma_1 = (1234)$, $\sigma_2 = (1432)$. Let $V = \Bbbk X = V_1 \oplus V_2$, where $V_1 = \Bbbk\mathcal{D}_4$ is spanned by $(x_i)_{i\in\mathbb{I}_4}$, and V_2 is spanned by x_5, x_6. Let $q_1, q_2 \in \Bbbk^\times$, $\zeta_1, \zeta_2 \in \mathbb{G}_4$. Define a cocycle on V by

$$(\mathfrak{q}_{ij})_{i,j\in\mathbb{I}_4} = \begin{pmatrix} -1 & -\zeta_1^2 & -\zeta_1^2 & -\zeta_1^2 \\ -1 & -1 & -1 & -\zeta_1^2 \\ -\zeta_1^2 & -1 & -1 & -1 \\ -\zeta_1^2 & -\zeta_1^2 & -\zeta_1^2 & -1 \end{pmatrix}, \quad (\mathfrak{q}_{ij})_{i,j\in\mathbb{I}_{5,6}} = \begin{pmatrix} -1 & -\zeta_2^3 \\ -\zeta_2^3 & -1 \end{pmatrix},$$
$$(\mathfrak{q}_{ij})_{i\in\mathbb{I}_{5,6},j\in\mathbb{I}_4} = \begin{pmatrix} 1 & q_2\zeta_1^3 & 1 & q_2\zeta_1 \\ \zeta_1^2 & q_2\zeta_1^3 & 1 & q_2\zeta_1^3 \end{pmatrix}, \quad \mathfrak{q}_{ij} = \begin{cases} \zeta_2^{1-i}, & j=5, \\ q_1\zeta_2^{i-1} & j=6, \end{cases} \quad i \in \mathbb{I}_4. \tag{4.91}$$

Exercise 4.35 Check that (4.91) satisfies the cocycle relation.

Theorem 4.19 ([52, Theorem 5.4]) *Let* $(V, c^{\mathfrak{q}})$ *be the braided vector space with* $\mathfrak{q}$ *given by* (4.91). *Assume that* $\zeta_1\zeta_2 = q_1q_2$ *and* $\zeta_2 \in \mathbb{G}_4'$. *Then* $\dim \mathscr{B}(X, \mathfrak{q}) =$ 262144.

Example 4.42 Let $X = \mathscr{T} \coprod \{5\}$. Let $V = \Bbbk X = V_1 \oplus V_2$, where $V_1 = \Bbbk\mathscr{T}$ is spanned by $(x_i)_{i\in\mathbb{I}_4}$, and V_2 is $\Bbbk x_5$. Let $a, q_1, q_2 \in \Bbbk^\times$. Define a braiding on V by

$$\begin{aligned} c_{|V_1\otimes V_1} &= c^{-1}, & c_{|V_2\otimes V_2} &= a\,\mathrm{id}, \\ c(x_i \otimes x_5) &= q_1 x_5 \otimes x_i, & c(x_5 \otimes x_i) &= q_2 x_i \otimes x_5, \quad i \in \mathbb{I}_4. \end{aligned} \tag{4.92}$$

Thus $V = V_1 \oplus V_2$ is a decomposition of braided vector spaces where V_1 is $(\Bbbk\mathscr{T}, c^{-1})$ as in Example 4.33 and $V_2 = \Bbbk x_5$ has dimension 1.

Exercise 4.36 Check that (4.92) satisfies the braid equation.

Theorem 4.20 ([52, Theorem 2.8]) *Let* (V, c) *be the braided vector space with* c *given by* (4.92). *Assume that* $-q_1q_2 \in \mathbb{G}_3'$ *and* $aq_1q_2 = 1$. *Then* $\dim \mathscr{B}(X, \mathfrak{q}) =$ 80621568.

The following remarkable result is the culmination of the series of papers [48, 51–53].

Theorem 4.21 ([53]) *Let* G *be a finite non-abelian group and* $V = V_1 \oplus V_2 \in {}^{\Bbbk G}_{\Bbbk G}\mathscr{YD}$, *where* V_1 *and* V_2 *are simple, the support of* V *generates* G *and* $c^2_{|V_1\otimes V_2} \neq \mathrm{id}$. *Assume that* $\dim \mathscr{B}(V) < \infty$. *Then as a braided vector space,* V *is isomorphic to one of the Examples 4.37, 4.38, 4.39, 4.40, 4.41 or 4.42.*

This formulation is simplified for the sake of the exposition; the actual result gives precise information of the possible groups G, it does not require $\Bbbk$ to be algebraically closed, and it extends to all characteristics, with new examples in characteristics 2 and 3.

4.4.5.4 Decompositions with $\theta > 2$ Summands

The proof of Theorem 4.21 uses the Weyl groupoid and a detailed analysis of the subgroups of the enveloping group of the racks involved. With similar techniques, the same authors went on and obtained in [54] the analogous classification but for $\theta > 2$, again without restriction on the characteristic. The outcome is that essentially Dynkin diagrams of simple Lie algebras are, up to just a few exceptions, the main characters of the classification!

As the precise formulation of the main theorem of [54] requires a careful preparation beyond the scope of these notes, we refer the interested reader to the original source [54].

Acknowledgements I was invited to give a conference at the XX Congreso Colombiano de Matemáticas in Manizales, in July 2015 and as a prolongation of my visit, to the Geometric, Algebraic and Topological Methods for Quantum Field Theory Summer School, Villa de Leyva. I am grateful to César Galindo for his friendly hospitality and the organization of the visit, to Sylvie Paycha for the invitation to the School and the encouragement and advice to write these notes, and to the many Colombian colleagues that welcomed me as kindly as always. I would like also to thank João Matheus Jury Giraldi and Guillermo Sanmarco for comments on the first version of this text and Giovanna Carnovale for discussions on Remark 4.9.

References

1. N. Andruskiewitsch, About finite dimensional Hopf algebras. Contemp. Math. **294**, 1–54 (2002)
2. N. Andruskiewitsch, On finite-dimensional Hopf algebras, in *Proceedings of the ICM Seoul*, Vol. II, pp. 117–141 (2014)
3. N. Andruskiewitsch, I. Angiono, On Finite dimensional Nichols algebras of diagonal type, http://www.famaf.unc.edu.ar/~andrus/articulos.html
4. N. Andruskiewitsch, I. Angiono, I. Heckenberger, On finite GK-dimensional Nichols algebras over abelian groups, arXiv:1606.02521. Submitted
5. N. Andruskiewitsch, I. Angiono, I. Heckenberger, On finite GK-dimensional Nichols algebras of diagonal type. In preparation
6. N. Andruskiewitsch, G. Carnovale, G.A. García, Finite-dimensional pointed Hopf algebras over finite simple groups of Lie type I. Non-semisimple classes in $PSL_n(q)$. J. Algebra **442**, 36–65 (2015)
7. N. Andruskiewitsch, G. Carnovale, G.A. García, Finite-dimensional pointed Hopf algebras over finite simple groups of Lie type II. Unipotent classes in the symplectic groups, Commun. Contemp. Math. **18**(4), Article ID 1550053, 35 p. (2016)
8. N. Andruskiewitsch, G. Carnovale, G.A. García, Finite-dimensional pointed Hopf algebras over finite simple groups of Lie type III. Semisimple classes in $PSL_n(q)$, to appear in Rev. Mat. Iberoam
9. N. Andruskiewitsch, G. Carnovale, G.A. García, Finite-dimensional pointed Hopf algebras over finite simple groups of Lie type IV. Unipotent classes in Chevalley and Steinberg groups, arXiv:1612.07964
10. N. Andruskiewitsch, J. Cuadra, On the structure of (co-Frobenius) Hopf algebras. J. Noncommut. Geom. **7**, 83–104 (2013)
11. N. Andruskiewitsch, F. Fantino, M. Graña, L. Vendramin, Finite-dimensional pointed Hopf algebras with alternating groups are trivial. Ann. Math. Pura Appl. **190**, 225–245 (2011)
12. N. Andruskiewitsch, F. Fantino, M. Graña, L. Vendramin, Pointed Hopf algebras over the sporadic simple groups. J. Algebra **325**, 305–320 (2011)
13. N. Andruskiewitsch, F. Fantino, G.A. García, L. Vendramin, On Nichols algebras associated to simple racks. Contemp. Math. **537**, 31–56 (2011)
14. N. Andruskiewitsch, J.M.J. Giraldi, Nichols algebras that are quantum planes, Linear and Multilinear Algebra (to appear)

15. N. Andruskiewitsch, M. Graña, From racks to pointed Hopf algebras. Adv. Math. **178**, 177–243 (2003)
16. N. Andruskiewitsch, I. Heckenberger, H.-J. Schneider, The Nichols algebra of a semisimple Yetter-Drinfeld module. Am. J. Math. **132**, 1493–1547 (2010)
17. N. Andruskiewitsch, H.-J. Schneider, Lifting of quantum linear spaces and pointed Hopf algebras of order p^3. J. Algebra **209**, 658–691 (1998)
18. N. Andruskiewitsch, H.-J. Schneider, Finite quantum groups and Cartan matrices. Adv. Math. **154**, 1–45 (2000)
19. N. Andruskiewitsch, H.-J. Schneider, Pointed Hopf Algebras, New Directions in Hopf Algebras, MSRI series, pp. 1–68. Cambridge University Press, Cambridge (2002)
20. N. Andruskiewitsch, H.-J. Schneider, On the classification of finite-dimensional pointed Hopf algebras. Ann. Math. **171**, 375–417 (2010)
21. I. Angiono, A presentation by generators and relations of Nichols algebras of diagonal type and convex orders on root systems. J. Eur. Math. Soc. **17**, 2643–2671 (2015)
22. I. Angiono, On Nichols algebras of diagonal type. J. Reine Angew. Math. **683**, 189–251 (2013)
23. Y. Bazlov, Nichols-Woronowicz algebra model for Schubert calculus on Coxeter groups. J. Algebra **297**, 372–399 (2006)
24. G. Carnovale, A. García Iglesias, θ-semisimple twisted conjugacy classes of type D in $PSL_n(q)$. J. Algebra **442**, 36–65 (2015)
25. V. Chari, A. Pressley, *A Guide to Quantum Groups* (Cambridge University Press, Cambridge, 1994)
26. M. Cuntz, I. Heckenberger, Finite Weyl groupoids. J. Reine Angew. Math. **702**, 77–108 (2015)
27. Y. Doi, M. Takeuchi, Multiplication alteration by two-cocycles. The quantum version. Commun. Algebra **22**, 5715–5732 (1994)
28. V.G. Drinfeld, Quantum groups. Proc. Int. Cong. Math., Berkeley **1**, 789–820 (1986)
29. V.G. Drinfeld, Quasi-Hopf algebras. Leningrad Math. J. **1**, 1419–1457 (1990)
30. V.G. Drinfeld, On some unsolved problems in quantum group theory, in Quantum Groups (Leningrad, 1990), Lecture Notes in Math. **1510**, Springer, Berlin, 1–8 (1992)
31. P. Etingof, S. Gelaki, D. Nikshych, V. Ostrik, *Mathematical Surveys and Monographs*, vol. 205, Tensor Categories (American Mathematical Society, Providence, 2015). pp. 344
32. P. Etingof, T. Schedler, A. Soloviev, Set-theoretical solutions to the Quantum Yang-Baxter Equation. Duke Math. J. **100**, 169–209 (1999)
33. L.D. Faddeev, NYu. Reshetikhin, L.A. Takhtajan, Quantization of Lie groups and Lie algebras. Leningrad Math. J. **1**, 193–225 (1990)
34. F. Fantino, Conjugacy classes of p-cycles of type D in alternating groups. Comm. Algebra **42**, 4426–4434 (2014)
35. F. Fantino, L. Vendramin, On twisted conjugacy classes of type D in sporadic simple groups. Contemp. Math. **585**, 247–259 (2013)
36. S. Fomin, K.N. Kirillov, *Progress in Mathematics*, vol. 172 (Birkhauser, Boston, 1999), pp. 146–182
37. G.A. García, A. García Iglesias, Finite dimensional pointed Hopf algebras over $\mathbb{S}_4$. Isr. J. Math. **183**, 417–444 (2011)
38. G.A. García, J.M.J. Giraldi, On Hopf Algebras over quantum subgroups, arXiv:1605.03995
39. A. García Iglesias, C. Vay, Finite-dimensional pointed or copointed Hopf algebras over affine racks. J. Algebra **397**, 379–406 (2014)
40. A. García Iglesias, L. Vendramin, An explicit description of the second cohomology group of a quandle. Math. Z. **286**(3–4), 1041–1063 (2017)
41. M. Graña, On Nichols algebras of low dimension. Contemp. Math. **267**, 111–134 (2000)
42. D. Gurevich, Algebraic aspects of the Quantum Yang-Baxter equation. Leningrad J. Math. **2**, 801–828 (1991)
43. T. Hayashi, Quantum groups and quantum determinants. J. Algebra **152**, 146–165 (1992)
44. I. Heckenberger, The Weyl groupoid of a Nichols algebra of diagonal type. Invent. Math. **164**, 175–188 (2006)
45. I. Heckenberger, Classification of arithmetic root systems. Adv. Math. **220**, 59–124 (2009)

46. I. Heckenberger, A. Lochmann, L. Vendramin, Braided racks, Hurwitz actions and Nichols algebras with many cubic relations. Transform. Groups **17**, 157–194 (2012)
47. I. Heckenberger, A. Lochmann, L. Vendramin, Nichols algebras with many cubic relations. Trans. Am. Math. Soc. **367**, 6315–6356 (2015)
48. I. Heckenberger, H.-J. Schneider, Nichols algebras over groups with finite root system of rank two I. J. Algebra **324**, 3090–3114 (2010)
49. I. Heckenberger, H.-J. Schneider, Root systems and Weyl groupoids for Nichols algebras. Proc. Lond. Math. Soc. **101**(3), 623–654 (2010)
50. I. Heckenberger, H.-J. Schneider, Yetter-Drinfeld modules over bosonizations of dually paired Hopf algebras. Adv. Math. **244**, 354–394 (2013)
51. I. Heckenberger, L. Vendramin, Nichols algebras over groups with finite root system of rank two II. J. Group Theory **17**(6), 1009–1034 (2014)
52. I. Heckenberger, L. Vendramin, Nichols algebras over groups with finite root system of rank two III. J. Algebra **422**, 223–256 (2015)
53. I. Heckenberger, L. Vendramin, The classification of Nichols algebras with finite root system of rank two. J. Eur. Math. Soc. **19**(7), 1977–2017 (2017)
54. I. Heckenberger, L. Vendramin, A classification of Nichols algebras of semi-simple Yetter-Drinfeld modules over non-abelian groups. J. Eur. Math. Soc. **19**, 299–356 (2017)
55. J. Hietarinta, Solving the two-dimensional constant quantum Yang-Baxter equation. J. Math. Phys. **34**, 1725 (1993)
56. D. Joyce, Simple quandles. J. Algebra **79**, 307–318 (1982)
57. V. Kac, *Infinite-Dimensional Lie Algebras*, 3rd edn. (Cambridge University Press, Cambridge, 1990)
58. C. Kassel, *Graduate Texts in Mathematics*, vol. 155, Quantum Groups (Springer, New York, 1995)
59. G. Krause, T. Lenagan, Growth of algebras and Gelfand-Kirillov dimension, *Revised edition. Graduate Studies in Math*, vol. 22 (American Mathematical Society, Providence, RI, 2000), p. x+212
60. R. Larson, M. Sweedler, An associative orthogonal bilinear form for Hopf algebras. Am. J. Math. **91**, 75–94 (1969)
61. J.-H. Lu, M. Yan, Y.-C. Zhu, On Set-theoretical Yang-Baxter equation. Duke Math. J. **104**, 1–18 (2000)
62. G. Lusztig, Finite-dimensional Hopf algebras arising from quantized universal enveloping algebra. J. Am. Math. Soc. **3**, 257–296 (1990)
63. G. Lusztig, Quantum groups at roots of 1. Geom. Dedicata **35**, 89–113 (1990)
64. G. Lusztig, *Introduction to Quantum Groups* (Birkhäuser, Boston, 1993)
65. S. Majid, Free braided differential calculus, braided binomial theorem, and the braided exponential map. J. Geom. Phys. **34**, 4843–4856 (1993)
66. S. Majid, *Foundations of Quantum Group Theory* (Cambridge University Press, Cambridge, 1995)
67. S. Majid, Noncommutative differentials and Yang-Mills on permutation groups S_n, *Hopf Algebras in Noncommutative Geometry and Physics*, vol. 239, Lecture Notes in Pure and Applied Mathematics (Dekker, New York, 2005), pp. 189–213
68. S. Majid, R. Oeckl, Twisting of Quantum Differentials and the Planck Scale Hopf Algebra. Commun. Math. Phys. **205**, 617–655 (1999)
69. G. Malle, D. Testerman, Linear algebraic groups and finite groups of lie type, Camb. Stud. Adv. Math. **133** (2011)
70. A. Milinski, H.-J. Schneider, Pointed Indecomposable Hopf Algebras over Coxeter Groups. Contemp. Math. **267**, 215–236 (2000)
71. S. Montgomery, *Hopf Algebras and their Actions on Rings, CMBS*, vol. 82 (American Mathematical Society, Providence, 1993)
72. W.D. Nichols, Bialgebras of type one. Comm. Algebra **6**, 1521–1552 (1978)
73. D.E. Radford, Hopf algebras, Series on Knots and Everything 49. Hackensack, NJ: World Scientific. xxii, 559 p. (2012)

74. M. Rosso, Groupes quantiques et algebres of battage quantiques. C. R. A. S. (Paris) **320**, 145–148 (1995)
75. M. Rosso, Quantum groups and quantum shuffles. Inventiones Math. **133**, 399–416 (1998)
76. N. Reshetikhin, Multiparameter quantum groups and twisted quasitriangular Hopf algebras. Lett. Math. Phys. **20**, 331–335 (1990)
77. P. Schauenburg, *On coquasitriangular Hopf algebras and the quantum Yang-Baxter equation, Algebra-Berichte 67 Math* (Inst. Uni, München, 1992)
78. H.-J. Schneider, Lectures on Hopf algebras, Trab. Mat. **31/95**, FaMAF, Universidad Nacional de Córdoba, Córdoba (1995)
79. A.M. Semikhatov, I.Y. Tipunin, The Nichols algebra of screenings. Commun. Contemp. Math. **14**, 1250029, 66 (2012)
80. M.E. Sweedler, *Hopf algebras* (Benjamin, New York, 1969)
81. M. Takeuchi, A two-parameter quantization of $GL(n)$. Proc. Japan Acad. Ser. A Math. Sci. **66**, 112–114 (1990)
82. M. Takeuchi, Survey of braided Hopf algebras. Contemp. Math. **267**, 301–324 (2000)
83. L. Vendramin, Proc. Amer. Math. Soc. **140**, 3715–3723 (2012)
84. R.A. Wilson, The Finite Simple Groups. Graduate Texts Math. 251. Springer (2009)

Chapter 5
Quantum Field Theory in Curved Space-Time

Notes Based on Lectures by A. Ashtekar

Andrés F. Reyes Lega

Abstract These notes present an overview of the lectures held by Abhay Ashtekar on the theory of quantum fields in curved space-times and its applications to cosmology at the Summer School "Geometric, topological, and algebraic methods for quantum field theory", held at Villa de Leyva in 2015. The first part of the notes is pretty much self contained, assuming only a basic knowledge of quantum field theory in Minkowski space-time, while the second part—more in the style of a seminar—presents applications to the theory of inflation.

5.1 Introduction

Over the last 20 years, a good understanding of the origin of the large-scale structure of the universe has been obtained. Much of the theoretical basis for this understanding comes from the study of quantum field theory in curved space-times, which is the subject of this chapter. We are going to study *linear quantum field theories* which, although simpler than fully interacting ones, have a series of subtle and physically remarkable properties. There are several functional-analytic aspects that we will not cover, but which are discussed in references [5, 6, 9, 10]. One of the main differences between quantum field theories in flat (i.e., Minkowski) and curved space-times is that in the latter case translation invariance is lost and hence Fourier transforms are not available. One important consequence of this is the lack of uniqueness of the ground state, which in the Minkowski case is singled out by its invariance with respect to the isometry group. This is of course not new, as we know from the Hawking effect. But, even if the theory has been developed since the seventies, there are still fundamental open problems. That is the case, e.g., with applications to the very early universe.

A.F. Reyes Lega (✉)
Departamento de Física, Universidad de los Andes, Carrera 1 18A-10, Bloque Ip, Bogotá, Colombia
e-mail: anreyes@uniandes.edu.co

A. Cardona et al. (eds.), *Quantization, Geometry and Noncommutative Structures in Mathematics and Physics*, Mathematical Physics Studies,
DOI 10.1007/978-3-319-65427-0_5

To understand how quantum field theory fits in this context, it is important to have an idea of the different scales, in particular of the very extreme conditions in the early universe. For example, we know that matter density must have been of the order of

$$\rho \sim 10^{-12} \rho_P = 10^{81} \text{ gm/cc}.$$

This should be compared with the density of nuclear matter, which is of the order of 10^{14}gm/cc. Here ρ_P refers to the Planck density, a number that is obtained from the three fundamental physical constants c, G and $\hbar$. One combination of these constants gives a fundamental length, the Planck length, given by

$$l_P = \sqrt{\frac{G\hbar}{c^3}} \approx 10^{-33} \text{ cm}.$$

The corresponding density, the Planck density, is then given by

$$\rho_P = \frac{c^5}{G\hbar^2} \approx 10^{93} \text{ gm/cc}.$$

In the sequel, we will set $c = 1$. In physics, dimensions are important, in particular, dimensionful parameters set *scales*. The quantities l_P and ρ_P introduced above correspond to the Planck scale, where quantum gravity effects become important. But in these notes, we will stay away from this scale and so we will ignore quantum effects of nonlinear gravity. In our setting, gravity will be described by general relativity, but we will consider *quantum fields* in curved space-times. That means, in this chapter, space-time is classical (described by general relativity) with quantum fields on it.

The role of quantum fields for cosmology is very important. According to the current understanding, the early universe underwent a period of near exponential expansion called inflation. Given that the very early universe was extremely homogeneous and isotropic, the question of how did the large-scale structure arise is a very important one. There must have been very "tiny" fluctuations that led to cosmic inhomogeneities in the cosmic microwave background (CMB) and seeded the formation of large-scale structure in the universe. The answer to this question involves the theory of quantum fields in curved space-times, the subject of these lecture notes. We will start with a brief review of quantum field theory in Minkowski space-time, in order to appreciate the change in perspective that will be necessary when considering curved space-times.

5.2 Quantum Field Theory in Minkowski Space-Time

Minkowski space-time is a 4-manifold $M = \mathbb{R}^4$, endowed with a metric

$$ds^2 = -dt^2 + d\boldsymbol{x}^2.$$

The Riemann tensor corresponding to this metric vanishes, meaning that we are dealing with special relativity (and so, from the three fundamental constants c, $\hbar$, G, we are ignoring the last one). The simplest example of a field theory is provided by the real scalar field, ϕ, which is a (real) solution to the Klein–Gordon equation ($\Box \equiv -\partial_t^2 + \nabla^2$):

$$(\Box - \mu^2)\phi(x) = 0, \tag{5.1}$$

where μ is the mass of the scalar field. In quantum field theory (QFT), we replace the function $\phi(x)$ by an operator-valued distribution $\hat{\phi}(x)$. Denoting space-time points as $x = (x_0, \boldsymbol{x})$, we can use Fourier transform to write

$$\phi(x) = \frac{1}{(2\pi)^4} \int d^4k \tilde{\phi}(k) e^{ik\cdot x}, \tag{5.2}$$

with $k \cdot x = -k_0 x_0 + \boldsymbol{k} \cdot \boldsymbol{x}$. Since $\phi(x)$ should be a solution to the field equation, we obtain:

$$\tilde{\phi}(k) = \delta(-\omega_k^2 + k_0^2) a(k), \tag{5.3}$$

where ω_k is defined as

$$\omega_k := \sqrt{\boldsymbol{k}^2 + \mu^2}. \tag{5.4}$$

This means that the integral actually has support on the hyperboloid of mass μ.

Exercise 5.1 Defining $A(\boldsymbol{k}) := \frac{a(\omega_k, \boldsymbol{k})}{2\pi\sqrt{2\omega_k}}$, carry out the integral with respect to k_0 to get

$$\hat{\phi}(x) = \frac{1}{(2\pi)^3} \int \frac{d^3\boldsymbol{k}}{\sqrt{2\omega_k}} \left[e^{i(\boldsymbol{k}\cdot\boldsymbol{x} - \omega_k t)} \hat{A}(\boldsymbol{k}) + e^{i(-\boldsymbol{k}\cdot\boldsymbol{x} + \omega_k t)} \hat{A}^\dagger(\boldsymbol{k}) \right]. \tag{5.5}$$

So, upon quantization, it is the Fourier coefficient functions ($a(k)$) that take on the operator character of the field. These creation/annihilation operators obey the canonical commutation relations (CCR):

$$\left[\hat{A}(\boldsymbol{k}), \hat{A}^\dagger(\boldsymbol{k}')\right] = \frac{\hbar}{(2\pi)^3}\delta(\boldsymbol{k}, \boldsymbol{k}'), \quad \left[\hat{A}(\boldsymbol{k}), \hat{A}(\boldsymbol{k}')\right] = 0 = \left[\hat{A}^\dagger(\boldsymbol{k}), \hat{A}^\dagger(\boldsymbol{k}')\right]. \tag{5.6}$$

Exercise 5.2 Check that the CCR for the $\hat{A}(\boldsymbol{k})$ and $\hat{A}^\dagger(\boldsymbol{k})$ operators are equivalent to the following ones:

$$\left[\hat{\varphi}(\boldsymbol{x}), \hat{\pi}^\dagger(\boldsymbol{y})\right] = i\hbar\delta(\boldsymbol{x}, \boldsymbol{y}), \tag{5.7}$$

where

$$\begin{aligned} \hat{\varphi}(\boldsymbol{x}) &:= \hat{\phi}(t_0, \boldsymbol{x}), \\ \hat{\pi}(\boldsymbol{x}) &:= \frac{\partial}{\partial t}\hat{\phi}(t, \boldsymbol{x})|_{t=t_0}. \end{aligned} \tag{5.8}$$

Now, if we consider the spatial topology to be that of a 3-torus, that is if we replace $\mathbb{R}^4$ by $\mathbb{R} \times T^3$, then $\boldsymbol{k}$ becomes a discrete variable. Equivalently, we may consider periodic boundary conditions on space, of the form $x^i \equiv x^i + l$ $(i = 1, 2, 3)$. Then the volume of space is finite, $V_0 = l^3$, and integrals over momenta become discrete sums:

$$\frac{d^3\boldsymbol{k}}{(2\pi)^3} \longrightarrow \frac{1}{V_0} \sum_{k \in \frac{2\pi}{l} n_k} .$$

Switching to the more compact notation $\hat{A}_{\boldsymbol{k}} \equiv \hat{A}(\boldsymbol{k})$, we then obtain the following expression for the quantum field:

$$\hat{\phi}(x) = \frac{1}{V_0} \sum_{\boldsymbol{k}} e^{i\boldsymbol{k}\cdot\boldsymbol{x}} \left(\frac{e^{-i\omega_k t}}{\sqrt{2\omega_k}} \hat{A}_{\boldsymbol{k}} + \frac{e^{+i\omega_k t}}{\sqrt{2\omega_k}} \hat{A}^\dagger_{-\boldsymbol{k}} \right). \tag{5.9}$$

This makes it apparent that we can regard the free quantum field as an "assembly of simple harmonic oscillators." We can now proceed in the standard way and construct states of the field from the vacuum. As with the simple harmonic oscillator, the CCR imply that there must be a state vector $|0\rangle$ (the vacuum state) which is annihilated by all the operators $\hat{A}_{\boldsymbol{k}}$:

$$\hat{A}_{\boldsymbol{k}}|0\rangle = 0. \tag{5.10}$$

Then we can construct 1-particle states as linear combinations of vectors of the form

$$|\boldsymbol{k}\rangle = \hat{A}^\dagger_{\boldsymbol{k}}|0\rangle, \tag{5.11}$$

as well as 2-particle states,

$$|\boldsymbol{k}_1, \boldsymbol{k}_2\rangle = \hat{A}^\dagger_{\boldsymbol{k}_1}\hat{A}^\dagger_{\boldsymbol{k}_2}|0\rangle, \tag{5.12}$$

and so on. Notice that, due to the CCR, all N-particle states are *symmetric* under exchange of the particles' labels. For instance, we have $|\boldsymbol{k}_1, \boldsymbol{k}_2\rangle = |\boldsymbol{k}_2, \boldsymbol{k}_1\rangle$. This reflects the fact that we are dealing with a bosonic theory. The Hilbert space obtained this way is called Fock space. Let us have a closer look at how it is constructed.

One of the basic structures we have to consider (and one that will play an essential role in the case of QFT in curved space-times) is the linear structure of the space of (real) solutions of the classical field equations. So let us consider the vector space of all real solutions of the Klein–Gordon equation, let us call it V. As we have seen, the presence of Poincaré symmetry allows us to decompose any solution as the sum of positive and negative frequency parts, so that—on the complexified space $V_{\mathbb{C}}$—we can always write $\phi = \phi^+ + \phi^-$, with ϕ real, and ϕ^+ (resp. ϕ^-) containing only positive (resp. negative) frequencies. We therefore have a decomposition of the form $V_{\mathbb{C}} = V^+ \oplus V^-$. This decomposition is covariant, meaning that it does not depend on the frame of reference. It turns out that the space V^+ can actually be given the structure of a Hilbert space. This Hilbert space (the space of 1-particle states) will be denoted by h. So a general 1-particle state will be given by the restriction of a real solution ϕ to its positive frequency part:

$$h \ni \phi^+(x) = \frac{1}{V_0} \sum_{\boldsymbol{k}} e^{i\boldsymbol{k}\cdot\boldsymbol{x}} e^{-i\omega_k t} A(\boldsymbol{k}). \tag{5.13}$$

Notice that here $A(\boldsymbol{k})$ is a function, the "Fourier coefficients" of the field, and so it depends on the initial data. The inner product on h is, then, given by

$$\langle \phi_1^+, \phi_2^+ \rangle := \frac{i}{\hbar} \int_\Sigma d^3\boldsymbol{x} \left(\overline{\phi_1^+} \nabla_a \phi_2^+ - \phi_2^+ \nabla_a \overline{\phi_1^+} \right) n^a, \tag{5.14}$$

where the integral is taken over any space-like hypersurface Σ. Here, n^a denotes the (time-like) vector normal to Σ. The integrand in (5.14) is given by the conserved current of the Klein–Gordon field, and therefore it is independent of the chosen Σ.

Exercise 5.3 Check the properties of the Hermitian product defined by (5.14), namely,

- $\langle \phi_1^+, \phi_2^+ \rangle = \overline{\langle \phi_2^+, \phi_1^+ \rangle}$,
- $\langle \phi_1^+, \phi_1^+ \rangle \geq 0, \quad \langle \phi_1^+, \phi_1^+ \rangle = 0 \Leftrightarrow \phi_1^+ = 0.$

Having constructed the 1-particle Hilbert space h, we can now define the Fock space, as follows:

$$\mathscr{H} := \mathbb{C} \oplus h \oplus (h \otimes_s h) \oplus \cdots \oplus (\otimes_s^n h) \oplus \cdots \tag{5.15}$$

Here, the subscript "s" in the tensor product means that we are considering the symmetrized tensor product of h with itself (this reflects the bosonic character of the field operators). One can easily check that the operators $\hat{A}_{\boldsymbol{k}}$ and $\hat{A}_{\boldsymbol{k}}^\dagger$ act on $\mathscr{H}$ in the expected way, and also that $\hat{\phi}(x)$ has a well-defined action on $\mathscr{H}$. In particular, the Fock space contains "sectors" labeled by the number of particles, that is, by the (integer) eigenvalues of the number operator $\hat{N} = \sum_k \hat{N}_{\boldsymbol{k}}$, where $\hat{N}_{\boldsymbol{k}} = \hat{A}_{\boldsymbol{k}}^\dagger \hat{A}_{\boldsymbol{k}}$.

After having reviewed the basic properties of the scalar field in Minkowski space, let us comment on some of the limitations.

1. We have made use of the Fourier transform in order to find the decomposition into positive/negative frequency parts. But in a general space-time, we will not have this tool at our disposal. However, what is fundamental, and what we have at our disposal, is an operator algebra, generated by (smeared) operators $\hat{\pi}(x)$, $\hat{\phi}(x')$. This algebra will be called the *canonical algebra* and denoted by $\mathscr{A}_{\text{can}}$. The task is then to find representations of this algebra. A relevant question is, then, how many equivalence classes of representations are there? In quantum mechanics, where the number of degrees of freedom is finite, we have von Neumann's uniqueness theorem, so the answer in that case is that there is, up to unitary equivalence, only one representation. But in QFT, where we have infinitely many degrees of

freedom, there are many (in fact, infinitely many) inequivalent representations of the canonical algebra. The one that was presented here was selected by Poincaré invariance of $|0\rangle$.
2. In curved space-time, there is no Poincaré group, so we will be confronted with a fundamental lack of uniqueness for the choice of a vacuum state. The first question we will address is how can we find an analog of the decomposition into positive and negative frequency parts, when working with a more general space-time where this decomposition will not be available, at least not in the straightforward way we found for the Minkowski case.

5.3 Quantum Field Theory in Curved Space-Time

In the previous section, we discussed the quantization of a scalar field on Minkowski space. Now we will consider the generalization of the same problem to a more general class of space-time manifolds. The curved space-times we will consider will be given by globally hyperbolic spaces, endowed with a metric g_{ab} of Minkowski signature $(-,+,+,+,)$. Topologically, the condition of being globally hyperbolic means that the underlying manifold is of the form

$$M^{(4)} = \Sigma^{(3)} \times \mathbb{R}. \tag{5.16}$$

Thus, globally hyperbolic space-times have the structure of a foliation, for which the 3-manifold $\Sigma^{(3)}$ plays the role of a Cauchy surface; that is, every inextensible time-like curve intersects $\Sigma^{(3)}$ only once. This in turn means that, for the type of classical equation of motion we are considering, the initial value problem is well-posed.

We will follow the convention to denote points belonging to $\Sigma^{(3)}$ with a *bar*. Thus, $\bar{x}$ means $\bar{x} \in \Sigma^{(3)}$. Let us consider two functions $\varphi(\bar{x})$ and $\pi(\bar{x})$ (with appropriate regularity/support properties) that we regard as the *initial data* for the classical equation of motion, Eq. (5.1). Then we get a unique solution $\phi(x)$, $x \in M^{(4)}$, which is such that

$$\phi(x)\big|_{\Sigma^{(3)}} = \varphi(\bar{x}), \qquad n^a \nabla_a \phi(x)\big|_{\Sigma^{(3)}} = \pi(\bar{x}). \tag{5.17}$$

5.3.1 *Quantization*

In its more general form, the Stone–von Neumann theorem can be stated in terms of the uniqueness of (strongly continuous, unitary, irreducible) representations of an abstract Weyl algebra $\mathscr{W}(V,\sigma)$ which is associated to any finite-dimensional symplectic vector space (V,σ). This algebra is generated by elements of the form $W(u)$, $u \in V$, subject to the relations

$$W(u)W(v) = e^{-\frac{i}{2}\sigma(u,v)} W(u+v). \tag{5.18}$$

If we replace symplectic vector spaces by symplectic manifolds (i.e., if we give up the linear structure), then the theorem does not hold true, something that is well known in the context of geometric quantization. But even if we keep the linear structure, but consider infinite dimensional (symplectic) vector spaces, there will be many inequivalent representations. This kind of situation is very common in physics, and it is precisely this structure (that of an infinite dimensional symplectic vector space) the one that is relevant for the study of scalar quantum fields on curved space-times.

The (symplectic) vector space we are interested in is the space Γ of *real* solutions of the field Eq. (5.1). As every solution can be obtained from a suitable pair $\gamma \equiv (\varphi, \pi)$, we may as well consider the vector space $\Gamma_{\rm can}$ of initial data as the relevant vector space. Both spaces are related through an isomorphism

$$\begin{aligned} I_\Sigma : \Gamma &\longrightarrow \Gamma_{\rm can} \\ \phi &\longmapsto (\varphi, \pi). \end{aligned} \tag{5.19}$$

The symplectic form on $\Gamma_{\rm can}$ (which we will denote as Ω) is given by the following formula:

$$\Omega\left((\varphi_1, \pi_1), (\varphi_2, \pi_2)\right) := \int_\Sigma (\varphi_1 \pi_2 - \varphi_2 \pi_1) dV_\Sigma, \tag{5.20}$$

where Σ denotes *any* Cauchy surface. We can use the isomorphism I_Σ to obtain (by pull-back) the equivalent of Ω on Γ. It is given by $\boldsymbol{\Omega} = I_\Sigma^* \Omega$. Explicitly we have, for $\phi_1, \phi_2 \in \Gamma$,

$$\boldsymbol{\Omega}\,(\phi_1, \phi_2) := \int_\Sigma (\phi_1 \nabla_a \phi_2 - \phi_2 \nabla_a \phi_1) n^a dV_\Sigma. \tag{5.21}$$

As we know, a quantization map in the sense of Dirac does not in general exist. For this reason, we will focus on a special class of observables, for which quantization is well defined. To $\gamma_0 = (f, g) \in \Gamma_{\rm can}$, we associate the *linear observable* $\mathcal{O}_{\gamma_0}$, defined as the following function on Γ:

$$\mathcal{O}_{\gamma_0}(\gamma) := \Omega(\gamma_0, \gamma). \tag{5.22}$$

Notice that for $\gamma_0 = (f, g)$ and $\gamma = (\varphi, \pi)$ we then have

$$\mathcal{O}_{\gamma_0}(\gamma) = \int_\Sigma (f\, \pi - g\, \varphi) dV_\Sigma. \tag{5.23}$$

Upon quantization, we expect to obtain an operator $\hat{\mathcal{O}}_{\gamma_0}$, linear in γ_0, of the form

$$\hat{\mathcal{O}}_{\gamma_0}(\gamma) = \int_\Sigma (f(\bar{x})\, \hat{\pi}(\bar{x}) - g(\bar{x})\, \hat{\varphi}(\bar{x})) dV_\Sigma. \tag{5.24}$$

The commutation relations these operators should obey can be read off from the corresponding classical expression in terms of the Poisson bracket:

$$\{\mathscr{O}_{\gamma_1}, \mathscr{O}_{\gamma_2}\} = \Omega(\gamma_1, \gamma_2). \tag{5.25}$$

Hence, we impose the following CCR:

$$[\hat{\mathscr{O}}_{\gamma_1}, \hat{\mathscr{O}}_{\gamma_2}] = i\hbar\Omega(\gamma_1, \gamma_2)\hat{\mathbb{1}}. \tag{5.26}$$

Exercise 5.4 Check that the CCR in the form (5.26) are actually equivalent to

$$[\hat{\varphi}(\bar{x}_1), \hat{\pi}(\bar{x}_2)] = i\hbar\delta(\bar{x}_1, \bar{x}_2)\hat{\mathbb{1}}, \quad [\hat{\varphi}(\bar{x}_1), \hat{\varphi}(\bar{x}_2)] = [\hat{\pi}(\bar{x}_1), \hat{\pi}(\bar{x}_2)] = 0. \tag{5.27}$$

The CCR in the form (5.27) only involve (the quantized counterparts of) the initial data $\varphi(\bar{x})$ and $\pi(\bar{x})$. But it turns out that, because of the hyperbolic character of the field equation, if we assume that the quantized field $\hat{\phi}(x)$ also satisfies the field equation

$$(\Box - \mu^2)\hat{\phi}(x) = 0, \tag{5.28}$$

then using (5.27) we can obtain the commutation relations at arbitrary space-time points. The result is

$$[\hat{\phi}(x_1), \hat{\phi}(x_2)] = i\hbar\Delta(x_1, x_2)\hat{\mathbb{1}}, \tag{5.29}$$

where $\Delta(x_1, x_2) = G_A(x_1, x_2) - G_R(x_1, x_2)$ is the commutator function, defined as the difference between the advanced (G_A) and retarded (G_R) Green's functions of the Klein–Gordon operator. These are defined as distributional solutions to

$$(\Box - \mu^2)G_{A/R}(x, y) = \delta^{(4)}(x, y)$$

with advanced/retarded support. In a more abstract setting, we may consider the free algebra $\mathscr{A}_{\text{can}}$ generated by the smeared quantum fields

$$\hat{\varphi}(f) = \int_\Sigma \hat{\varphi}(\bar{x})f(\bar{x})dV_\Sigma \tag{5.30}$$

and

$$\hat{\pi}(g) = \int_\Sigma \hat{\pi}(\bar{x})g(\bar{x})dV_\Sigma, \tag{5.31}$$

where f and g are test functions on Σ. The smeared field operators satisfy the following form of the CCR:

$$[\hat{\varphi}(f), \hat{\pi}(g)] = i\hbar\Omega(f, g)\hat{\mathbb{1}}, \tag{5.32}$$

with other commutators vanishing. Notice that we are using the compact notation

$$\Omega(f, g) \equiv \Omega((f, 0), (0, g)).$$

The form (5.32) is easily obtained from (5.27), as we have:

$$\begin{aligned}[\hat{\varphi}(f), \hat{\pi}(g)] &= \int_\Sigma dV_\Sigma^{(1)} dV_\Sigma^{(2)} f(\bar{x}_1) g(\bar{x}_2) [\hat{\varphi}(\bar{x}_1), \hat{\pi}(\bar{x}_2)] \\ &= i\hbar \int_\Sigma dV_\Sigma f(x) g(x) \equiv \Omega((f, 0), (0, g)).\end{aligned}$$

The algebra $\mathscr{A}_{\text{can}}$ can also be given the structure of a $*$-algebra, with an involution ($*$-operation) given, for f, g real test functions, by

$$\hat{\pi}^*(f) = \hat{\pi}(f), \quad \hat{\varphi}^*(g) = \hat{\varphi}(g).$$

Although it is not possible to endow $\mathscr{A}_{\text{can}}$ with a norm (the field operators are unbounded), we can consider the exponentiated form of the CCR and instead focus on the resulting Weyl algebra ($\mathscr{W}(\Gamma, \boldsymbol{\Omega})$, in the notation introduced above). For such an algebra, it is always possible to find a suitable norm.

Now, in view of the isomorphism I_Σ, we may also consider the covariant algebra $\mathscr{A}_{\text{cov}}$, generated as a free $*$-algebra by the quantum fields $\hat{\phi}(x)$, which are solutions to the field Eq. (5.28), are subject to the CCR (5.29), and are Hermitian: $\hat{\phi}(x) = \hat{\phi}^*(x)$.

5.3.2 *Representations*

The representation of the field algebra in the case of Minkowski space-time made crucial use of a decomposition into positive and negative frequency modes coming directly from the Fourier transform. As in the general case we will not have this possibility, we will instead make use of a *polarization* defined on $\Gamma_\mathbb{C}$. The idea is to be able to express every real solution ϕ of the field equation in the form

$$\phi = \phi^+ + \phi^-, \tag{5.33}$$

where ϕ^+ and ϕ^- are complex fields such that $\bar{\phi}^- = \phi^+$. Since the phase space Γ is a real vector space, the way to achieve this is through a (compatible) complex structure J. This is a real linear map $J : \Gamma \to \Gamma$ such that $J^2 = -\mathbb{1}$. A polarization on $\Gamma_\mathbb{C}$ is then obtained by defining, for each $\phi \in \Gamma$,

$$\phi^+ := \frac{1}{2}(\phi - iJ\phi), \qquad \phi^- := \frac{1}{2}(\phi + iJ\phi). \tag{5.34}$$

It then follows that $J\phi^{\pm} = \pm i\phi^{\pm}$. The compatibility conditions that we must require between the symplectic form and the complex structure are:

(i) $\Omega(J\gamma_1, J\gamma_2) = \Omega(\gamma_1, \gamma_2)$, for all $\gamma_1, \gamma_2 \in \Gamma$.
(ii) $\Omega(\gamma, J\gamma) \geq 0$, for all $\gamma \in \Gamma$ and $\Omega(\gamma, J\gamma) = 0 \Leftrightarrow \gamma = 0$.

These conditions allow us to define the following Hermitian inner product on Γ_J, the complexification of Γ obtained from the complex structure J:

$$\langle \gamma_1, \gamma_2 \rangle_J := \frac{1}{2\hbar}\left[\Omega(\gamma_1, J\gamma_2) + i\Omega(\gamma_1, \gamma_2)\right]. \tag{5.35}$$

Exercise 5.5 Check that (5.35) indeed defines a Hermitian inner product on Γ_J.

An equivalent description is obtained if instead we work directly on $\Gamma_{\mathbb{C}}$, using the polarization and restricting $\langle \cdot, \cdot \rangle$ to $\Gamma^+ := \{\phi^+ = \frac{1}{2}(1 - iJ)\phi \mid \phi \in \Gamma\}$. From $J\phi^+ = i\phi^+$, we obtain

$$\begin{aligned} \langle \phi_1^+, \phi_2^+ \rangle &= \frac{i}{\hbar}\Omega(\bar{\phi}_1^+, \phi_2^+) \\ &= \frac{i}{\hbar}\int_{\Sigma} dV_{\Sigma} n^a (\bar{\phi}_1^+ \nabla_a \phi_2^+ - \phi_2^+ \nabla_a \bar{\phi}_1^+). \end{aligned} \tag{5.36}$$

Furthermore, a simple calculation shows that $\langle \phi_1^+, \phi_2^+ \rangle = \langle \phi_1, \phi_2 \rangle_J$. Upon Cauchy completion, what we obtain is an isomorphism of Hilbert spaces, between $(\Gamma_J, \langle\ ,\ \rangle_J)$ and $(\Gamma^+, \langle\ ,\ \rangle)$.

We will therefore define the 1-particle Hilbert space to be the Cauchy completion of Γ^+, that is we set

$$h := (\overline{\Gamma^+}, \langle\ ,\ \rangle). \tag{5.37}$$

The corresponding Fock space is then given (as in (5.15)) by

$$\mathscr{H} := \mathbb{C} \oplus h \oplus (h \otimes_s h) \oplus \cdots \oplus (\otimes_s^n h) \oplus \cdots \tag{5.38}$$

Now we have an algebra (either $\mathscr{A}_{\text{cov}}$ or $\mathscr{A}_{\text{can}}$) and a Hilbert space $\mathscr{H}$. The next question to be answered is: How do we construct a representation map? Assuming we have already made a choice of complex structure J, we just need to introduce an orthonormal basis on h. That is, we must choose a set $\{e_n(x)\}_n$ of solutions of the Klein–Gordon equation such that

(i) $Je_n = ie_n$
(ii) $\langle e_n, e_m \rangle = \delta_{n,m}$, i.e., $\boldsymbol{\Omega}(\overline{e}_n, e_m) = -i\hbar\delta_{n,m}$.

It is because of condition (i) that we call them "positive frequency" solutions. The choice of this set of solutions then gives rise to a representation, via the following definition:

$$\mathscr{R}_J(\hat{\phi}(x)) := \sum_n \left(e_n(x)\hat{A}_n + \overline{e}_n(x)\hat{A}_n^\dagger \right). \tag{5.39}$$

Exercise 5.6 Check that, for $\mathscr{R}_J$ to be a representation, we must have $[\hat{A}_n, \hat{A}_m^\dagger] = \delta_{n,m}$ and $[\hat{A}_n, \hat{A}_m] = 0$.

One of the most prominent features of the theory of quantum fields on curved space-times is the non-uniqueness of the vacuum state. As mentioned before, in the case of Minkowski space-time, there is a clear physical condition that singles out the vacuum as the unique Poincaré invariant state of the theory. But in more general situations, we do not have such symmetry properties at our disposal. Finding physically relevant conditions that might single out a vacuum state becomes a most important problem.

In the remaining part of this section, we are going to discuss one such condition, the significance of which is better appreciated if we first discuss some of the properties of *coherent states*. It is convenient to first discuss coherent states in the context of the simple harmonic oscillator, described by the Hamiltonian

$$\hat{H} = \frac{\hat{P}^2}{2m} + \frac{1}{2}m\omega^2\hat{X}^2, \tag{5.40}$$

where $\hat{X}$ and $\hat{P}$ satisfy the usual canonical commutation relations. Defining the dimensionless operators

$$\hat{x} = \sqrt{\frac{m\omega}{\hbar}}\hat{X}, \qquad \hat{p} = \frac{1}{\sqrt{m\omega\hbar}}\hat{P}, \tag{5.41}$$

as well as creation and annihilation operators

$$\hat{a} = \frac{1}{\sqrt{2}}(\hat{x} + i\hat{p}), \qquad \hat{a}^\dagger = \frac{1}{\sqrt{2}}(\hat{x} - i\hat{p}), \tag{5.42}$$

we obtain the following simple expression for the Hamiltonian:

$$\hat{H} = \hbar\omega\left(\hat{a}^\dagger\hat{a} + \frac{1}{2}\right). \tag{5.43}$$

The spectrum of $\hat{H}$ is readily obtained by making use of the commutation relations $[\hat{a}, \hat{a}^\dagger] = 1$. The ground state $|0\rangle$ is characterized by the condition $\hat{a}|0\rangle = 0$ and gives the lowest energy eigenvalue. The remaining energy eigenstates are obtained from $|0\rangle$ by repeated application of the creation operator:

$$|n\rangle = \frac{(\hat{a}^\dagger)^n}{\sqrt{n!}}|0\rangle. \tag{5.44}$$

An equivalent description of these states can be given in terms of holomorphic functions, in the following way. First let us consider the classical phase space Γ (which here is just the 2-dimensional space $\mathbb{R}^2$), the elements of which we denote as $\gamma = (x, p)$. Upon identification of Γ with $\mathbb{C}$, γ becomes $z = x + ip$. The natural complex structure of $\mathbb{C}$ can then be mapped to a complex structure J on Γ. Explicitly, we have (for the mapping $z \mapsto iz$ seen as a real map on Γ):

$$\begin{array}{rccc} J: & \Gamma & \longrightarrow & \Gamma \\ & \begin{pmatrix} x \\ p \end{pmatrix} & \longmapsto & \begin{pmatrix} 0 & -1 \\ 1 & 0 \end{pmatrix}\begin{pmatrix} x \\ p \end{pmatrix}. \end{array} \tag{5.45}$$

Exercise 5.7 Check that J as defined above is compatible with the symplectic form $\Omega = dx \wedge dp$.

We now introduce a Hilbert space $\mathscr{H}$ of holomorphic functions on $\Gamma \equiv \mathbb{C}$, with inner product given by

$$\langle \psi_1, \psi_2 \rangle := \frac{i}{\pi}\int_\Gamma dz \wedge d\bar{z}\, e^{-z\bar{z}}\, \overline{\psi_1(z)}\psi_2(z). \tag{5.46}$$

A general state vector in this space is written as

$$\psi = \psi_0 + \psi_1 z + \psi_2 \frac{z^2}{\sqrt{2!}} + \cdots + \psi_n \frac{z^n}{\sqrt{n!}} + \cdots \tag{5.47}$$

In terms of the basis (5.44), the above state is written as

$$\psi = \psi_0|0\rangle + \psi_1|1\rangle + \psi_2|2\rangle + \cdots \tag{5.48}$$

As can be readily checked, in this representation creation and annihilation operators are given by the following expressions:

$$\hat{a}^\dagger \psi(z) = z\psi(z), \qquad \hat{a}\psi(z) = \frac{d\psi(z)}{dz}. \tag{5.49}$$

Notice that in this representation (which is known as the *Bargmann–Segal representation* [7]), ψ is a function on phase space, and not only on configuration space.

Now let us turn to the definition of coherent states. For $\zeta \in \mathbb{C}$, we put

$$|\psi_\zeta\rangle := N e^{\bar{\zeta}\hat{a}^\dagger}|0\rangle, \tag{5.50}$$

where $N = e^{-1/2|\zeta|^2}$ is a normalization constant. Such a state is a coherent superposition of *all* basis states of the harmonic oscillator:

$$|\psi_\zeta\rangle = N\left(|0\rangle + \bar{\zeta}|1\rangle + \frac{\bar{\zeta}^2}{2!}|2\rangle + \cdots\right).$$

Coherent states play a fundamental role in physics (e.g., in quantum optics), because they are states of minimum uncertainty. In fact, it turns out that the state $|\psi_\zeta\rangle$ is sharply peaked around the point (x_0, p_0), where $\zeta = x_0 + ip_0$. Discussions of coherent states are much easier in the Bargmann–Segal (holomorphic) representation. For instance notice that, in this representation, the coherent state defined in (5.50) is given by

$$\psi_\zeta(z) = e^{-1/2|\zeta|^2} e^{\bar{\zeta} z}. \tag{5.51}$$

The probability distribution corresponding to this state is easily computed and found to be given by a Gaussian distribution centered around ζ:

$$\mathscr{P}(z) = \overline{\psi_\zeta}(z)\psi_\zeta(z)e^{-z\bar{z}} = N^2 e^{\zeta\bar{z}+\bar{\zeta}z}e^{-z\bar{z}} = e^{-|z-\zeta|^2}.$$

A straightforward computation then shows that the uncertainties in x and p are indeed the minimum possible:

$$\Delta x = \frac{1}{\sqrt{2}}, \qquad \Delta p = \frac{1}{\sqrt{2}}.$$

Exercise 5.8 Starting from the requirement of minimum uncertainty, obtain the general form (5.50) for a coherent state.

A salient feature of coherent states following from this sharp localization is that the quantum evolution of the state follows very closely the classical path.

Let us now consider quantum fields: Starting with a real solution of the classical field equation, say $\phi \in \Gamma_{cov}$, and assuming we have already made a choice of complex structure J, we obtain a 1-particle state ϕ^+ in h, from which a creation operator $\hat{A}^\dagger_{\phi^+}$ can be constructed. Acting by exponentiation with this operator on the vacuum, we then obtain a coherent state $|\Psi_{\phi^+}\rangle$, that will be sharply peaked around the classical solution we started with.

Let ϕ_0 denote a specific 1-particle state. For any choice of a Cauchy surface Σ, we will have, as in (5.17), initial data given by functions $\varphi_0(\bar{x})$ and $\pi_0(\bar{x})$. Then one obtains, for the expectation value of the corresponding quantum operators on the coherent state $|\Psi_{\phi^+}\rangle$,

$$\langle\Psi_{\phi^+}|\hat{\varphi}(\bar{x})|\Psi_{\phi^+}\rangle = \varphi_0(\bar{x}) \tag{5.52}$$

$$\langle\Psi_{\phi^+}|\hat{\pi}(\bar{x})|\Psi_{\phi^+}\rangle = \pi_0(\bar{x}). \tag{5.53}$$

Furthermore, in the context of the standard Fock quantization of scalar fields on Minkowski space-time, it is possible to show that the expectation value of the (quantum) Hamiltonian with respect to a coherent state $|\Psi_{\phi^+}\rangle$ precisely coincides with the energy of the classical field ϕ. This property can in turn be used to characterize the Fock representation [3].

From the above discussion, we have learned: (i) that, in general, there is no unique choice of vacuum sate and (ii) a choice of vacuum state is related to a choice of complex structure on Γ. Thus, some questions that arise are: Is it possible to use a physical criterion (like the one discussed above) in order to single out a vacuum state in more general situations? What happens if we consider two different complex structures? We will not answer these questions in full generality in these notes, but will rather consider specific classes of space-times for which we can state relevant results. Let us, therefore, consider the following two classes of space-times:

- Space-times are "trivial spatially." These include solutions to Einstein's equations of the FLRW type, which are relevant for cosmology.
- Space-times that are "trivial in time," meaning that there exists a Killing vector field t^a on M (the space-time manifold) such that the metric g_{ab} is invariant with respect to the diffeomorphisms generated by i_{t^a}, i.e., $i_t^* g_{ab} = g_{ab}$. Or equivalently (infinitesimal form), the metric is such that $\mathscr{L}_t g_{ab} = 2\nabla_{[a} t_{b]} = 0$. In general relativity, such stationary space-times include the ones produced by stars (or even black holes) in equilibrium.

A choice of complex structure for stationary space-times, which is a natural generalization from static space-times, can be obtained making use of an energy requirement similar to the one described above. Under the assumption that there exists ε such that $\lambda := -t^a t_a \geq \varepsilon > 0$, we have [2]:

Theorem There exists a unique complex structure J on $\Gamma_{\text{cov}} \ni \phi$, $(\Box - \mu^2)\phi = 0$ which respects the isometry i_t and is compatible with Ω, in the sense that

$$\Omega(i_t\phi_1, i_t\phi_2) = \Omega(\phi_1, \phi_2), \tag{5.54}$$

as well as

$$[J, i_t] = 0 \tag{5.55}$$

are satisfied, and such that

$$E_t(\phi) = \langle \psi_{\phi_0} | \hat{H}_t | \psi_{\phi_0} \rangle, \tag{5.56}$$

where $E_t(\phi)$ stands for the classical energy of the field.

5.4 Cosmology

5.4.1 General Remarks

In this last section, we will consider applications of the theory discussed so far to problems in cosmology. As mentioned before, in cosmology we often consider metrics for which "space is trivial." By this we mean that the space-time manifold (M, g_{ab}) is foliated by a 1-parameter family of Cauchy surfaces such that each leaf, as a metric manifold, is maximally symmetric. Physically this means that space is *homogeneous* (there is no preferred point) and isotropic (there is no preferred direction). In this case, we will have six independent Killing vector fields and, correspondingly, a six-dimensional isometry group.

A fundamental question in cosmology is the one related to the origin of structure. What is the mechanism behind the formation of, say, density fluctuations that eventually gave rise to the formation of galaxies? As we will see, the origin of these fluctuations is of a quantum nature. In theoretical models, we therefore start with a homogeneous and isotropic background metric and include inhomogeneities as perturbations to this metric. The requirements of homogeneity and isotropy lead to a metric of the following form:

$$g_{\mu\nu}dx^\mu dx^\nu = -dt^2 + q_{ab}dx^a dx^b, \tag{5.57}$$

where the metric q_{ab}, corresponding to the spatial part Σ, is of constant curvature. The Riemann tensor, then, takes the following form,

$$\mathscr{R}_{abcd} = \frac{1}{6}\mathscr{R} q_{a[c}q_{b]d},$$

with $\mathscr{R}$ (Ricci scalar) constant. Depending on the value of $\mathscr{R}$, we distinguish three cases:

1. $\mathscr{R} > 0$. In this case, Σ has the topology of a 3-sphere.
2. $\mathscr{R} = 0$. The manifold Σ is $\mathbb{R}^3$ with flat metric $q_{ab}dx^a dx^b = dx_1^2 + dx_2^2 + dx_3^2$.
3. $\mathscr{R} < 0$: Σ is a constant-vacuum surface of negative curvature.

So far, the second case is the one compatible with observations [8]. The evolution of the universe (i.e., of the metric g_{ab}) is determined by Einstein's equations, for a given matter/energy source. From the symmetries discussed, and also from observational evidence, we are led to propose a metric of the form

$$g_{ab} = -dt^2 + a^2(t)d\bar{x}^2. \tag{5.58}$$

The factor $a(t)$ will play the role of a dynamical variable. Due to the symmetries of the space-time, Einstein's equations reduce to the following two equations:

$$\left(\frac{\dot{a}}{a}\right)^2 = \frac{8\pi G}{3}\rho \qquad \text{(Friedmann equation).} \tag{5.59}$$

$$\left(\frac{\ddot{a}}{a}\right) = -\frac{4\pi G}{3}(\rho + 3p) \qquad \text{(Raychaudhuri equation),} \tag{5.60}$$

where ρ is the energy density and p is the pressure of the matter field.

Exercise 5.9 Use Einstein's field equations in order to obtain the previous two equations.

In cosmology, there is a quite interesting interplay between particle and nuclear physics on one hand and general relativity on the other. This is so because, according to Einstein's field equations, the metric (and hence the dynamical variable $\tilde{a}(t)$) is determined by the matter and energy content of the universe. On the other hand, a dynamical universe, one where space itself is changing over time, will determine when do certain "particle species" dominate, according to the specific cross sections, energy scales, etc. Nowadays, we have a fairly clear picture of the way our universe has evolved after the big bang. So, for instance, right after inflation, until about 4×10^4 years after the big bang, we have a "radiation-dominated era," followed by a "matter-dominated era." During the latter, the energy density of matter took over the energy density of radiation. Some 300.000 years after the big bang, the universe had cooled enough, so that hydrogen atoms could form. Light eventually decoupled from matter, giving rise to the cosmic microwave background (CMB) radiation that we observe today at a temperature of about 2.7 K. In our current time ($t \approx 13.8 \times 10^9$ years), it is the cosmological constant Λ the one dominating the evolution of the universe. How do we explain this? The energy densities for matter and radiation in an expanding universe depend on the scale factor a as follows [11]:

$$\rho_{\text{dust}} \sim \frac{1}{a^3}, \qquad \rho_{\text{rad}} \sim \frac{1}{a^4}. \tag{5.61}$$

This means that they *decrease* in time. On the other hand, ρ_Λ is time independent. So when ρ_{dust} and ρ_{rad} become vanishingly small, ρ_Λ becomes dominant (this leading to an accelerated universe).

Important information about inhomogeneities that are crucial in order to understand the dynamics of structure formation is imprinted in the CMB radiation. What is the origin of these fluctuations? In order to answer this question, we should focus on the physics of the very early universe. The basic idea is that inhomogeneities have their origin in *quantum fluctuations*.

In the remaining part of this section, we will describe the basic assumptions behind inflationary models, emphasizing the applications of the concepts of quantum field theory in curved space-times that we have discussed so far. One of the main assumptions we make is about the existence of a "matter" scalar field Φ that is subject to a suitable potential $V(\Phi)$. One possible choice for the potential is

$$V(\Phi) = \frac{1}{2}\mu^2\Phi^2.$$

Already at the CMB epoch, the universe looks extremely homogeneous and isotropic, but there should be "tiny" fluctuations that our model should account for. At the onset of inflation, space-time is homogeneous and isotropic. Now, perturbations in homogeneities would be zero classically. But in quantum field theory, these perturbations have their origin in the properties of the vacuum state and—because of the uncertainty principle—there will be fluctuations, of a quantum nature. We will, therefore, consider quantum field theory in a FLRW, spatially flat, universe. We shall consider the wave equation for a scalar field on this background space-time. The metric is given by

$$ds^2 = -dt^2 + \tilde{a}^2(t)d\bar{x}^2, \tag{5.62}$$

with t denoting proper time. If we now go to conformal time, that is, if we introduce the change of variables

$$d\eta = \frac{dt}{a(t)}, \qquad a(\eta) = \tilde{a}(t), \tag{5.63}$$

then the metric takes the form

$$ds^2 = a^2(\eta)(-d\eta^2 + d\bar{x}^2). \tag{5.64}$$

Exercise 5.10 Using the general form of the d'Alembertian in a curved space-time,

$$\Box\phi = \frac{1}{\sqrt{-g}}\partial_a\sqrt{-g}\, g^{ab}\partial_b\phi, \tag{5.65}$$

obtain the field equation

$$\phi''(x) - 2\frac{a'}{a}\phi'(x) - \nabla^2\phi(x) + \mu^2 a^2\phi(x) = 0, \tag{5.66}$$

where ϕ' denotes the derivative of ϕ with respect to η.

We can now perform a Fourier transformation with respect to space:

$$\phi(\bar{x}, \eta) = \frac{1}{V_o}\sum_{\bar{k}} e^{i\bar{k}\cdot\bar{x}}\phi_{\bar{k}}(\eta), \tag{5.67}$$

obtaining

$$\phi''_{\bar{k}} + 2\frac{a'(\eta)}{a(\eta)}\phi'_{\bar{k}} + \left(\bar{k}^2 + \mu^2 a^2(\eta)\right)\phi_{\bar{k}} = 0. \tag{5.68}$$

For each $\bar{k}$, we obtain two independent solutions. Denoting them by $e_k(\eta), \overline{e}_k(\eta)$, we then have

$$\phi_{\bar{k}}(\eta) = e_k(\eta) A_{\bar{k}} + \overline{e}_k(\eta) \overline{A}_{-\bar{k}} \tag{5.69}$$

Let us recall that in the flat space-time case we had

$$e_k(\eta) = \frac{e^{-i\omega\eta}}{\sqrt{2\omega}}.$$

This choice was singled out by time-translation symmetry. But in our present example, there is no natural splitting into positive and negative frequency solutions. A choice of basis has to be made, as well as a choice of a splitting into "positive" and "negative" frequency parts. Let us explain this in more detail. First of all we choose, for each mode, a solution $e_k(\eta)$. We will call these the positive frequency solutions. Then we *define* a complex structure on the space of solutions by means of the following formula:

$$J\phi(x,t) := \frac{1}{V_0} \sum_{\bar{k}} e^{i\bar{k}\cdot\bar{x}} \left(i e_k(\eta) A_{\bar{k}} - i\, \overline{e}_k(\eta) \overline{A}_{-\bar{k}} \right). \tag{5.70}$$

That is, on basis elements we define $Je_k(\eta) = ie_k(\eta)$ and $J\bar{e}_k(\eta) = -i\bar{e}_k(\eta)$. Now we need to check the compatibility of J with the symplectic form Ω. Recall that the symplectic form is given by the following expression:

$$\Omega(\phi_1, \phi_2) = \int d^3v (\phi_1 n^a \nabla_a \phi_2 - \phi_2 n^a \nabla_a \phi_1). \tag{5.71}$$

The compatibility conditions are

$$\Omega(J\phi_1, J\phi_2) = \Omega(\phi_1, \phi_2), \tag{5.72}$$

$$\Omega(\phi, J\phi) \geq 0 \; (= 0 \Leftrightarrow \phi = 0). \tag{5.73}$$

It can be easily checked that (5.72) and (5.73) are satisfied if we impose the following normalization conditions:

$$\bar{e}_k e_k' - \bar{e}_k' e_k = -\frac{i}{a^2(\eta)} \quad \text{and} \quad e_k e_{-k}' - e_k' e_{-k} = 0. \tag{5.74}$$

Notice that if these conditions are satisfied at $\eta = \eta_0$, then they satisfied at all η, because the e_k are solutions of the wave equation, i.e.,

$$e_k'' + 2\frac{a'}{a} e_k' + (k^2 + a^2\mu^2) e_k = 0. \tag{5.75}$$

Upon quantization, this will guarantee that the operators $\hat{A}^{(\dagger)}_{\bar{k}}$ fulfill the canonical commutation relations. We then get a Hilbert space $\mathscr{H}_J$, which is generated by the J-vacuum (i.e., a state vector annihilated by all the $\hat{A}_{\bar{k}}$) together with the state vectors generated from the vacuum by successive application of creation operators, obtaining in this manner a representation of the field operators on this Hilbert space:

$$\mathscr{R}_J(\hat{\phi}(x)) = \frac{1}{V_0}\sum_{\bar{k}} e^{i\bar{k}\cdot\bar{x}}\left(e_k(\eta)\hat{A}_{\bar{k}} + \bar{e}_k(\eta)\hat{A}^{\dagger}_{-\bar{k}}\right). \tag{5.76}$$

For the construction of the representation, we have made explicit use of a complex structure J. An important question is, therefore, the one related to the dependence of results on different choices of J. Since the way we have defined the complex structure makes use of a choice of basis, let us consider a different choice of basis, related to the first choice in the following way,

$$e^{(2)}_k = \alpha_k e^{(1)}_k + \beta_k \bar{e}^{(1)}_k, \tag{5.77}$$

with α_k, β_k complex (Bogoliubov) coefficients.

Exercise 5.11 Check that the basis vectors $e^{(2)}_k$ and $e^{(1)}_k$ both satisfy the normalization conditions if and only if

$$|\alpha_k|^2 - |\beta_k|^2 = 1. \tag{5.78}$$

The simplest situation we can consider is given by the choice $\beta_k = 0$. In this case, (5.78) reduces to $\alpha_k = e^{i\theta_k}$ and therefore $e^{(2)}_k = e^{i\theta_k} e^{(1)}_k$. So, in this case, creation and annihilation operators do not mix, and we have:

$$\hat{A}^{(2)}_{\bar{k}}|0\rangle_1 = 0 \Leftrightarrow |0\rangle_1 = |0\rangle_2.$$

In this case, the resulting complex structures are equivalent, and so are the corresponding Hilbert space representations. The non-trivial cases correspond to choices for which $\beta_k \neq 0$. Here, the "new" annihilation operators are given by a linear superposition of the "old" creation *and* annihilation operators:

$$\hat{A}^{(2)}_{\bar{k}} = \bar{\alpha}_k A^{(1)}_{\bar{k}} - \bar{\beta}_k A^{(1)\dagger}_{-\bar{k}}. \tag{5.79}$$

Now we have two *a priori* different vacua, determined by the conditions

$$\hat{A}^{(1)}_{\bar{k}}|0\rangle_1 = 0, \quad \hat{A}^{(2)}_{\bar{k}}|0\rangle_2 = 0. \tag{5.80}$$

Even though the two vacua $|0\rangle_1$ and $|0\rangle_2$ are different, it could be possible that they correspond to equivalent representations. Let us explore this possibility.

Given two representations $\mathscr{R}_{J_1}$ and $\mathscr{R}_{J_2}$, we want to know under which conditions on the corresponding complex structures J_1 and J_2 can we guarantee the existence of a unitary operator $U : \mathscr{H}_1 \to \mathscr{H}_2$ such that

$$\mathscr{R}_{J_1}(\hat{\phi}(x)) = U^{-1}\mathscr{R}_{J_2}(\hat{\phi}(x))U. \tag{5.81}$$

It turns out that the existence of such an operator follows from the condition

$$\sum_k |\beta_k|^2 < \infty. \tag{5.82}$$

Although we will not discuss it here, this is equivalent to a mathematical characterization in terms of Hilbert–Schmidt operators [9]. Instead, let us comment on the physical interpretation of the condition. As can be easily checked, what (5.82) means is that the expectation value of the number operator $\hat{N}^{(2)}$ with respect to the vacuum $|0_1\rangle$ is finite ($\hat{N}_k = \hat{A}_k^\dagger \hat{A}_k$):

$$\sum_k \langle 0_1|U^{-1}\hat{N}_k^{(2)}U|0_1\rangle = \sum_k |\beta_k|^2 < \infty. \tag{5.83}$$

Hence, the Fock vacuum corresponding to J_1 looks as an excited state in the vacuum corresponding to J_2 (with a finite number of particles) if and only if $\sum_k |\beta_k|^2 < \infty$.

We therefore obtain a partition of the set of complex structures into equivalence classes. How do we select a preferred equivalence class? A physically well-motivated answer to this question is given by the requirement of UV regularity [1, 5, 6, 10]. Generically, physical observables are constructed out of the basic quantum field operator $\hat{\phi}(x)$. But a composite field operator, like $\hat{\phi}(x)^2$, leads to divergences, due to the local character of the field operators. On Minkowski space-time, the divergencies of such a product are taken care of by introducing normal ordering. But in a general curved space-time, this requires a regularization procedure and this, in turn, can be well defined only if certain regularity conditions are demanded. Consider, for instance, the two-point function

$$\langle 0_J|\phi(\overline{x}_1,\eta)\phi(\overline{x}_2,\eta)|0_J\rangle = \int \frac{d^3k}{(2\pi)^3} e^{i\overline{k}\cdot(\overline{x}_1-\overline{x}_2)}|e_k(\eta)|^2, \tag{5.84}$$

evaluated with respect to a given vacuum $|0_J\rangle$. Now we let $\overline{x}_1 \to \overline{x}_2$. As is evident from the above expression, a divergence will appear, but it is possible to demand that this be a "controlled divergence" (UV regularity). There are several conditions that can be imposed on the vacuum state and that guarantee such a behavior, like the so-called Hadamard condition, adiabatic regularity. [1, 4, 10].

The imposition of UV regularity, then, serves a double purpose. First of all, for states with this property it is possible to properly define fundamental physical objects like the energy–momentum tensor $\hat{T}_{ab}$. On the other hand, it helps in selecting an equivalence class, based on physical criteria.

5.4.2 *Inflation*

In this last section, we will briefly discuss some of the main assumptions underlying inflationary models, as well as the way in which theory and observation are connected. We will emphasize those points that are most closely related to the subject of the present notes.

Regarding inflation, the basic assumptions we make are the following:

1. Some time, soon after the big bang, the universe is well described by a FLRW model satisfying Einstein's equations, with a scalar field Φ in a suitable (very shallow) potential $V(\Phi)$ (e.g., $V(\Phi) = \frac{1}{2}\mu\Phi^2$). Although it is possible to devise models with several fields, here we will only consider the case of a single (scalar) field.
2. The potential is to be chosen in such a way that, starting at a finite energy configuration, the field induces a period of exponential expansion, while "slowly rolling down the potential." The slow-roll condition implies that the kinetic energy is small compared with the potential energy. The reason why this works is the following. Consider the Friedmann equation (in the spatially flat case):

$$\left(\frac{\dot{a}}{a}\right)^2 = \frac{8\pi G}{3}\rho. \tag{5.85}$$

   It is evident that, for ρ "almost" constant, we will have a solution of the form $a(t) = a(0)\, e^{Ht}$, i.e., a period of exponential expansion which, if assumed, allows us to solve at once three problems of the original big bang model, namely the *flatness problem*, the *horizon problem*, and the *monopole problem* [11]. From the explicit form of the energy-momentum tensor for the scalar field, one finds that the slow-roll condition provides the required mechanism. It is therefore customary to define the Hubble "parameter" H as the nearly constant (during slow-roll) function

$$H := \frac{\dot{a}}{a}. \tag{5.86}$$

   For H exactly constant, we have

$$ds^2 = -dt^2 + e^{2Ht}d\overline{x}^2 = \frac{1}{(H\eta)^2}(-d\eta^2 + d\overline{x}^2). \tag{5.87}$$

3. We first consider a homogeneous and isotropic universe model. Perturbations leading to inhomogeneities are then introduced. A basic assumption of the theory is that perturbations have their origin in quantum fluctuations due to the presence of quantum fields. Mathematically (in linearized gravity), there will be three modes: two tensor modes ($\tau^{(1)}$ and $\tau^{(2)}$, related to gravitational waves) and a scalar mode ($\mathscr{R}$, related to density fluctuations).

4. For a de Sitter geometry, we have a "preferred" vacuum for the quantum field $\hat{\phi}(x)$, which is the *Bunch–Davies vacuum* $|0_{BD}\rangle$. In this case, by requiring the vacuum state to be invariant under the (de Sitter) isometry group, we fix the choice of complex structure.

Based on these 4 assumptions, we want to make predictions on the behavior of observables. The basic observable, called the *power spectrum* and denoted $\mathscr{P}_\phi(k)$, is defined (up to a delta distribution) as the Fourier transform of the 2-point function:

$$\langle 0_{BD}|\hat{\phi}(k_1,\eta)\hat{\phi}(k_2,\eta)|0_{BD}\rangle = (2\pi)^3\delta(\overline{k}_1+\overline{k}_2)2\pi^2 P_\phi(k). \tag{5.88}$$

But we know, from direct calculation, that

$$\langle 0_{BD}|\hat{\phi}(k_1,\eta)\hat{\phi}(k_2,\eta)|0_{BD}\rangle = (2\pi)^3\delta(\overline{k}_1+\overline{k}_2)|e_k(\eta)|^2, \tag{5.89}$$

from which

$$\mathscr{P}_\phi(k) = \frac{\hbar k^3}{2\pi^2}|e_k(\eta)|^2 \tag{5.90}$$

follows. The next question is, then, what are the $e_k(\eta)$ for the Bunch–Davies vacuum? The wave Eq. (5.75) in the free case takes the form

$$e_k'' + 2\frac{a'}{a}e_k' + k^2 e_k = 0. \tag{5.91}$$

Solving this equation with $a(\eta) = -1/(H\eta)$, we obtain:

$$S_k^{(1)}(\eta) = \frac{e^{-ik\eta}}{\sqrt{2k}}\left(\frac{iH}{k} - H\eta\right) \equiv e_k(\eta), \tag{5.92}$$

$$S_k^{(2)}(\eta) = \overline{S}_k^{(1)}(\eta). \tag{5.93}$$

As the wavelength of the mode grows larger, it starts to "feel" the curvature. At the end of inflation, the power spectrum becomes $(k|\eta| \ll k)$

$$\mathscr{P}_\phi(k) = \frac{\hbar k^3}{2\pi^2}|e_k(\eta)|^2 \approx \frac{\hbar H^2}{4\pi^2}. \tag{5.94}$$

It is independent of k (scale invariant). But, what is the relation to the scalar and tensor perturbations? We are assuming that the (quantum) fluctuations on the inflaton field are responsible for the origin of structure in the universe. Therefore, all other fluctuations should be related to those of the inflaton. As long as the inhomogeneities produced by the quantum fluctuations remain "small," we may assume that these relations are of linear type. These relations are given in terms of *transfer functions*.

For instance, for the scalar perturbation, we have

$$\hat{\mathscr{R}}(x) = \sqrt{\frac{4\pi G}{(\dot{H}/H^2)}}\hat{\phi}(x), \tag{5.95}$$

where, going beyond the de Sitter approximation, we have $H(t) = H_0 + \dot{H}t$. If we define the slow-roll parameters

$$\varepsilon = \frac{\dot{H}}{H^2}, \quad \delta = -\frac{\ddot{H}}{2H\dot{H}}, \tag{5.96}$$

then we obtain (with $|\eta_k|\, k = 1$),

$$\mathscr{P}_{\mathscr{R}}(k) = \frac{\hbar 4\pi G}{\varepsilon}\left(\frac{H(\eta_k)}{2\pi}\right)^2, \tag{5.97}$$

where η_k is the time at which the mode k exits the Hubble horizon during inflation.

Equation (5.97) is the expression for the power spectrum ($\mathscr{P}_{\mathscr{R}}$) of the scalar perturbations at the end of inflation. Note that, $\mathscr{P}_{\mathscr{R}}$ for a Fourier mode k depends on the background evolution via the slow-roll parameter (ε) and the value of the Hubble rate ($H(\eta_k)$) at time η_k, when the mode k exits the Hubble horizon during inflation. That is, when the physical wavelength of the mode becomes equal to the Hubble horizon during inflation. While ε can be taken to be constant up to first order, the Hubble rate during inflation decreases slowly. Since the modes with higher k exit the horizon at a later time, the amplitude of the power spectrum also decreases with increasing k. As a result, the power spectrum has a nonzero slope which is characterized by the spectral index n_s defined in terms of the logarithmic derivative of the natural logarithm of the power spectrum: $n_s - 1 = \frac{d\log(\mathscr{P}_{\mathscr{R}})}{d\log(k)}$. Using this definition and the expression of the slow-roll parameters, one can obtain: $n_s - 1 = -4\varepsilon + 4\delta$. In the slow-roll approximation, i.e., assuming ε and δ are approximately constant during inflation, inflation predicts a spectral index $n_s < 1$. This deviation of the power spectrum from the exact scale invariance and the form of the power spectrum are robust predictions of the inflationary paradigm which have been tested with observational data from missions such as WMAP and *Planck*. Similarly, inflation also predicts tensor power spectrum with small deviation from the exact scale invariance, which can again be obtained using the framework of linear quantum perturbations on cosmological space-time described above. So far, the CMB measurements have only been able to put an upper bound on the amplitude of the tensor perturbations. Several future observational missions have been planned to measure tensor modes in the CMB. Such a detection of tensor mode in the CMB would be a huge leap in favor of the inflationary paradigm and can be used to select and/or reject various inflationary models. This is a powerful example of how the abstract mathematical framework of quantum fields in curved space-time can lead to concrete physical insight to the early universe.

Acknowledgements These notes represent the written version of the lectures held by Abhay Ashtekar at Villa de Leyva in 2015. I thank Abhay Ashtekar for having accepted the invitation, and for his lectures, that were very much appreciated by the participants of the school. I am also grateful to Mateo Restrepo and Juan Carlos Sanabria for their comments on earlier versions of this document. Special thanks to Brajesh Gupt, whose comments and suggestions helped me improve the presentation in a significant manner.

References

1. Ivan Agullo, Abhay Ashtekar, Unitarity and ultraviolet regularity in cosmology. Phys. Rev. D **91**, 124010 (2015)
2. A. Ashtekar, Anne Magnon, Quantum fields in curved space-times. Proc. R. Soc. Lond. Math. Phys. Eng. Sci. **346**(1646), 375–394 (1975)
3. Abhay Ashtekar, Anne Magnon-Ashtekar, A curiosity concerning the role of coherent states in quantum field theory. Pramana **15**(1), 107–115 (1980)
4. R. Paul, Anderson and Leonard Parker, Adiabatic regularization in closed Robertson-Walker universes. Phys. Rev. D **36**, 2963–2969 (1987)
5. Nicholas David Birrell, Paul Charles William Davies, *Quantum Fields in Curved Space*, no. 7 (Cambridge University Press, London, 1984)
6. Stephen A Fulling, *Aspects of Quantum Field Theory in Curved Spacetime*, vol. 17 (Cambridge University Press, London, 1989)
7. Brian Hall, *Perspectives on quantization*, Contemp. Math., vol. 214, ch. Quantum mechanics in phase space, 47–62, Amer. Math. Soc., Providence, RI, (1998)
8. Gary Hinshaw, D. Larson, Eiichiro Komatsu, D.N. Spergel, C.L. Bennett, J. Dunkley, M.R. Nolta, M. Halpern, R.S. Hill, N. Odegard et al., Nine-year Wilkinson microwave anisotropy probe (wmap) observations: Cosmological parameter results. Astrophys. J. Suppl. Ser. **208**(2,19), (2013)
9. David Shale, Linear symmetries of free boson fields. Trans. Am. Math. Soc. **103**(1), 149–167 (1962)
10. Robert M Wald, *Quantum Field Theory in Curved Spacetime and Black Hole Thermodynamics* (University of Chicago Press, Chicago, 1994)
11. Steven Weinberg, *Cosmology* (Oxford University Press, 2008)

Chapter 6
An Introduction to Pure Spinor Superstring Theory

Nathan Berkovits and Humberto Gomez

Abstract In these lecture notes, we give an introduction to superstring theory. We begin by studying the particle and superparticle in order to get a better understanding on the superstring side. Afterward, we review the pure spinor formalism and end by computing the scattering amplitude for three gravitons at tree level.

6.1 Introduction

For more than a decade, a manifestly super-Poincaré covariant formulation for the superstring, known as the pure spinor formalism [1], has shown to be a powerful framework in two branches. The first one is the computation of scattering amplitudes, and the second one is the quantization of the superstring in curved backgrounds which can include Ramond–Ramond flux. The strength of the pure spinor formalism resides precisely in the fact that it can be quantized in a manifestly super-Poincaré manner, so this covariance is not lost neither in the scattering amplitudes computation nor in the quantization of the superstring in curved backgrounds.

One key ingredient in this formalism is a bosonic ghost λ^α, constrained to satisfy Cartan's pure spinor condition in 10 space-time dimensions [2]. The prescription for computing multiloop amplitudes was given in [3], where as in the RNS formalism, it was necessary to introduce picture changing operators (PCOs) in order to absorb the zero modes of the pure spinor variables. Up to two loops, various amplitudes were computed in [4]. Later on, by introducing a set of non-minimal variables $\bar{\lambda}_\alpha$ and r_α,

N. Berkovits (✉)
ICTP South American Institute for Fundamental Research, Instituto de Física Teórica
UNESP-Universidade Estadual Paulista, Rua Dr. Bento T. Ferraz 271, Sao Paulo, SP 01140-070, Brazil
e-mail: nberkovi@ift.unesp.br

H. Gomez
Facultad de Ciencias Básicas, Universidad Santiago de Cali,
Calle 5 N°, 62-00 Barrio Pampalinda, Cali, Valle, Colombia
e-mail: humgomzu@gmail.com

A. Cardona et al. (eds.), *Quantization, Geometry and Noncommutative Structures in Mathematics and Physics*, Mathematical Physics Studies,
DOI 10.1007/978-3-319-65427-0_6

an equivalent prescription for computing scattering amplitudes was formulated in [5, 6]. This last superstring description is known as the "non-minimal" pure spinor formalism, in order to distinguish it from the former "minimal" pure spinor formalism. With the non-minimal formalism also were computed scattering amplitudes up to three loops [7, 8]. Because of its topological nature, in the non-minimal version it is not necessary to introduce PCOs. Nevertheless, it is necessary to use a regulator. The drawback of having to introduce this regulator appears beyond three loops, since it gets more complicated due to the divergences coming from the poles contribution of the b-ghost [9].

In this short note, we give an introduction to superstring theory in the pure spinor formalism. We are going to start with very general comments about the superparticle in 10 dimensions.

6.2 Particle and Superparticle

We begin this note with a brief introduction to the relativistic point particle and superparticle, please review the Refs. [10, 12, 13, 17, 18].

A relativistic particle is described by a point in a flat space-time[1] $(1, D-1)$, whose evolution over time is described by a curve (worldline) (Fig.6.1).

The simplest Poincaré and τ-reparameterization invariant action is proportional to the worldline length

$$S = -M \int ds, \tag{6.1}$$

where M is the mass of the particle. The "−" sign is introduced in order to guarantee that the S functional is going to have a local minimum, i.e., a stable classical trajectory. Let us recall that the space-time induces a metric on the worldline; thus, the ds line element is just given by the square root of the induced metric

$$ds = \sqrt{-\eta_{\mu\nu}\dot{X}^\mu \dot{X}^\nu}\, d\tau. \tag{6.2}$$

Since the worldline is a causal trajectory (see Fig. 6.2), i.e., the velocity vector (tangent vector) is a timelike vector

$$\eta_{\mu\nu}\dot{X}^\mu \dot{X}^\nu < 0, \tag{6.3}$$

one must introduce the "−" sign into the square root so as to obtain a positive number.

Nevertheless, although the action in (6.1) seems simple, it is too hard to quantize because we do not know how to perform a path integral with a square root. In addition, this action only describes massive particles, so to compute scattering of

[1]The notation $(1, D-1)$ means the metric of the space-time is given by $\eta_{\mu\nu} = \mathrm{diag}(-, +, +, \cdots, +)$.

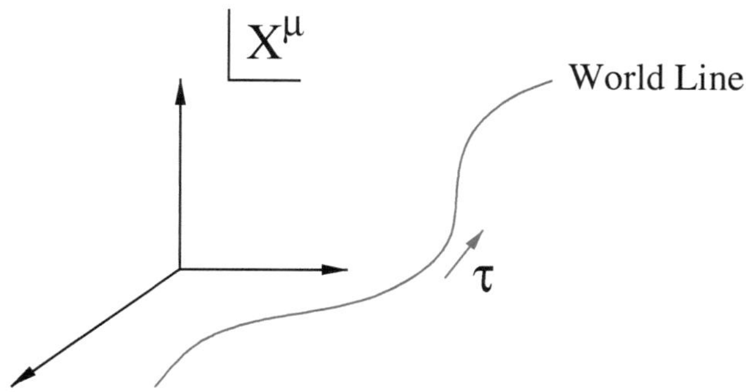

Fig. 6.1 Point particle evolving over time. The worldline is parameterized by τ, and the $X^\mu = (X^0, X^i) = (t, \vec{r})$ are the space-time coordinates

Fig. 6.2 Causal trajectory

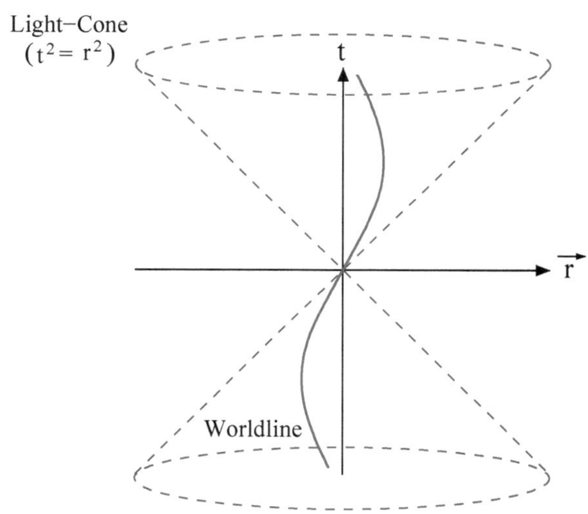

photons, gluons, or gravitons we need to modify it. In order to solve these problems, the following first-order action can be proposed

$$S = -\int d\tau \left[P^\mu \dot{X}_\mu + \frac{e}{2}(P^\mu P_\mu + M^2) \right]. \tag{6.4}$$

This action is classically equivalent to (6.1), i.e., using the P_μ equations of motion. Furthermore, it supports massless particles and its quantization is easier than (6.1).

Note that (6.4) is invariant, up to total derivative, by the local (gauge) transformation

$$\delta P^\mu = \xi \, \dot{P}^\mu, \qquad \delta X^\mu = \xi \, \dot{X}^\mu, \qquad \delta e = \xi \, \dot{e} + \dot{\xi} \, e, \tag{6.5}$$

where $\xi = \xi(\tau)$ is a local parameter. Using this gauge symmetry, one can fix the Lagrange multiplier, "e," and perform a BRST quantization. Nevertheless, from (6.5) it is clear that the e field is a 1-form on the worldline, i.e., $e \in H^1_{\rm dR}(C)$, where C is the worldline and $H^1_{\rm dR}(C)$ is the first de-Rham cohomology group over C [15]. Therefore, the choice of the gauge fixing depends on the worldline topology. Since here we are not focused on this issue, for more details see [15].

6.2.1 Brink–Schwarz Superparticle

As the main topic of this note is to give an introduction to superstring theory, we will center in a space-time of 10 dimensions. So, we begin with a superparticle in 10 dimensions. The main references for this section are [12, 17, 18].

Brink–Schwarz Superparticle

The Brink–Schwarz(BS) action for the 10-dimensional (massless) superparticle is given by

$$S = \int d\tau \left(\Pi^\mu P_\mu + e P^\mu P_\mu\right), \tag{6.6}$$

with

$$\Pi^\mu := \dot{X}^\mu - \frac{1}{2}\dot{\theta}^\alpha \gamma^\mu_{\alpha\beta}\theta^\beta, \quad \mu = 0, \dots, 9, \quad \alpha, \beta = 1, \dots 16,$$

where P^μ is the canonical momentum of X^μ, e is the Lagrange multiplier to impose the massless condition, $P^2 = 0$, and θ^α is a Grassmann or fermionic coordinate,[2] i.e., $\theta^\alpha\theta^\beta = -\theta^\beta\theta^\alpha$. The gamma matrices, $\gamma^\mu_{\alpha\beta}$ and $\gamma_\mu^{\alpha\beta}$, are 16×16 symmetric matrices which satisfy the Clifford algebra, $(\gamma^\mu)_{\alpha\beta}(\gamma^\nu)^{\beta\rho} + (\gamma^\nu)_{\alpha\beta}(\gamma^\mu)^{\beta\rho} = 2\eta^{\mu\nu}\delta^\rho_\alpha$. In the Weyl representation, $(\gamma^\mu)_{\alpha\beta}$ and $(\gamma^\mu)^{\alpha\beta}$ are the off-diagonal blocks of the 32×32 Dirac γ^μ matrices, i.e.,

$$\gamma^\mu = \begin{pmatrix} 0 & (\gamma^\mu)^{\alpha\beta} \\ (\gamma^\mu)_{\alpha\beta} & 0 \end{pmatrix}, \quad \text{where} \quad \{\gamma^\mu, \gamma^\nu\} = 2\eta^{\mu\nu}. \tag{6.7}$$

Besides being invariant by reparameterization

$$\delta P^\mu = \xi \dot{P}^\mu, \qquad \delta X^\mu = \xi \dot{X}^\mu, \qquad \delta\theta^\alpha = \xi \dot{\theta}^\alpha, \qquad \delta e = \xi \dot{e} + \dot{\xi} e, \tag{6.8}$$

the BS action is invariant under the global transformation

$$\delta\theta^\alpha = \varepsilon^\alpha, \qquad \delta X^\mu = \frac{1}{2}\theta^\alpha\gamma^\mu_{\alpha\beta}\varepsilon^\beta, \qquad \delta P^\mu = \delta e = 0, \tag{6.9}$$

[2] α is a (Weyl) spinorial index. When the space-time has D-dimensions, where D is an even integer number, then a Weyl spinor has $2^{\frac{D}{2}-1}$ components.

where ε^{α} is a constant Grassmann parameter. Using Noether's theorem, this global symmetry is generated by the charge

$$Q_{\alpha} := p_{\alpha} - \frac{1}{2}\gamma^{\mu}_{\alpha\beta}\theta^{\beta}P_{\mu}, \tag{6.10}$$

where

$$p_{\alpha} := \frac{\partial L}{\partial\dot{\theta}^{\alpha}} = -\frac{1}{2}\gamma^{\mu}_{\alpha\beta}\theta^{\beta}P_{\mu}, \tag{6.11}$$

is the canonical momentum of θ^{α}, namely[3] $\{p_{\beta}, \theta^{\alpha}\}_{\text{PB}} = -i\delta^{\alpha}_{\beta}$. It is simple to check

$$\{Q_{\alpha}, Q_{\beta}\} = i\gamma^{\mu}_{\alpha\beta}P_{\mu}. \tag{6.12}$$

The charge Q_{α} is known as the *supercharge*, and the transformations in (6.9) are the *supersymmetry transformations*.

The BS action is also invariant under the local transformation

$$\delta\theta^{\alpha} = P^{\mu}\gamma^{\alpha\beta}_{\mu}\kappa_{\beta}, \qquad \delta X^{\mu} = -\frac{1}{2}\theta^{\alpha}\gamma^{\mu}_{\alpha\beta}\delta\theta^{\beta}, \qquad \delta P^{\mu} = 0, \qquad \delta e = \dot{\theta}^{\alpha}\kappa_{\alpha}, \tag{6.13}$$

where $\kappa_{\alpha} = \kappa_{\alpha}(\tau)$ is a local Grassmann parameter. This local symmetry is known as the *Kappa symmetry*. This symmetry is going to be used to perform the light-cone gauge.

From the canonical momentum p_{α} obtained in (6.11), we obtain a constraint system given by the conditions

$$d_{\alpha} := p_{\alpha} + \frac{1}{2}\gamma^{\mu}_{\alpha\beta}\theta^{\beta}P_{\mu} = 0. \tag{6.14}$$

The algebra of these constraints is given by

$$\{d_{\alpha}, d_{\beta}\}_{\text{PB}} = -i\gamma^{\mu}_{\alpha\beta}P_{\mu}. \tag{6.15}$$

Because $P^2 = 0$, then one has 8 first-class constraints and eight second-class constraints. To see this, we choose a frame where $P^{\mu} = (E, 0, \ldots, E)$, and later, we define the light-cone coordinates and $\gamma-$matrices as

$$X^{\pm} = \frac{1}{\sqrt{2}}(X^0 \pm X^9), \qquad P^{\pm} = \frac{1}{\sqrt{2}}(P^0 \pm P^9), \qquad \gamma^{\pm} = \frac{1}{\sqrt{2}}(\gamma^0 \pm \gamma^9). \tag{6.16}$$

[3]PB means Poisson bracket. Let us remember that when the two variables are Grassmann numbers, then this commutator becomes an anticommutator, "$-\ \rightarrow\ +$."

It is clear that $P^- = 0$; therefore, the algebra in (6.15) becomes

$$\{d_\alpha, d_\beta\}_{\text{PB}} = -i\gamma^-_{\alpha\beta}P^+ \propto \begin{pmatrix} \mathbb{1}_{8\times 8} & \mathbb{O}_{8\times 8} \\ \mathbb{O}_{8\times 8} & \mathbb{O}_{8\times 8} \end{pmatrix}. \tag{6.17}$$

Gauge Fixing

Let us recall that to quantize a theory with second-class constraints the Poisson bracket must be replaced by the Dirac bracket, which is defined as

$$\{A, B\}_{\text{DB}} := \{A, B\}_{\text{PB}} - \{A, \phi_i\}_{\text{PB}}\, C^{-1}_{ij}\, \{\phi_j, B\}_{\text{PB}}, \tag{6.18}$$

where ϕ_i's are the second-class constraints and C^{-1}_{ij} is the inverse matrix of the second-class constraints algebra, $C_{ij} := \{\phi_i, \phi_j\}_{\text{PB}}$.

For the BS superparticle, it is not possible to separate, in a Lorentz covariant way, the first- and second-class constraints, in order to obtain the C_{ij} matrix. However, as shown in (6.17), there is a frame where the first- and second-class constraints are disjoint, which is known as the light-cone gauge.

To be more precise, the light-cone gauge consists in choosing a θ^α field such that $(\gamma^+\theta)_\alpha = 0$, which is possible by the Kappa symmetry. Since $P^- = 0$ and $P^i = 0$, $i = 1, ..., 8$, on the frame $P^\mu = (E, 0, ..., E)$, one can fix $\kappa_\beta = \frac{1}{2P^+}(\gamma^+\theta)_\beta$. Using the κ transformation given in (6.13), it is straightforward to check

$$\theta'^\alpha = \theta^\alpha + \delta\theta^\alpha = -\frac{1}{2}(\gamma^+\gamma^-\theta)^\alpha, \tag{6.19}$$

where we have used $\{\gamma^+, \gamma^-\} = -1$. So, it is clear that $(\gamma^+\theta')_\alpha = 0$. In this gauge, the BS action becomes

$$\begin{aligned} S &= \int d\tau\, \left(\Pi^\mu P_\mu + e\, P^\mu P_\mu\right) \\ &= \int d\tau \left[\dot{X}^\mu P_\mu - \frac{1}{2}(-\dot\theta\gamma^+\theta P^- - \dot\theta\gamma^-\theta P^+ + \dot\theta\gamma^i\theta P^i) + e\, P^\mu P_\mu\right] \\ &= \int d\tau \left(\dot{X}^\mu P_\mu - \frac{1}{2}\dot{S}_a S_a + e\, P^\mu P_\mu\right), \quad a = 1, ..., 8, \end{aligned} \tag{6.20}$$

where we have utilized $\dot\theta\gamma^+\theta = \dot\theta\gamma^i\theta = 0$ and defined $S^a = 2^{1/4}\sqrt{P^+}\theta^a$. It is useful to remember that a Weyl spinor in a 10-dimensional space-time can be decomposed in two Weyl spinors in a eight-dimensional space-time, namely

$$\theta^\alpha = \begin{pmatrix} \theta^a \\ \theta^{\dot a} \end{pmatrix}, \qquad a, \dot a = 1, 2, ..., 8. \tag{6.21}$$

In addition, there is a representation where the $\gamma^{-}_{\alpha\beta}$ matrix looks

$$\gamma^{-}_{\alpha\beta} = -\sqrt{2}\begin{pmatrix} \mathbb{1}_{8\times 8} & \mathbb{0}_{8\times 8} \\ \mathbb{0}_{8\times 8} & \mathbb{0}_{8\times 8} \end{pmatrix}, \tag{6.22}$$

hence[4] $\frac{1}{2}\dot{\theta}\gamma^{-}\theta P^{+} = -\frac{1}{2}\dot{S}_a S_a$.

The BS action in the light-cone gauge is more friendly than the original one, but we have lost the Lorentz covariance since an action with eight-dimensional space-time spinor fields.

Quantization

From the BS action in (6.20), the canonical momentum of S_a, i.e., $\{p_a, S_b\}_{\mathrm{PB}} = -\delta_{ab}$, is given by

$$p_a := \frac{\partial L}{\partial \dot{S}^a} = -\frac{1}{2}S_a; \tag{6.23}$$

therefore, there are eight constraints, $d_a = p_a + \frac{1}{2}S_a = 0$. The algebra of these constraints is straightforward to compute

$$\{d_a, d_b\}_{\mathrm{PB}} = -\delta_{ab}, \tag{6.24}$$

which implies that these constraints are of second class. Thus, using Dirac's method (see (6.18)), we get the anticommutator

$$\begin{aligned} \{S_a, S_b\}_{\mathrm{DB}} &= \{S_a, S_b\}_{\mathrm{PB}} - \{S_a, d_c\}_{\mathrm{PB}}\, \{d^c, d^e\}^{-1}_{\mathrm{PB}}\, \{d_e, S_b\}_{\mathrm{PB}} \\ &= 0 - (-\delta_{ac})(-\delta_{ce})(-\delta_{eb}) \\ &= \delta_{ab}, \end{aligned} \tag{6.25}$$

which is the Clifford algebra. A representation of this algebra gives us the quantum states of the theory.

In order to build a representation of (6.25), it is convenient to keep in mind the $SO(8)$ Pauli matrices,[5] which satisfy

$$\mathbb{S}^i_{a\dot{a}}\mathbb{S}^j_{\dot{a}b} + \mathbb{S}^j_{a\dot{a}}\mathbb{S}^i_{\dot{a}b} = 2\delta_{ab}\delta^{ij}, \qquad i, j, a, \dot{a}, b = 1, ..., 8, \tag{6.26}$$

where i, j are vector indices (space-time) and a, b, $\dot{a}$ are spinor indices.[6] Following the Pauli matrices properties, we can represent the algebra in (6.25) using the definitions

[4]The 8-dimensional space-time spinor metric is just the identity, $S_a = \delta_{ab}S^b$.

[5]$SO(8)$ means special orthogonal group in 8-dimensional space-time.

[6]The a label is known as a chiral index and $\dot{a}$ as an antichiral.

$$S_a|\dot{a}\rangle = \frac{1}{\sqrt{2}}\mathbb{S}^j_{a\dot{a}}|j\rangle, \tag{6.27}$$

$$S_a|i\rangle = \frac{1}{\sqrt{2}}\mathbb{S}^i_{a\dot{b}}|\dot{b}\rangle. \tag{6.28}$$

Clearly, $\{S_a, S_b\}|\dot{a}\rangle = \delta_{ab}|\dot{a}\rangle$ and $\{S_a, S_b\}|i\rangle = \delta_{ab}|i\rangle$; therefore, the physical spectrum is a $SO(8)$ vector, given by $|i\rangle$, and a $SO(8)$ antichiral spinor, given by $|\dot{a}\rangle$, which are massless by the equation of motion, $P^2 = 0$. This is the same spectrum of $D = 10$ super-Yang-Mills (SYM), eight degree of freedom (d.o.f) for the gluon and eight d.o.f for the gluino.

6.2.2 Pure Spinor Superparticle

This section is based on the Refs. [12, 13].

As shown above, the BS action is read as

$$S = \int d\tau \left(\dot{X}^\mu P_\mu - \frac{1}{2}\dot{S}_a S_a + e P^\mu P_\mu\right), \tag{6.29}$$

in the light-cone gauge. Nevertheless, we can think that this action comes from a bigger theory, different from the one given in (6.6), such that after fixing the symmetries, one obtains (6.29).

Let us consider the following action

$$S = \int d\tau \left(\dot{X}^\mu P_\mu - \frac{1}{2}\dot{S}_a S_a + e P^\mu P_\mu + \dot{\theta}^\alpha p_\alpha + f^\alpha \hat{d}_\alpha\right), \tag{6.30}$$

where (θ^a, p_a) are independent fermionic fields,[7] f^α is a fermionic Lagrange multiplier, and $\hat{d}_\alpha$ are the fermionic constraints[8]

$$\hat{d}_\alpha := d_\alpha + \frac{1}{\sqrt{P^+}} P_\mu(\gamma^\mu\gamma^+ S)_\alpha, \qquad \text{with } \; d_\alpha := p_\alpha + \frac{1}{2}P_\mu(\gamma^\mu\theta)_\alpha. \tag{6.32}$$

From the algebra $\{S_a, S_b\} = i\delta_{ab}$ and $\{d_\alpha, d_\beta\} = -iP_\mu\gamma^\mu_{\alpha\beta}$, it is not hard to check

$$\{\hat{d}_\alpha, \hat{d}_\beta\} = -\frac{i}{2P^+}P^2(\gamma^+)_{\alpha\beta}, \tag{6.33}$$

[7] The θ^α field is not related to S_a as in (6.20).

[8] It is useful to recall the notation

$$(\gamma^+)^{\beta\rho}S_\rho = \sqrt{2}\begin{pmatrix} \mathbb{1}_{8\times 8} & \mathbb{0}_{8\times 8} \\ \mathbb{0}_{8\times 8} & \mathbb{0}_{8\times 8}\end{pmatrix}\begin{pmatrix} S_a \\ 0 \end{pmatrix} = (\gamma^+)^{\beta a}S_a \tag{6.31}$$

.

where the identities, $(\gamma^{+})^{\delta a}(\gamma^{+})^{\mathbb{S}a} = \sqrt{2}\,(\gamma^{+})^{\delta\mathbb{S}}$ and $\{\gamma^{\mu}, \gamma^{\nu}\}_{\alpha}^{\beta} = 2\eta^{\mu\nu}\delta_{\alpha}^{\beta}$, have been used. Clearly, the $\hat{d}_{\alpha}$s are first-class constraints, which generate a gauge symmetry. Using this gauge symmetry, one can fix $\theta^{\alpha} = 0$, and so (6.30) becomes the BS action. But the idea is to use the BRST method to quantize this new action (for details of the BRST quantization in superstring theory, one can review the Ref. [10]).

From the BRST method, we know that for each gauge symmetry there are ghost and antighost fields with inverse statistics. For example, using the reparameterization gauge symmetry we can fix $e = 1/2$, so

$$\begin{array}{lll} \text{Gauge fixing} & \text{Fermionic} - (\text{ghost, antighost}) & \text{First} - \text{class constraint} \\ e = \frac{1}{2} & (c, b) & P^{2} = 0. \end{array} \tag{6.34}$$

So, using the gauge symmetry generated by the first-class constraints, $\hat{d}_{\alpha} \approx 0$, we can fix

$$\begin{array}{lll} \text{Gauge Fixing} & \text{Bosonic} - (\text{ghost, antighost}) & \text{First} - \text{class constraint} \\ f^{\alpha} = 0 & (\hat{\lambda}^{\alpha}, \hat{\omega}_{\alpha}) & \hat{d}_{\alpha} = 0, \end{array} \tag{6.35}$$

and the action in (6.30) becomes

$$S = \int d\tau \left(\dot{X}^{\mu} P_{\mu} - \frac{1}{2}\dot{S}_{a}S_{a} - \frac{1}{2} P^{\mu} P_{\mu} + \dot{\theta}^{\alpha} p_{\alpha} + \dot{c}\, b + \dot{\hat{\lambda}}^{\alpha} \hat{\omega}_{\alpha} \right). \tag{6.36}$$

After fixing the local symmetries and introducing the ghost fields, the gauge symmetries turn into global symmetries; thus, using Noether's procedure one can obtain the conserved charge. That charge is known as the *BRST charge*, which is denoted by Q, and in general, it can be written as the ghost field times its corresponding constraint (it is a fermionic charge). In addition, that charge must be nilpotent, i.e., $\{Q, Q\} = Q^{2} = 0$. Therefore, following those ideas one may suspect that the charge should have the form

$$\hat{Q} = \hat{\lambda}^{\alpha}\hat{d}_{\alpha} + cP^{2}, \tag{6.37}$$

but this charge is not nilpotent, $\hat{Q}^{2} = -\frac{i}{2P^{+}}P^{2}(\hat{\lambda}\gamma^{+}\hat{\lambda})$. In order to realize a nilpotent BRST charge, we must add the term

$$\hat{Q} = \hat{\lambda}^{\alpha}\hat{d}_{\alpha} + cP^{2} + \frac{i}{4P^{+}}b(\hat{\lambda}\gamma^{+}\hat{\lambda}), \tag{6.38}$$

which, in fact, arises naturally from Noether's method.

Pure Spinor Condition

Since the BRST charge is nilpotent, $Q^{2} = 0$, one can wonder about its cohomology [15], i.e., the coset space defined as

$$H(Q) := \text{Ker}Q/\text{Im}Q \tag{6.39}$$

where

$$\text{Ker}Q := \{\Psi \in \mathscr{C}^\infty : Q\Psi = 0\}, \qquad \text{Im}Q := \{\Psi \in \mathscr{C}^\infty : \Psi = Q\Omega\}. \tag{6.40}$$

Clearly, $\text{Im}Q \subset \text{Ker}Q$.

In the BRST language, the physical states are defined as the states which are in the BRST cohomology, i.e.,

$$H(Q) = \text{Physical states.} \tag{6.41}$$

So, to compute the physical states of the action in (6.36), we must find the $\hat{Q}$ cohomology of the operator in (6.38). But, furthermore to being a complicated operator, it is not Lorentz covariant. In [12], it was shown that the $\hat{Q}$−cohomology is actually equivalent to the cohomology of the simple operator

$$Q = \lambda^\alpha d_\alpha, \tag{6.42}$$

which is independent of $\{S_a, c\}$. Thus, the action in (6.36) can be modified to the new and simpler action

$$S^{\text{PS}} = \int d\tau \left(\dot{X}^\mu P_\mu - \frac{1}{2} P^\mu P_\mu + \dot{\theta}^\alpha p_\alpha + \dot{\lambda}^\alpha \omega_\alpha\right). \tag{6.43}$$

As d_α is not a really first-class constraint, $\{d_a, d_b\} = -iP_\mu \gamma^\mu_{\alpha\beta}$, the BRST charge in (6.42) is nilpotent if and only if the λ^α field satisfies the condition

$$\frac{1}{2}Q^2 = \{Q, Q\} = -iP_\mu(\lambda\gamma^\mu\lambda) \Rightarrow (\lambda\gamma^\mu\lambda) = 0, \quad \mu = 0, \ldots, 9. \tag{6.44}$$

This condition is known as the pure spinor condition for spinors in 10 dimensions. This condition implies that λ^α is a complex spinor. For example, let us consider $\mu = 0$, i.e.,

$$(\lambda\gamma^0\lambda) = -[(\lambda^1)^2 + (\lambda^1)^2 + \cdots + (\lambda^{16})^2] = 0, \tag{6.45}$$

thus, in order to obtain a non-trivial solutions λ^α must be a complex spinor.[9]

In addition, the pure spinor action given in (6.43) is invariant under the global transformation

$$\lambda^\alpha \rightarrow e^{iz}\lambda^\alpha, \quad \omega_\alpha \rightarrow e^{-iz}\omega_\alpha. \tag{6.47}$$

[9]We have used a representation of the Dirac matrices where

$$\gamma^0_{\alpha\beta} = -\begin{pmatrix} \mathbb{1}_{8\times 8} & \mathbb{O}_{8\times 8} \\ \mathbb{O}_{8\times 8} & \mathbb{1}_{8\times 8} \end{pmatrix}. \tag{6.46}$$

By Noether's procedure, the conserved charge is

$$J := \lambda^\alpha w_\alpha, \tag{6.48}$$

which is known as the *ghost number*. Clearly, λ^α and Q have ghost number 1, and ω_α has ghost number -1.

Quantization

In order to find the Q–cohomology, it is useful to write the d_α constraint as an operator. From the canonical momentum representation, $p_\alpha \to \frac{\partial}{\partial\theta^\alpha}$ and $P_\mu \to \frac{\partial}{\partial X^\mu}$, we map the d_α constraint to the operator

$$d_\alpha = p_\alpha + \frac{1}{2}P_\mu(\gamma^\mu\theta)_\alpha \quad \to \quad D_\alpha = \frac{\partial}{\partial\theta^\alpha} + \frac{1}{2}(\gamma^\mu\theta)_\alpha\frac{\partial}{\partial X^\mu}. \tag{6.49}$$

The D_α operator is known as the super-covariant derivative, and its algebra is just given by $\{D_\alpha, D_\beta\} = -i\gamma^\mu_{\alpha\beta}\frac{\partial}{\partial X^\mu}$.

Now, we write the most general super-Poincaré covariant wavefunction that can be constructed from $(X^\mu, \theta^\alpha, \lambda^\alpha)$

$$\begin{aligned}\Psi(X,\theta,\lambda) = C(X,\theta) + \lambda^\alpha A_\alpha(X,\theta) + (\lambda\gamma^{\mu_1,\dots,\mu_5}\lambda)A^*_{\mu_1,\dots,\mu_5}(X,\theta)\\ + \lambda^\alpha\lambda^\beta\lambda^\gamma C^*_{\alpha\beta\gamma}(X,\theta) + \cdots,\end{aligned} \tag{6.50}$$

where we have expanded around the bosonic variable, λ^α. The terms in $\cdots$ include superfields with more than three powers of λ^α (ghost number greater than 3), which are in the trivial cohomology.

For example, $Q\Psi = -i\lambda^\alpha D_\alpha C - i\lambda^\alpha\lambda^\beta D_\alpha A_\beta + \dots$, so $Q\Psi = 0$ implies that $A_\alpha(x,\theta)$ satisfies the equation of motion $\lambda^\alpha\lambda^\beta D_\alpha A_\beta = 0$. But since $\lambda^\alpha\lambda^\beta$ are pure spinors (see appendix), they are proportional to

$$(\lambda\gamma^{\mu_1\mu_2\mu_3\mu_4\mu_5}\lambda)\gamma^{\alpha\beta}_{\mu_1\mu_2\mu_3\mu_4\mu_5},$$

this implies that $D\gamma^{\mu_1\mu_2\mu_3\mu_4\mu_5}A = 0$, which is the linearized version of the super-Yang-Mills equation of motion. Furthermore, if one defines the gauge parameter by $\Omega = i\Lambda + \lambda^\alpha\omega_\alpha + \dots$, the gauge transformation $\delta\Psi = Q\Omega$ implies $\delta A_\alpha = D_\alpha\Lambda$ which is the linearized super-Yang-Mills gauge transformation.

So, $A_\alpha(X,\theta)$ contains the on-shell super-Yang-Mills gluon and gluino, $a_\mu(X)$ and $\chi^\alpha(X)$, which satisfy the linearized equations of motion and gauge invariances

$$\partial^\mu\partial_{[\mu}a_{\nu]} = \gamma^\mu_{\alpha\beta}\partial_\mu\chi^\beta = 0, \quad \delta a_\mu = \partial_\mu s.$$

Since gauge invariances of antifields correspond to equations of motion of fields, one expects to have antifields $a^{*\mu}(x)$ and $\chi^*_\alpha(x)$ in the cohomology of Q which satisfy the linearized equations of motion and gauge invariances

$$\partial_\mu a^{*\mu} = 0, \quad \delta a^{*\mu} = \partial_\nu(\partial^\nu s^\mu - \partial^\mu s^\nu), \quad \delta\chi^*_\alpha = \gamma^\mu_{\alpha\beta}\partial_\mu\kappa^\beta, \tag{6.51}$$

where s^μ and κ^β are gauge parameters. Indeed, these antifields $a^{*\mu}$ and χ^*_α appear in components of the ghost number +2 superfield $A^*_{\mu_1\mu_2\mu_3\mu_4\mu_5}$ of (6.50). Using $Q\Psi = 0$ and $\delta\Psi = Q\Omega$, $A^*_{\mu_1\mu_2\mu_3\mu_4\mu_5}$ satisfies the linearized equation of motion $\lambda^\alpha(\lambda\gamma^{\mu_1\mu_2\mu_3\mu_4\mu_5}\lambda)D_\alpha A^*_{\mu_1\mu_2\mu_3\mu_4\mu_5} = 0$ with the linearized gauge invariance $\delta A^*_{\mu_1\mu_2\mu_3\mu_4\mu_5} = \gamma^{\alpha\beta}_{\mu_1\mu_2\mu_3\mu_4\mu_5}D_\alpha\omega_\beta$. Expanding ω_α and $A^*_{\mu_1\mu_2\mu_3\mu_4\mu_5}$ in components, one learns that $A^*_{\mu_1\mu_2\mu_3\mu_4\mu_5}$ can be gauged to the form

$$\begin{aligned} &A^*_{\mu_1\mu_2\mu_3\mu_4\mu_5} = \\ &= (\theta\gamma_{[\mu_1\mu_2\mu_3}\theta)(\theta\gamma_{\mu_4\mu_5]})^\alpha\chi^*_\alpha(x) + (\theta\gamma_{[\mu_1\mu_2\mu_3}\theta)(\theta\gamma_{\mu_4\mu_5]s}\theta)a^{*s}(x) + \cdots \end{aligned} \tag{6.52}$$

where χ^*_α and a^{*s} satisfy the equations of motion and residual gauge invariances of (6.51), and "$\cdots$" involves terms higher order in θ^α which depend on derivatives of χ^*_α and a^{*s}.

In addition to these fields and antifields, one also expects to find the Yang-Mills ghost $c(X)$ and antighost $c^*(X)$ in the cohomology of Q. The ghost $c(x)$ is found in the $\theta = 0$ component of the ghost number 0 superfield,

$$C(X,\theta) = c(X) + \cdots,$$

and the antighost $c^*(x)$ is found in the $(\theta)^5$ component of the ghost number +3 superfield,

$$C^*_{\alpha\beta\gamma}(X,\theta) = \cdots + c^*(X)(\gamma^{\mu_1}\theta)_\alpha(\gamma^{\mu_2}\theta)_\beta(\gamma^{\mu_3}\theta)_\gamma(\theta\gamma_{\mu_1\mu_2\mu_3}\theta) + \cdots$$

It was proven in [13] that the above states are the only states in the cohomology of Q, and therefore, although Ψ of (6.50) contains superfields of arbitrarily high ghost number, only superfields with ghost number between 0 and 3 contain states in the cohomology of Q.

The linearized equations of motion and gauge invariances $Q\Psi = 0$ and $\delta\Psi = Q\Omega$ are easily generalized to the nonlinear equations of motion and gauge invariances

$$Q\Psi + g\Psi\Psi = 0, \quad \delta\Psi = Q\Omega + g[\Psi,\Omega], \tag{6.53}$$

where Ψ and Ω transform in the adjoint representation of the gauge group. For the superfield $A_\alpha(X,\theta)$, (6.53) implies the super-Yang-Mills equations of motion and gauge transformations. Furthermore, the equation of motion and gauge transformation of (6.53) can be obtained from the space-time action[10]

$$\mathscr{S} = \mathrm{Tr}\int d^{10}X\,\langle\frac{1}{2}\Psi Q\Psi + \frac{g}{3}\Psi\Psi\Psi\rangle, \tag{6.54}$$

[10]This space-time action was first proposed by Edward Witten [19].

using the normalization (measure) definition[11] that

$$\langle (\lambda\gamma^{\mu_1}\theta)(\lambda\gamma^{\mu_2}\theta)(\lambda\gamma^{\mu_3}\theta)(\theta\gamma_{\mu_1\mu_2\mu_3}\theta)\rangle = 1. \tag{6.55}$$

Although (6.55) may seem strange, it is the only one scalar in the $Q-$ cohomology with ghost number 3. This measure becomes important in the superstring scattering amplitudes context. After expressing (6.54) in terms of component fields and integrating out auxiliary fields, it is possible to show that (6.54) reduces to the standard Batalin–Vilkovisky action for super-Yang-Mills,

$$\mathscr{S} = \mathrm{Tr}\int d^{10}X(\frac{1}{4}f_{\mu\nu}f^{\mu\nu} + \chi^\alpha\gamma^\mu_{\alpha\beta}(\partial_\mu\chi^\beta + ig[a_\mu,\chi^\beta]) \tag{6.56}$$

$$+ ia^{*\mu}(\partial_\mu c + ig[a_\mu, c]) - g\chi^*_\alpha\{\chi^\alpha, c\} - gccc^*). \tag{6.57}$$

6.3 Pure Spinor Superstring

In this section, we give an introduction to superstring theory using the pure spinor formalism. Our main objective is to compute, explicitly, the scattering amplitude of gravitons for three points at tree level. This section is based from the Refs. [1, 11, 12, 16].

6.3.1 *General Issues*

From the superparticle pure spinor action found in (6.30), one may integrate out the P^μ field, so the pure spinor superparticle action becomes

$$S^{\mathrm{PS}} = \int d\tau\left(\frac{1}{2}\dot{X}^\mu\dot{X}_\mu + \dot{\theta}^\alpha p_\alpha + \dot{\lambda}^\alpha\omega_\alpha\right), \tag{6.58}$$

and the BRST charge stays the same.

The most natural and simplest generalization from superparticle to superstring is just to consider a surface instead of worldline curve, i.e.,

$$(\tau) \rightarrow (z,\bar{z}), \tag{6.59}$$
$$\{X(\tau),\theta(\tau),p(\tau),\lambda(\tau),\omega(\tau)\} \rightarrow \{X(z,\bar{z}),\theta(z,\bar{z}),p(z,\bar{z}),\lambda(z,\bar{z}),\omega(z,\bar{z})\},$$

[11] This definition will become clear in the following section.

and the pure spinor superstring action becomes

$$S^{\mathrm{PS}} = \frac{1}{2\pi\alpha'}\int d^2z\left(\frac{1}{2}\partial X^{\mu}\,\bar{\partial}X_{\mu} + p_{\alpha}\bar{\partial}\theta^{\alpha} + \omega_{\alpha}\bar{\partial}\lambda^{\alpha} + \hat{p}_{\alpha}\partial\hat{\theta}^{\alpha} + \hat{\omega}_{\alpha}\partial\hat{\lambda}^{\alpha}\right), \quad (6.60)$$

where we have denoted $d^2z = dz\,d\bar{z}$, $\partial = \partial_z$, $\bar{\partial} = \partial_{\bar{z}}$ and we introduced the global factor $\frac{1}{2\pi\alpha'}$, which is the string tension. Furthermore, λ^{α} and $\hat{\lambda}^{\alpha}$ are pure spinors, $(\lambda\gamma^{\mu}\lambda) = (\hat{\lambda}\gamma^{\mu}\hat{\lambda}) = 0$.

Clearly, the complex coordinates parameterize the surface or worldsheet, which is always possible locally. We have also introduced more fields (the hat fields), in order to obtain a real action. Nevertheless, the fermion spinors, (p_a, θ^{α}) and $(\hat{p}_a, \hat{\theta}^{\alpha})$, and the bosonic ones, $(\lambda^a, \omega_{\alpha})$ and $(\hat{\lambda}^a, \hat{\omega}_{\alpha})$, may have different chirality, which will define the type of the string. In addition, since the fields are on a surface, they can have different boundary conditions. The boundary conditions depend on whether the surface is open or closed.

For the open string, the boundary conditions are given by[12]

$$\begin{aligned} \partial X^{\mu} &= \bar{\partial}X^{\mu} \\ \theta^{\alpha}(z) &= \hat{\theta}^{\alpha}(\bar{z}) \\ p_{\alpha}(z) &= \hat{p}_{\alpha}(\bar{z}), \qquad \text{when } z = \bar{z}. \\ \lambda^{\alpha}(z) &= \hat{\lambda}^{\alpha}(\bar{z}) \\ \omega_{\alpha}(z) &= \hat{\omega}_{\alpha}(\bar{z}) \end{aligned} \quad (6.61)$$

It is useful to remember that the equations of motion of the pure spinor superstring action are

$$\begin{aligned} \partial\bar{\partial}X^{\mu} &= 0, \\ \bar{\partial}\theta^{\alpha} &= \bar{\partial}p_{\alpha} = \bar{\partial}\lambda^{\alpha} = \bar{\partial}\omega_{\alpha} = 0, \\ \partial\hat{\theta}^{\alpha} &= \partial\hat{p}_{\alpha} = \partial\hat{\lambda}^{\alpha} = \partial\hat{\omega}_{\alpha} = 0. \end{aligned} \quad (6.62)$$

Therefore, the holomorphic fields, $\{\theta^{\alpha}, p_{\alpha}, \lambda^{\alpha}, \omega_{\alpha}\}$, are known as the left sector and the antiholomorphic fields, $\{\hat{\theta}^{\alpha}, \hat{p}_{\alpha}, \hat{\lambda}^{\alpha}, \hat{\omega}_{\alpha}\}$, are the right sector.

The boundary conditions of the closed string are just given by the periodicity, for example,

$$\bar{\partial}X^{\mu}(z+2\pi) = \bar{\partial}X^{\mu}(z), \;\; \theta^{\alpha}(z+2\pi) = \theta^{\alpha}(z), \;\; \lambda^{\alpha}(z+2\pi) = \lambda^{\alpha}(z), \;\; \ldots \quad (6.63)$$

[12]In the open string, in order to preserve the supersymmetry, it is necessary that the spinors have the same chirality. This string is known as Type I.

In the closed string, when the fields, $\{\theta^\alpha, p_\alpha, \lambda^\alpha, \omega_\alpha\}$ and $\{\hat{\theta}^\alpha, \hat{p}_\alpha, \hat{\lambda}^\alpha, \hat{\omega}_\alpha\}$, have the same chirality, it is called *string type IIB*. When the fields, $\{\theta^\alpha, p_\alpha, \lambda^\alpha, \omega_\alpha\}$ and $\{\hat{\theta}_\alpha, \hat{p}^\alpha, \hat{\lambda}_\alpha, \hat{\omega}^\alpha\}$, have the opposite chirality, and then, this string is called *string type IIA*.

The BRST charge looks very similar to the one found in the superparticle

$$Q := \int dz\,(\lambda^\alpha d_\alpha), \qquad \bar{Q} := \int d\bar{z}\,(\hat{\lambda}^\alpha \hat{d}_\alpha). \tag{6.64}$$

We have now two BRST charges, holomorphic and antiholomorphic, which are independent in the closed string. The $d_\alpha\,(\hat{d}_\alpha)$ constraint is little different than the one obtained in superparticle, which is written as

$$d_\alpha := p_\alpha - \frac{1}{2}(\gamma^\mu\theta)_\alpha \partial X_\mu - \frac{1}{8}(\gamma^\mu\theta)_\alpha(\theta\gamma_\mu\partial\theta), \tag{6.65}$$

and its algebra is $\{d_\alpha, d_\beta\} = -\gamma^\mu_{\alpha\beta}\Pi_\mu$, where $\Pi^\mu = \partial X_\mu + \frac{1}{2}(\theta\gamma_\mu\partial\theta)$ is known as the supersymmetric momentum.[13] This constraint arises naturally from the Green-Schwarz action for superstring, but we will not consider it here.[14]

6.3.2 Some Symmetries

It is very useful to remember that in the superparticle case we had gauged the reparameterization invariance by fixing $e = -1/2$. On the worldline, the e-field is interpreted as its metric. Therefore, on the string side the generalization of the e-field is the two-dimensional metric, $g_{ab}, a, b = 1, 2$, but in the superstring pure spinor action it is not very well understood. In addition, the action in (6.60) has the remnant symmetry which is known as conformal symmetry (holomorphic transformations)

$$z \;\rightarrow\; z' = z'(z), \quad \text{Holomorphic transformation.} \tag{6.66}$$

Since the fields, $(X^\mu, \theta^\alpha, \lambda^\alpha)$, are scalars on the worldsheet and $(p_\alpha, \omega_\alpha)$ are $(1, 0)$ differential forms, the current conserved is

$$T(z) = -\frac{1}{2}\partial X^\mu \partial X_\mu - p_\alpha \partial\theta^\alpha + \omega_\alpha \partial\lambda^\alpha, \tag{6.67}$$

that is known as the holomorphic stress tensor. Its antiholomorphic counterpart is just given by the fields with hat.

[13]The definition of $\hat{d}_\alpha$ is $\hat{d}_\alpha := \hat{p}_\alpha - \frac{1}{2}(\gamma^\mu\hat{\theta})_\alpha \bar{\partial} X_\mu - \frac{1}{8}(\gamma^\mu\hat{\theta})_\alpha(\hat{\theta}\gamma_\mu\bar{\partial}\hat{\theta})$.

[14]In the rest of the document, we only work with the left sector.

The pure spinor superstring action has also the global symmetries

$$\begin{array}{ll} \text{space} - \text{timesupersymmetry} & \text{ghost} - \text{number} \\ \delta\lambda^\alpha = \delta\omega_\alpha = 0 & \delta\lambda^\alpha = e^{i\alpha}\lambda^\alpha \\ \delta X^\mu = \frac{1}{2}(\varepsilon\gamma^\mu\theta) & \delta\omega_\alpha = e^{-i\alpha}\omega_\alpha \\ \delta\theta^\alpha = \varepsilon^\alpha & \delta X^\mu = 0 \\ \delta p_\alpha = -\frac{1}{2}(\varepsilon\gamma^\mu)_\alpha\partial X_\mu + \frac{1}{8}(\varepsilon\gamma^\mu\theta)(\partial\theta\gamma_\mu)_\alpha & \delta\theta^\alpha = \delta p_\alpha = 0 \end{array} \tag{6.68}$$

These symmetries give us the charges

$$q_\alpha = -\int dz\left(p_\alpha + \frac{1}{2}(\varepsilon\gamma^\mu)_\alpha\partial X_\mu + \frac{1}{24}(\theta\gamma^\mu\partial\theta)(\theta\gamma_\mu)_\alpha\right), \qquad \text{Supercharge} \tag{6.69}$$

$$G = \int dzJ(z) = \int dz\,(\lambda^\alpha\omega_\alpha), \qquad \text{Ghost} - \text{number}. \tag{6.70}$$

Poincaré invariance, which can be written as

$$X'^\mu = \Lambda^\mu_\nu X^\nu + a^\mu, \tag{6.71}$$

$$\theta'^\alpha = \frac{1}{4}\Lambda_{\mu\nu}(\gamma^{\mu\nu}\theta)^\alpha, \qquad p'_\alpha = \frac{1}{4}\Lambda_{\mu\nu}(\gamma^{\mu\nu}p)_\alpha, \tag{6.72}$$

$$\lambda'^\alpha = \frac{1}{4}\Lambda_{\mu\nu}(\gamma^{\mu\nu}\lambda)^\alpha, \qquad \omega'_\alpha = \frac{1}{4}\Lambda_{\mu\nu}(\gamma^{\mu\nu}\omega)_\alpha, \tag{6.73}$$

where $\Lambda_{\mu\nu} = -\Lambda_{\nu\mu}$, it is generated by the currents

$$P^\mu = \partial X^\mu, \qquad L^{\mu\nu} = X^\mu\partial X^\nu - X^\nu\partial X^\mu, \tag{6.74}$$

$$\Sigma^{\mu\nu} = \frac{1}{2}(p\gamma^{\mu\nu}\theta) \tag{6.75}$$

$$N^{\mu\nu} = \frac{1}{2}(\omega\gamma^{\mu\nu}\lambda). \tag{6.76}$$

Finally, the pure spinor action in (6.60) has an extra local symmetry as a consequence of the pure spinor constraint, $(\lambda\gamma^\mu\lambda) = 0$, which is given by

$$\delta\omega_\alpha = \Lambda_\mu(\gamma^\mu\lambda)_\alpha. \tag{6.77}$$

The pure spinor constraint implies that the number of degrees of freedom of λ^α is just 11 (see Appendix 6.4); in addition, using the local symmetry in (6.77) one can fix 5 of the 16 components of ω_α. Hence, the number of degrees of freedom of λ^α and ω_β is the same, 11.

6.3.3 OPEs and Anomaly

In two-dimensional theories, particularly in conformal theories, one often has to compute the operator product expansion (OPE) among different physical operators. The OPE gives us much information about the theory, such as the topology of the target space, anomalies, symmetries, and amplitudes. For this section, one can review [1, 3, 6, 20].

Roughly speaking, the OPEs are defined just as the correlation function between operators. In addition, as it is well known from quantum field theory, a correlation function is just a Green function of some operator. For example, from the pure spinor action, it is simple to see that the correlation function among X^{μ} with itself is just the Green function of Laplace the operator $\partial\bar{\partial}$, namely (on the sphere)

$$\langle X^{\mu}(z)X_{\nu}(y)\rangle := X^{\mu}(z)X^{\nu}(y) = -\frac{\eta^{\mu\nu}}{2}\ln|z-y|^2 + \text{reg}, \tag{6.78}$$

where "reg" means regular terms in $(z-y)$. In the similar way, OPEs among the others fields are

$$p_{\alpha}(z)\theta^{\beta}(y) = \frac{\delta_{\alpha}^{\beta}}{z-y} + \text{reg}, \tag{6.79}$$

$$\omega_{\alpha}(z)\lambda^{\beta}(y) = \frac{\delta_{\alpha}^{\beta}}{z-y} + \frac{\text{correction from the pure spinor condition}}{z-y} + \text{reg}, \tag{6.80}$$

where the correction from the pure spinor condition is a little complicated and for more details see [1].

Using the previous fundamental OPEs and applying the Wick theorem, we can compute the OPEs among the different currents. For instance, let us consider the followings two OPEs

$$T(z)T(y) = \frac{2}{(z-y)^2}T(y) + \frac{1}{(z-y)}\partial T(y) + \text{reg} \tag{6.81}$$

$$T(z)J(y) = \frac{8}{(z-y)^3} + \frac{1}{(z-y)^2}J(y) + \frac{1}{(z-y)}\partial J(y). \tag{6.82}$$

The first one means that the pure spinor formalism is free of conformal anomaly. In general, the OPE among the stress tensor with itself is given by

$$T(z)T(y) = \frac{c}{2\,(z-y)^4} + \frac{2}{(z-y)^2}T(y) + \frac{1}{(z-y)}\partial T(y) + \text{reg}. \tag{6.83}$$

The first term is the anomalous term, and the "c" constant is known as the central charge. Theories with nonzero central charge have a conformal anomaly, i.e., at quantum level the conformal symmetry in these theories is broken. Furthermore, since the stress tensor is the generator of the conformal transformation, the quadratic pole of the OPE with $T(z)$ gives the conformal weight, namely how the field transforms under a conformal transformation. Clearly, $T(z) = T_{zz}(z)$ has conformal weight 2, to wit $T'(z') = (\mathfrak{p}_{z'}z)^2 T(z)$.

From the second OPE in (6.82), one can see that the ghost current has conformal weight 1, as it was expected that λ^α is a worldsheet scalar and ω_α is an holomorphic form; in addition, this current has an anomaly given by the number 8 in the cubic pole. As $J(z)$ just depends on the pure spinor and its conjugate momentum, then this anomaly gives us topological properties of the pure spinor space. Since that the pure spinor action is invariant under the ghost number transformation (see (6.68)), this implies that the anomaly is presented in the integration measure of the path integral, i.e.,

$$[\mathscr{D}\lambda][\mathscr{D}\omega] \;\rightarrow\; \text{ghost number } 8. \tag{6.84}$$

The ω field, which is a differential form over the worldsheet, can be expanded as a linear combination of the eigenfunctions of the operator $\bar{\partial}$, namely

$$\omega_\beta = \sum_i \omega^i_\beta f_i(z,\bar{z}), \qquad \text{where } \bar{\partial} f_i(z,\bar{z}) = \gamma_i f_i(z,\bar{z}). \tag{6.85}$$

Let us recall that on the sphere there are no global holomorphic forms, so there is no an eigenfunction with eigenvalue zero, $\gamma_i \neq 0$. The eigenfunctions with eigenvalue zero are called the *zero modes*, so the ω_α field does not have zero modes on the sphere and the measure $[\mathscr{D}\omega]$ reads

$$[\mathscr{D}\omega] = \prod_{i=1}[d\omega^i_\beta], \tag{6.86}$$

where $[d\omega^i_\beta]$ is the ω_β measure over the phase space $(\lambda^\alpha, \omega_\beta)$. Now, as the λ^α is a scalar field over the worldsheet, then it can be expanded as a linear combination of the eigenfunctions of the operator $\bar{\partial}\partial$, i.e.,

$$\lambda^\alpha = \sum_i \lambda^\alpha_i\, h_i(z,\bar{z}), \qquad \text{where } \bar{\partial}\partial\, h_i(z,\bar{z}) = \rho_i\, h_i(z,\bar{z}). \tag{6.87}$$

Clearly, the $f_i(z,\bar{z})$ functions are related to the $h_i(z,\bar{z})$ by, $f_i(z,\bar{z}) = \partial\, h_i(z,\bar{z})$. Nevertheless, the operator $\bar{\partial}\partial$ has zero modes on the sphere, so the measure $[\mathscr{D}\lambda]$ becomes

$$[\mathscr{D}\lambda] = [d\lambda^\alpha_0] \prod_{i=1}[d\lambda^\alpha_i], \tag{6.88}$$

where $[d\lambda^\alpha_i]$ is the holomorphic measure of the pure spinor space. Therefore, the total measure can be written as

$$[\mathscr{D}\lambda][\mathscr{D}\omega] = [d\lambda^\alpha_0] \prod_{i=1}[d\lambda^\alpha_i]\,[d\omega^i_\beta]. \tag{6.89}$$

Since λ^α has ghost number 1 and ω_β has ghost number -1, the measure $\prod_{i=1}[d\lambda_i^\alpha]$ $[d\omega_\beta^i]$ has ghost number 0; thus, we conclude that the ghost number anomaly is just given by the measure of the zero modes

$$[d\lambda_0^\alpha] \rightarrow \text{ ghost number } 8. \tag{6.90}$$

In order to compute scattering amplitudes, we must build a top holomorphic form, $[d\lambda_0^\alpha]$, to wit a 11−form, with ghost number 8. This top holomorphic form can be written in the following covariant way[15]

$$[d\lambda^\alpha](\lambda\gamma^{\mu_1})_{\alpha_1}(\lambda\gamma^{\mu_2})_{\alpha_2}(\lambda\gamma^{\mu_3})_{\alpha_3}(\gamma_{\mu_1\mu_2\mu_3})_{\alpha_4\alpha_5} = \varepsilon_{\alpha_1\ldots\alpha_5\beta_1\ldots\beta_{11}} d\lambda^{\beta_1} \wedge \cdots \wedge d\lambda^{\beta_{11}}, \tag{6.91}$$

where $\varepsilon_{\alpha_1\ldots\alpha_5\beta_1\ldots\beta_{11}}$ is the 16-dimensional totally antisymmetric tensor (Levi-Civita symbol) and we have removed the zero modes subindex "0." Using the pure spinor constraint and the γ−matrices algebra, it is not hard to check that, in fact, the term $(\lambda\gamma^{\mu_1})_{\alpha_1}(\lambda\gamma^{\mu_2})_{\alpha_2}(\lambda\gamma^{\mu_3})_{\alpha_3}(\gamma_{\mu_1\mu_2\mu_3})_{\alpha_4\alpha_5}$ is totally antisymmetric in the spinorial labels. Clearly, the left side and right side of the equality in (6.91) have ghost number 11, and the term on the left hand is the same one which appears in (6.55).

6.3.4 *Massless States*

In order to give a prescription to compute scattering amplitudes in the pure spinor formalism, it is needed to introduce the vertex operators, namely to find the BRST cohomology. This section is going to be brief due to the long computations to check the results, for more details review [1, 12, 14, 16]

The physical states in the pure spinor formalism are defined as ghost number 1 states in the BRST cohomology of $Q = \int dz(\lambda^\alpha d_\alpha)$. In addition, since we are just interested in massless states, they must have conformal weight zero by the relation $(mass)^2 = k^2 = \frac{n}{2}$, where n is the conformal weight and k^μ is the momentum vector. So, the most general massless operator at ghost number 0 is

$$V(z) = \lambda^\alpha A_\alpha(X, \theta). \tag{6.92}$$

which is known as the unintegrated vertex operator. From the BRST cohomology condition, $QV = 0$, one obtains the constraint

$$(\gamma_{\mu_1\mu_2\mu_3\mu_4\mu_5})^{\alpha\beta} D_\alpha A_\beta = 0,$$

which is the equation of motion for the spinor prepotential of super-Yang-Mills. Furthermore, the gauge transformation

[15]It is useful to see the Appendix.

$$\delta V = Q\Omega(X,\theta) = \lambda^\alpha D_\alpha \Omega(X,\theta),$$

reproduces the usual super-Yang-Mills gauge transformation $\delta A_\alpha = D_\alpha \Omega(X,\theta)$, where $\Omega(X,\theta)$ is a generic scalar superfield. So, the ghost number 1 cohomology of Q for the massless sector reproduces the desired super-Yang-Mills spectrum.

It is possible to show there is a gauge such that

$$A_\alpha(X,\theta) = \tag{6.93}$$
$$= \frac{1}{2}a_\mu(X)(\gamma^\mu\theta)_\alpha - \frac{1}{3}(\xi(X)\gamma_\mu\theta)(\gamma^\mu\theta)_\alpha - \frac{1}{16}\partial_{[\mu}a_{\nu]}(\theta\gamma^{\delta\mu\nu}\theta)(\gamma_\delta\theta)_\alpha + \ldots$$

where $a_\mu(X) = e_\mu\, e^{ik\cdot X}$ and $\xi^\alpha(X) = \chi^\alpha\, e^{ik\cdot X}$ are the gluon and gluino fields of the SYM theory; e_μ and χ^α are the polarization vectors; and "$[\mu,\nu]$" is the antisymmetrization of the indices.

The unintegrated vertex operators are needed to fix the global symmetry over the Riemann surface. For example, on the sphere (tree-level amplitude) the global symmetry group is $PSL(2,\mathbb{C})$, which has three generators. So, in order to fix this global symmetry, one must use three unintegrated vertex operators in the scattering amplitudes prescription, which can be fixed at any point. The other vertex operators in the scattering amplitudes prescription are integrated vertex operators. The integrated vertex operator, which we will call as $U(z)$, associated with the unintegrated vertex operator V is defined to satisfy

$$QU(z) = \partial_z V(z). \tag{6.94}$$

Note that, $Q(\int U(z)) = 0$. From this definition, one can check that the integrated vertex operator associated with $V(z) = \lambda^\alpha A_\alpha(X,\theta)$ is

$$U(z) = \partial_z\theta^\alpha A_\alpha(X,\theta) + \Pi^\mu B_\mu(X,\theta) + d_\alpha W^\alpha(X,\theta) + \frac{1}{2}N_{\mu\nu}\mathcal{F}^{\mu\nu}(X,\theta) \tag{6.95}$$

where the superfields, $\{B_\mu(X,\theta), W^\alpha(X,\theta), F^{\mu\nu}(X,\theta)\}$, satisfy the constraints

$$D_\alpha A_\beta + D_\beta A_\alpha - \gamma^\mu_{\alpha\beta}B_\mu = 0, \tag{6.96}$$
$$D_\alpha B_\mu - \partial_\mu A_\alpha - (\gamma_\mu)_{\alpha\beta}W^\beta = 0, \tag{6.97}$$
$$D_\alpha W^\beta - \frac{1}{4}(\gamma_{\mu\nu})_\alpha{}^\beta\mathcal{F}^{\mu\nu} = 0, \tag{6.98}$$
$$\lambda^\alpha\lambda^\beta(\gamma_{\mu\nu})_\beta{}^\gamma D_\alpha\mathcal{F}^{\mu\nu} = 0, \tag{6.99}$$

which imply the super-Maxwell equations of motion.

For the closed string, the vertex operators are just the tensorial product of operators from the left and right sector, to wit

$$V_{\text{closed}} = V(z)\otimes\hat{V}(\bar{z}) = \lambda^\alpha\hat{\lambda}^\beta A_\alpha(\theta)\otimes\hat{A}_\beta(\hat{\theta})\, e^{ik\cdot X}, \tag{6.100}$$

$$U_{\text{closed}} = U(z) \otimes \hat{U}(\bar{z}), \tag{6.101}$$

where the graviton, $g_{\mu\nu}$, is identified with $e_\mu \otimes \hat{e}_\nu$ and the gravitino, ψ^α_μ ($\hat{\psi}^\alpha_\mu$), with $e_\mu \otimes \hat{\chi}^\alpha$ ($\chi^\alpha \otimes \hat{e}_\mu$).

6.3.5 Tree-Level Scattering Amplitudes

For more details of this section, one can review [1, 3, 5].

In this section, we give an example how to compute scattering amplitudes at tree level using the pure spinor formalism, in particular for the closed string, i.e., on a sphere.

In general, the scattering amplitude prescription on a sphere is given by the expression

$$\mathscr{M}_n := \prod_{i=4}^{n} \int d^2 z_i \left| \langle V(z_1)\, V(z_2)\, V(z_3)\, U(z_4) \cdots U(z_n) \rangle \right|^2, \tag{6.102}$$

where the power two is due to left and right sectors. The three unintegrated vertex operators fix the $PSL(2, \mathbb{C})$ global symmetry, and the points $\{z_1, z_2, z_3\}$ are arbitrary on the sphere, which often are chosen to be $z_1 = 1, z_2 = 0, z_3 = \infty$. The triangular bracket, $\langle \cdots \rangle$, means integration by all fields, i.e.,

$$\langle \cdots \rangle = \int [\mathscr{D}X][\mathscr{D}\lambda][\mathscr{D}\omega][\mathscr{D}\theta][\mathscr{D}d] \ \cdots, \tag{6.103}$$

where we have replaced the $[\mathscr{D}p]$ integration by $[\mathscr{D}d]$.

Since λ^α and ω_α are complex variables, the integration by these variables must be a contour integral. The contour can be fixed introducing the Cauchy kernel (delta Dirac function), which is known as the picture changing operator. Nevertheless, in 2005 Berkovits introduced a new set of fields, the complex conjugate of $(\lambda^\alpha, \omega_\alpha)$, i.e., $(\bar{\lambda}_\alpha, \bar{\omega}^\alpha)$, in order to integrate over whole pure spinor space. In addition, so as to keep the central charge, $c = 0$ (see (6.83)), two more fermionic fields must be introduced, (r_α, s^α), where r_α is constrained to satisfy,[16] $(\bar{\lambda}\gamma^\mu r) = 0$, $\mu = 0, ..., 9$. The BRST charge is also modified[17]

$$Q = \int dz(\lambda^\alpha d_\alpha) \quad \longrightarrow \quad \tilde{Q} = \int dz(\lambda^\alpha d_\alpha + \bar{\omega}^\alpha r_\alpha), \tag{6.104}$$

but the cohomology of Q and $\tilde{Q}$ is the same.

---

[16]Note that the r_α field can be interpreted as an antiholomorphic form over the pure spinor space, to wit $r_\alpha \equiv d\bar{\lambda}_\alpha$.

[17]Clearly, the operator $\int(r_\alpha \bar{\omega}^\alpha)$ can be identified with the Dolbeault operator $d\bar{\lambda}_\alpha \frac{\partial}{\partial \bar{\lambda}_\alpha}$. So, $\tilde{Q}$ is an equivariant operator.

In this new version, the ghost anomaly is -3, i.e.,

$$[\mathscr{D}\theta][\mathscr{D}d][\mathscr{D}\lambda][\mathscr{D}\bar{\lambda}][\mathscr{D}\omega][\mathscr{D}\bar{\omega}][\mathscr{D}r][\mathscr{D}s] \longrightarrow \text{ghost number } -3, \tag{6.105}$$

where the ghost current is given by $J(z) = (\omega_\alpha \lambda^\alpha) - (\bar{\omega}^\alpha \bar{\lambda}_\alpha)$. But the total integral given in (6.102) has ghost number 0, to wit

$$[\mathscr{D}\theta][\mathscr{D}d][\mathscr{D}\lambda][\mathscr{D}\bar{\lambda}][\mathscr{D}\omega][\mathscr{D}\bar{\omega}][\mathscr{D}r][\mathscr{D}s]V(z_1)V(z_2)V(z_3)U(z_4)\cdots U(z_n) \longrightarrow \text{ghost number } 0.$$

It is not hard to check that the integration

$$\int [\mathscr{D}\theta][\mathscr{D}d][\mathscr{D}\lambda][\mathscr{D}\bar{\lambda}][\mathscr{D}\omega][\mathscr{D}\bar{\omega}][\mathscr{D}r][\mathscr{D}s]\cdots$$

is equivalent to the bracket

$$\int [\mathscr{D}\theta][\mathscr{D}d][\mathscr{D}\lambda][\mathscr{D}\bar{\lambda}][\mathscr{D}\omega][\mathscr{D}\bar{\omega}][\mathscr{D}r][\mathscr{D}s]\cdots \longrightarrow \big\langle (\lambda\gamma^{\mu_1}\theta)(\lambda\gamma^{\mu_2}\theta)(\lambda\gamma^{\mu_3}\theta)(\theta\gamma_{\mu_1\mu_2\mu_3}\theta)\big\rangle = \mathrm{C}, \tag{6.106}$$

where C is a constant. In general, this constant is normalized to be $\mathrm{C} = 1$, so as in (6.55).

6.3.5.1 Three Gravitons at Tree Level

In this example, we compute a scattering amplitude at tree level for three gravitons. This is the simplest one because it is not needed to introduce an integrated vertex operator. This section is based on C. Mafra's master thesis [17]

The amplitude is given by

$$\mathscr{M}_3 = \Big| \langle V(z_1)V(z_2)V(z_3)\rangle \Big|^2, \tag{6.107}$$

with

$$V(z_j) = \frac{1}{2} e^j_\mu (\lambda\gamma^\mu\theta) e^{ik_j\cdot X} - \frac{1}{16} k^j_\mu e^j_\nu (\lambda\gamma_\rho\theta)(\theta\gamma^{\mu\nu\rho}\theta) e^{ik_j\cdot X} \tag{6.108}$$

where we have just considered the bosonic contribution, i.e., the polarization vector e^j_μ, where j is the label of the corresponding particle.

From the integration given in (6.106), the only nonzero contributions are those in which there are five θs. So, following the expansion in (6.108), there are just three possibilities to distribute the θ field, (1, 1, 3), (1, 3, 1), (3, 1, 1).

The first contribution is given by

$$\mathcal{M}_3^1 = e^1_{\mu_1} e^2_{\mu_2} k^3_{\nu_3} e^3_{\mu_3} \left\langle (\lambda\gamma^{\mu_1}\theta)(\lambda\gamma^{\mu_2}\theta)(\lambda\gamma_\rho\theta)(\theta\gamma^{\rho\nu_3\mu_3}\theta) \right\rangle \times \\ \times \left\langle e^{ik_1\cdot X(z_1)} e^{ik_2\cdot X(z_2)} e^{ik_3\cdot X(z_3)} \right\rangle . \tag{6.109}$$

The integration by the X^μ field is simple, and the answer is

$$\left\langle e^{ik_1\cdot X(z_1)} e^{ik_2\cdot X(z_2)} e^{ik_3\cdot X(z_3)} \right\rangle = \int [\mathscr{D}X] e^{-\int d^2z\, \partial X\cdot\bar\partial X} e^{ik_1\cdot X(z_1)} e^{ik_2\cdot X(z_2)} e^{ik_3\cdot X(z_3)} \\ = |z_{12}|^{2k_1\cdot k_2} |z_{13}|^{2k_1\cdot k_3} |z_{23}|^{2k_2\cdot k_3}, \tag{6.110}$$

where $z_{ij} := z_i - z_j$. From the on-shell condition, $k_j^2 = 0$, and the momentum conservation constraint, $k_1^\mu + k_2^\mu + k_3^\mu = 0$, it is trivial to check $k_1 \cdot k_2 = k_1 \cdot k_3 = k_2 \cdot k_3 = 0$; therefore,

$$\left\langle e^{ik_1\cdot X(z_1)} e^{ik_2\cdot X(z_2)} e^{ik_3\cdot X(z_3)} \right\rangle = 1. \tag{6.111}$$

Note that, we have not introduced the functional determinant, $\det(\partial\bar\partial)$, in (6.110), because it will be canceled out by the other functional determinants.

Up to an overall factor, it is not hard to check[18]

$$\left\langle (\lambda\gamma^{\mu_1}\theta)(\lambda\gamma^{\mu_2}\theta)(\lambda\gamma_\rho\theta)(\theta\gamma^{\rho\nu_3\mu_3}\theta) \right\rangle \propto \eta^{\mu_1\nu_3}\eta^{\mu_2\mu_3} - \eta^{\mu_1\mu_3}\eta^{\mu_2\nu_3}. \tag{6.112}$$

Finally, the contribution (1, 1, 3) becomes

$$\mathcal{M}_3^1 = (e^1 \cdot k^3)(e^2 \cdot e^3) - (e^2 \cdot k^3)(e^1 \cdot e^3). \tag{6.113}$$

In a similar way, the contributions (1, 3, 1) and (3, 1, 1) are given by

$$\mathcal{M}_3^2 = -(e^1 \cdot k^2)(e^2 \cdot e^3) + (e^3 \cdot k^2)(e^1 \cdot e^2), \tag{6.114}$$
$$\mathcal{M}_3^3 = (e^2 \cdot k^1)(e^1 \cdot e^3) - (e^3 \cdot k^1)(e^1 \cdot e^2). \tag{6.115}$$

Therefore, the total amplitude reads

$$\mathcal{M}_3 = \left| \mathcal{M}_3^1 + \mathcal{M}_3^2 + \mathcal{M}_3^3 \right|^2 \\ = \left| 2(e^1 \cdot e^2)(e^3 \cdot k^2) + 2(e^1 \cdot e^3)(e^2 \cdot k^1) + 2(e^2 \cdot e^3)(e^1 \cdot k^3) \right|^2, \tag{6.116}$$

where we have used the momentum conservation, $k_1^\mu + k_2^\mu + k_3^\mu = 0$, and the transversality condition, $e^j \cdot k^j = 0$. Up to overall constant, this is the right result [10].

[18] For more details, see appendix of [7].

6.4 Appendix

6.4.1 Cartan and Chevalley Definitions

This appendix is based in the lectures on beta-gamma system given in [20].

The $SO(2d)$ pure spinor λ^α is constrained to satisfy [2]

$$\lambda^\alpha (\gamma^{\mu_1 \dots \mu_j})_{\alpha\beta} \lambda^\beta = 0, \qquad \text{for} \qquad 0 \leq j < d \tag{6.117}$$

where $\mu = 1$ to $2D$, $\alpha = 1$ to 2^{d-1}, and $\gamma_{\alpha\beta}^{\mu_1 \dots \mu_j}$ is the antisymmetrized product of j Pauli matrices, i.e.,

$$\gamma^{\mu_1 \dots \mu_j} := \frac{1}{j!} \gamma^{[\mu_1} \gamma^{\mu_2} \dots \gamma^{\mu_j]}. \tag{6.118}$$

This implies that $\lambda^\alpha \lambda^\beta$ can be written as

$$\lambda^\alpha \lambda^\beta = \frac{1}{n!\, 2^d} \gamma^{\alpha\beta}_{\mu_1 \dots \mu_d} (\lambda^\rho \gamma_{\rho\delta}^{\mu_1 \dots \mu_d} \lambda^\delta) \tag{6.119}$$

where $\lambda \gamma^{\mu_1 \dots \mu_d} \lambda$ defines an d-dimensional complex plane $\mathbb{C}^d \subset \mathbb{R}^{2d} \otimes \mathbb{C}$. This d-dimensional complex plane is preserved by a $U(d)$ subgroup of $SO(2d)$ rotations. Also, multiplying λ by a nonzero complex number does not change this plane. So, if we consider the space of λs obeying up to rescalings, the space of *projective pure spinors*, $\mathbb{P}\mathrm{PS}_{2d}$ in $D = 2d$ Euclidean dimensions, then:

$$\mathbb{P}\mathrm{PS}_{2d} = SO(2d)/U(d) \tag{6.120}$$

The real dimension of this space is $d(d-1)$. The space PS_{2d} of pure spinors is a cone over $\mathbb{P}\mathrm{PS}_{2d}$. The space X_{2d}, which is PS_{2d} with the point $\lambda = 0$ deleted, can be thought of the moduli space of Calabi–Yau complex structures on $\mathbb{R}^{2d}$, i.e., the space of pairs

$$(\text{identification } \mathbb{C}^d \approx \mathbb{R}^{2d}, \Omega \in \Lambda^d \mathbb{C}^d)$$

This is an important space in the context of B-type topological strings.

6.4.2 Pure Spinor Parameterization

In order to solve the 10-dimensional pure spinor constraints, it is useful to write them in terms of the $U(5)$ variables.

A vector in 10 dimensions, $\tilde{V}^\mu$, can be written as a direct sum of two five-dimensional vectors

$$V^a := \frac{1}{\sqrt{2}} (\tilde{V}^a + i \tilde{V}^{a+5}), \qquad a = 1, 2, \dots, 5 \tag{6.121}$$

$$V_a := \frac{1}{\sqrt{2}}(\tilde{V}^a - i\tilde{V}^{a+5}), \tag{6.122}$$

i.e., we have broken the 10-dimensional vector representation of $SO(10)$ as a sum of two vectorial representations of $U(5)$, $10 = 5 \oplus \bar{5}$. In the 10-dimensional gamma matrices, we have

$$b^a := \frac{1}{\sqrt{2}}(\gamma^a + i\gamma^{a+5}), \qquad a = 1, 2, ..., 5 \tag{6.123}$$

$$b_a := \frac{1}{\sqrt{2}}(\gamma^a - i\gamma^{a+5}), \tag{6.124}$$

where the gamma algebra becomes $\{b_a, b^c\} = \delta_a^c$. Now, the (b_a, b^c) matrices satisfy a ladder algebra, and we can construct a finite representation.

We define the fundamental state such that $b_a|0\rangle = 0$, $a = 1, ..., 5$, so all states are created applying the b^α matrix on $|0\rangle$. Since the pure spinor is a chiral spinor and the chiral operator just counts the number of b^a matrices which acts on $|0\rangle$, the most general positive chiral spinor is written as

$$|\lambda^\alpha\rangle = \lambda^+|0\rangle + \frac{1}{2}\lambda_{ab}b^a b^b|0\rangle + \frac{1}{24}\lambda^a \varepsilon_{abcde} b^b b^c b^d b^e|0\rangle, \tag{6.125}$$

where positive chirality means the number of b^a is even and $\lambda_{ab} = -\lambda_{ba}$. Clearly, we have broken the λ^α spinor as $\lambda^\alpha = (\lambda^+, \lambda_{ab}, \lambda^a)$, where the number of degrees of freedom of λ^+ is one, of λ_{ab} is 10, and the λ^α is 5, namely $16 \rightarrow (1, \bar{10}, 5)$.

Finally, using the $U(5)$ representation the pure spinor constraints become

$$\lambda^+\lambda^a + \frac{1}{8}\varepsilon^{abcde}\lambda_{bc}\lambda_{de} = 0, \qquad a = 1, ..., 5, \tag{6.126}$$

$$\lambda^b\lambda_{ba} = 0. \tag{6.127}$$

Choosing the chart where $\lambda^+ \neq 0$ and using the parameterization $\lambda^+ = \gamma$, $\lambda_{ab} = \gamma\, u_{ab}$, the solution of the equations in (6.126) is straightforward

$$\lambda^a = -\frac{\gamma}{8}\varepsilon^{abcde}u_{bc}u_{de}, \tag{6.128}$$

and the equations in (6.127) become trivial.

As a final remark, because the pure spinor has ghost number 1, then obviously γ has ghost number 1 and u_{ab} has ghost number 0. Therefore, we can write an holomorphic top form over the pure spinor space with ghost number 8 as

$$[d\lambda^\alpha] = \gamma^7\, d\gamma \wedge du_{12} \wedge du_{13} \wedge \cdots \wedge du_{45}, \tag{6.129}$$

which matches with the one written in (6.91).

Acknowledgements We thank S. Mizera, G. Zhang, and C. Mafra for reading the manuscript. H.G. would like to thank M. Guillen for discussions about superparticle. H.G. is very grateful to the Perimeter Institute for hospitality during this work. The work of H.G. is supported by USC Grant DGI-COCEIN-No 935-621115-N22.

References

1. N. Berkovits, Super-Poincaré Covariant Quantization of the Superstring. JHEP **04**, 018 (2000), arXiv:hep-th/0001035
2. E. Cartan, *Lecons sur la Theorie des Spineurs* (Hermann, Paris, 1937)
3. N. Berkovits, Multiloop amplitudes and vanishing theorems using the pure spinor formalism for the superstring. JHEP **0409**, 047 (2004), arXiv:hep-th/0406055
4. N. Berkovits, Super-Poincaré covariant two-loop superstring amplitudes, JHEP 0601, 005 (2006). doi:10.1088/1126-6708/2006/01/005, arXiv:hep-th/0503197; N. Berkovits, C.R. Mafra, Equivalence of two-loop superstring amplitudes in the pure spinor and RNS formalisms, Phys. Rev. Lett. 96, 011602 (2006) 2006. doi:10.1103/PhysRevLett.96.011602, arXiv:hep-th/0509234; C.R. Mafra, Four-point one-loop amplitude computation in the pure spinor formalism, JHEP 0601, 075. doi:10.1088/1126-6708/2006/01/075, arXiv:hep-th/0512052
5. N. Berkovits, Pure spinor formalism as an N = 2 topological string. JHEP **0510**, 089 (2005), arXiv:hep-th/0509120
6. N. Berkovits, N. Nekrasov, Multiloop superstring amplitudes from non-minimal pure spinor formalism. JHEP **0612**, 029 (2006), arXiv:hep-th/0609012
7. N. Berkovits, C.R. Mafra, Some Superstring Amplitude Computations with the Non-Minimal Pure Spinor Formalism. JHEP **0611**, 079 (2006). doi:10.1088/1126-6708/2006/11/079, arXiv:hep-th/0607187
8. H. Gomez, C.R. Mafra, The closed-string 3-loop amplitude and S-duality, JHEP 1310, 217 (2013). doi:10.1007/JHEP10(2013)217, arXiv:1308.6567 [hep-th]
9. P.A. Grassi, P. Vanhove, Higher-loop amplitudes in the non-minimal pure spinor formalism. JHEP **0905**, 089 (2009). doi:10.1088/1126-6708/2009/05/089 [arXiv:0903.3903 [hep-th]]. Y. Aisaka and N. Berkovits, Pure Spinor Vertex Operators in Siegel Gauge and Loop Amplitude Regularization, JHEP 0907, 062 (2009). doi:10.1088/1126-6708/2009/07/062, arXiv:0903.3443 [hep-th]
10. J. Polchinski, String Theory vol I, (Cambridge, Published 1998), M. Green, J. Schwarz, E. Witten, Superstring Theory: vol 1, Introduction, (Cambridge,Published 1987)
11. E. D'Hoker, D.H. Phong, The geometry of string perturbation theory. Rev. Mod. Phys. **60**, 917 (1988)
12. N. Berkovits, ICTP lectures on covariant quantization of the superstring, arXiv:hep-th/0209059
13. N. Berkovits, Covariant quantization of the superparticle using pure spinors. JHEP **0109**, 016 (2001). doi:10.1088/1126-6708/2001/09/016, arXiv:hep-th/0105050
14. N. Berkovits, O. Chandia, Massive superstring vertex operator in D = 10 superspace, JHEP **0208**, 040 (2002). doi:10.1088/1126-6708/2002/08/040, arXiv:hep-th/0204121
15. Griffiths and Harris, Principles of Algebraic Geometry, (Wiley Classics Library Edition Published 1994)
16. C.R. Mafra, Superstring scattering amplitudes with the pure spinor formalism, arXiv:0902.1552 [hep-th]
17. C.R. Mafra, Os Formalismos da Supercorda, March of 2005, IFT-São Paulo-Brasil
18. Max Guillen, D = 10 Super Yang-Mills, D = 11 Supergravity and the Pure Spinor Superfield Formalism, March of 2016, IFT-São Paulo-Brasil
19. E. Witten, Non-commutative geometry and string field theory. Nuclear Physics B **268**(2), 253–294 (1986). doi:10.1016/0550-3213(86)90155-0. ISSN 0550-3213
20. N.A. Nekrasov, Lectures on curved beta-gamma systems, pure spinors, and anomalies, arXiv:hep-th/0511008

Chapter 7
Introduction to Elliptic Fibrations

Mboyo Esole

Abstract The modern study of elliptic fibrations started in the early 1960s with seminal works by Kodaira and by Néron. Elliptic fibrations play a central role in the classification of algebraic surfaces, in many aspects of arithmetic geometry, theoretical physics, and string geometry. In these notes, we introduce the reader to basic geometric properties of elliptic fibrations over the complex numbers. We start with an introduction to the geometry of elliptic curves defined over the complex numbers. We then discuss Weierstrass models, Kodaira's classification of singular fibers of elliptic surfaces, Tate's algorithm, and Miranda's regularization of elliptic threefolds.

7.1 Introduction

The theory of elliptic curves is an elegant and vast subject in mathematics that can be traced back to ancient Greece and beyond. An elliptic curve is a non-singular projective curve of genus one, with a choice of a rational point. The chosen rational point plays the role of the neutral element of the Mordell–Weil group of the elliptic curve. An elliptic fibration is the relative case of an elliptic curve. Intuitively, an elliptic fibration is the variety swapped by an elliptic curve moving over a base variety. The study of elliptic fibrations started in 1962–1963 with Kodaira's work on compact complex analytic surfaces [12] followed in 1964 by Néron's paper on minimal models of Abelian varieties [19].

Elliptic curves are a pillar of number theory; they are instrumental in cryptography and geometric modeling. Elliptic curves have also invaded many branches of theoretical physics through their modular properties. Elliptic fibrations are at the heart of F-theory, the theory that describes (among other things) the non-perturbative regime

M. Esole (✉)
Department of Mathematics, Northeastern University, 360 Huntington Ave., Boston, MA 02115, USA
e-mail: j.esole@northeastern.edu

A. Cardona et al. (eds.), *Quantization, Geometry and Noncommutative Structures in Mathematics and Physics*, Mathematical Physics Studies,
DOI 10.1007/978-3-319-65427-0_7

of type IIB string theory. Elliptic fibrations also provide geometric constructions of certain superconformal field theories including some that do not have a Lagrangian description.

In these notes, we will focus on the basic properties of elliptic fibrations over the complex numbers. We do not have space for complete proofs, but we will give appropriate references. In Sect. 7.2, we review the theory of elliptic curves over the complex numbers. In Sect. 7.3, we study the theory of elliptic fibrations. In particular, we start in Sect. 7.3.1 by reviewing the Riemann–Roch argument to derive the Weierstrass model of an elliptic curve. In Sects. 7.3.2 and 7.3.3, we explain in detail how the Riemann–Roch argument is combined with an appropriate base change theorem to obtain Weierstrass models for an elliptic fibrations. In Sect. 7.4, we introduce the Kodaira–Néron classification of singular fibers of a minimal elliptic surface and discuss Tate's algorithm. In Sect. 7.5, we study Miranda's regularization of elliptic threefolds and the notion of collisions of singularities.

There are many important questions that we will not address. As an apology, we give the following reading list for elliptic fibrations and Weierstrass models:

- The classical reference for Weierstrass models is the original paper of Deligne (in French) known as the "Formulaire" [4]. Deligne beautifully explains how to derive a Weierstrass model for an elliptic fibration with a ration section. It also introduces Tate's notation widely used today. The construction of Weierstrass models is also discussed in detail by Mumford and Suominen in [16, Chap. 3] and Nakayama [17, 18].
- The original paper of Kodaira on elliptic surfaces [12], Néron [19], and Tate [28] contain significant details not usually covered in reviews.
- In [21], Schütt and Shioda give a short introduction to the theory of elliptic surfaces.
- Chapter 3 of the book of Mumford and Suominen on the theory of moduli [16] has a self-contained section on elliptic curves and elliptic fibrations where the authors carefully derive the existence of a Weierstrass model for an elliptic fibration with a rational section.
- For more advanced topics, we refer to Liu's book on arithmetic geometry [13].
- Miranda's lecture notes on elliptic surfaces [15] are another classic review for the study of elliptic surfaces over an algebraically closed field.
- Nakayama analyzes the global and local structure of elliptic fibrations [17, 18]. He takes the interesting point of view of the variation of Hodge structure to describe elliptic fibrations. He shows that a polarized variation of Hodge structures of rank two, weight one over a base B is equivalent to a Weierstrass model.
- In [14], Miranda studies the problem of finding regular models for Weierstrass models over a smooth surface. He discusses the phenomena of collisions of Kodaira fibers and classifies the singular fibers that appear over codimension two points after the specific regularization that he considers. These are some of the first examples of non-Kodaira singular fibers.
- In his Ph.D. thesis [26], Szydlo generalizes the regularization of Miranda to the case of elliptic n-folds under the same assumptions as Miranda. He also considers the arithmetic case, when the field is not of characteristic zero and provides a

generalization of Tate's algorithm to the case of a complete discrete valuation ring with non-perfect residue field [27].

- Dolgachev and Gross have computed the Ogg–Shfarevich Theory of elliptic threefolds using Miranda's models [7].
- Conrad has an elegant unpublished paper on minimal models for elliptic curves with a strong EGA flavor in which he promises to "free the theory of elliptic curves from the curse of Weierstrass equations" [2]. However, before doing this, he presents a systematic derivation of the Weierstrass equation over $Spec(R)$.
- In Chapter IX of [1], Beauville gives a short introduction to the theory of elliptic surfaces from the point of view of the Kodaira dimension. Cossec and Dolgachev study genus-one fibration in Chap. 5 of [3].

7.2 Elliptic Curves over $\mathbb{C}$

In this section, we collect basic facts about elliptic curves over the complex numbers. This topic is elegantly covered in numerous books. For the proofs, we refer to Chap. 1 (Sects. 1–6) of [23], Chap. 3 of [11], and Chap. VII of [22].

We denote by $\mathbb{C}$ the field of complex numbers and by $\mathbb{Z}$ the ring of integers.

7.2.1 Modular Group and Complex Tori

Modulo similitude transformations, an elliptic curve over the complex numbers is equivalent to a complex torus $\mathbb{C}/(\mathbb{Z}+\tau\mathbb{Z})$, that is, the quotient of the complex plane by the double-lattice $\mathbb{Z}+\tau\mathbb{Z}$ generated by 1 and the complex number τ (the *period*). The Abelian group structure on the elliptic curve is then induced from the addition in $\mathbb{C}$. Geometrically, the period τ characterizes the shape of the complex torus. By convention, τ is restricted to be in the upper-half plane:

$$\mathscr{H} = \{\tau \in \mathbb{C} | \operatorname{Im}(\tau) > 0\}. \tag{7.1}$$

More generally, for a complex torus $\mathbb{C}/(\omega_1\mathbb{Z}+\omega_2\mathbb{Z})$ with periods (ω_1, ω_2), after a rescaling by ω_1^{-1}, we get $\mathbb{C}/(\mathbb{Z}+\tau\mathbb{Z})$ with $\tau = \frac{\omega_2}{\omega_1}$. We can permute ω_1 and ω_2 if necessary to ensure that $\operatorname{Im}(\tau) > 0$ (Fig. 7.1).

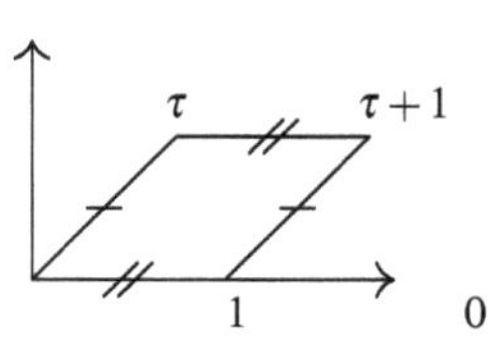

Fig. 7.1 Torus seen as the quotient $\mathbb{C}/(\mathbb{Z}+\tau\mathbb{Z})$

Theorem 7.1 *Two 2-tori are equivalent modulo similitudes if and only if their periods are related by a modular transformation:*

$$\begin{pmatrix} a & b \\ c & d \end{pmatrix} \cdot \tau = \frac{a\tau + b}{c\tau + d}, \quad \begin{pmatrix} a & b \\ c & d \end{pmatrix} \in SL(2, \mathbb{Z}). \tag{7.2}$$

Proof See Lemmas 1.1 and 1.2 in Chap. 1. Sect. 1 of [23].

In particular, if $g = \begin{pmatrix} a & b \\ c & d \end{pmatrix} \in \mathrm{SL}(2, \mathbb{Z})$, $Im(\tau)$ (the imaginary part of τ) transforms as

$$\mathrm{Im}(g \cdot \tau) = \frac{\mathrm{Im}\tau}{|c\tau + d|^2}. \tag{7.3}$$

We denote by I_2 the 2×2 identity matrix. Since the matrix $(-\mathrm{I}_2)$ acts trivially on τ, to have a faithful action, we consider the *modular group* to be the quotient

$$\Gamma(1) := \mathrm{SL}(2, \mathbb{Z})/\{\pm \mathrm{I}_2\}. \tag{7.4}$$

We use the same symbol for a matrix in $\mathrm{SL}(2, \mathbb{Z})$ and its projection to $\Gamma(1)$.

Theorem 7.2 *The group $SL(2, \mathbb{Z})$ is generated by the following two elements:*

$$S := \begin{pmatrix} 0 & -1 \\ 1 & 0 \end{pmatrix}, \quad T := \begin{pmatrix} 1 & 1 \\ 0 & 1 \end{pmatrix}. \tag{7.5}$$

They act as

$$S \cdot \tau = \frac{-1}{\tau}, \quad T \cdot \tau = \tau + 1. \tag{7.6}$$

S and T satisfy the following relations in $SL(2, \mathbb{Z})$:

$$S^2 = (ST)^3 = -\mathrm{I}_2. \tag{7.7}$$

Proof See Remark 1.3 on p. 10 of [23].

When S and T are considered as elements of $\Gamma(1)$, we have $S^2 = (ST)^3 = \mathrm{Id}$ so that $\Gamma(1)$ can be considered as the free group $\mathbb{Z}/2\mathbb{Z} * \mathbb{Z}/3\mathbb{Z}$:

$$\Gamma(1) \cong \langle a, b : a^2 = b^3 = 1 \rangle. \tag{7.8}$$

7.2.2 The Weierstrass Equation

The Weierstrass $\wp$-function provides a natural description of a complex torus $\mathbb{C}/(\mathbb{Z} + \tau\mathbb{Z})$ as a cubic curve in $\mathbb{P}^2$ in Weierstrass form. It is defined as follows:

$$\wp(z,\tau)=\frac{1}{z^2}+\sum_{\substack{w\in\Lambda_\tau\\ w\neq 0}}\Big(\frac{1}{(z-w)^2}-\frac{1}{w^2}\Big),\tag{7.9}$$

where $\Lambda_\tau=\mathbb{Z}+\tau\mathbb{Z}$. The Weierstrass $\wp$-function is a meromorphic function with double poles at the lattice points $w\in\Lambda_\tau$, and doubly periodic:

$$\wp(z+1,\tau)=\wp(z,\tau),\quad \wp(z+\tau,\tau)=\wp(z,\tau).\tag{7.10}$$

The Weierstrass $\wp$-function has a pole of order 2 at the origin, while its derivative $\wp'$ (with respect to z) has a pole of order 3. Together, they satisfy the Weierstrass equation

$$(\wp')^2=4\wp^3-g_2\wp-g_3,\quad \text{where } g_2(\tau):=60G_4(\tau) \text{ and } g_3(\tau):=140G_6(\tau).\tag{7.11}$$

For a given lattice Λ_τ, the Eisenstein series G_{2k} of weight $2k$ are by definition

$$G_{2k}(\tau)=\sum_{\substack{w\in\Lambda_\tau\\ w\neq 0}} w^{-2k}\tag{7.12}$$

Theorem 7.3 *The map*

$$\mathbb{C}/(\mathbb{Z}+\tau\mathbb{Z})\to\mathbb{P}^2: z\mapsto\left[\wp:\frac{1}{2}\wp':1\right],\tag{7.13}$$

provides an analytic isomorphism between the complex torus $\mathbb{C}/(\mathbb{Z}+\tau\mathbb{Z})$ and the following cubic in $\mathbb{P}^2$:

$$E: zy^2=x^3+fxz^2+gz^3,\tag{7.14}$$

with $f=-g_2/4$, $g=-g_3/4$, $y=\wp'/2$, $x=\wp$. For a regular curve $E: y^2=x^3+fx+g$, there is a unique lattice Λ_τ (up to modular transformation on τ) such that E and $\mathbb{C}/(\mathbb{Z}+\tau\mathbb{Z})$ are analytic isomorphic as complex Lie groups through the previous map.

Proof Corollary 4.3 on p. 35 of [23].

7.2.3 Moduli Space of Smooth Elliptic Curves

To classify smooth elliptic curves up to isomorphisms, we introduce the *Klein j-invariant* (also called the modular invariant).The j-invariant is a rational function of G_4^3/G_6^2, which ensures that it is a modular invariant. Any modular invariant is a rational function of G_4^3/G_6^2 or equivalently a rational function of the j-invariant. The j-invariant maps bijectively the moduli space of complex tori modulo similitudes (and therefore the moduli space of smooth elliptic curves) to the complex plane $\mathbb{C}$. Two elliptic curves over $\mathbb{C}$ are isomorphic if and only if they have the same j-invariant.

Definition 7.1 The j-invariant of a Weierstrass equation $y^2 = x^3 + fx + g$ is defined as follows:

$$j(\tau) = 1728\frac{4f^3}{4f^3 + 27g^2} = 1728 - \frac{27g^2}{4f^3 + 27g^2}. \tag{7.15}$$

The coefficient 1728 is chosen to ensure that the j-invariant has residue 1 at infinity.

Theorem 7.4 *The j-invariant can be expressed (as a function of τ) by a Laurent series in $q = \exp(2\pi i\tau)$ of the form:*

$$j(\tau) = \frac{1}{q} + 744 + \sum_{n>0} c_n q^n, \quad c_n \in \mathbb{N}. \tag{7.16}$$

Since $\operatorname{Im}\tau > 0, q = \exp(2\pi i\tau)$ is in the unit disk $|q| < 1$. The modular group admits as a *fundamental domain* the closure of the open region:

$$R_\Gamma = \{\tau \in \mathscr{H} : |\tau + \overline{\tau}| < 1 \text{ and } |\tau| > 1\}, \tag{7.17}$$

with a $\mathbb{Z}/2\mathbb{Z}$ identification on the boundary given by $\tau \cong -\overline{\tau}$. When we have to make a choice between two points on the boundary, we will take the one with negative real part. We recall some additional properties of the j-invariant:

$$j(\mathrm{i}) = 1728, \quad j(\mathrm{e}^{\frac{2\pi}{3}\mathrm{i}}) = 0, \quad j(-\overline{\tau}) = \overline{j(\tau)}, \quad \lim_{Im(\tau)\to+\infty} |j(\tau)| = \infty. \tag{7.18}$$

Geometrically, the moduli space of complex tori modulo similitude is the orbifold

$$Y(1) := \mathscr{H}/\Gamma(1). \tag{7.19}$$

If is useful to also include tori admitting an infinite value for the j-invariant. This corresponds to allowing an infinite value for the imaginary part of τ. By the action of the modular group, we should then also include all the rational points of the real line. This defines the *extended upper-half plane*

$$\mathscr{H}^* := \mathscr{H} \cup \mathbb{P}^1(\mathbb{Q}) = \mathscr{H} \cup \mathbb{Q} \cup \{\infty\} \quad \text{and } X(1) := \mathscr{H}^*/\Gamma(1). \tag{7.20}$$

$X(1)$ is called the *modular curve*. The points of $X(1)\backslash Y(1)$ are called the *cups*. They are the orbit of $\tau = i\infty$ under the action of $\Gamma(1)$. The name cusp can be confusing as $\tau = i\infty$ actually corresponds to a nodal elliptic curve, but the name cusp in this context refers to the singularities of $X(1)$ and not to a singular elliptic curve.

The following classical theorem is proven, for example, in Sect. 4.1 of Chap. 1 of [23].

Theorem 7.5 *The j-invariant is an isomorphism between $X(1)$ and the Riemann sphere $\mathbb{P}^1$:*

$$j : X(1) \to \mathbb{P}^1 : \tau \mapsto [U : V] = [1728 \cdot 4f^3 : 4f^3 + 27g^2], \tag{7.21}$$

where $[U : V]$ denotes the projective coordinates of $\mathbb{P}^1$. The value of the j-invariant at $\tau = i\infty$ is the point at infinity $[1 : 0]$.

One can define an appropriate topology and complex structure on the modular curve $X(1)$. This is explained in Chap. 1 of [23]. Every meromorphic function on $X(1)$ is then a rational function of j. For this reason, the j-invariant is also called the *modular invariant*.

Theorem 7.6 *There is an elliptic curve with a given j-invariant for any $j_0 \in \mathbb{C}$:*

$$y^2 z = x^3 + gz^3, \quad g \neq 0, \, j_0 = 0, \tag{7.22}$$

$$y^2 z = x^3 + fxz^2, \quad f \neq 0, \, j_0 = 1728, \tag{7.23}$$

$$y^2 z = x^3 - \frac{27 j_0 \lambda^2}{4(j_0 - 1728)} xz^2 - \frac{27\lambda^3 j_0}{4(j_0 - 1728)} z^3, \quad \lambda \neq 0, \, j_0 \neq 0, 1728. \tag{7.24}$$

Proof Direct computation using the definition of the j-invariant.

Remark 7.1 A Weierstrass equation with a nodal singularity is given, for example, by the following equation:

$$zy^2 = x^3 - 3xz^2 + 2z^3. \tag{7.25}$$

The presence of a node can be seen by factorizing the r.h.s. to get

$$zy^2 = (x - z)^2 (x + 2z). \tag{7.26}$$

Such a curve has an infinite j-invariant since f and g are nonzero, while the discriminant vanishes.

Theorem 7.7 (Automorphism of an elliptic curve) *The group of automorphisms of an elliptic curve E_j with invariant j is $\mathbb{Z}/2\mathbb{Z}$ for $j \neq 0, 1728$. It is $\mathbb{Z}/4\mathbb{Z}$ for $j = 1728$ and $\mathbb{Z}/6\mathbb{Z}$ for $j = 0$.*

Proof See [11, Chap. 3 Sect. 4].

Remark 7.2 (*Ramifications and Automorphisms of elliptic curves*) Since the j-invariant can be expressed as $j = 1728(4f^3)/(4f^3 + 27g^2)$, it has a ramification of degree 3 at $f = 0$ for which $j = 0$. As we can also write $j - 1728 = -27g^2/(4f^3 + 27g^2)$, there is also a ramification of degree 2 at $g = 0$ for which

$j = 1728$. An elliptic curve with invariant $j = 0$ is given by $E : y^2 = x^3 + g$ with $g \neq 0$. An elliptic curve with invariant $j = 1728$ is $E : y^2 = x^3 + fx$ with $f \neq 0$. The $\mathbb{Z}/2\mathbb{Z}$ automorphism of an elliptic curve with invariant $j \neq 0, 1728$ is given by $(x, y) \rightarrow (x, -y)$. It is the inverse of the group law. For $j = 0$, it is induced by $(x, y) \rightarrow (\omega x, -y)$ where ω is a choice of a cubic root of the unit ($\omega^3 = 1$). For $j = 1728$, it is generated by $(x, y) \rightarrow (-x, \mathrm{i}y)$ where $\mathrm{i}^2 = -1$.

Remark 7.3 (*Cusps and jump phenomena*) Consider a regular elliptic curve in Weierstrass form $E : y^2 = x^3 + fx + g$ defined over a field k. For any nonvanishing $\lambda \in k$, we can define the curve $E_\lambda : y^2 = x^3 + f\lambda^4 x + g\lambda^6$. For $\lambda \neq 0$, E_λ is isomorphic to E after the redefinition $(x, y) \mapsto (\lambda^2 x, \lambda^3 y)$. However, at $\lambda = 0$ we always have the cusp $E_0 : y^2 = x^3$. It follows that an elliptic curve with an arbitrary j-invariant can jump to a cusp. For this reason, cusps are excluded in the moduli space of elliptic curves. When considering only smooth curves, the j-invariant maps the space of elliptic curve modulo isomorphism onto $\mathbb{C}$. This space can be compactified by allowing curves of arithmetic genus one with a nodal singularity.

7.3 Elliptic Fibrations

We work over an algebraically closed field k of characteristic zero. The reader is welcome to think of the base field k as the field of complex numbers $\mathbb{C}$. Most of the results do not require the field to be algebraically closed nor of characteristic zero. But we still assume it out of convenience. We denote by $\mathbb{Z}$ the ring of relative integers. By a variety we mean a reduced and irreducible algebraic scheme [10]. Given a variety X, we denote by $\mathscr{O}_X$ the sheaf of regular functions of X. Given a Cartier divisor D in a normal variety X, we denote by $\mathscr{O}_X(-D)$ the normal bundle of D in X. The sheaf $\mathscr{O}_X(nD)$ ($n \in \mathbb{Z}$) is the sheaf of rational functions with a pole of degree n over the divisor D. The dual sheaf of $\mathscr{O}_X(nD)$ is denoted $\mathscr{O}_X(-nD)$. In particular, D is the vanishing locus of a section of $\mathscr{O}_X(D)$.

Definition 7.2 (*Genus-one fibration*) A genus-one fibration is a surjective proper morphism $\varphi : Y \rightarrow B$ between algebraic varieties such that the generic fiber is a regular projective curve of genus one. The variety B is called the *base of the fibration*.

Definition 7.3 (*Discriminant locus*) The locus of singular fibers of the fibration $\pi : Y \rightarrow B$ is called the *discriminant locus* of π and is denoted Δ.

To avoid trivialities, we assume that a genus-one fibration has a non-trivial discriminant locus (there is at least one singular fiber).

Definition 7.4 (*Rational section*) A *rational section* of a fibration $\varphi : Y \rightarrow B$ is a rational map $\sigma : B \rightarrow Y$ such that $\varphi \circ \sigma$ is the identity away from a Zariski closed set of B.

Definition 7.5 (*Elliptic fibration*) An elliptic fibration is a genus-one fibration endowed with a *rational section*.

7.3.1 Weierstrass Models for Elliptic Curves

Before discussing elliptic fibrations, we first review the classical argument to get a Weierstrass equation for a regular curve of arithmetic genus 1 with a choice of a rational point S. We follow Mumford and Suominen [16].

Let Y be a non-singular projective curve of genus one over k. Denote by S the divisor associated with a fixed base k-rational point O of Y. The Riemann–Roch theorem asserts that $\dim_k H^0(Y, \mathscr{O}_Y(nS)) = n$ for $n > 0$. Hence, the vector space $H^0(Y, \mathscr{O}_Y(2S))$ has dimension two. Since the only rational functions with at most a pole of degree one on an elliptic curve are the constants, there exists a rational function x with a double pole at O such that $\{1, x\}$ is a basis of $H^0(Y, \mathscr{O}_Y(2S))$. In the same way, since $\dim_k H^0(Y, \mathscr{O}_Y(3S)) = 3$, there is a rational function $y \in H^0(Y, \mathscr{O}_Y(3S))$ with a triple pole at O. Using the basis $\{1, x, y\}$ of $H^0(Y, \mathscr{O}_Y(3S))$, we can prove the following lemma:

Lemma 7.1 *The set* $\{1, x, x^2, \ldots, x^m, y, xy, yx^2, \ldots, yx^{m-2}\}$ *is a basis of* $H^0(Y, \mathscr{O}_Y(nS))$ *for* $n = 2m$. *We get a basis for* $H^0(Y, \mathscr{O}_Y(nS))$ *for* $n = 2m + 1$ *by adding the monomial* yx^{m-1}.

Proof By Riemann–Roch, $H^0(Y, \mathscr{O}_Y(nS))$ has dimension n for $n > 0$. The basis presented in the lemma contains n elements that are linearly independent since each function has a pole at the origin with a different order.

For a curve of genus g, any divisor of degree $2g + 1$ or bigger is a very ample divisor. For a curve of genus 1, any divisor of degree 3 is very ample. It follows that the divisor $3S$ provides a closed embedding of the elliptic curve into $\mathbb{P}^2$. All is left is to give the equation of that curve. Since $H^0(Y, \mathscr{O}_Y(3S))$ is generated by $\{1, x, y\}$, there is a unique embedding $Y \to \mathbb{P}^2$ such that $\mathscr{O}_Y(3S)$ is the pullback of the tautological line bundle $\mathscr{O}_{\mathbb{P}^2}(1)$ and $(x, y, 1)$ are the affine coordinates. The punch line of the proof of the existence of an isomorphic cubic curve in Weierstrass form for an elliptic curve is the following. Since $y^2 \in H^0(Y, \mathscr{O}_Y(6S))$, there are constants $a_0, a_1, a_3, a_4, a_6 \in k$ such that

$$y^2 + a_1xy + a_3y = a_0x^3 + a_2x^2 + a_4x + a_6. \tag{7.27}$$

Finally, we have to show that a_0 cannot be zero. If $a_0 = 0$, $\{y^2, xy, y, x^2, x, 1\}$ would be linearly dependent. But this is not possible since there is no terms to cancel out the pole (of order 6) of y^2. We can then redefine $(x, y, a_1, a_2, a_3, a_4, a_6) \to (a_0x, a_0^2y, a_0a_1, a_0^2a_3, a_2, a_0^3a_4, a_0^4a_6)$ and eliminate the overall factor of a_0^4 to get the Weierstrass equation in Tate form:

$$y^2 + a_1xy + a_3y = x^3 + a_2x^2 + a_4x + a_6. \tag{7.28}$$

7.3.2 Preparation for the Relative Case

Given an elliptic fibration $\varphi : Y \longrightarrow B$ with a section $\sigma : B \to Y$, we construct a Weierstrass model birational to Y. We assume the following conditions:

1. φ is a flat projective morphism between (quasi)-projective varieties.
2. Y is normal, and the base B is smooth.
3. The section σ is a morphism.
4. All fibers are irreducible projective curves.

As varieties, Y and B are in particular Noetherian schemes. Hence, the projectivity of φ implies that φ is also a proper morphism by [10, Chap. II, Theorem 4.9]. Since we work over an algebraically closed field, φ is flat if and only if φ is equidimensional (every fiber has the same dimension). Hence, the assumption (4) implies that φ is also a flat morphism.

Since $\varphi : Y \to B$ is a proper morphism, φ is in particular separated and the section σ defines a closed immersion of B in Y (an isomorphism from B onto a closed subscheme of Y) by [5, Corollary 5.4.6]. Let $\mathscr{I}$ denote the ideal sheaf of that subscheme; its support is a Cartier divisor S of Y. We denote by $\mathscr{N}_{S/Y}$ the normal sheaf of S in Y.

Using the Riemann–Roch Theorem, we can write a Weierstrass equation for each smooth fiber Y_p as in the previous section by studying the cohomology of the fiber. The challenge is now to understand how the cohomology along the fiber varies as a function of the fiber. This is a question of cohomology and base change, an important topic in algebraic geometry covered, for example, in Chap. 3 of [10].

In algebraic geometry, a family of schemes is simply a morphism $f : X \to Y$ and the members of the family are the fibers $X_y = X \times_Y \operatorname{Spec} k(y)$, where $k(y)$ is the residue field at the point $y \in Y$. To study the cohomology of family of schemes, the higher direct image functors $R^i f_*$ are introduced. They describe the "relative cohomology of X over Y".

Definition 7.6 Let X be any topological space, we denote by $\mathscr{U}(X)$ the category of sheaves of Abelian groups on X. Given a continuous function $f : X \to Y$ between topological spaces, for any integer $i \geq 0$, we define $R^i f_* : \mathcal{U}(X) \to \mathscr{U}(Y)$ as the right derived functors of the direct image function f_*.

The following theorem gives a local description of $R^i f_*(\mathscr{F})$:

Theorem 7.8 (Chapter III. Proposition 8.1 of [10]) *For each $i \geq 0$ and each $\mathscr{F} \in \mathscr{U}(X)$, $R^i f_*(\mathscr{F})$ is the sheaf associated to the presheaf $V \mapsto H^i(f^{-1}(V), \mathscr{F}|_{f^{-1}(V)})$ on Y.*

Given a fibration $f : X \to Y$, one would like to find some relation between the fiber cohomology groups $H^i(X_y, \mathscr{F}_y)$ and the globally defined sheaves $R^i f_*(\mathscr{F})$. The Proper Base Change Theorem is discussed by Mumford in Sect. 5 of Chap. 2 of "Abelian Varieties."

In the case of an elliptic fibration $\varphi : Y \to B$, the crucial step is the introduction of the fundamental line bundle $\mathscr{L}$ (over the base of the elliptic fibration) which provides a splitting of $\varphi_* \mathscr{O}_Y(nS)$ where S is a Cartier divisor defined by the section. When this is established, one can just follow the usual Riemann–Roch argument and define an embedding of the fibration in a projective bundle $\mathbb{P}^2 \to B$ using the fact that $\varphi_* \mathscr{O}_Y(3S)$ is very ample relatively to the base B. The projective bundle will be defined by the projectivation of $\varphi_* \mathscr{O}_Y(3S)$.

Lemma 7.2 (See [16] Chap. 3 Lemma 2)

1. *The ideal sheaf* $\mathscr{I}$ *corresponding to the subscheme* S *defined by the section is an invertible sheaf.*
2. *The sheaf of functions* $\mathscr{O}_Y(nS)$ *with* n*-fold poles along* S *is isomorphic to* $\mathscr{I}^{\otimes(-n)}$ *for any integer* $n > 0$.

Working fiber by fiber, we get the following lemma summarizing the (cohomological) properties of the pushforward of $\mathscr{O}_Y(nS)$:

Lemma 7.3 *For an elliptic fibration* $\varphi : Y \to B$ *with a section* $\sigma : B \to Y$ *defining a closed subscheme* S *of* Y, $R^1\varphi_*(nS)$ *and* $\varphi_* \mathscr{O}_Y(nS)$ *are both locally free for all* n *and we have:*

1. $\varphi_* \mathscr{O}_Y = \mathscr{O}_B$
2. $\varphi_* \mathscr{O}_Y(nS)$ *is locally free of rank* n *for all* $n > 0$.
3. $R^1\varphi_* \mathscr{O}_Y(nS) = 0, \quad$ *for all* $n > 0$, *and locally free for of rank one for* $n = 0$.
4. $R^1\varphi_* \mathscr{O}_Y \cong \varphi_* \mathscr{N}_{S/Y}$ *is an invertible sheaf.*
5. $R^i\varphi_* \mathscr{O}_Y(nS) = 0, \quad$ *for all* $i > 1$, *and all integers* n.

Proof See Mumford–Suominen [16, Chap. 3], Deligne [4], or Miranda [15, Lecture II Sect. 3].

The line bundle $R^1\varphi_* \mathscr{O}_Y$ is a fundamental invariant of the elliptic fibration $\varphi : Y \longrightarrow B$. This motivates the following definition [15].

Definition 7.7 (*Fundamental line bundle of an elliptic fibration*) The fundamental line bundle of an elliptic fibration $\varphi : Y \to B$ is the invertible sheaf $\mathscr{L}$ defined as:

$$\mathscr{L} := \left(R^1\varphi_* \mathscr{O}_Y\right)^{-1}. \tag{7.29}$$

Remark 7.4 The fundamental line bundle $\mathscr{L}$ is often defined as $(\varphi_* \mathscr{N}_{S/Y})^{-1}$. By the previous Lemma, the two definitions agree since $\varphi_* \mathscr{N}_{S/Y} \simeq R^1\varphi_* \mathscr{O}_Y$. It also follows from the Lemma that the sheaf $\mathscr{N}_{S/Y}$ does not depend on the section S.

For an elliptic fibration with a section, the fundamental line bundle provides a splitting of $\varphi_* \mathscr{O}_Y(nS)$ for $n > 1$:

Theorem 7.9 *For* $n > 1$, *we have*

$$\varphi_* \mathscr{O}_Y(nS) \cong \mathscr{O}_B \oplus \mathscr{L}^{-2} \oplus \mathscr{L}^{-3} \oplus \cdots \oplus \mathscr{L}^{-n}. \tag{7.30}$$

Proof See Lemma II.4.3 of [15].

Equipped with Theorem 7.9, we can now apply the familiar Riemann–Roch argument to derive the Weierstrass equation in the relative case.

7.3.3 Weierstrass Models for Elliptic Fibrations

Lemma 7.4 (Deligne [4]) *Given an invertible section μ of $\mathscr{L}$, there exists locally for Zariski topology a basis $\{1, x, y\}$ of $\varphi_*\mathscr{O}_Y(3S)$ such that:*

1. *1 is a generator of $\mathscr{O}_B$.*
2. *$\{1, x\}$ is a basis of $\varphi_*\mathscr{O}_Y(2S)$, and the image of x along $\mathscr{L}^{-2}$ is μ^{-2}.*
3. *y belongs to $\varphi_*\mathscr{O}_Y(3S)$ and the image of y along $\mathscr{L}^{-3}$ is μ^{-3}.*
4. *$\{1, x, x^2, \dots, x^n, y, yx, \dots, yx^{n-2}\}$ is a basis for $\varphi_*\mathscr{O}_Y(mS)$ if $m = 2n$ with $n > 1$.*
5. *$\{1, x, x^2, \dots, x^n, y, yx, \dots, yx^{n-2}, yx^{n-1}\}$ is a basis for $\varphi_*\mathscr{O}_Y(mS)$ if $m = 2n + 1$ with $n > 1$.*

Given a different choice μ' of an invertible section of $\mathscr{L}$, there exists well-defined u, r, s, t such that the new basis $(1, x', y')$ is related to the previous one as follows:

$$\begin{cases} x' = u^2 x + r \\ y' = u^3 y + s u^2 x + t \\ \mu' = u\mu \end{cases} \tag{7.31}$$

These transformations (7.31) will be called *admissible transformations of a Weiestrass model*. For $\varphi_*\mathscr{O}_Y(6S)$, we have the basis $\{1, x, x^2, x^3, y, yx\}$ composed of six generators, but the space of monomials generated by $\{1, x, y\}$ in $\varphi_*\mathscr{O}_Y(6S)$ is seven-dimensional. The missing monomial is y^2, and its image along $\mathscr{L}^{-\otimes 6}$ is μ^6, which matches the image of x^3. It follows that $y^2 - x^3$ can be uniquely written as a linear combination of generators of $\varphi_*\mathscr{O}_Y(5S)$. This gives the Weierstrass equation in Tate form:

$$y^2 + a_1 xy + a_3 = x^3 + a_2 x^2 + a_4 x + a_6. \tag{7.32}$$

For each index i, the coefficient a_i is a section of $\mathscr{L}^{\otimes i}$. The line bundle $\mathscr{O}_Y(3S)$ is very ample relatively to the base B. The basis $(1, x, y)$ can be seen as affine coordinates of a $\mathbb{P}^2$ in which each fiber is embedded. We have an immersion of the elliptic fibration Y into a $\mathbb{P}^2$ projective bundle over the base B:

$$Y \to \mathbb{P}(\mathscr{E}) := \mathrm{Proj}(\mathrm{Sym}\,\mathscr{E}^*), \quad \text{where } \mathscr{E} := \varphi_*\mathscr{O}_Y(3S), \tag{7.33}$$

and $(x, y, 1)$ are the affine coordinates as they generate $\varphi_*\mathscr{O}_Y(3S)$. When the characteristic is different from 2 and 3, the Weierstrass equation can be reduced to the shorter form: $y^2 = x^3 + fx + g$ where f and g are, respectively, sections of $\mathscr{L}^{\otimes 4}$ and $\mathscr{L}^{\otimes 6}$. We quickly review our conventions for projective bundles.

Remark 7.5 (*Conventions for projective bundles*)

- We use the classical convention for the projectivization $\pi : \mathbb{P}(\mathscr{E}) \to B$ of a locally free sheaf $\mathscr{E}$ over B: The fibers of $\mathbb{P}(\mathscr{E})$ are the lines of $\mathscr{E}$ passing through the origin and not the hyperplanes. In our conventions $\mathbb{P}(\mathscr{E}) := \text{Proj}(\text{Sym } \mathscr{E}^*)$. In other words, what we call $\mathbb{P}(\mathscr{E})$ corresponds to $\mathbb{P}(\mathscr{E}^*)$ in the convention of EGA II.4.1.1 or Hartshorne.
- We denote the tautological line bundle of the projective bundle $\mathbb{P}(\mathscr{E})$ by $\mathscr{O}_{\mathbb{P}(\mathscr{E})}(-1)$. Its dual is the canonical line bundle $\mathscr{O}_{\mathbb{P}(\mathscr{E})}(1)$. By an abuse of notation, we will write $\mathscr{O}(-1)$ and $\mathscr{O}(1)$, respectively, for $\mathscr{O}_{\mathbb{P}(\mathscr{E})}(-1)$ and $\mathscr{O}_{\mathbb{P}(\mathscr{E})}(1)$. We also write $\mathscr{O}(-n)$ (for $n > 0$) for the nth tensor product of $\mathscr{O}(-1)$. Its dual is $\mathscr{O}(n)$, the nth tensor product of $\mathscr{O}(1)$. In particular, in our notation $\pi_*\big(\mathscr{O}_{\mathbb{P}(\mathscr{E})}(1)\big) = \mathscr{E}^*$.
- Given a locally free sheaf $\mathscr{E} = \mathscr{O}_B \oplus \mathscr{L}^{\otimes a} \oplus \mathscr{L}^{\otimes b}$, there are natural embeddings $\mathscr{O}_B \hookrightarrow \mathscr{E}$, $\mathscr{L}^{\otimes a} \hookrightarrow \mathscr{E}$, and $\mathscr{L}^{\otimes b} \hookrightarrow \mathscr{E}$. We use these embeddings to define projective coordinates $[z : x : y]$ for $\mathbb{P}(\mathscr{E})$:

$$\begin{cases} z \text{ is a section of } \mathscr{O}(1) \\ x \text{ is a section of } \mathscr{O}(1) \otimes \pi^*\mathscr{L}^{\otimes a} \\ y \text{ is a section of } \mathscr{O}(1) \otimes \pi^*\mathscr{L}^{\otimes b} \end{cases}$$

We can now introduce the definition of a Weierstrass model.

Definition 7.8 (*Weierstrass models*) Given a base B endowed with a line bundle $\mathscr{L}$, the Weierstrass model $\mathscr{W}_B(\mathscr{L}|f, g)$ defines an elliptic fibration $Y \to B$ where Y is the zero-scheme of a section of $\mathscr{O}(3) \otimes \pi^*\mathscr{L}^{\otimes 6}$ in $\mathbb{P}[\mathscr{O}_B \oplus \mathscr{L}^{\otimes 2} \oplus \mathscr{L}^{\otimes 3}]$ cuts by the Weierstrass normal equation:

$$y^2 z = x^3 + fxz^2 + gz^3. \tag{7.34}$$

In the previous equation, $[z : x : y]$ are projective coordinates of the projective bundle as explained earlier. The coefficient f is a section of $\mathscr{L}^{\otimes 4}$ and g a section of $\mathscr{L}^{\otimes 6}$. It is assumed that the discriminant $\Delta := -16(4f^3 + 27g^2)$ is not identically zero and defines a Cartier divisor in the base B.

Definition 7.9 (*Canonical section of a Weierstrass model*) A Weierstrass model admits a section given by $x = z = 0$ which is always in the smooth locus of the elliptic fibration. It is called the *canonical section.*

Definition 7.10 (*Discriminant locus*) The discriminant locus of the Weierstrass model $\mathscr{W}_B(\mathscr{L}|f, g)$ is given by the zero-scheme of the following section of $\mathscr{L}^{\otimes 12}$:

$$\Delta = -16(4f^3 + 27g^2), \quad \Delta \in H^0(B, \mathscr{L}^{\otimes 12}). \tag{7.35}$$

Remark 7.6 The factor of (-16) is there to match the definition of the discriminant for a Weierstrass model in Tate form as it is given in the formulaire of Deligne and Tate. It also matches the definition $\Delta(\tau) = g_2^3 - 27g_3^2$ of the cusp form associated to the Weierstrass equation $(\wp')^2 = 4\wp^3 - g_2\wp - g_3$.

Theorem 7.10 (Equivalence of Weierstrass models) *Two Weierstrass models $\mathscr{W}_B(\mathscr{L}_1|f_1, g_1)$ and $\mathscr{W}_B(\mathscr{L}_2|f_2, g_2)$ over the same base B are equivalent if and only if there is a nowhere vanishing $u \in H^0(B, \mathscr{L}_2 \otimes \mathscr{L}_1^{-1})$ such that $f_2 = u^4 f_1$ and $g_2 = u^6 g_1$.*

We have proven the following.

Theorem 7.11 ([4, 15, 16]) *Let $\varphi : Y \to B$ be a smooth elliptic fibration admitting a section $\sigma : B \to Y$. Then, there exists a triplet $(\mathscr{L}, f, g)$ and an isomorphism $\mu : Y \to \mathscr{W}_B(\mathscr{L}|f, g)$ over B such that $\mu \circ \sigma$ is the canonical section and $\mathscr{L}^{-1} \simeq R^1\varphi_*\mathscr{O}_Y$. Moreover, the discriminant Δ is invertible over the locus of regular fibers.*

One can have a similar result in the presence of mild singularities.

Theorem 7.12 (Nakayama) *An elliptic fibration $\varphi : X \to B$ with a section $\sigma : B \to X$ is birationally equivalent to a Weierstrass model $\mathscr{W}_B(\mathscr{L}|f, g)$ with canonical singularities and such that $\mathscr{L}$ is the fundamental line bundle associated to the elliptic fibration.*

As a direct consequence of the adjunction formula, we have the following theorem.

Theorem 7.13 *The canonical bundle of a smooth Weierstrass model $\mathscr{W}_B(\mathscr{L}|f, g)$ for a smooth elliptic fibration $\varphi : Y \to B$ is*

$$\omega_Y \cong \varphi^*(\omega_B \otimes \mathscr{L}). \tag{7.36}$$

Lemma 7.5 (Elliptically fibered Calabi–Yau) *A Weierstrass model $Y = \mathscr{W}_B(\mathscr{L}|f, g)$ has a trivial canonical divisor if and only the dual of its fundamental line bundle is the canonical line bundle of the base. That is*

$$K_Y = 0 \iff \mathscr{L}^{-1} = \omega_B.$$

7.3.4 The *j*-Invariant

Given a Weierstrass model $\mathscr{W}_B(\mathscr{L}|f, g)$, for any nonvanishing section μ of $\mathscr{O}_B$, we can rescale $(f, g) \mapsto (u^4 f, u^6 g)$ and get an equivalent Weierstrass model with the same fundamental line bundle $\mathscr{L}$. It follows that there is a unique invariant f^3/g^2 that we can write. However, it is more convenient to use $f^3/(4f^3 + 27g^2)$ since $4f^3 + 27g^2$ is nonvanishing over regular fibers.

Definition 7.11 (*j-invariant*) To a Weierstrass model $\mathscr{W}_B(\mathscr{L}|f,g)$, we associate the j-invariant:

$$j(f,g) := 1728\frac{4f^3}{4f^3+27g^2} \in H^0(B,\mathscr{O}_B). \tag{7.37}$$

Remark 7.7 Kodaira uses the normalization $j = 4f^3/(4f^3+27g^2)$. The one we use here with the extra factor of $1728 = 12^3$ is the normalization used by number theorists. It matches the conventions of Deligne and Tate.

Remark 7.8 (*The j-map is not injective*) Let K be a field containing a nonzero element λ which has no square root in K. Two elliptic curves with the same j-invariant are isomorphic in a quadratic or cubic extension of the field. The elliptic curves $E_1 : y^2 = x^3 + fx + g$ and $E_2 : y^2 = x^3 + \lambda^2 fx + \lambda^3 g$ are not isomorphic over K even though they have the same j-invariant. They become isomorphic over any field extension K' of K containing a square root of λ.

7.3.5 Deligne's Formulaire

In this section, we would like to collect important definitions and formulas for an elliptic curve in Weierstrass form:

$$y^2z + a_1xyz + a_3yz^2 = x^3 + a_2x^2z + a_4xz^2 + a_6z^3. \tag{7.38}$$

We follow Tate's notation [4, 29]. Geometrically, the marked point of the Weierstrass form of an elliptic curve is its intersection point with the line at infinity $z=0$, namely the point $[x:y:z]=[0:1:0]$, which is a point of inflection and the only point at infinity of the curve. The curve is called a Weierstrass normal form since (in characteristic different from 2 and 3) after the change of variables:

$$\wp = x + \frac{1}{12}(a_1^2+4a_2), \quad \wp' = 2y + a_1x + a_3, \tag{7.39}$$

it reduces to the traditional cubic equation satisfied by the Weierstrass $\wp$-function and its derivative:

$$E : (\wp')^2 = 4\wp^3 - g_2\wp - g_3. \tag{7.40}$$

The *Néron differential* associated to the elliptic curve is the the differential invariant under translations in the group law and defined as follows:

$$\omega = \frac{dx}{2y+a_1x+a_3} = \frac{dy}{3x^2+2a_2x+a_4-a_1y} = \Big(= \frac{d\wp(z)}{\wp'(z)} = dz\Big). \tag{7.41}$$

A curve given by a Weierstrass equation is singular if and only if its discriminant Δ is zero. If we denote by $\bar{k}$ the algebraic closure of k, two smooth elliptic curves are isomorphic over $\bar{k}$ if and only if they have the same j-invariant. We recall the formulaire of Deligne and Tate which is useful to express the discriminant Δ, the j-invariant and to reduce the Weierstrass equation into simpler forms:

$$b_2 = a_1^2 + 4a_2, \quad b_4 = a_1 a_3 + 2a_4, \quad b_6 = a_3^2 + 4a_6, \tag{7.42}$$

$$b_8 = b_2 a_6 - a_1 a_3 a_4 + a_2 a_3^2 - a_4^2, \tag{7.43}$$

$$c_4 = b_2^2 - 24 b_4, \quad c_6 = -b_2^3 + 36 b_2 b_4 - 216 b_6, \tag{7.44}$$

$$\Delta = -b_2^2 b_8 - 8 b_4^3 - 27 b_6^2 + 9 b_2 b_4 b_6, \tag{7.45}$$

$$j = \frac{c_4^3}{\Delta} \tag{7.46}$$

These quantities are related by the following relations:

$$4b_8 = b_2 b_6 - b_4^2 \quad \text{and} \quad 1728\Delta = c_4^3 - c_6^2. \tag{7.47}$$

The variables b_2, b_4, b_6 are used to express the Weierstrass equation after completing the square in y by a redefinition

$$y \mapsto y - \frac{1}{2}(a_1 x + a_3 z), \tag{7.48}$$

which gives

$$zy^2 = x^3 + \frac{1}{4} b_2 x^2 z + \frac{1}{2} b_4 x z^2 + \frac{1}{4} b_6 z^3. \tag{7.49}$$

The variables c_2, c_4, and c_6 are then obtained after eliminating the term in x^2 by the redefinition

$$x \mapsto x - \frac{1}{12} b_2 z, \tag{7.50}$$

which finally gives the short form of the Weierstrass equation:

$$zy^2 = x^3 - \tfrac{1}{48} c_4 x z^2 - \tfrac{1}{864} c_6 z^3. \tag{7.51}$$

We will use the following normalization of the short Weierstrass equation (obtained by introducing $f = -\frac{1}{48} c_4$ and $g = -\frac{1}{864} c_6$):

$$E : zy^2 = x^3 + f x z^2 + g z^3, \quad \Delta = -16(4f^3 + 27g^2), \quad j = 1728 \frac{4f^3}{4f^3 + 27g^2}. \tag{7.52}$$

A Weierstrass equation is unique up to the following admissible coordinate transformation - (with $r, s, t, u \in k$ and $u \neq 0$):

$$x = u^2 x' + r, \quad y = u^3 y' + su^2 x' + t, \tag{7.53}$$

under which we have

$$\begin{aligned}
ua_1' &= a_1 + 2s,\\
u^2 a_2' &= a_2 - sa_1 + 3r - s^2,\\
u^3 a_3' &= a_3 + ra_1 + 2t,\\
u^4 a_4' &= a_4 - sa_3 + 2ra_2 - (t + rs)a_1 + 3r^2 - 2st,\\
u^6 a_6' &= a_6 + ra_4 + r^2 a_2 + r^3 - ta_3 - rta_1 - t^2,
\end{aligned} \tag{7.54a}$$

$$\begin{aligned}
u^2 b_2' &= b_2 + 12r,\\
u^4 b_4' &= b_4 + rb_2 + 6r^2,\\
u^6 b_6' &= b_6 + 2rb_4 + r^2 b_2 + 4r^3,\\
u^8 b_8' &= b_8 + 3rb_6 + 3r^2 b_4 + r^3 b_2 + 3r^4,
\end{aligned} \tag{7.54b}$$

$$u^4 c_4' = c_4, \quad u^6 c_6' = c_6, \tag{7.54c}$$

$$u^{12}\Delta' = \Delta, \quad u\omega' = \omega, \qquad j' = j. \tag{7.54d}$$

7.4 Kodaira–Néron Classification of Singular Fibers

For an elliptic fibration $\varphi : Y \to B$, a smooth fiber is isomorphic to a torus $\mathbb{C}/(\mathbb{Z} + \tau\mathbb{Z})$ where τ lives in the upper-half plane $\mathscr{H}$. Two elliptic curves with period τ and τ' are isomorphic if and only if they are related by a modular transformation:

$$\tau' = \frac{a\tau + b}{c\tau + d}, \quad \begin{pmatrix} a & b \\ c & d \end{pmatrix} \in \mathrm{SL}(2, \mathbb{Z}). \tag{7.55}$$

The elliptic fibration admits a discriminant locus over which the fibers are singular. Let B^* be the locus of points p of B such that the fiber Y_p over p is a smooth curve. By considering the ambiguity, we have a period mapping function $\tau : U \to \mathscr{H}$ from the universal covering space U of B^* into the upper-half plane $\mathscr{H}$ and a monodromy representation

$$\mu : \pi_1(B^*) \to \mathrm{SL}(2, \mathbb{Z}), \tag{7.56}$$

such that for $\gamma \in \pi_1(B^*)$ and $p \in U$

$$\tau(\gamma p) = \frac{a_\gamma \tau + b_\gamma}{c_\gamma \tau + d_\gamma}, \quad \mu(\gamma) = \begin{pmatrix} a_\gamma & b_\gamma \\ c_\gamma & d_\gamma \end{pmatrix} \in \mathrm{SL}(2, \mathbb{Z}). \tag{7.57}$$

Table 7.1 Quasi-unipotent matrices in SL(2, $\mathbb{Z}$)

I_a ($a \in \mathbb{Z}$)	II	III	IV
$\begin{pmatrix} 1 & a \\ 0 & 1 \end{pmatrix} = T^{\otimes a}$	$\begin{pmatrix} 1 & 1 \\ -1 & 0 \end{pmatrix} = -(ST)^2$	$\begin{pmatrix} 0 & 1 \\ -1 & 0 \end{pmatrix} = -S$	$\begin{pmatrix} 0 & 1 \\ -1 & -1 \end{pmatrix} = -ST$
I_b^* ($b \in \mathbb{Z}$)	II*	III*	IV*
$\begin{pmatrix} -1 & -b \\ 0 & -1 \end{pmatrix} = -T^{-b}$	$\begin{pmatrix} 0 & -1 \\ 1 & 1 \end{pmatrix} = ST$	$\begin{pmatrix} 0 & -1 \\ 1 & 0 \end{pmatrix} = S$	$\begin{pmatrix} -1 & -1 \\ 1 & 0 \end{pmatrix} = (ST)^2$

7.4.1 Monodromy

For a proper map $\varphi : Y \to B$ between smooth projective varieties, the monodromy around a point of the discriminant locus with at most normal crossing singularity is a quasi-unipotent matrix by Borel's lemma [20]. We recall the definition of quasi-unipotent and give a classification for SL(2, $\mathbb{Z}$) following Kodaira.

Definition 7.12 A matrix M is said to be *quasi-unipotent* if all its eigenvalues are roots of the unit. That is, there are integers $n, k \geq 1$ such that $(M^k - \mathrm{Id})^n = 0$.

In the case of SL(2, $\mathbb{Z}$), quasi-unipotent matrices up to conjugation form eight different classes:

Lemma 7.6 (Kodaira [12]) *A quasi-unipotent matrix in SL(2, $\mathbb{Z}$) is conjugated exactly to one of the matrices in Table 7.1.*

These eight conjugation classes provide a classification of the type of singular fibers over a general point of a component of the discriminant locus assuming that the singularity at that point is at most a normal crossing singularity.

7.4.2 Fiber Type

The local ring of a subvariety S of X is denoted $\mathscr{O}_{X,S}$, its maximal ideal is $\mathscr{M}_{X,S}$, and the quotient field is the residue field $\kappa(V) = \mathscr{O}_{X,S}/\mathscr{M}_{X,S}$. The local ring $\mathscr{O}_{X,S}$ is the stalk of the structure sheaf of X at the generic point η_S of S, and $\kappa(S)$ is the function field of S. If S is a divisor, $\mathscr{O}_{X,S}$ is a one-dimensional local domain. In case X is non-singular along S, $\mathscr{O}_{X,S}$ is a discrete valuation ring and the order of vanishing is given by the usual valuation.

Definition 7.13 (*Fiber over a point*) Let $\varphi : Y \longrightarrow B$ be a morphism of schemes. For any $p \in B$, the fiber over p is denoted Y_p and defined using a fibral product[1] as

[1]Given three sets (A_1, A_2, and S) and two maps $\varphi_1 : A_1 \to B$ and $\varphi_2 : A_2 \to B$, we define the fibral product $A_1 \times_S A_2$ as the subset of $A_1 \times A_2$ composed of couples (a_1, a_2) such that $\varphi_1(a_1) = \varphi_2(a_2)$.

Table 7.2 Allowed collisions of a Miranda model

$j=\infty$	$j=0$	$j=1728$
$I_{M_1}+I_{M_2}$	$\mathrm{II}+\mathrm{IV}$	$\mathrm{III}+\mathrm{I}_0^*$
$I_{M_1}+I^*_{M_2}$	$\mathrm{II}+I_0^*$	
	$\mathrm{II}+\mathrm{IV}^*$	
	$\mathrm{IV}+I_0^*$	

$$Y_p = Y \times_B \operatorname{Spec} \kappa(\mathrm{p}).$$

The first projection $Y_p \longrightarrow Y$ induces an homeomorphism from Y_p onto $f^{-1}(p)$ [13, Sect. 3.1 Proposition 1.16]. The second projection gives Y_p the structure of a scheme over the residue field $\kappa(p)$.

If p is not a closed point,[2] the residue field $\kappa(p)$ is not necessarily algebraically closed. Certain components of Y_p could be $\kappa(p)$-irreducible (i.e., irreducible when defined over $\kappa(p)$), while they become reducible after an appropriate field extension. An irreducible scheme over a field k is said to be *geometrically irreducible* when it stays irreducible after any field extension. The most refined description of the fiber Y_p is always the one corresponding to the algebraic closure $\overline{\kappa(p)}$ of $\kappa(p)$. This motivates the following definition (Table 7.2).

Definition 7.14 The geometric fiber over p is the fiber $Y_p \times_{\kappa(p)} \overline{\kappa(p)}$, the fiber Y_p after the base change induced by the field extension $\kappa(p) \to \overline{\kappa(p)}$ to the algebraic closure of $\kappa(p)$.

By construction, a geometric fiber is always composed of geometrically irreducible components.

Definition 7.15 We say that the type of a fiber Y_p is *geometric* if it does not change after a field extension.

For an elliptic n-fold, the Kodaira fibers are also the *geometric generic fibers* of the irreducible components of the reduced discriminant locus.

Definition 7.16 (*Algebraic cycle*) An algebraic cycle of a Noetherian scheme X is a finite formal sum $\sum_i n_i V_i$ of subvarieties V_i with integer coefficients n_i. If all the subvarieties V_i have the same dimension d, the cycle is called a d-cycle. The free group generated by subvarieties of dimension d is denoted $Z_d(X)$. The group of all cycles, denoted $Z(X) = \bigoplus_d Z_d(X)$, is the free group generated by subvarieties of X.

Definition 7.17 (*Degree of a zero-cycle* [9, Chap. 1, Definition 1.4, p. 13]) Let X be a complete scheme. The *degree* of a zero-cycle $\sum n_i p_i$ of X is $\deg(\sum_i n_i p_i) = \sum_i n_i [\kappa(p_i) : k]$, where $[\kappa(p_i) : k]$ is the degree of the field extension $\kappa(p_i) \to k$.

Let Θ be an algebraic one-cycle with irreducible decomposition $\Theta = \sum_i m_i \Theta_i$. We denote by $\Theta_i \cdot \Theta_j$ the zero-cycle defined by the intersection of Θ_i and Θ_j for

[2] For example, if p is the generic point of a subvariety of B.

$i \neq j$. A *n-point* of an algebraic one-cycle Θ is a point in $\bigcup_i \Theta_i$, which belongs to exactly n distinct irreducible components Θ_i. An algebraic one-cycle Θ is said to be a *tree* if it does not have n-points for $n > 2$. Two curves intersect transversally if their intersection consists of isolated reduced closed points.

Following Kodaira [12], we introduce the following definition:

Definition 7.18 (*Fiber type*) By the *type* of an algebraic one-cycle $\Theta \in Z_1(X)$ with irreducible decomposition $\Theta = \sum_i m_i\Theta_i$, we mean the isomorphism class of each irreducible curve Θ_i, together with the topological structure of the reduced polyhedron $\sum \Theta_i$ (that is, the collection of zero-cycles $\Theta_i \cdot \Theta_j$ $(i \neq j)$), and the homology class of $\Theta = \sum_i m_i\Theta_i$ in the Chow group $A_1(X)$.

Example 7.1 For instance, $\Theta_1 \cdot \Theta_2 = 2p_1 + 3p_2$ indicates that the two curves Θ_1 and Θ_2 meet at two points p_1 and p_2 with respective intersection multiplicity 2 and 3.

Definition 7.19 (*Dual graph*) To an algebraic one-cycle Θ with irreducible decomposition $\Theta = \sum_i m_i\Theta_i$, we associate a weighted graph (called the *dual graph* of Θ) such that:

- The vertices are the irreducible components of the fiber.
- The weight of a vertex corresponding to the irreducible component Θ_i is its multiplicity m_i. When the multiplicity is one, it can be omitted.
- The vertices corresponding to the irreducible components Θ_i and Θ_j $(i \neq j)$ are connected by $\hat{\Theta}_{i,j} = \deg(\Theta_i \cdot \Theta_j)$ edges.

Definition 7.20 (*Kodaira symbols, See* [12]) Kodaira has introduced the following symbols characterizing the type of one-cycles appearing in the study of minimal elliptic surfaces. See Table 7.3 for a visualization of these fibers.

1. Type I_0: a smooth curve of genus 1.
2. Type I_1: an irreducible nodal rational curve.
3. Type II: an irreducible cuspidal rational curve.
4. Type I_2: $\Theta = \Theta_1 + \Theta_2$ and $\Theta_1 \cdot \Theta_2 = p_1 + p_2$: two smooth rational curves intersecting transversally at two distinct points p_1 and p_2. The dual graph of I_2 is $\tilde{A}_1$.
5. Type III: $\Theta = \Theta_1 + \Theta_2$ and $\Theta_1 \cdot \Theta_2 = 2p$: two smooth rational curves intersecting at a double point. Its dual graph is $\tilde{A}_1$.
6. Type IV: $\Theta = \Theta_1 + \Theta_2 + \Theta_3$ and $\Theta_1 \cdot \Theta_2 = \Theta_1 \cdot \Theta_3 = \Theta_2 \cdot \Theta_3 = p$: a 3-star composed of smooth rational curves. Its dual graph is $\tilde{A}_2$.
7. Type I_n $(n \geq 3)$: $\Theta = \Theta_0 + \cdots \Theta_n$ with $\Theta_i \cdot \Theta_{i+1} = p_i$ $i = 0, \ldots, n-1$ and $\Theta_n \cdot \Theta_0 = p_n$. Its dual graph is the affine Dynkin diagram $\tilde{A}_{n-1}$.
8. Type I_n^* $(n \geq 0)$: $\Theta = \Theta_0 + \Theta_1 + 2\Theta_2 + \cdots + 2\Theta_{n+2} + \Theta_{n+3} + \Theta_{n+4}$, with $\Theta_i \cdot \Theta_{i+1} = p_i$ $(i = 1, \ldots, n+2)$, $\Theta_0 \cdot \Theta_2 = p_0$, $\Theta_{n+4} \cdot \Theta_{n+2} = p_{n+4}$. The dual graph is the affine Dynkin diagram $\tilde{D}_{4+n}$.
9. Type IV^*: $\Theta = \Theta_0 + \Theta_1 + 2\Theta_2 + 2\Theta_3 + 3\Theta_4 + 2\Theta_5 + \Theta_6$ with $\Theta_i \cdot \Theta_{i+1} = p_i$ $(i = 3, \ldots, 6)$, $\Theta_1 \cdot \Theta_3 = p_1$, $\Theta_0 \cdot \Theta_2 = p_0$, $\Theta_2 \cdot \Theta_4 = p_2$. The dual graph is the affine Dynkin diagram $\tilde{E}_6$.

Table 7.3 Kodaira–Néron classification of geometric fibers over codimension-one points of the base of an elliptic fibration [12, 19]. The type of the fiber is given by its Kodaira symbol. In the second, third, and fourth column, $\nu(A)$ is the valuation of A. The j-invariant of the I_0^* is never ∞ and can take any finite value

Type	$\nu(c_4)$	$\nu(c_6)$	$\nu(\Delta)$	j	Monodromy	Fiber	Dual Graph
I_0	≥ 0	≥ 0	0	$\mathbb{C}$	I_2	Smooth Elliptic Curve	-
I_1	0	0	1	∞	$\begin{pmatrix}1&1\\0&1\end{pmatrix}$	 (curve of arithmetic genus 1 with a nodal singularity)	$\widetilde{A}_0$
II	≥ 1	1	2	0	$\begin{pmatrix}1&1\\-1&0\end{pmatrix}$	 (curve of arithmetic genus 1 with a cuspidal singularity)	$\widetilde{A}_0$
III	1	≥ 2	3	1728	$\begin{pmatrix}0&1\\-1&0\end{pmatrix}$	 Two rational curves intersecting at a double point	$\widetilde{A}_1$
IV	≥ 2	2	4	0	$\begin{pmatrix}0&1\\-1&-1\end{pmatrix}$	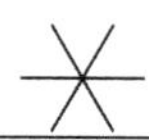	$\widetilde{A}_2$
I_n	0	0	$n>1$	∞	$\begin{pmatrix}1&n\\0&1\end{pmatrix}$	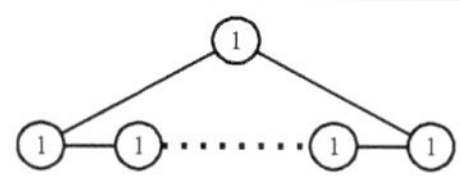 n nodes	$\widetilde{A}_{n-1}$
I_n^*	2	≥ 3	$n+6$	∞	$\begin{pmatrix}-1&-n\\0&-1\end{pmatrix}$	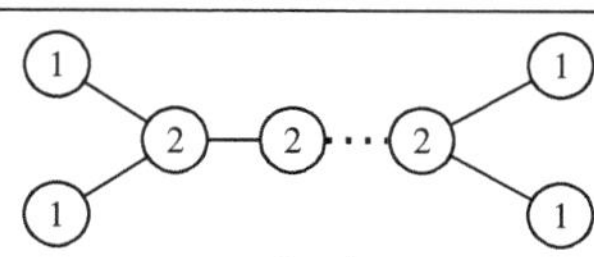 $n+5$ nodes	$\widetilde{D}_{n+4}$
	≥ 2	3	$n+6$				
IV^*	≥ 3	4	8	0	$\begin{pmatrix}-1&-1\\1&0\end{pmatrix}$	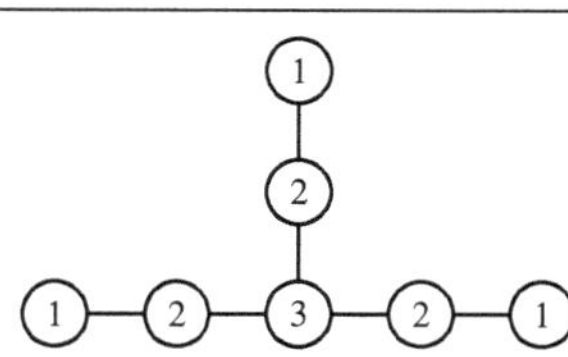	$\widetilde{E}_6$
III^*	3	≥ 5	9	1728	$\begin{pmatrix}0&-1\\1&0\end{pmatrix}$	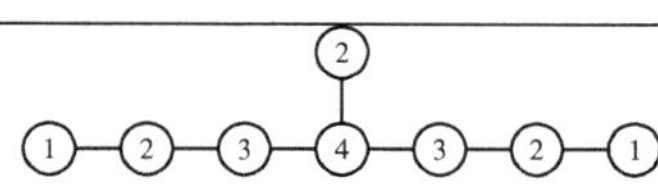	$\widetilde{E}_7$
II^*	≥ 4	5	10	0	$\begin{pmatrix}0&-1\\1&1\end{pmatrix}$	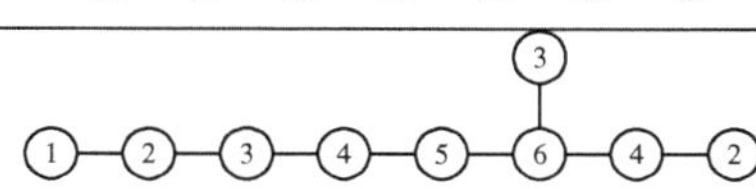	$\widetilde{E}_8$

10. Type III*: $\Theta = \Theta_0 + 2\Theta_1 + 2\Theta_2 + 3\Theta_3 + 4\Theta_4 + 3\Theta_5 + 2\Theta_6 + \Theta_7$ with $\Theta_i \cdot \Theta_{i+1} = p_i$ $(i = 3, \dots, 6)$, $\Theta_1 \cdot \Theta_3 = p_1$, $\Theta_0 \cdot \Theta_1 = p_0$, $\Theta_2 \cdot \Theta_4 = p_2$. The dual graph is the affine Dynkin diagram $\tilde{E}_7$.
11. Type II*: $\Theta = 2\Theta_1 + 3\Theta_2 + 4\Theta_3 + 6\Theta_4 + 5\Theta_5 + 4\Theta_6 + 3\Theta_7 + 2\Theta_8 + \Theta_0$, with $\Theta_i \cdot \Theta_{i+1} = p_i$ $(i = 3, \dots, 7)$, $\Theta_1 \cdot \Theta_3 = p_1$, $\Theta_8 \cdot \Theta_0 = p_8$, and $\Theta_2 \cdot \Theta_4 = p_2$. The dual graph is the affine Dynkin diagram $\tilde{E}_8$.

While the dual graph of a Kodaira fiber is an affine Dynkin diagram of type $\tilde{A}_k$, $\widetilde{D}_{4+k}$, $\widetilde{E}_6$, $\widetilde{E}_7$, or $\widetilde{E}_8$, the dual graph of the generic (arithmetic) fiber itself can also be a twisted Dynkin diagram of type $\widetilde{B}^t_{3+k}$, $\widetilde{C}^t_{2+k}$, $\widetilde{G}^t_2$, or $\widetilde{F}^t_4$. This is reviewed in Table 7.5. These dual graphs are not geometric in the sense that after an appropriate base change, they become $\widetilde{D}_{4+n}$, $\widetilde{A}_{2+2k}$ or $\widetilde{A}_{1+2k}$, and $\widetilde{E}_6$, respectively. The Kodaira fibers of the following type never need a field extension: I_1, II, III, III*, and II*.

The remaining Kodaira fibers (IV, $\mathrm{I}_{n>1}$, I^*_n, and IV*) can come from fibers Y_p whose types are not geometric and require at least a field extension of degree 2 to describe a fiber with a geometric type. When the fiber Y_p has a geometric type, the type of the fiber is said to be *split*. Otherwise, the type of Y_p is said to be non-split. When that is the case we mark the fiber with an "ns" superscript: $\mathrm{IV}^{\mathrm{ns}}$, $\mathrm{I}^{\mathrm{ns}}_n$, $\mathrm{I}^{*\mathrm{ns}}_n$, $(n \geq 2)$ and $\mathrm{IV}^{*\mathrm{ns}}$. When a field extension is not needed, the fibers are marked with an "s" superscript ("split"): IV^{s}, $\mathrm{I}^{\mathrm{s}}_n$, $\mathrm{I}^{*\mathrm{s}}_n$, $(n \geq 2)$ and $\mathrm{IV}^{*\mathrm{s}}$. The fiber of type I^*_0 can be split, semi-split, or non-split if the Kodaira types require no field extension, at field extension of degree 2, or a field extension of degree 3. The corresponding dual graphs are, respectively, $\widetilde{G}^t_2$, $\widetilde{B}^t_3$, and $\widetilde{D}_4$.

7.4.3 *Tate's Algorithm*

Let R be a complete discrete valuation ring with valuation v, uniformizing parameter s, and perfect residue field $\kappa = R/(s)$. We are interested in the case where κ has characteristic zero. We recall that a discrete valuation ring has only three ideals, the zero ideal, the ring itself, and the principal ideal sR. It follows that the scheme $\mathrm{Spec}(R)$ has only two points[3]: the generic point (defined by the zero ideal) and the closed point (defined by the principal ideal sR).

Let E/R be an elliptic curve over R with Weierstrass equation

$$y^2 + a_1xy + a_3y = x^3 + a_2x^2 + a_4x + a_6, \quad a_i \in R.$$

The generic fiber is a regular elliptic curve. After a resolution of singularities, we have a regular model $\mathscr{E}$ over R and the *special fiber* is the fiber over the closed point of Spec R.

Tate's algorithm determines the type of the geometric special fiber over the closed point of $\mathrm{Spec}(R)$ by manipulating the valuations of the coefficients and the discriminant and the arithmetic properties of some auxiliary polynomials. The type of the

[3]As usual we take the convention in which the ring itself is not a prime ideal.

geometric fiber is given by its Kodaira's symbol. The special fiber becomes geometric after a quadratic or a cubic field extension κ'/κ. Keeping track of the field extension used gives a classification of the special fiber as a κ-scheme as discussed, for example, in [13, Sect. 10.2]. The information on the required field extension needed to have geometrically irreducible components is already carefully encoded in Tate's original algorithm, as it is needed to compute the local index.

Tate's algorithm consists of the following eleven steps (see [28], [23, Sect. IV.9], [6]).

Step 1. $v(\Delta) = 0 \implies \mathrm{I}_0$.

Step 2. If $v(\Delta) \geq 1$, change coordinates so that $v(a_3) \geq 1$, $v(a_4) \geq 1$, and $v(a_6) \geq 1$.
If $v(b_2) = 0$, the type is $\mathrm{I}_{v(\Delta)}$. To have a fiber with geometric irreducible components, it is enough to work in the splitting field κ' of the following polynomial of $\kappa[T]$:

$$T^2 + a_1 T - a_2.$$

The discriminant of this quadric is b_2. If b_2 is a square in κ, then $\kappa' = \kappa$, otherwise $\kappa' \neq \kappa$:
(a) $\kappa' = \kappa \implies \mathrm{I}_n^{\mathrm{s}}$ (b) $\kappa' \neq \kappa \implies \mathrm{I}_n^{\mathrm{ns}}$

Step 3. $v(b_2) \geq 1$, $v(a_3) \geq 1$, $v(a_4) \geq 1$, and $v(a_6) = 1 \implies$ II.

Step 4. $v(b_2) \geq 1$, $v(a_3) \geq 1$, $v(a_4) = 1$, and $v(a_6) \geq 2 \implies$ III.

Step 5. $v(b_2) \geq 1$, $v(a_3) \geq 1$, $v(a_4) \geq 2$, $v(a_6) \geq 2$, and $v(b_6) = 2 \implies$ IV.
The fiber has geometric irreducible components over the splitting field κ' of the polynomial

$$T^2 + a_{3,1} T - a_{6,2}$$

Its discriminant is $b_{6,2}$. If $b_{6,2}$ is a square in κ, then $\kappa' = \kappa$ otherwise $\kappa' \neq \kappa$.
(a) $\kappa' = \kappa \implies \mathrm{IV}^{\mathrm{s}}$ (b) $\kappa' \neq \kappa \implies \mathrm{IV}^{\mathrm{ns}}$

Step 6. $v(b_2) \geq 1$, $v(a_3) \geq 1$, $v(a_4) \geq 2$, $v(a_6) \geq 3$, $v(b_6) \geq 3$, $v(b_8) \geq 3$. Then, make a change of coordinates such that $v(a_1) \geq 1$, $v(a_2) \geq 1$, $v(a_3) \geq 2$, $v(a_4) \geq 2$, and $v(a_6) \geq 3$. Let

$$P(T) = T^3 + a_{2,1} T^2 + a_{4,2} T + a_{6,3}$$

If $P(T)$ is a separable polynomial in κ, that is, if $P(T)$ has three distinct roots in a field extension of κ, then the type is I_0^*. The geometric fiber is defined over the splitting field κ' of $P(T)$ in κ. The type of the special fiber before to go to the splitting field depends on the degree of the field extension $\kappa' \to \kappa$:

- $[\kappa' : \kappa] = 3$ or $6 \implies \mathrm{I}_0^{*\mathrm{ns}}$ with dual graph $\tilde{G}_2^t$.
- $[\kappa' : \kappa] = 2 \implies \mathrm{I}_0^{*\mathrm{ss}}$ with dual graph $\tilde{B}_3^t$.
- $[\kappa' : \kappa] = 1 \implies \mathrm{I}_0^{*\mathrm{s}}$ with dual graph $\tilde{D}_4$.

where "ns", "ss", and "s" stand, respectively, for "non-split", "semi-split", and "split". In the notation of Liu, these fibers are, respectively, $\mathrm{I}_{0,3}^*$, $\mathrm{I}_{0,2}^*$, and I_0^*.

Step 7. If $P(T)$ has a double root, then the type is I_n^*.
Make a change of coordinates such that the double root is at the origin. Then $v(a_1) \geq 1$, $v(a_2) = 1$, $v(a_3) \geq 2$, $v(a_4) \geq 3$, $v(a_6) \geq 4$, and $v(\Delta) = n + 6$ $(n \geq 1)$.

Step 8. If $P(T)$ has a triple root, change coordinates such that the triple root is zero. Then, $v(a_1) \geq 1$, $v(a_2) \geq 2$, $v(a_3) \geq 2$, $v(a_4) \geq 3$, $v(a_6) \geq 4$.
Let

$$Q(T) = T^2 + a_{3,2}T - a_{6,4}$$

If Q has two distinct roots ($v(b_6) = 4$ or equivalently $v(\Delta) = 8$), the type is IV*. The split type depends on the rationality of the roots. If $b_{6,4}$ is a perfect square modulo s, the fiber is $\mathrm{IV}^{*\mathrm{s}}$, otherwise the fiber is $\mathrm{IV}^{*\mathrm{ns}}$.
The split form can be enforced with $v(a_6) \geq 5$ and hence $v(a_3) = 2$ to ensure that $v(b_6) = 4$.

Step 9. If Q has a double root, we change coordinates so that the double root is at the origin. Then:
$v(a_1) \geq 1$, $v(a_2) \geq 2$, $v(a_3) \geq 3$, $v(a_4) = 3$, $v(a_6) \geq 5 \implies$ type III*.

Step 10. $v(a_1) \geq 1$, $v(a_2) \geq 2$, $v(a_3) \geq 3$, $v(a_4) \geq 4$, $v(a_6) = 5 \implies$ type II*.

Step 11. Else $v(a_i) \geq i$ and the equation is not minimal. Divide all the a_i by s^i and start again with the new equation.

7.5 Miranda Models

The theory of elliptic surfaces has been treated by Kodaira. The geometry of the singular fibers is specially elegant. Singular fibers appear over isolated points on the base where their positions are given by the zeros scheme of the reduced discriminant. The complete list of singular fibers encompassed two infinite series (I_n and I_n^*) and six exceptional cases (II, III, IV, IV*, III*, II*). They can also be classified by their monodromies, and they can be attributed a well-defined value for the j-invariant. Namely $j = 0$ for the two infinite series (I_n, I_n^* with $n > 0$), arbitrary for I_0^*, $j = 0$ for II, IV, IV*, II* and $j = 1728$ for III and III*. If a Weierstrass model is given, the singular fibers can also be classified purely algebraically by Tate's algorithm.

If the base of the fibration is higher dimensional, we can still use Kodaira results and Tate's algorithm over codimension-one loci in the base. But there is a new challenge in determining the structure of singular fibers over higher codimensional loci in the base, for example, at the collisions of several components of the discriminant locus.

Miranda has given an explicit algorithm for finding a resolution of an elliptic threefold given by a (singular) Weierstrass models. Blow up the base until the reduced discriminant locus has simple normal crossings. Continue further so that only one of a small list of possible collisions between component of the discriminant locus occurs. Namely the following seven possibilities:

This list of collisions is obtained by requiring three conditions:

1. The reduced discriminant has simple normal crossing.
2. Only fibers with the same j-invariant are allowed to collide. This ensures that after the resolution, the j-invariant is a morphism.
3. Collisions that do not admit a small resolution are excluded. This ensures that the resolution gives a flat fibration.

The only places where one leaves the category of schemes in Miranda's resolution is in using a small resolution of an ordinary double point in resolving the collision $I_{M_1} + I_{M_2}$ when M_1 and M_2 are both odd. One has to contract a ruled surface to a $\mathbb{P}^1$ to ensure that the fibers are unidimensional. However, if one blows up such a collision point, one obtains over the exceptional curve a fiber of type $I_{M_1+M_2}$. Since $M_1 + M_2$ would be even, we can avoid collisions $I_{M_1} + I_{M_2}$ with M_1 and M_2 odd. Miranda's construction is purely local. But he also shows that it is well defined globally.

Following Dolgachev and Gross [7], we define a Miranda elliptic fibrations as follows:

Definition 7.21 (*Miranda elliptic fibrations*) A Miranda elliptic fibration is an elliptic fibration $\varphi : Y \to B$ such that (1) Y and B are regular and φ is flat and admits a section. (2) The discriminant locus has simple normal crossing. (3) All collisions are of the following seven types $I_{M_1} + I_{M_2}$, $I_{M_1} + I^*_{M_2}$, II+IV, II+I_0^*, II+IV*, IV+I_0^* or III+I_0^*.

7.5.1 Fibers at the Collisions of a Miranda Model

In Miranda models, in addition to the usual Kodaira fibers, there are new fibers that appear in higher codimensions. For an elliptic threefold, we have fibers in codimension-2 that could be one of the five exceptional types that are essentially sequences chains of 3, 4 or 5 rational curves with multiplicities (see Table 7.4). There is also the fibers I_n and I_n^* that can appear and a new infinite family called I_n^{*+} which admits as a dual graph the Dynkin diagram of D_{n+5} (we recall that a fiber I_n^* has a dual graph $\tilde{D}_{n+4}$. It consists of two rational curves of multiplicity one connected to a chain of $n+2$ rational curves of multiplicity 2. One can think of a I_n^{*+} fiber as a I_n^* fiber in the limit in which one of the two pairs of curves of multiplicity one is identified. Since there is a section, it is necessary the pair that does not intersect the section.

Table 7.4 Colliding singularities in an elliptic threefold as constructed by Miranda. The non-Kodaira fiber I_n^{*+} has the shape of a diagram of type D_{n+4}. The last column shows the fiber that would be obtained for an elliptic with base a smooth curve passing through the point of collision. The last column is what would be predicted by "applying" Tate algorithm in higher codimension

j-inv	Collision	Dual graph	if the base was a smooth curve through the collision point
∞	$I_{M_1}+I_{M_2}$	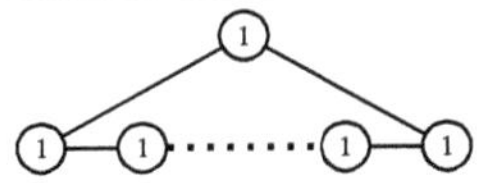 $\mathrm{I}_{M_1+M_2}$ nodes	same
∞	$I_{2n}+I_m^*$	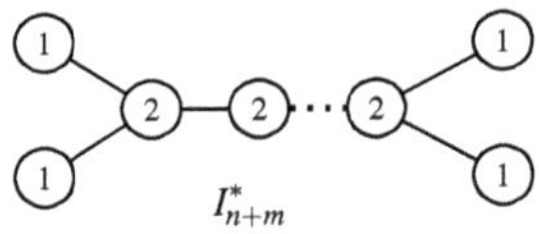 I_{n+m}^*	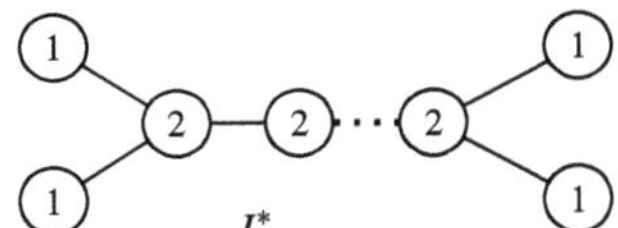 I_{2n+m}^*
∞	$I_{2n+1}+I_m^*$	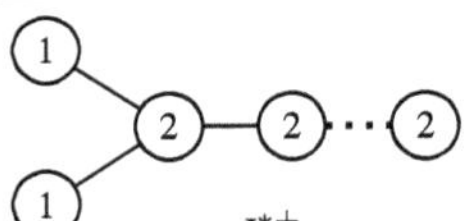 I_{n+m}^{*+} ($n+m+4$ components)	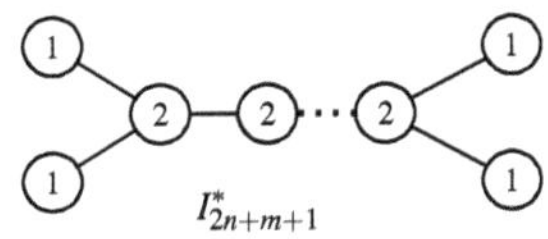 I_{2n+m+1}^*
0	$II+IV$	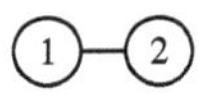	I_0^* 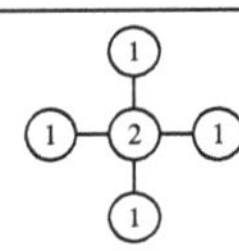
0	$II+I_0^*$	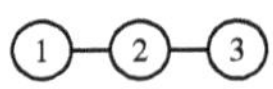	IV^* 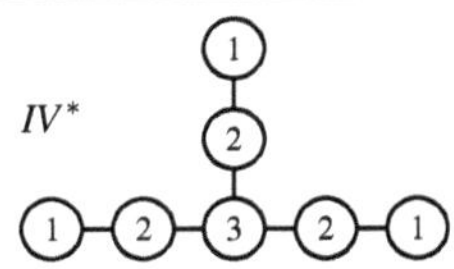
0	$II+IV^*$	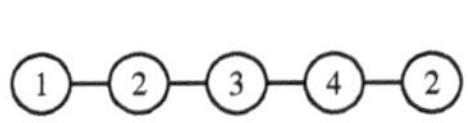	II^* 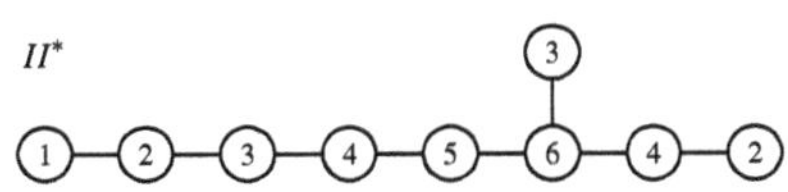
0	$IV+I_0^*$	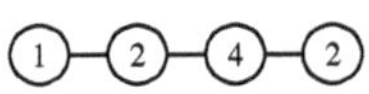	II^* 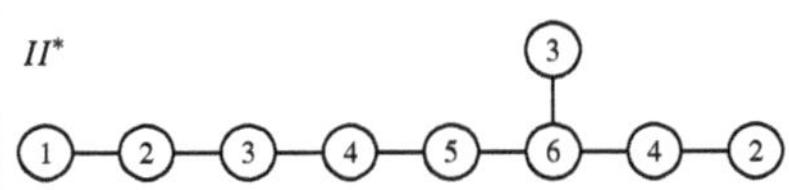
1728	$III+I_0^*$	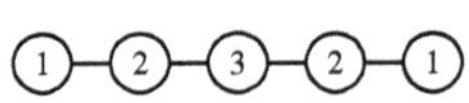	III^* 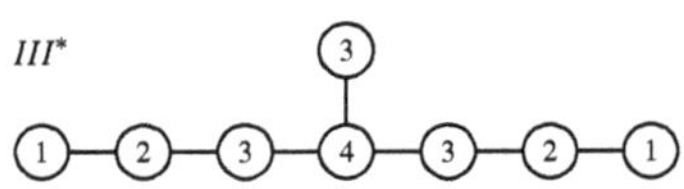

Table 7.5 Dual graphs for elliptic fibrations. The fiber type follows the notation of Liu [13, Sect. 10.2]. A fiber type is called T_d if the corresponding geometric fiber has Kodaira type T and a field extension of at least degree d is necessary to make all the components of the fiber geometrically irreducible. This indicates some nodes are not geometrically irreducible and split into d geometrically irreducible curves after a field extension of degree d

Fiber Type	Dual graph	Dual graph of the geometric fibre
IV_2 $\widetilde{A}_1$	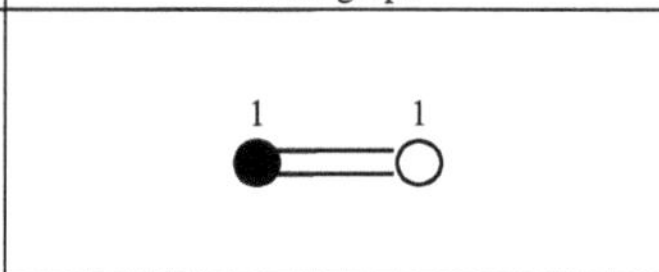	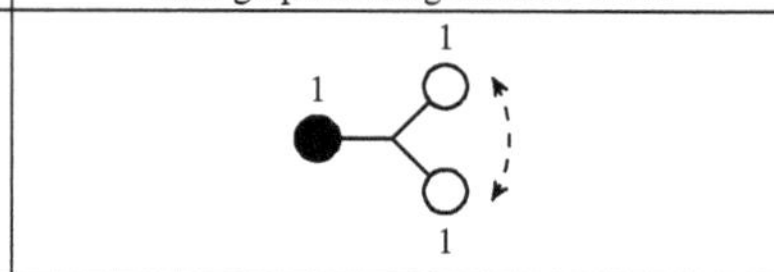
$I^*_{\ell-3,2}$ $\widetilde{B}^t_\ell$ $(\ell \geq 3)$	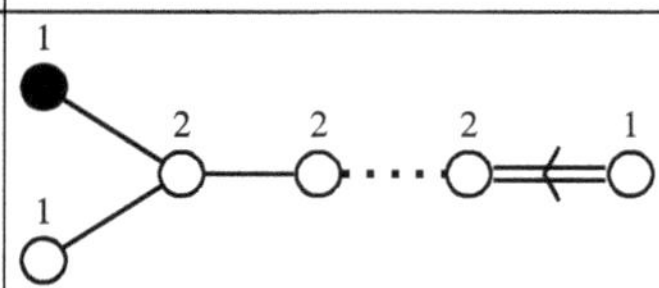	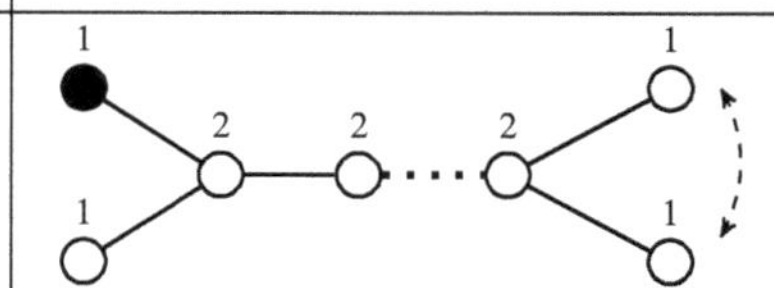
$I_{2\ell+2,1}$ $\widetilde{C}^t_{\ell+1}$ $(\ell \geq 1)$	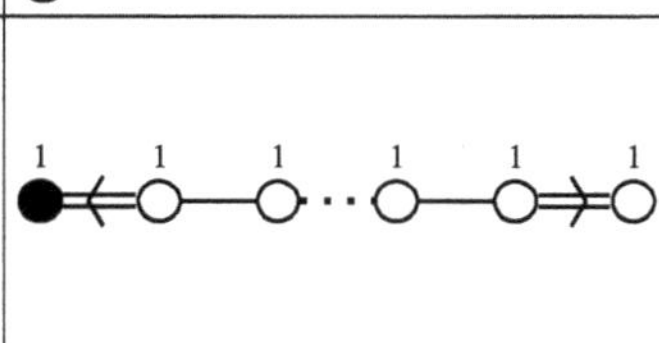	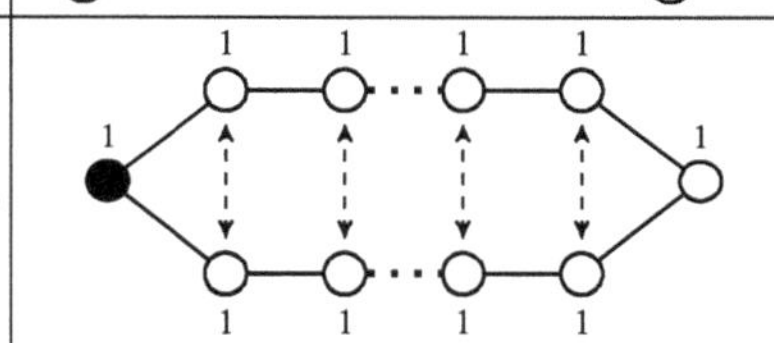
$I_{2\ell+3,2}$ $\widetilde{C}^t_{\ell+1}$ $(\ell \geq 1)$	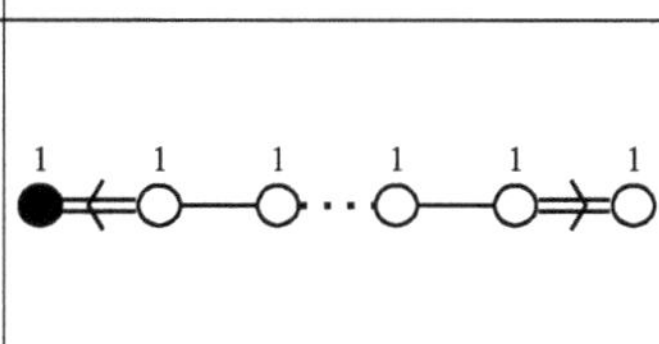	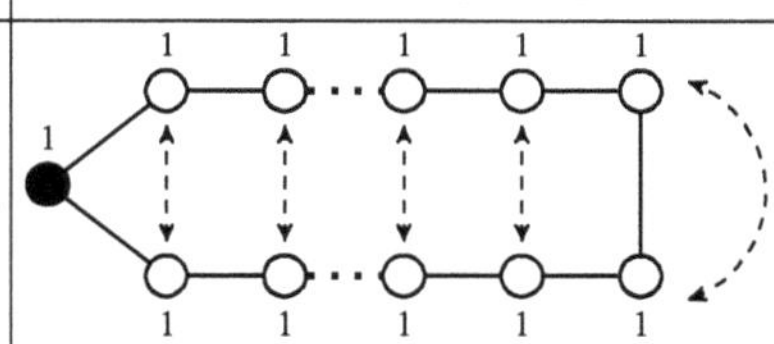
IV^*_2 $\widetilde{F}^t_4$	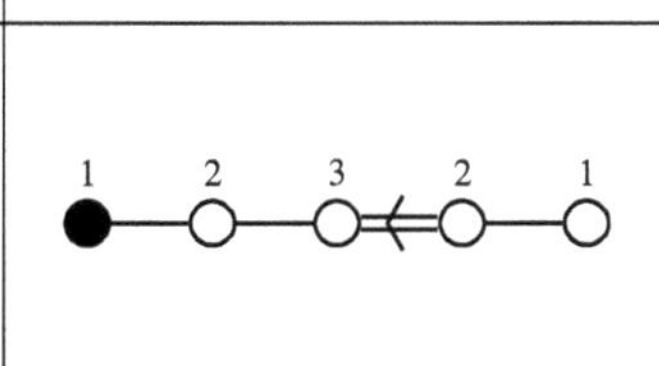	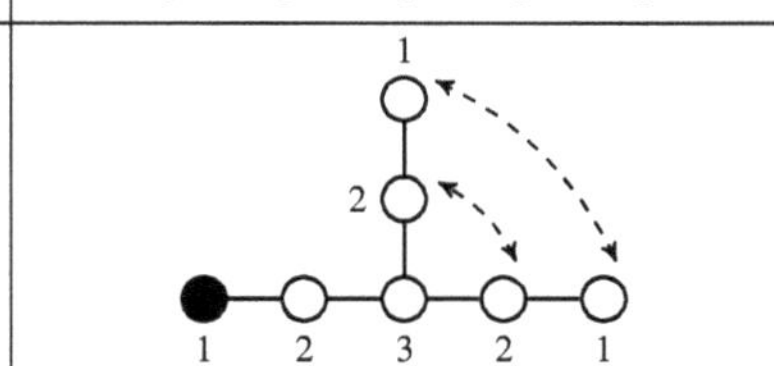
I^*_2 $\widetilde{B}^t_3$	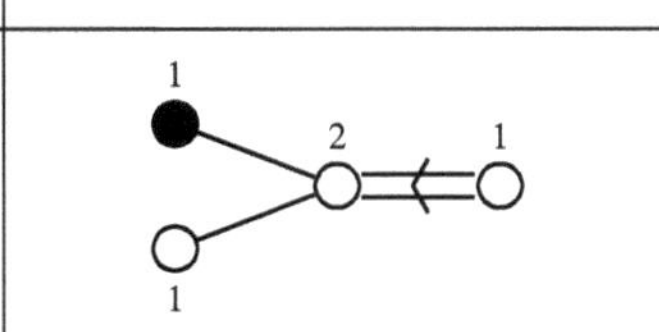	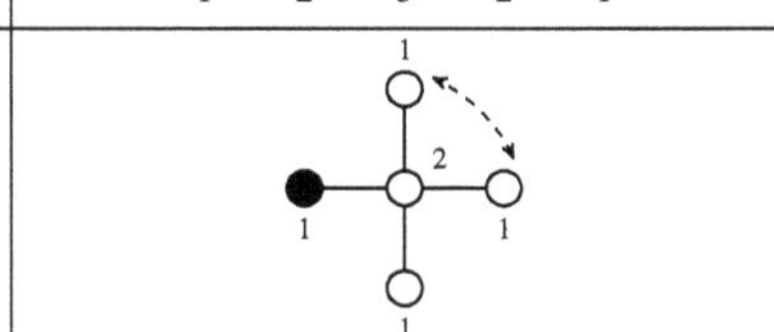
$I^*_{0,3}$ $\widetilde{G}^t_2$	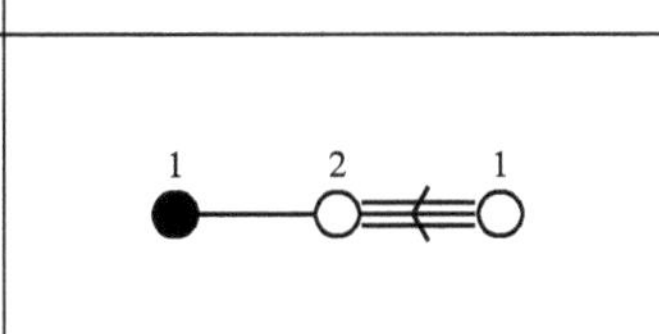	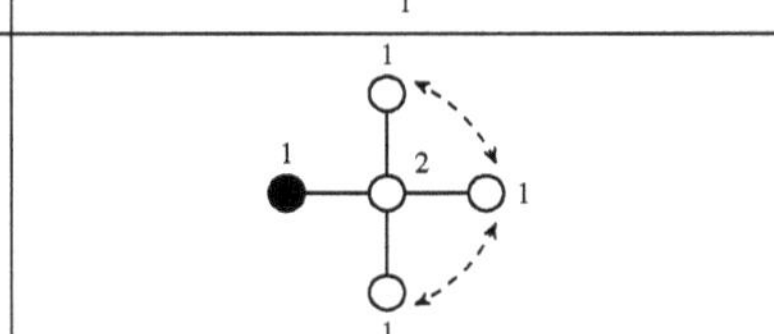

7.5.2 Szydlo's Generalization of Miranda Models

Assuming the same conditions as Miranda, Szydlo has analyzed the general case of collisions in higher codimensions [26]. He assumes that the base scheme of the fibration is Noetherian, n-dimensional, regular, integral, and separated. He also allows mixed characteristic.

Szydlo does not assume that the residue field is perfect, it follows that an irreducible polynomial can have roots with multiplicity so that the roots only exist in non-separable extension of the residue field. The translation needed in Tate's algorithm translates the singular point of a Weierstrass model to the origin and the multiple root of certain quadratic polynomial to the origin (Table 7.5).

Interestingly, starting from codimension-three, the only collisions possible are those with $J = \infty$ (type I_n and I_n^*) with the following restrictions: There are at most one fiber of type I_n^* and at most one fiber of type I_{2m+1}, and the number of fiber of type I_{2n} is bounded by the codimension of the collision. Taking this into account, we have the following four types of collisions:

$$\begin{aligned} J = \infty : I_{2n_1} + \cdots I_{2n_k} & \longrightarrow I_{2n}, \quad n = n_1 + \cdots + n_k \\ I_{2n_1} + \cdots + I_{2n_k} + I_{2r+1} & \longrightarrow I_{2n+2r+1}, \\ I_{2n_1} + \cdots + I_{2n_k} + I_m^* & \longrightarrow I_{n+m+1}^*, \\ I_{2n_1} + \cdots + I_{2n_k} + I_{2r+1} + I_m^* & \longrightarrow I_{n+r+m+1}^{*+}. \end{aligned} \tag{7.58}$$

The resolution of the singularities at the collisions depends on some discrete choices. In particular, the order in which the blowups are performed is crucial for the final result. For example, Miranda and Szydlo don't have the same results for the collision $IV + I_0^*$ and the justification can be traced back to different conventions on how to order the blowups:

(Miranda)	(Szydlo)
$IV + I_0^*$: (1)–(2)–(4)–(2)	$I_0^* + IV$: (1)–(2)–(3)–(2)

(7.59)

Acknowledgements I would like to thank the organizers and the participants of the Villa de Leyva Summer School of 2015. I am deeply grateful to my collaborators Ravi Jagadeesan, Patrick Jefferson, Monica Kang, Sabrina Pasterski, Julian Salazar, Shu-Heng Shao, and Shing-Tung Yau for helpful discussions. I am also thankful to Paolo Aluffi and Matilde Marcolli who have introduced me to the school. This work is partly supported by the National Science Foundation (NSF) grants DMS-1406925 and DMS-1701635 "Elliptic Fibrations and String Theory."

References

1. A. Beauville, *Complex Algebraic Surfaces*. London Mathematical Society Lecture Note, vol. 68 (Cambridge Univ. Press, Cambridge, 1983)
2. B. Conrad, *Minimal Models for Elliptic Curves*, unpublished work available on his website
3. F.E. Cossec, I.V. Dolgachev, *Enriques Surfaces I*. Progress in Math. vol. 76 (Birkhäuser, 1989)
4. P. Deligne, Courbes elliptiques: formulaire d'après J. Tate, Modular functions of one variable, IV (Proc. Internat. Summer School, Univ. Antwerp, Antwerp, 1972), Lecture Notes in Mathematics, vol. 476 (Springer, Berlin, 1975), pp. 53–73
5. J. Dieudonné, A. Grothendieck, *Eléments de Géometrie Algébrique. I. Le Language des Schémas* (French) Inst. Hautes Etudes Sci. Publ. Math. vol. 4 (1960), p. 228
6. T. Dokchitser, V. Dokchitser, A remark on Tate's algorithm and Kodaira types. Acta Arithmetica **160**, 95–100 (2013)
7. I. Dolgachev, M. Gross, Elliptic three-folds I: Ogg-Shafarevich theory. J. Algebraic Geom. **3**, 39–80 (1994). MR 95d:14037
8. I.V. Dolgachev, On the purity of the degeneration loci of families of curves. Invent. Math. **8**, 34–54 (1969)
9. W. Fulton, *Intersection theory*, volume 2 of *Ergebnisse der Mathematik und ihrer Grenzgebiete. 3. Folge. A Series of Modern Surveys in Mathematics [Results in Mathematics and Related Areas. 3rd Series. A Series of Modern Surveys in Mathematics]*, 2 edn. (Springer, Berlin, 1998)
10. R. Hartshorne, *Algebraic Geometry*. Graduate Texts in Mathematics, vol. 52 (Springer, Berlin, 1977)
11. D. Husemöller, *Elliptic Curves* (Springer, Berlin, 2004)
12. K. Kodaira, On compact analytic surfaces II–III. Ann. Math. **77**, 563–626 (1963); **78**, 1–40 (1963)
13. Q. Liu, *Algebraic Geometry and Arithmetic Curves*. Oxford Graduate Texts in Mathematics, vol. 6 (Oxford University Press, Oxford, 2002). Translated from the French by Reinie Erné, Oxford Science Publications
14. R. Miranda, *Smooth Models for Elliptic Threefolds,* in: R. Friedman, D.R. Morrison (Eds.), The Birational Geometry of Degenerations, Progress in Mathe-matics 29, Birkhauser (1983), pp. 85–133
15. R. Miranda, *The Basic Theory of Elliptic Surfaces, Dottorato di Ricerca in Matematica* (Dipartimento di Matematica dell' Universita di Pisa, ETS Editrice Pisa, 1989)
16. D. Mumford, K. Suominen, Introduction to the theory of moduli, in Algebraic Geometry, Oslo, 1970, Proceedings of the 5th Nordic summer school in mathematics. Wolters-Noordhoff, 171–222 (1972)
17. N. Nakayama, *Local Structure of an Elliptic Fibration,* Higher dimensional birational geometry. Advanced Studies in Pure Mathematics, vol. 35 (Mathematical Society, Kyoto, 1997), pp. 185–295
18. N. Nakayama, Global structure of an elliptic fibration. Publ. Res. Inst. Math. Sci. **38**, 451–649 (2002)
19. A. Néron, Modèles Minimaux des Variétés Abeliennes sur les Corps Locaux et Globaux, Publ. Math. I.H.E.S. 21, 361–482 (1964)
20. W. Schmid, Variation of Hodge structure: the singularities of the period mapping. Invent. Math. **22**, 211–319 (1973)
21. M. Schütt, T. Shioda, Elliptic surfaces, Algebraic geometry in East Asia - Seoul 2008. Adv. Stud. Pure Math. **60**, 51–160 (2010)
22. J.P. Serre, *A Course in Arithmetic* (Springer, New York, 1973)
23. J. Silverman, *Advanced topics in the Arithmetic of Elliptic Curves* (Springer, Berlin, 1995)
24. J. Silverman, *The Arithmetic of Elliptic Curves* (Springer, Berlin, 1986)
25. V. Snyder, C.H. Sisam, *Analytic Geometry of Space* (Henry Holt and Company, 1914)
26. M. Szydlo, *Flat Regular Models of Elliptic Schemes*, Ph.D thesis, Harvard University (1999)
27. M. Szydlo, Elliptic fibers over non-perfect residue fields. J. Number Theory **104**, 75–99 (2004)

28. J.T. Tate, Algorithm for determining the type of a singular fiber in an elliptic pencil, in *Modular Functions of One Variable IV*. Lecture Notes in Mathematics, vol. 476 (Springer, Berlin, 1975)
29. J.T. Tate, The arithmetics of elliptic curves. Invent. Math. **23**, 170–206 (1974)

Chapter 8
Batalin–Vilkovisky Formalism as a Theory of Integration for Polyvectors

Pierre J. Clavier and Viet Dang Nguyen

Abstract The Batalin–Vilkovisky (BV) formalism is a powerful generalization of the BRST approach of gauge theories and allows to treat more general field theories. We will see how, starting from the case of a finite dimensional configuration space, we can see this formalism as a theory of integration for polyvectors over the shifted cotangent bundle of the configuration space and arrive at a formula that admits a generalization to the infinite dimensional case. The process of gauge fixing and the observables of the theory will be presented.

8.1 Motivations and Program

If you ask your best experimental physicist friend what the Universe is, there is a good chance that he will talk about particles; gluons and quarks, photons and fermions if he is studying the small structures of matter at high energies. On the other hand, if he is doing experimental astrophysics, he might discuss stars, black holes, and so on.

These objects are described by very different (and, to date, incompatible) theories: quantum field theory for the former, general relativity for the latter. However, quantum field theory and general relativity have a point in common: they happen to be gauge theories. The goal of the formalism devised by Batalin and Vilkovisky in [1, 2] is to deal with such theories and some of their generalizations. For the sake of completeness let us briefly, in non-technical terms, recall what a gauge theory is.

P.J. Clavier (✉)
Institut für Mathematik, Potsdam Universität, Potsdam, Deutschland
e-mail: clavier@math.uni-potsdam.de

V.D. Nguyen
Institut Camille Jordan, Université Claude Bernard Lyon 1, UMR 5208, 69622 Lyon, France
e-mail: dang@math.univ-lyon1.fr

V.D. Nguyen
Institut Camille Jordan, CNRS, UMR 5208, 69622 Lyon, France

A. Cardona et al. (eds.), *Quantization, Geometry and Noncommutative Structures in Mathematics and Physics*, Mathematical Physics Studies,
DOI 10.1007/978-3-319-65427-0_8

Let us assume that you have a d-dimensional space-time $\mathscr{M}^d$. A field living on that space is a function from this space-time to a target space which depends on the theory under consideration: It is $\mathbb{R}$ for a scalar field theory, a vector space for a vector field theory, and so on. The guiding principle of a gauge theory is to reparametrize your field by a "rotation"

$$\Psi(x) \longrightarrow e^{i\theta}\Psi(x).$$

This is rather similar to the invariance under rotation of the wavefunctions solutions of the Schrödinger equation in usual quantum mechanics. This reparametrization is, however, a generalization of quantum mechanics in two important ways.

- The reparametrization can be made into a group more general than the group $U(1)$ of rotations. We will typically say that the reparametrization parameter θ is an element of some semi-simple Lie algebra $\mathfrak{g}$. This algebra will be called the gauge algebra.
- This $\theta \in \mathfrak{g}$ has a value that depends on the point of space-time at which we evaluate the field Ψ: $\theta = \theta(x)$.

We call gauge theory a theory that has the invariance under such a reparametrization, plus the usual properties of any nice physical theory: Lorentz invariance, locality, renormalizability (in the case of quantum field theories).

Now, one of the arguably most elegant formulations of physics is known as the path-integral formalism. It was devised by Feynman in [7]. It rests upon the observation that the basic principles of quantum mechanics forbid us to determine by what slit a photon goes in the double slots experiment, or more precisely, it states that this is a meaningless question. Then, one can increase the number of screens with two slots on them between the photon's emission and detection points. For each of the screens, there is no meaning to ask to which slot the photon went through. Therefore, in the limit of an infinite number of screens, we conclude that we cannot tell which *path* a photon follows from a to b: One has to make a (weighted) average over all possible paths, that is to perform an integration over the space of paths.

Two obvious difficulties arise. The first one is that the space of paths is huge, typically having an uncountable number of dimensions. Integrals are in general ill-defined in such spaces. The second one comes from the gauge freedom: If two paths can be mapped to each other by a gauge transformation, they correspond to the same physical path, and we shall not count both.

Let us briefly explain the program behind the BV formalism. First, we take a finite dimensional configuration space M, $N = \dim(M)$. Then, naively, we can imagine our observables as elements of $\mathscr{C}^\infty(M)$. Path integral has to be an evaluation map

$$<>: \mathscr{C}^\infty(M) \longrightarrow \mathbb{R}.$$

The most natural way to do this is to choose a volume form $\Omega \in \Lambda^N T^*M$ and to define, for an observable $f \in \mathscr{C}^\infty(M)$

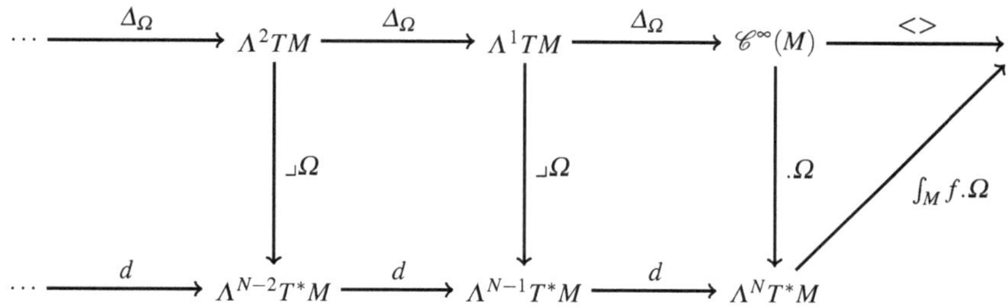

Fig. 8.1 A summary of the BV program

$$< f > = \int_M f\Omega.$$

This has two assets. First, the integration of forms is a powerful, very well understood tool. Second, it provides an equivalence relation between elements of $\mathscr{C}^\infty(M)$ which have the same evaluation: If $f\Omega$ and $f'\Omega$ differ by a d-exact term (where d is the de Rham differential), then they give rise to the same measured values. Hence, we understand that the true observables are rather elements of the de Rham cohomology group. This remark gives some hope about our ability to treat gauge theories.

However, it also suffers from two serious drawbacks: The notion of a top form does not have a clear meaning in the infinite dimensional case, and even less so if the dimension is uncountably infinite, as it will be in the cases of interest. Moreover, Ω might not exist.

The BV formalism offers an escape road to these drawbacks. The idea is to work with polyvectors rather than with forms, since a $N-1$ form is the contraction of a 1-polyvector with a N form. We gain in that we can work with polyvectors even in the infinite dimensional case! The usual concepts of integration can be lifted to the level of polyvectors and arrived at a formulation that admits a natural generalization to the infinite dimensional case. This project is summarized in Fig. 8.1.

The plan of this paper is the following: In the subsequent second section, we will define the BV integral and state some of its properties.[1] The third section will make explicit how to deal with the gauge freedom within the BV formalism and will introduce the Schouten–Nijenhuis brackets when working in simple (but still interesting) gauge theories. The fourth section will present the famous master equations and give a few consistency checks on the set of observables built within the BV framework. The conclusion contains a discussion on the reasons why one might be interested in studying the BV formalism.

[1]In most of the case, only hints of proofs will be given, first to keep this text within a reasonable size, and second to clarify the idea of the construction. In practice, this will make the next section quite short.

8.2 BV Integral

From now on, we will use the so-called shift isomorphism. Let M be a finite dimensional vector space. The space of polyvectors is isomorphic (as a vector space) to the space of functions on the shifted cotangent space:

$$\Lambda TM \simeq \mathscr{C}^{\infty}(\Pi T^* M). \tag{8.1}$$

We denote by ΠM the vector space M shifted by 1. Here shifted means that we reverse the parity of coordinates: Fermionic coordinates are now bosonic and vice versa. Let $(e^a)_a$ be a basis of M. Then, the e^a are linear coordinates on M^*, and the isomorphism is given in term of the basis by

$$e^{a_1} \wedge \cdots \wedge e^{a_k} \mapsto \left(e^{a_1} \dots e^{e_k}\right),$$

(extended by linearity). The left-hand side of the above map is a polyvector, while the right-hand side is a polynomial. The basic idea behind the proof is that both sides have the same parity (that is why we have $\Pi T^* M$) and that all functions on a shifted space (which we call superfunctions) are polynomials. Hence, in the following, we will freely use this isomorphism, and in particular we will often write polyvectors as superfunctions.

8.2.1 BV Laplacian

In Fig. 8.1, we were imprecise when writing a polyvector field as an element of $\Lambda^p TM$, while a polyvector is instead a section of the bundle $\Lambda^p TM$ above M, exactly like a form is a section of the cotangent bundle $\Lambda^q T^* M$. This distinction is often neglected in Physics, for going from one to the other boils down to specifying a basis on M. However, we have $\alpha \in \Gamma(M, \Lambda^p(TM))$. Then the contraction operator $\lrcorner$ is a map

$$\lrcorner : \Gamma(M, \Lambda^p(TM)) \times \Gamma(M, \Lambda^q(T^*M)) \to \Gamma(M, \Lambda^{q-p}(T^*M)) \tag{8.2}$$

defined for $p = 1$ by $(X \lrcorner \omega)(X_1, \dots, X_{q-1}) := (\iota_X \omega)(X_1, \dots, X_{q-1}) = \omega(X, X_1, \dots, X_{q-1})$ and is then extended by $(\alpha \wedge \beta) \lrcorner \omega = \alpha \lrcorner (\beta \lrcorner \omega)$.

Now, let α be a polyvector[2] and Ω a well behaved (i.e., nowhere vanishing) volume form. Then Δ_Ω is the operator from $\Gamma(M, \Lambda^p(TM))$ to $\Gamma(M, \Lambda^{p-1}(TM)$ defined by

$$(\Delta_\Omega \alpha) \lrcorner \Omega = d(\alpha \lrcorner \Omega). \tag{8.3}$$

[2]A polyvector field, to be precise but in the following we will not bother writing field everywhere.

If we denote by $\mathscr{F}_\Omega$ the isomorphism

$$\mathscr{F}_\Omega : \alpha \in \Gamma(M, \Lambda TM) \longmapsto \alpha \lrcorner \Omega \in \Gamma(M, \Lambda T^*M) \tag{8.4}$$

then $\mathscr{F}_\Omega^{-1}$ is well defined since Ω was assumed to vanish nowhere. Thus

$$\Delta_\Omega = \mathscr{F}_\Omega^{-1} \circ d \circ \mathscr{F}_\Omega \tag{8.5}$$

where d is the usual de Rham differential. The homology of this operator is described in the subsequent lemma.

Lemma 1 *Let Ω be a volume form and Δ_Ω be its associated BV Laplacian. Then the operator Δ_Ω defined above satisfies $\Delta_\Omega^2 = 0$.*

In the remaining parts of this text, we will use the usual terminology and call a polyvector α to be Δ_Ω-closed when $\Delta_\Omega \alpha = 0$ and Δ_Ω-exact when there exists a polyvector β such that $\alpha = \Delta_\Omega \beta$.

While Δ_Ω depends on the chosen volume form, we will show later that the observables, built as elements of the homology complex of Δ_Ω do not. Moreover one can relate the various BV Laplacians to each others, as we will see with (8.15).

8.2.2 Definition of the Integral

Given a well-behaved volume form Ω and its associated BV Laplacian Δ_Ω, we define the BV integral of a polyvector field α as

$$\int_{\Pi N^* \Sigma}^{BV} \alpha = \int_\Sigma \alpha \lrcorner \Omega$$

where $N^* \Sigma$ is the conormal of Σ, which means the part of the cotangent space T^*M which vanishes on the tangent of Σ. It is a space of dimension $N = \dim(M)$. Indeed, if $\dim(\Sigma) = p$, then $N^* \Sigma$ has $N - p$ dimensions in the fiber. This definition can be justified in the finite dimensional case by working out its right-hand side with the shift isomorphism until we reach the left-hand side. To make the link with the language of symplectic geometry, let us notice that conormal spaces are examples of Lagrangian submanifolds of T^*M endowed with its natural symplectic form $dx^i \wedge dx_i^*$.

With this definition of the BV integral at hand, most of the important theorems of integration of forms can be translated as results on the integration of polyvectors. The first is the following counterpart to the Stokes theorem.

Theorem 2 *Let Ω be a volume form and Δ_Ω be its associated BV Laplacian. Let Σ be a smooth submanifold with smooth boundary $\partial \Sigma$. Then for any polyvector $\alpha \in C^\infty(\Pi T^* M)$:*

$$\int_{\Pi N^* \partial \Sigma}^{BV} \alpha = \int_{\Pi N^* \Sigma}^{BV} \Delta_\Omega \alpha.$$

Proof

$$\begin{aligned}
\int_{\Pi N^*\partial\Sigma}^{BV} \alpha &= \int_{\partial\Sigma} \alpha\lrcorner\Omega \quad \text{by definition of the BV integral,}\\
&= \int_{\Sigma} d(\alpha\lrcorner\Omega) \quad \text{by the usual Stokes theorem, and with } d \text{ the de Rham differential,}\\
&= \int_{\Sigma} (\Delta_\Omega\alpha)\lrcorner\Omega \quad \text{by definition of } \Delta_\Omega,\\
&= \int_{\Pi N^*\Sigma}^{BV} \Delta_\Omega\alpha. \quad \text{by definition of the BV integral.}
\end{aligned}$$

This result has an obvious corollary.

Corollary 3 *Let Σ_1, Σ_2 be two smooth submanifolds belonging to the same homology class. Then for any Δ_Ω-closed polyvector $\alpha \in C^\infty(\Pi T^*M)$:*

$$\int_{\Pi N^*\Sigma_1}^{BV} \alpha = \int_{\Pi N^*\Sigma_2}^{BV} \alpha.$$

Finally, there is also a very simple but important lemma coming from the previous definitions.

Lemma 4 *If a polyvector α is Δ_Ω-exact, then*

$$\int_{\Pi N^*\Sigma}^{BV} \alpha = 0$$

This lemma is the equivalent to the Poincaré lemma for forms.

As for the Theorem 2, the proofs of this corollary and of this lemma are fairly easy and follow simply from playing with the definition of the BV integral and the usual integration of forms.

8.2.3 Advantages of the BV Formalism

Having defined the BV integral, we may start answering a simple question: What are the advantages of the BV formalism over other approaches such as the Faddeev–Popov determinant or the BRST formalism. We will give the historical answer here. Other arguments in favor of the BV formalism will be given in the conclusion.

A symmetry is said to be open when it is fulfilled only on-shell, that is on the critical domain of the action S_0, i.e., on the submanifold of the configuration space where the fields are solutions to the usual equations of motion. The archetypal example of a physical theory with open symmetries is supergravity without auxiliary fields. As first noticed in [11], when working in a theory with open symmetries, we might end up with quartic ghost terms in the gauge-fixed lagrangian.

In the Faddeev–Popov formalism, ghosts are interpreted as fermionic variables coming from the restriction of the domain of integration. This restriction is performed with delta functions, and brings a determinant, written as an integral over fermionic variables: the ghosts. Therefore, we do not have many freedom on the ghost terms that can be treated in the Faddeev–Popov formalism. In particular, quartic terms are not allowed, thus the Faddeev–Popov formalism is not adapted to the treatment of theories with open symmetries.

On the other hand, once again in the case of a theory with open symmetries, one can show that the BRST differential does not square to zero off-shell, hence making the BRST cohomology ill-defined. This is clearly explained in [13], where more references on the subject can be found. However, in sharp constrast with the Faddeev–Popov and BRST formalisms, since the BV formalism is a theory of integration, one can treat much more general functions than in the Faddeev–Popov formalism, and the well-definiteness of the integral does not depend on the precise form of these functions. Therefore, this formalism is particularly adapted to the treatment of theories with open symmetries.

Another reason to prefer the BV formalism over other approaches is that some questions are easier to answer with it. In particular, Jean Zinn-Justin and Laurent le Guillou first developed an equivalent formalism (which corresponds to the classical version of the formalism presented here) to study anomalies in gauge theories. Nowadays, anomalies are still often studied within the framework of the BV formalism.

Other arguments in favor of the BV formalism will be presented in the conclusion, for they involve contemporary concerns that can be treated with it.

8.3 Gauge Fixing

8.3.1 Gauge Fixing in BV Formalism

We are now ready to handle gauge freedom. We have two types of freedom: the choice of the volume form Ω, and the choice of the surface Σ to integrate over. Which one corresponds to the gauge freedom, and which one is a spurious choice, in which case we will make sure that have the evaluation of observables does not depend on it?

To answer this question, it is useful to realize that, in the finite dimensional case, we have

$$\int_M e^{iS/\hbar}\Omega = \int_{\underline{0}}^{BV} e^{iS/\hbar}.$$

Here $\underline{0}$ is the conormal of M, which has its N dimensions on the base space M, and none on the fiber i.e., it is the zero section of the shifted cotangent bundle ΠT^*M.

But $\underline{0}$ is a very bad choice of integration domain in ΠT^*M! Indeed, in the infinite dimension limit, this integral will diverge, due to the gauge freedom. The reason for this is that $\underline{0}$ is identified (as a Lagrangian submanifold of ΠT^*M) with M, and that the gauge group $\mathfrak{g}$ is a symmetry of the theory, the integrand is left unchanged under the action of the group $\mathfrak{g}$. Hence, in the infinite dimensional limit, we find an integral over a non-compact domain of a constant.

Fortunately, Corollary 3 tells us that if an integrand is Δ_Ω-closed, then one can change the domain of integration in ΠT^*M without changing the value of the integral. More precisely, if Σ is a smooth submanifold of M and if the integration domain in ΠT^*M is $\Pi N^*\Sigma$ (the conormal space of Σ), then without changing the value of the integral, we can integrate over $\Pi N^*\Sigma'$, provided that Σ' is in the same homology class of M than Σ. Hence, the choice of a gauge will be the choice of a surface Σ to integrate over in ΠT^*M. And, of course, saying that a quantity A is gauge invariant amounts to requiring that such a change of the integration surface will not change the value of the integral. Hence, the Δ_Ω-closedness condition is the gauge-invariance condition.

Now, a question a reader may raise is to which extent can we explicitly describe the allowed integration surfaces? Homology is a notoriously difficult subject, so is this definition of any practical interest? As a partial answer, we will show that we can describe a non-trivial set of submanifolds on which it is legitimate to integrate. We read this observation in [8], although the idea was already present in the seminal work of Batalin and Vilkovisky [2]. Take Ψ_1, Ψ_2 two smooth functions on M. Consider the submanifolds $\mathscr{L}_{\Psi_j}$ of ΠT^*M defined by

$$(x^i, x_i^*) \in \mathscr{L}_{\Psi_j} \Longleftrightarrow \Psi_j(x^i) = 0, x_i^* = \frac{\partial \Psi_j}{\partial x^i} \tag{8.6}$$

for $j \in \{1, 2\}$. Since we can build the homotopy

$$\Psi_t = t\Psi_1 + (1-t)\Psi_2,$$

we find that $\mathscr{L}_{\Psi_1}$ and $\mathscr{L}_{\Psi_2}$ are in the same homology class (if M is connected). Finally, since the zero section is defined by $x_i^* = 0$, we see that, for a smooth function Ψ, the submanifold $\mathscr{L}_\Psi$ defined in the same way as $\mathscr{L}_{\Psi_i}$ is an admissible submanifold to be integrated over: It has the same homology class as the zero section.

With this picture in mind, we see that Lemma 4 allows to say that two gauge-invariant quantities that differ only by a Δ_Ω-exact term will give the same observed values, and thus actually lead to the same observables. In more rigorous terms, the observables will be elements of the homology group of Δ_Ω.

To summarize, the choice of the surface to integrate over is the choice of a gauge in the usual formulations of gauge theories, and the condition of gauge invariance of a quantity A is translated into $\Delta_\Omega(A) = 0$. Hence, we will have to check that the set of observables that we get at the end of this procedure does not depend on the choice of the volume form Ω. This will be carried out at the very end of this presentation.

8.3.2 Schouten–Nijenhuis Bracket

Before deriving the quantum master equations, let us present the so-called Schouten–Nijenhuis bracket for a special case on which we will focus from now on. We will take our configuration space M to be

$$M = X \times \Pi \mathfrak{g} \tag{8.7}$$

where X is a space of fields (which is assumed to be bosonic), and $\Pi \mathfrak{g}$ is the space of ghosts of the theory. We will write $(x^i) = (\phi^i, c^\alpha)$ for a coordinate basis of M. Then, the associated basis of ΠT^*M is

$$(\phi_i^*, c_\alpha^*, \phi^i, c^\alpha). \tag{8.8}$$

There is a natural grading on this space, defined by

$$|c_\alpha^*| = -2 \quad |\phi_i^*| = -1 \quad |\phi^i| = 0 \quad |c^\alpha| = +1 \tag{8.9}$$

and $|f.g| = |f| + |g|$. Hence, antighosts c_α^* are bosonic, while the antifields ϕ_i^* and the ghosts are fermionic.

Let us notice that if we had ghosts of ghosts[3] they would be of degree +2, ghosts of ghosts of ghosts would be of degree +3, and so on ...

The Schouten–Nijenhuis bracket of two polyvectors F and G, seen as superfunctions over ΠT^*M, is defined as

$$\{F, G\} = \frac{\overleftarrow{\partial} F}{\partial x^i}\frac{\overrightarrow{\partial} G}{\partial x_i^*} - \frac{\overleftarrow{\partial} F}{\partial x_i^*}\frac{\overrightarrow{\partial} G}{\partial x^i} \tag{8.10}$$

where the right derivative $\overrightarrow{\partial}$ is the usual derivative and the left derivative $\overleftarrow{\partial}$ is defined by

$$\frac{\overleftarrow{\partial} F}{\partial y} = (-1)^{|y|(|F|+1)}\frac{\partial F}{\partial y}. \tag{8.11}$$

This bracket has many good properties (e.g., $\{F, .\}$ is a graded derivation, and it obeys a graded Jacobi identity), but they are tedious to show (essentially because of the powers of -1) and left as an exercise. However, one of their important features is their link to the BV Laplacian when written in coordinates.

---

[3] Such objects are needed when the Koszul homology is not a resolution, i.e., has non-trivial homology groups other than the zeroth one. This will make the zeroth BRST cohomology not isomorphic to the set of observables of the theory (see [8] for a clear proof of this statement), and has to be cured with the Tate procedure [12] which will produce ghosts of ghosts and so on. A presentation of BRST formalism coherent with the current notations can be found in [5].

8.4 Master Equations

8.4.1 BV Laplacians in Coordinate

In order to derive the link discussed above, we take the configuration space of fields X to be finite dimensional, with $\dim(X) = N$. Let $(\phi_i)_{i=1,\ldots,N}$ be a basis of X and $\Omega_0 = \mathrm{d}\phi^1 \wedge \ldots \wedge \mathrm{d}\phi^N$ the associated canonical volume form. The BV Laplacian associated to this volume form (which we will write Δ from now on) reads

$$\Delta = \frac{\partial}{\partial \phi^i}\frac{\partial}{\partial \phi^*_i}. \tag{8.12}$$

The proof of this result follows directly from the definition of the BV Laplacian (given by 8.3). Then, one can compute the right-hand side and the left-hand side of (8.3) with the proposed Δ. It is very easy to see that both sides lie in the same directions on ΠT^*X, but the hard part is to check that they are actually equal.

We can then play the same game in the space $\Pi T^*\mathfrak{g}$, leading to the same result. In order to generalize this to $M = X \times \Pi\mathfrak{g}$, we have to merge our two partial Laplacians. It turns out that the right way of doing so is to define the BV Laplacian Δ on M to be

$$\Delta = \frac{\partial}{\partial \phi^i}\frac{\partial}{\partial \phi^*_i} - \frac{\partial}{\partial c^\alpha}\frac{\partial}{\partial c^*_\alpha}. \tag{8.13}$$

Before going further, let us notice that we defined objects that admit a natural generalization to the infinite dimensional case! Indeed, one would just have to replace the derivatives in the Schouten–Nijenhuis bracket and in this BV Laplacian by functional derivatives to have operators on the configuration space of a quantum field theory. We will nevertheless go on working in the finite dimensional setup to derive the quantum master equations, but we have to keep in mind that what we are doing has a meaning in the infinite dimensional case.

The Laplacian (8.13) has many interesting properties. An essential property is that the Schouten–Nijenhuis bracket measures the obstruction preventing the BV Laplacian Δ from being a (graded) derivation. More precisely

$$\Delta(fg) = (\Delta f)g + (-1)^{|f|}f(\Delta g) + (-1)^{|f|}\{f, g\}. \tag{8.14}$$

This property is important for it can be used as the starting point of another approach of the BV formalism, developed in particular by Owen Gwilliam and Kevin Costello. The reader may want to look at the thesis [9] and the references therein for a detailed presentation of this approach.

Lemma 5 (Gwilliam [9]) *Δ is the unique translation invariant second-order differential operator that decreases the degree of the polyvectors it acts on by exactly* 1 *(for the grading defined above) and satisfies* (8.14) *for ΠT^*X.*

The proof of this lemma relies on the study of the constraints set by (8.14) on the free parameters of the most general translation invariant second-order differential operator that decreases the degree of the polyvectors it acts on of exactly 1, which is

$$\Delta = a^i{}_j(x)\frac{\partial}{\partial x^i}\frac{\partial}{\partial x^*_j} + b_i(x)\frac{\partial}{\partial x^*_i} + c_{ij}{}^k(x)x^*_k\frac{\partial}{\partial x^*_i}\frac{\partial}{\partial x^*_j}.$$

In this alternative approach, everything derives from first principles, which might make it more elegant than our down-to-earth approach. However, we believe that the approach presented here is more constructive in the sense that every definition is justified from physical considerations relative to the problem at hand.

We are now almost ready to derive the set of observables of our theory. We only need one more detail, which will be important when having to show that this set of observables is independent of the chosen volume form Ω. Any well-behaved volume form Ω is still linked to the canonical volume form Ω_0 by a conformal factor:

$$\Omega = e^f\Omega_0.$$

From the Definition (8.3), we see that the quantity that we have to compute to get Δ_Ω is

$$d(\alpha \lrcorner \Omega) = d(e^f\alpha \lrcorner \Omega_0) = e^f\left(d(\alpha \lrcorner \Omega_0) + df \wedge (\alpha \lrcorner \Omega_0)\right).$$

Since $e^f d(\alpha \lrcorner \Omega_0) = (\Delta\alpha)\lrcorner \Omega$ while $e^f df \wedge (\alpha \lrcorner \Omega_0) = \{f, \alpha\}\lrcorner \Omega$, we have

$$\Delta_\Omega = \Delta + \{f, -\}. \tag{8.15}$$

8.4.2 Quantum Master Equations

In order to derive the quantum master equations, let us consider the canonical volume form on the finite dimensional space M. Recall that a quantity is gauge invariant if, and only if, it is Δ-closed. The very first thing to do is to check when the zero-point function

$$\int_M e^{iS/\hbar}\Omega = \int_{\underline{0}}^{BV} e^{iS/\hbar}.$$

is gauge invariant. This is therefore equivalent to solving the equation

$$\Delta\left(e^{i\frac{S}{\hbar}}\right) = 0 \tag{8.16}$$

for S. This can be simplified by expanding the exponential, using the fact that Δ is a derivation and showing by induction that $\Delta S^n = nS^{n-1}\Delta S + \frac{n(n-1)}{2}S^{n-2}\{S, S\}$.

After some work, we reach a very nice equation, which is generally called *the* quantum master equation:

$$\{S, S\} - i\hbar\Delta S = 0. \tag{8.17}$$

The classical master equation obtained from the quantum master equation by taking the limit $\hbar \to 0$ reads $\{S, S\} = 0$. This is actually an important equation, but we will not discuss it further.

We have already stated that any solution of this equation defines an action such that the zero-point function of the theory arising from this action is gauge invariant. Then, one can study the observables of the theory, and so on ... In other words, the solution of this equation gives all the possible theories with a given gauge symmetry. This is why the study of the solutions of this Quantum Master Equation is still nowadays an active domain of research, see e.g., [10].

Before deriving the observables of a theory, an essential consistency check of this construction is to make sure that the usual BRST action (when written in the language of the BV formalism) defines a gauge-invariant zero-point function. This will be stated in a forthcoming theorem, but for which we need a few definitions.

Let S_0 be an action functional over X invariant under a gauge group $\mathfrak{g}$ (let us recall that this means that $\mathfrak{g}$ is represented on the vector fields of X), that is, for any generator e_α

$$\rho(e_\alpha)S_0 = 0,$$

where $\rho : \mathfrak{g} \mapsto \Gamma(X, TX)$ is the representation function of $\mathfrak{g}$. Then, the BV action is

$$S = S_0 + S_E + S_R = S_0 + \underbrace{c^\alpha c^\beta C^\gamma_{\alpha\beta} c^*_\gamma}_{S_E} + \underbrace{\rho^i_\alpha c^\alpha x^*_i}_{S_R}. \tag{8.18}$$

One can view S as a perturbation of S_0 in ΠT^*M.[4]

To state the link between this action and the BRST formalism, let us say that $S_E + S_R$ is the symbol of the BRST differential (seen as a vector field) in ΠT^*M. The symbol of a vector field Q on a smooth manifold V of cotangent bundle $\pi : T^*V \mapsto V$ is defined as a function $\sigma(Q) \in T^*V$ such that the Hamiltonian vector field $\{\sigma(Q), .\}$ is a vector field on T^*V whose horizontal part is just Q. In other words, $\forall f \in C^\infty(V), \{\sigma(Q), \pi^* f\} = Qf$.

Theorem 6 *Let $S = S_0 + S_E + S_R$ the action defined above and Δ the BV Laplacian corresponding to the canonical volume form Ω_0, if*

- *the Lie algebra $\mathfrak{g}$ acts on X in such a way that it preserves the measure Ω_0*
- *the Lie algebra $\mathfrak{g}$ is unimodular,*

then S is a solution of the Quantum Master Equation.

---

[4]Someone working in homology theory may ponder about this terminology.

The proof of this theorem is quite long and far beyond the scope of this introduction. It can be found in many places in the literature, and in particular in [5]. However, interestingly, we have $\Delta S = \{S, S\} = 0$. Hence, S is a solution of the classical master equation as well as the quantum master equation: We say that S is non-anomalous.

Indeed, the classical and quantum master equations encode, in a very deep sense, the symmetries of the theory. Hence, the classical and the quantum version of the theory have the same symmetries, which is exactly the definition of what non-anomalous mean. For completeness, let us point out that in the BV formalism, there is a weaker version of non-anomalous, which is just that we can find a completion (order by order in $\hbar$) of a solution of the classical master equation such that this completion fulfills the quantum master equation.

8.4.3 Observables in BV Formalism

Assuming that we have found an action S solution of the Quantum Master Equation, what are the observables of the theory? First of all, they have to be gauge-invariant quantities so, according to the previous explanations $\int^{BV} \mathscr{F} e^{iS/\hbar}$ has to be well defined, that is

$$\Delta \left(\mathscr{F} e^{iS/\hbar}\right) = 0.$$

As for the Quantum Master equation, expanding the exponential we eventually found that this amounts to $\mathscr{F}$ being a solution of

$$\{S, \mathscr{F}\} - i\hbar\Delta\mathscr{F} = 0. \tag{8.19}$$

This equation is the other quantum master equation and justifies the plural form in the title of this section. Now, observables will be equivalent classes of solutions of this equation. Indeed, if $\Delta\left(\mathscr{F} e^{iS/\hbar}\right) = \Delta\left(\mathscr{F}' e^{iS/\hbar}\right) = 0$ and if $(\mathscr{F} - \mathscr{F}')e^{iS/\hbar} = \Delta\mathscr{G}$ then for any Lagrangian submanifold Σ

$$\int_{\Pi N^*\Sigma}^{BV} (\mathscr{F} - \mathscr{F}')e^{iS/\hbar} = 0,$$

i.e., $\mathscr{F}$ and $\mathscr{F}'$ lead to the same measured values. Hence, we see that the observables of the theory are instead elements of the zeroth homology group of

$$\mathscr{O} = \{S, -\} - i\hbar\Delta. \tag{8.20}$$

Finally, our last task is to check that these are independent of the choice of the volume form Ω.

So, let us assume that we have chosen an action S solution of the Quantum Master Equation. We want to know what are the observables of the theory, deriving them for instance from the volume form

$$\Omega = e^{iS/\hbar}\Omega_0.$$

With respect to this volume form, we rewrite the integral to be computed

$$\int_{\Pi N^*\Sigma}^{BV,\Delta} \mathscr{F} e^{iS/\hbar} = \int_{\Pi N^*\Sigma}^{BV,\Delta_\Omega} \mathscr{F} := \int_{\Sigma} \mathscr{F} e^{iS/\hbar} \lrcorner \Omega_0,$$

where we have explicitly written the BV Laplacian with which every BV integral is built. Therefore, for the right-hand side of this equation, the gauge-invariance condition for $\mathscr{F}$ reduces to $\Delta_\Omega \mathscr{F} = 0$.

Before we check that this is equivalent to the quantum master equation (8.19), let us notice that until now, we have made sure that the two integrals are the same in the finite dimensional case, and we just have to check that this equality is preserved when on the level of the quantum master equations. This is motivated by the fact that these quantum master equations are the starting point (in the BV approach) for defining observables in quantum field theory.

From the relation (8.15), we see that

$$\Delta_\Omega \mathscr{F} = 0 \Leftrightarrow \Delta\mathscr{F} + \{iS/\hbar, \mathscr{F}\} = 0 \Leftrightarrow \{S, \mathscr{F}\} - i\hbar\Delta\mathscr{F} = 0. \tag{8.21}$$

Hence, the set of observables is independent of the choice of a volume form! Now, we could have taken a more general volume form

$$\Omega = e^f \Omega_0,$$

in which case the zero-point function would be

$$\int^{BV,\Delta_\Omega} e^{-f} e^{iS/\hbar} := \int e^{iS/\hbar}\Omega_0 = \int^{BV,\Delta_{\Omega_0}} e^{iS/\hbar}.$$

Then one can check that

$$\Delta_\Omega \left(e^{-f} e^{iS/\hbar}\right) = 0 \Leftrightarrow \{S, S\} - i\hbar\Delta S = 0.$$

One moreover checks that the set of observables is independent of f, that is

$$\Delta_\Omega \left(e^{-f} e^{iS/\hbar}\mathscr{F}\right) = 0 \Leftrightarrow \{S, \mathscr{F}\} - i\hbar\Delta\mathscr{F} = 0$$

if S is a solution of the Quantum Master Equation. This is only slightly more cumbersome than the case we treated in more details above.

8.5 Conclusion

We have now presented the basics idea behind our approach of the BV formalism. Many technical details were left behind the curtain, but we hope that the main philosophy was somehow conveyed to the reader. Let us briefly summarize it here, before saying a word on a few questions that are left unanswered in this text.

The BV formalism aims to deal with theories having a gauge symmetry. In such theories, the action is invariant under the action of a group (the gauge group). Thus, the integrand is constant over non-compact submanifolds of the configuration space, making the path integral ill-defined. The BRST solution to this problem is to quotient out the orbits of the gauge group and to declare that the real physically relevant configuration space is the quotiented space.

The BV approach is quite different. The idea is to work in some extended configuration space, that is on the shifted cotangent space of the initial configuration space, and to view the path integral (in the finite dimensional case) as an integral over some lagrangian submanifold of this extended space. One can change the integration domain without changing the value of the integral if the integrand has a nice property (which is just gauge invariance) and if the new integration domain is in the same homology class than the former.

The possible gauge-invariant actions and the possible gauge-invariant observables of the theory defined by such an action are shown to be independent of the choice of a volume form in the finite dimensional case. The remarkable point is that formulas derived in this way admit a natural generalization when the configuration space is infinite dimensional, which is of course the case of interest in quantum field theory.

These infinite dimensional generalizations can be transposed to be the definition of the observables in quantum field theory. This is the same philosophy as for the BRST approach of quantum field theory: In the finite dimensional case, one shows that the observables are the elements of a certain cohomology group. This cohomology group is still well defined in the infinite dimensional case, and we take it to be the definition of the observables in quantum field theory.

Now, a reader may ask: What happens in practice when moving to the infinite dimensional case? Well, first, the coordinates x^i will become fields, so that any derivative (for example in the BV Laplacian) will become a functional derivative. Whereas in the finite dimensional case, repeated indices meant a summation over the basis elements of the configuration space, in the infinite dimensional case, this has to be understood as an integral over the space-time one is dealing with as well as a discrete summation over the gauge group generators in the case of $c^\alpha c^*_\alpha$. Then, other usual features of quantum field theories can be treated with these objects. For example, the issue of renormalization has been dealt with since the beginning of BV formalism, but a modern point of view on the subject can be found in [6].

We have given above some arguments in favor of the BV formalism, namely that it allows to treat more general theories than the Faddeev–Popov and BRST formalisms, and that it is more adapted to some questions, e.g., the study of anomalies in QFT.

Moreover, we have recently observed a revival of the BV formalism, which has found applications in various areas.

In particular, it has been used (together with its hamiltonian formulation: the Batalin–Fradkin–Vilkovisky (BFV) formalism) to compute the Chern–Simons invariants of manifolds with boundaries. The article [4] of the present volume gives a detailed account of this line of research. Moreover, it has also recently been argued in [3] that the BV formalism written in the language of category theory might allow to work out perturbative quantum gravity as a perturbative quantum field theory. However, we would like to finish this text stressing the intrinsic elegance of the BV theory, which is one of the main reasons that motivated our interest for it.

Acknowledgements We would like to thank the organizers of this very interesting and pleasant school as well as all the participants and speakers who have clearly contributed to the quality of the school. Moreover, the authors would like to thank Christian Brouder and Frédéric Hélein for many enlightening discussions.

References

1. I.A. Batalin, G.A. Vilkovisky, Relativistic S-matrix of dynamical systems with boson and fermion constraints. Phys. Lett. B **69**(3), 309–312 (1977). ISSN 0370-2693
2. I.A. Batalin, G.A. Vilkovisky, Gauge algebra and quantization. Phys. Lett. B **102**(1), 27–31 (1981). ISSN 0370-2693
3. R. Brunetti, K. Fredenhagen, K. Rejzner, Quantum gravity from the point of view of locally covariant quantum field theory (2013)
4. A.S. Cattaneo, K. Wernli, P. Mnev, Split Chern–Simons theory in the BV-BFV formalism (2016)
5. P.J. Clavier, Analytic and geometrical approaches of non-perturbative quantum field theories. Ph.D. thesis, Université Pierre et Marie Curie, 2015
6. K. Costello, *Renormalization and Effective Field Theory* (AMS, Providence, 2011)
7. R.P. Feynman, Relativistic cut-off for quantum electrodynamics. Phys. Rev. **74**, 1430–1438 (1948)
8. J. Figueroa-O'Farrill, BRST cohomology (2006)
9. O. Gwilliam, Factorization algebra and free field theories. Ph.D. thesis, 2012
10. Y. Igarashi, K. Itoh, H. Sonoda, Quantum master equation for QED in exact renormalization group. Prog. Theor. Phys. **118**, 121–134 (2007)
11. R.E. Kallosh, Modified Feynman rules in supergravity. Nucl. Phys. B **141**(1), 141–152 (1978). ISSN 0550-3213
12. J. Tate, Homology of Noetherian rings and local rings. Ill. J. Math. **1**(1), 14–27 (1957). ISSN 0370-2693
13. S. Weinberg, *The Quantum Theory of Fields*, vol. 2 (Cambridge University Press, Cambridge, 1996). ISBN 0-521-55001-7

Chapter 9
Split Chern–Simons Theory in the BV-BFV Formalism

Alberto S. Cattaneo, Pavel Mnev and Konstantin Wernli

Abstract The goal of this note is to give a brief overview of the BV-BFV formalism developed by the first two authors and Reshetikhin in (Cattaneo et al., Commun Math Phys 332(2), 535–603, 2014) [9], (Cattaneo et al., Perturbative Quantum Gauge Theories on Manifolds with Boundary, 2015) [10] in order to perform perturbative quantisation of Lagrangian field theories on manifolds with boundary, and present a special case of Chern–Simons theory as a new example.

9.1 Introduction

Since the proposal of functorial quantum field theory by Atiyah and Segal [1, 25] mathematical research in this topic has progressed far and in many directions (see e.g. the books [17, 27], or the review article [22]). Recently, the first two authors together with Reshetikhin introduced the BV-BFV formalism, which can be seen either as an extension of functorial QFT to perturbative quantisation or, from another viewpoint, as a method to perturbatively quantise gauge theory in the presence of a boundary. The main idea is to unify the Lagrangian Batalin-Vilkovisky (BV) formalism [2, 4] in the bulk and the Hamiltonian Batalin-Fradkin-Vilkovisky (BFV) formalism [3] on the boundary.

A.S. Cattaneo (✉) · K. Wernli
Institut für Mathematik, Universität Zürich, Winterthurerstrasse 190, 8053 Zürich, Switzerland
e-mail: cattaneo@math.uzh.ch

K. Wernli
e-mail: konstantin.wernli@math.uzh.ch

P. Mnev
Max-Planck-Institut für Mathematik, Vivatsgasse 7, 53111 Bonn, Germany
e-mail: pmnev@pdmi.ras.ru

P. Mnev
St. Petersburg Department of V.A., Steklov Institute of Mathematics of the Russian Academy of Sciences, Fontanka 27, St. Petersburg 191023, Russia

A. Cardona et al. (eds.), *Quantization, Geometry and Noncommutative Structures in Mathematics and Physics*, Mathematical Physics Studies,
DOI 10.1007/978-3-319-65427-0_9

One possible application is to shed new light on the relation between perturbative techniques and mathematical ideas that are concepts of non-perturbative quantisation, like the Reshitikhin-Turaev invariants ([21], see also [14]), and thus ultimately about non-perturbative results to the path integral itself. In this note, a very first step on this road is taken by applying the formalism to a special form of Chern–Simons theory.

The note is structured as follows: Sect. 9.2 delivers a short overview of the relevant formal concepts via the example of abelian BF theory. Section 9.3 discusses a variant of Chern–Simons theory known as split Chern–Simons theory, in its BV-BFV formulation. Section 9.4 computes the state of this theory explicitly in lowest orders on the solid torus, which is a first step towards constructing the Chern–Simons invariant for lens spaces.

9.2 Overview of the BV and BV-BFV Formalisms

The goal of this section is to give a very brief introduction to the BV-formalism on manifolds without boundary, see also [11] in the present volume, and the BV-BFV formalism on manifolds with boundary, for two special examples. For the technical details we will refer to the papers [9, 10] where the Classical and Quantum BV-BFV formalisms were discussed in depth.

9.2.1 *Perturbative Quantisation of Lagrangian Field Theories*

Fix a dimension d. A Lagrangian field theory assigns to every closed d-dimensional manifold a *space of fields* F_M and an *action functional* $S_M\colon F_M \to \mathbb{R}$. This action functional is required to be local, i.e. of the form

$$S_M[\phi] = \int_M \mathscr{L}[\phi(x), \partial\phi(x), \ldots],$$

where $\mathscr{L}$, the so-called *Lagrangian density*, should depend only on the fields ϕ and finitely many of their derivatives. The critical points of the action functional are called the classical solutions of the theory, and are obtained by solving the Euler-Lagrange equations, also called equations of motion.

One way of quantising such a theory, suggested by the path integral from quantum mechanics, is to compute "integrals" of the form

$$\int_{F_M} \mathscr{O}[\phi] e^{\frac{i}{\hbar} S_M[\phi]} \mathscr{D}\phi,$$

where $\mathscr{O}$ is an "observable", over the space of fields F_M (these integrals are usually also called path integrals, even though they do not involve any paths). In this note we are only interested in the so-called *vaccuum state* or *partition function*

$$\psi = \int_{F_M} e^{\frac{i}{\hbar}S_M[\phi]}\mathscr{D}\phi. \tag{9.1}$$

However, in almost all relevant examples the spaces of fields have infinite dimension, and there is no sensible integration theory at hand. One way to still make sense of such expressions in the limit $\hbar \to 0$ is to use (formally) the principle of stationary phase. This produces an expansion in powers of $\hbar$ around critical points of the action. The terms in such an expansion can conveniently be labelled by diagrams, which after their inventor are called Feynman diagrams. A concise introduction can be found in [19].

Remark 1 (*Perturbative expansion*) We will only consider actions of the form $S = S_0 + S_{\text{int}}$ where S_0 is the quadratic part (also called "free" or "kinetic" part). In this case one usually considers the interaction or perturbation term to be small ("weak coupling") so we can expand the action around critical points of S_0 in powers of the interaction ("coupling constant"), and the integral then can be formally computed from the theory of Gaussian moments,[1] usually referred to as *Wick's theorem* in quantum field theory. Details can be found e.g. in the Book by Peskin and Schroeder [18] or lecture notes such as [5, 26].

9.2.2 Perturbative Quantisation of Gauge Theories

In many cases important for physics and mathematics, the Lagrangian is actually *degenerate*, i.e. its critical points are not isolated, and we cannot apply the stationary phase expansion, see e.g. [20]. This is usually due to the presence of symmetries on the space of fields that leave the action invariant.

This problem can often be solved by so-called *gauge-fixing* procedures (a thorough introduction to gauge theories from a physical viewpoint can be found in [13], a concise introduction to the mathematical formalisms in [15]). The common idea is to add more fields, corresponding to the generators of those symmetries, to remove the degeneracies in the Lagrangian. The most powerful gauge-fixing procedure (in the sense that it deals with the most general situation) is the Batalin-Vilkovsky formalism ([2, 4], for a short introduction to the mathematics see [12]). We will not discuss it in full generality, but rather explain the idea using the example of abelian BF theory, which will be important later in this note.

[1] Slight abuse of language as we are actually considering *Fresnel* integrals, i.e. with complex exponent.

9.2.2.1 Abelian BF Theory

Let M be a closed manifold, i.e. a compact manifold without boundary. Abelian BF theory has the space of fields

$$F_M = \Omega^1(M, \mathbb{R}) \oplus \Omega^{d-2}(M, \mathbb{R}) \ni (A, B).$$

Here $\Omega^p(M, \mathbb{R})$ denotes the vector space of real-valued differential p-forms on M. The action functional is

$$S_M[A, B] = \int_M B \wedge \mathrm{d}A$$

and the critical points are simply closed forms $\mathrm{d}A = 0, \mathrm{d}B = 0$. Clearly, the critical points are not isolated. In fact, adding any exact form to either A or B will leave the action invariant by Stokes' theorem. Therefore, the symmetries of the theory are generated by $\mathscr{A} := C^\infty(M) \oplus \Omega^{d-3}(M)$. An element $(c, \tau) \in \mathscr{A}$ acts on F_M by $(A, B) \mapsto (A + \mathrm{d}c, B + \mathrm{d}\tau)$. Since both the space of fields and the space of symmetries are linear here, the space of symmetries can be identified with the space of generators of the symmetries. We then declare the new space of fields to be

$$F_M^1 := F_M \oplus \mathscr{A}[1].$$

Here $\mathscr{A}[1]$ means that we give the fields in $\mathscr{A}$ ghost number 1.

Remark 2 (*Reducible symmetries*) In this note we will only be concerned with dimension $d = 3$, which we fix from now. However, in dimension $D \geq 4$, the symmetries of BF theory are *reducible*, that is, "the symmetries have some symmetries themselves": We do not change the symmetry of the action given by (c, τ) if we add to τ the differential of a $D - 4$-form τ_2. In this case one has to introduce the so-called "ghosts-for-ghosts" of ghost number 2, which amounts to adding to the space of fields $\Omega^{D-4}(M)[2]$, and continue all the way until we reach $\Omega^{D-D}(M)[D - 2]$.

Remark 3 (*Total degree*) Forms commute or anticommute according to their form degree, i.e. if ω is a p-form and τ is a q-form we have $\omega \wedge \tau = (-1)^{pq} \tau \wedge \omega$. If we introduce ghost fields, fields commute or anticommute according to their *total degree*, which is defined to be the form degree plus the ghost number. In BF theory in 3 dimensions, all fields have total degree 1, so all fields anticommute.

These new fields are not enough to make the action nondegenerate. One way to resolve the situation is to pass to the *BV space of fields*

$$\mathscr{F}_M := T^*[-1]F_M^1 = F_M^1 \oplus (F_M^1)^*[-1] = F_M \oplus \mathscr{A}[1] \oplus F_M^*[-1] \oplus \mathscr{A}^*[-2].$$

Table 9.1 The fields involved in BV version of abelian BF theory in dimension 3, with their form degree, ghost number and total degree

Field	Form degree	Ghost number	Total degree = ghost number + form degree
A	1	0	1
B	d–2 = 1	0	d–2 = 1
c	0	1	1
τ	d–3 = 0	1	d–2 = 1
A^+	d–1 = 2	–1	1
B^+	2	–1	1
c^+	d = 3	–2	1
τ^+	3	–2	1

The prescription to use cotangent bundle comes from finite dimensions where the dual of a vector space is always unique. Here we will not use the real dual spaces of differential forms (i.e. currents[2]), but use the Poincaré pairing

$$(\cdot,\cdot)\colon \Omega^p(M,\mathbb{R}) \times \Omega^{D-p}(M,\mathbb{R}) \to R$$
$$(\alpha,\beta) \mapsto \int_M \alpha \wedge \beta$$

to set $F(M)^* = (\Omega^1(M) \oplus \Omega^{d-2}(M))^* = \Omega^{d-1}(M) \oplus \Omega^2(M)$ and $\mathscr{A}^* = (\Omega^0(M) \oplus \Omega^{d-3}(M))^* = \Omega^d(M) \oplus \Omega^3(M)$. Denoting the dual fields with a $^+$, we summarise the fields and their degrees in Table 9.1.

The new (BV) action is then

$$\mathscr{S}_M = \int_M B \wedge \mathrm{d}A + A^+ \wedge \mathrm{d}c + B^+ \wedge \mathrm{d}\tau$$

which leads to Euler-Lagrange equations

$$\mathrm{d}A = \mathrm{d}B = \mathrm{d}c = \mathrm{d}A^+ = \mathrm{d}B^+ = \mathrm{d}c^+ = 0.$$

Of course, right now it seems we only introduced more degeneracy, but this is where the gauge fixing comes into play. First, however, we shall need a couple of remarks.

Remark 4 (*Superfields*) At this point it is very convenient to introduce the "superfields"

$$\mathsf{A} = c + A + B^+ + \tau^+ \in \Omega^\bullet(M),$$
$$\mathsf{B} = \tau + B + A^+ + c^+ \in \Omega^\bullet(M).$$

[2]This leads to another theory with a larger space of fiels called *canonical BF theory*, see [15].

The action now simply reads

$$\mathscr{S}_M = \int_M \mathsf{B} \wedge \mathsf{dA},$$

where only the integral of the top-degree part is non-zero, and the Euler-Lagrange equations can be summarised as

$$\mathsf{dB} = \mathsf{dA} = 0.$$

Remark 5 (*Structure of the space of fields*) The grading by ghost number endows $\mathscr{F}_M$ with the structure of a *graded vector space*. The pairing of fields and anti-fields endows $\mathscr{F}_M$ with a so-called *odd symplectic structure* (odd because it pairs fields whose degrees add up to -1, rather than to 0). If δ denotes the de Rham differential on $\mathscr{F}_M$, it is given by

$$\omega_M = \int_M \delta\mathsf{A} \wedge \delta\mathsf{B}. \tag{9.2}$$

As every odd symplectic structure it induces an odd Poisson bracket on $\mathrm{Fun}(\mathscr{F}_M)$, which in this case is called the *BV bracket*. It is well defined on the subspace of local functionals (see the discussion of BV formalism in [7]). Also, one has the *BV Laplacian*

$$\Delta = \sum_{k=0}^{3} (-1)^{k+1} \int_M \frac{\delta^2}{\delta\mathsf{A}^{(k)}(x)\delta\mathsf{B}^{(k)}(x)},$$

where $\mathsf{A}^{(k)}$ denotes the k-form part of A. Together with the BV bracket, it gives $\mathrm{Fun}(\mathscr{F}_M)$ the structure of a so-called *BV algebra*. However, in the infinite dimensional setting this expression for the BV Laplacian is very singular and needs to be regularised carefully.

The BV formalism to compute integral (9.1) now proceeds as follows: one picks a Lagrangian subspace $\mathscr{L}$ of $\mathscr{F}_M$ such that the BV action has isolated critical points on $\mathscr{L}$. This is the gauge fixing in the BV formalism. The integral

$$\psi = \int_{\mathscr{L}} e^{\frac{i}{\hbar}\mathscr{S}[\phi]} \mathscr{D}\phi$$

can be computed by methods of Feynman diagrams. If the BV action satisfies the *Quantum Master Equation* $\Delta(e^{\frac{i}{\hbar}S}) = 0$, then under deformations of $\mathscr{L}$, the result changes by a Δ-exact term.

Remark 6 (*Quantum and Classical Master Equations*) The Quantum Master equation $\Delta(e^{\frac{i}{\hbar}\mathscr{S}}) = 0$ is equivalent to $(\mathscr{S}, \mathscr{S}) - 2i\hbar\Delta\mathscr{S} = 0$, where $(\cdot, \cdot)$ is the BV bracket. Expanding $\mathscr{S}$ as a power series in $\hbar$, the degree 0 part S_0 has to satisfy $(S_0, S_0) = 0$. This is called the *Classical Master Equation*.

Remark 7 The statements above can be made entirely precise and rigorously proven for finite-dimensional spaces of fields (see e.g. [11] in this volume, [12] or [10], Chap. 2). In the infinite-dimensional setting, the BV formalism produces a number of postulates that one has to prove *a posteriori*. An example (for the extension to manifolds with boundary, the BV-BFV formalism) in this note is the mQME (9.4) which is proven for abelian BF theory. In [10] a general procedure to prove the mQME is described.

Remark 8 (*Perturbative expansion of interacting gauge theories*) Abelian BF theory is an example for a free theory (i.e. $S_{\text{int}} \equiv 0$). For theories that are perturbations of free theories, the gauge-fixing for the free part of the theory can be used to compute the expansion in powers of the coupling constant. We will call theories that are perturbations of abelian BF theory "BF-like". Examples are the Poisson Sigma model and non-abelian BF theory, and, most importantly for this note, split Chern–Simons theory.

9.2.2.2 Residual Fields

It can happen that the degeneracy in the quadratic part of the action does not stem from the gauge symmetries alone. This is the case when the operator in the quadratic part of the action has non-trivial "zero modes" i.e. it has zeros that are not related under gauge symmetries. In the case of abelian BF theory, the operator in question is the de Rham differential, while the gauge symmetries are given by shifting the fields by exact forms. It follows that the space of inequivalent zero modes is precisely the de Rham cohomology of M.

In this case the procedure is as follows. One splits the space of fields $\mathscr{F}_M = \mathscr{Y}' \times \mathscr{Y}''$ into a space of *residual fields*[3] $\mathscr{Y}'$, consisting of representatives of the zero modes, and a complement $\mathscr{Y}''$ that we will call *fluctuations*.[4] Then one only integrates over a Lagrangian subspace $\mathscr{L}$ of $\mathscr{Y}''$, so that the result depends on the residual fields. This yields the definition of the *effective action*:

$$e^{\frac{i}{\hbar}\mathscr{S}^{\text{eff}}(\phi')} = \int_{\phi'' \in \mathscr{L} \subset \mathscr{Y}''} e^{\frac{i}{\hbar}S(\phi',\phi'')}\mathscr{D}\phi''.$$

To be compatible with the BV formalism, $\mathscr{Y}'$ and $\mathscr{Y}''$ should be odd symplectic themselves, such that $\mathscr{F}_M$ has the product structure. In this case, one can prove that in the finite-dimensional case, the QME for the action on $\mathscr{F}$ induces the QME for the effective action. In the case at hand of abelian BF theory, we choose a finite-dimensional space of residual fields, the de Rham cohomology, and one can prove explicitly that the effective action satisfies the QME. Therefore $\mathscr{Y}'$ should be given by representatives of the de Rham cohomology of M. Such a splitting (and a suitable

[3] Also known as background fields, slow fields, infrared fields.

[4] Otherwise known as fast fields or ultraviolet fields.

choice of Lagrangian) can then be found e.g. by Hodge decomposition. Choosing a Riemannian metric g, the space $\mathcal{Y}'$ is given by g-harmonic forms and the Lagrangian $\mathcal{L}$ by the Lorentz gauge condition $d^*\phi = 0$. On this space d has no kernel and therefore the restriction of the BV extension of the abelian BF action to this Lagrangian subspace is non-degenerate.

9.2.3 On Manifolds with Boundary

We will now consider the case of manifolds with boundary. The strategy that is compatible with the mathematical idea of gluing of manifolds along boundary components is not to fix boundary conditions, but instead to think of the state as a functional on the possible boundary fields.

Consider first the case of a theory without gauge symmetries. Under some assumptions, one can show that a d-dimensional field theory induces a space of fields F^{∂}_{Σ} on $(d-1)$-dimensional manifolds Σ that has a natural even symplectic structure. The space of states should be a quantisation of this symplectic manifold. In many examples, F^{∂}_{Σ} is actually an affine space, and one can define a quantisation from a Lagrangian polarisation[5] with a smooth leaf space (examples of this are the position or momentum space) B_{Σ}. In this case, the space of states is the space of functionals on B_{Σ}. If $\Sigma = \partial M$, there is a surjective submersion $F_M \to F^{\partial}_{\partial M}$ given by restriction of fields to the boundary. If we denote by p the composition of this map with the projection $F^{\partial}_{\partial M} \to B_{\partial M}$, we can define the state by the "integral"

$$\widehat{\psi}_M(\beta) = \int_{p^{-1}(\beta)} e^{\frac{i}{\hbar}S[\phi]}\mathcal{D}\phi$$

for $\beta \in B_{\partial M}$.

9.2.4 The BV-BFV Formalism

Now we want to combine this with the method used to deal with gauge theories discussed above. Given a space of BV fields $\mathcal{F}_M$ for every d-dimensional manifold M, there is again an induced space of fields $\mathcal{F}^{\partial}_{\Sigma}$ on $d-1$-dimensional manifolds endowed with what is called a BFV structure (see [24] for a mathematical discussion of BFV structure). The result is what is called a *BV-BFV manifold*, whose definition we will now recall.

Definition 9.2.1 (*BFV manifold*) A *BFV manifold* is a triple $(\mathcal{F}, \omega, Q)$, where

- $\mathcal{F}$ is a $\mathbb{Z}$-graded manifold,

---

[5]This is basically a choice of coordinates and canonically conjugate momenta, similar to the p and q variables in quantum mechanics.

- $\omega = \delta\alpha$ is an exact degree 0 symplectic form on $\mathscr{F}$,
- Q is a degree $+1$ vector field on $\mathscr{F}$,

such that

- Q is symplectic for ω, i.e. $L_Q\omega = 0$,
- Q is cohomological, i.e. $Q^2 = 0$ or equivalently $[Q, Q] = 0$.

For degree reasons this implies the existence of a degree 1 Hamiltonian function S for Q, i.e. $\iota_Q\omega = \delta S$ (and the datum of such function specifies a cohomological symplectic vector field) and this function S automatically satisfies the Classical Master Equation $(S, S) = 2\iota_Q\iota_Q\omega = 0$. The $\mathbb{Z}$-grading of the manifold is the ghost number we briefly explained above.

Definition 9.2.2 (*BV-BFV manifold*) A *BV-BFV manifold* over a given BFV manifold $(\mathscr{F}^\partial, \omega^\partial = \delta\alpha^\partial, Q^\partial$ is a quintuple $(\mathscr{F}, \omega, Q, \mathscr{S}, \pi)$ where

- $\mathscr{F}$ is a $\mathbb{Z}$-graded manifold,
- ω is a degree -1 symplectic form,
- Q is a degree $+1$ cohomological vector field,
- $\mathscr{S}$ is a degree 0 function on $\mathscr{F}$,
- π is a surjective submersion $\mathscr{F} \to \mathscr{F}^\partial$,

such that[6]

- $\delta\pi(Q) = Q^\partial$,
- $\iota_Q\omega = \delta\mathscr{S} + \pi^*\alpha^\partial$.

The axioms imply the *modified Classical Master Equation* (mCME)

$$\frac{1}{2}\iota_Q\iota_Q\omega - \pi^*\mathscr{S}^\partial = 0. \tag{9.3}$$

Remark 9 (*Shifting* α) Given a BV-BFV theory and a functional f on the space of boundary fields, we can define a new BV-BFV theory by $\alpha^\delta \mapsto \alpha^\partial + \delta f$, $\mathscr{S} \mapsto \mathscr{S} - \pi^* f$. It will coincide with the previous theory on closed manifolds.

In many cases, the BV structure on the bulk and the BFV structure on the boundary look very similar in the superfield formalism.

Let us look at the example of abelian BF theory on a 3-manifold M with boundary ∂M that is included via $\iota\colon \partial M \to M$. Let $\mathscr{F}_M$ be the space of BV fields $\Omega^\bullet(M)[1] \oplus \Omega^\bullet(M)[1] \ni (\mathsf{A}, \mathsf{B})$. Denote by $\mathsf{A}^\partial := \iota^*\mathsf{A}$, $\mathsf{B}^\partial := \iota^*\mathsf{B}$ the restrictions of these fields to the boundary. Then the space of boundary BFV fields is $\mathscr{F}^\partial_{\partial M} = \Omega^\bullet(\partial M)[1] \oplus \Omega^\bullet(\partial M)[1] \ni (\mathsf{A}^\partial, \mathsf{B}^\partial)$. The symplectic form and action have the same form as before

[6] This definition differs from the one in [9] by a purely conventional sign $(-1)^n$ in front of δS.

$$\omega^{\partial}_{\partial M} = \int_{\partial M} \delta \mathsf{A}^{\partial} \wedge \delta \mathsf{B}^{\partial},$$

$$S^{\partial}_{\partial M} = \int_{\partial M} \mathsf{B}^{\partial} \wedge \mathsf{d}\mathsf{A}^{\partial},$$

and the corresponding Hamiltonian vector field on $\mathscr{F}^{\partial}_{\partial M}$ is

$$Q^{\partial}_{\partial M} = \int_{\partial M} \mathsf{d}\mathsf{A}^{\partial} \frac{\delta}{\delta \mathsf{A}^{\partial}} + \mathsf{d}\mathsf{B}^{\partial} \frac{\delta}{\delta \mathsf{B}^{\partial}}.$$

However, considering Table 9.1 and that the dimension of ∂M is 2, notice that $\omega^{\partial}_{\partial M}$ pairs fields of *opposite* ghost number, and thus has degree 0. I.e., $(\mathscr{F}^{\partial}_{\partial M}, \omega^{\partial}_{\partial M}, Q^{\partial}_{\partial M})$ is a BFV manifold.

Claim 1 *If we denote*

$$Q_M = \int_M \mathsf{d}\mathsf{A} \frac{\delta}{\delta \mathsf{A}} + \mathsf{d}\mathsf{B} \frac{\delta}{\delta \mathsf{B}}$$

and $\pi_M = \iota^* \colon \mathscr{F}_M \to \mathscr{F}^{\partial}_{\partial M}$ *the restriction of fields to the boundary, then in abelian BF theory the quintuple* $(\mathscr{F}_M, \omega_M, Q_M, S_M, \pi_M)$ *is a BV-BFV manifold over the BFV manifold* $(\mathscr{F}^{\partial}_{\partial M}, \omega^{\partial}_{\partial M}, Q^{\partial}_{\partial M})$.

Proof We will just prove the central BV-BFV identity $\iota_{Q_M} \omega_M = \delta S_M + \pi^* \alpha^{\partial}_{\partial M}$. Notice that the de Rham differential on $\mathscr{F}_M$ is given by

$$\delta = \int_M \delta \mathsf{A} \frac{\delta}{\delta \mathsf{A}} + \delta \mathsf{B} \frac{\delta}{\delta \mathsf{B}}$$

and one choice of $\alpha^{\partial}_{\partial M}$ is

$$\alpha^{\partial}_{\partial M} = \int_{\partial M} \mathsf{B} \wedge \delta \mathsf{A}.$$

On the one hand,

$$\iota_{Q_M} \omega_M = \int_M \mathsf{d}\mathsf{A} \wedge \delta \mathsf{B} + \delta \mathsf{A} \wedge \mathsf{d}\mathsf{B}.$$

On the other hand, integrating by parts yields

$$\begin{aligned} \delta \mathscr{S}_M &= \delta \int_M \mathsf{B} \wedge \mathsf{d}\mathsf{A} = \int_M \mathsf{B} \wedge \mathsf{d}\delta A + \int_M \delta B \wedge \mathsf{d}\mathsf{A} \\ &= \int_M \mathsf{d}\mathsf{B} \wedge \delta \mathsf{A} + \int_M \delta B \wedge \mathsf{d}\mathsf{A} - \int_{\partial M} \mathsf{B} \wedge \delta \mathsf{A} = \iota_{Q_M} \omega_M - \pi^*_M \alpha^{\partial}_{\partial M}. \end{aligned}$$

9.2.5 The Quantum BV-BFV Formalism

We now explain the data of a quantum BV-BFV theory and show how to quantise in the example of abelian BF theory, before turning to the example of Chern–Simons theory. The perturbative quantisation of a BV-BFV theory consists of the following data:

1. A cochain complex $(\mathcal{H}_\Sigma^{\mathcal{P}}, \Omega_\Sigma^{\mathcal{P}})$ for every $(d-1)$-manifold Σ with a choice of polarisation in $\mathcal{F}_\Sigma^\partial$.
2. A finite-dimensional BV manifold $(\mathcal{V}_M, \Delta_{\mathcal{V}_M})$ - called the space of *residual fields* - associated to every d-manifold M and polarisation $\mathcal{P}$ on $\mathcal{F}_{\partial M}^\partial$.
3. Let $\widehat{\mathcal{H}_M^{\mathcal{P}}} := \mathcal{H}_{\partial M}^{\mathcal{P}} \hat{\otimes} C^\infty(\mathcal{V}_M)$ and endow it with the two commuting coboundary operators $\widehat{\Omega}_M^{\mathcal{P}} := \Omega_{\partial M}^{\mathcal{P}} \otimes \mathrm{id}$ and $\widehat{\Delta}_M^{\mathcal{P}} = \mathrm{id} \otimes \Delta_{\mathcal{V}_M}$. Then we require the existence of a *state* $\widehat{\psi}_M$ satisfying the *modified Quantum Master Equation* (mQME)
$$(\hbar^2 \widehat{\Delta}_M^{\mathcal{P}} + \widehat{\Omega}_M^{\mathcal{P}})\widehat{\psi}_M = 0, \tag{9.4}$$
the quantum counterpart of the mCME (9.3).

Some comments are in order. The cochain complex $(\mathcal{H}_\Sigma^{\mathcal{P}}, \Omega_\Sigma^{\mathcal{P}})$ is to be constructed as a sort of geometric quantisation of the symplectic manifold $\mathcal{F}_{\partial M}^\partial$ with the polarisation $\mathcal{P}$ and the action $\mathcal{S}_{\partial M}^\partial$. The general construction of the boundary quantisation is not important in this note. More important is the idea of residual fields that was explained in Sect. 9.2.2.2. The state is then computed by combining the methods of Sects. 9.2.2 and 9.2.3. Again, assume we have a polarisation $\mathcal{P}$ of $\mathcal{F}_{\partial M}^\partial$ with smooth leaf space $\mathcal{B}_{\partial M}^{\mathcal{P}}$. In this case $\mathcal{H}_\Sigma^{\mathcal{P}} \subset \mathrm{Fun}(\mathcal{B}_{\partial M}^{\mathcal{P}})$ is a certain subspace of functionals on boundary conditions defined in detail in Sects. 3.5.1 and 4.1.1 to 4.1.3 in [10].[7] We will further assume that actually $\mathcal{F}_M = \mathcal{B}_{\partial M}^{\mathcal{P}} \times \mathcal{Y}$ so that the fibers of the projection $p\colon \mathcal{F}_M \to \mathcal{B}_{\partial M}^{\mathcal{P}}$ are just $\{b\} \times \mathcal{Y}$. Moreover, we assume there is a functional $f_{\partial M}^{\mathcal{P}}$ such that $\alpha_{\partial M} - \delta f_{\partial M}^P$ vanishes when restricted to the fibers, i.e. on $\mathcal{Y}$, and then adapt the bulk action as in Remark 9. We then split $\mathcal{Y} = \mathcal{V}_M \times \mathcal{Y}''$ into a space of residual fields and fluctuations $\mathcal{Y}''$. Then we can finally define the state $\widehat{\psi}_M$ by

$$\widehat{\psi}_M(b,\phi) = \int_{\mathcal{L} \subset \mathcal{Y}''} e^{\frac{i}{\hbar} S_M(b,\phi,\phi'')} \mathcal{D}\phi'' \in \widehat{\mathcal{H}_M^{\mathcal{P}}} = \mathcal{H}_{\partial M}^{\mathcal{P}} \hat{\otimes} C^\infty(\mathcal{V}_M).$$

Again, we define the BV effective action by

$$\widehat{\psi}_M(b,\phi) = e^{\frac{i}{\hbar} S^{\mathrm{eff}}(b,\phi)}.$$

Instead of entering a general discussion of the above, let us continue the example of abelian BF theory.

[7] There are some subtleties arising from the regularisation of higher functional derivatives that would be too much for the purpose of this note.

9.2.6 *Abelian BF Theory in the Quantum BV-BFV Formalism*

9.2.6.1 Polarisations

Here there are two easy polarisations on $\mathscr{F}^{\partial}_{\partial M} = \Omega^\bullet(\partial M)[1] \oplus \Omega^\bullet(\partial M)[1]$, namely the ones given by $\frac{\delta}{\delta \mathsf{A}^\partial}$ (whose leaf space can be identified with the B^∂ fields) and $\frac{\delta}{\delta \mathsf{B}^\partial}$ (whose leaf space can be identified with the A^∂ fields).

Let now M be a manifold with boundary $\partial M = \partial_1 M \sqcup \partial_2 M$. We then define the polarisation $\mathscr{P}$ to be the $\frac{\delta}{\delta \mathsf{B}^\partial}$-polarisation on $\partial_1 M$ and the $\frac{\delta}{\delta \mathsf{A}^\partial}$-polarisation on $\partial_2 M$, so that we have the leaf space $\mathscr{B}^{\mathscr{P}}_{\partial M} = \Omega^\bullet(\partial_1 M)[1] \oplus \Omega^\bullet(\partial_2 M)[1]$, we denote the coordinates on it by $(\mathbb{A}, \mathbb{B})$. The correct way to adapt the boundary 1-form is to subtract the differential $f^{\mathscr{P}}_{\partial M} = \int_{\partial_2 M} \mathsf{B}^\partial \wedge \mathsf{A}^\partial$ from it.

9.2.6.2 Choosing a Splitting

We now split the space of fields $\mathscr{F}_M$ by choosing extensions $\widetilde{\mathbb{A}}, \widetilde{\mathbb{B}}$ of $\mathbb{A}$ and $\mathbb{B}$ from the boundary to the bulk of the manifold and splitting $\mathsf{A} = \widetilde{\mathbb{A}} + \widehat{\mathsf{A}}$, $\mathsf{B} = \widetilde{\mathbb{B}} + \widehat{\mathsf{B}}$ where $\widehat{\mathsf{A}}$ and $\widehat{\mathsf{B}}$ restrict to 0 on $\partial_1 M$ resp. $\partial_2 M$. As discussed in [10], one needs to require the extensions to be discontinuous extensions by 0 outside of the boundaries. One way to make this more precise is to work with a family of regular decompositions approximating this singular one, resulting a family of states that only in the limit will satisfy the mQME. We will therefore choose these extensions and identify $\widetilde{\mathbb{A}} = \mathbb{A}, \widetilde{\mathbb{B}} = \mathbb{B}$. This is our splitting $\mathscr{F}_M = \mathscr{B}^{\mathscr{P}}_{\partial M} \times \mathscr{Y}$.

9.2.6.3 Residual Fields and Fluctuations, Gauge Fixing

We now want to split $\mathscr{Y}$ into residual fields and fluctuations. As discussed above, in abelian BF theory the residual fields should contain the de Rham cohomology of M. In the case with boundary, for our polarisation, the space of residual fields is

$$\mathscr{V}_M = H^\bullet(M, \partial_1 M)[1] \oplus H^\bullet(M, \partial_2 M)[1].$$

We choose representatives $\chi_i \in \Omega^\bullet_{closed}(M, \partial_1 M)$ and $\chi^j \in \Omega^\bullet_{closed}(M, \partial_2 M)$ such that their cohomology classes form a basis of $H^\bullet(M, \partial_1 M)$ resp. $H^\bullet(M, \partial_2 M)$ and $\int_M \chi_i \wedge \chi^j = \delta_i^j$. Then, we write $\mathsf{a} = \sum_i z^i \chi_i$, $\mathsf{b} = \sum_i z_i^+ \chi^i$ for elements of $\mathscr{V}_M \subset \mathscr{F}_M$. The BV Laplacian $\Delta_{\mathscr{V}_M}$ is then

$$\Delta_{\mathscr{V}_M} = \sum_i -\frac{\partial}{\partial z^i}\frac{\partial}{\partial z_i^+}.$$

A possible way to choose such a basis, a complement $\mathscr{Y}''$ and a Lagrangian $\mathscr{L} \subset \mathscr{Y}''$ is to pick a Riemannian metric and use Hodge decomposition on manifolds with boundary (see [6]). This is the choice of gauge fixing (it is a variant of the Lorentz Gauge Fixing mentioned earlier). Its most important feature is that the gauge-fixing Lagrangian does not depend the boundary and background fields. We will avoid the details of this lengthy discussion, referring the interested reader again to [10] (Sect. 3.3 and Appendix A), and simply assume we can decompose the fields $\widehat{\mathsf{A}} = \mathsf{a} + \alpha$, $\widehat{\mathsf{B}} = \mathsf{b} + \beta$ into residual fields and fluctuations.

Remark 10 (*Decomposition of the action*) The decomposition of the fields also induces a decomposition of the adapted action

$$\mathscr{S}^{\mathscr{P}}_{\mathscr{M}} = \widehat{\mathscr{S}}_{M,0} + \mathscr{S}^{\text{back}}_{M} + \mathscr{S}^{\text{source}}_{M}, \tag{9.5}$$

where

$$\begin{aligned}
\widehat{\mathscr{S}}_{M,0} &= \int_M \beta \wedge \mathrm{d}\alpha, \\
\mathscr{S}^{\text{back}}_{M} &= -\left(\int_{\partial_2 M} \mathbb{B} \wedge \mathsf{a} + \int_{\partial_1 M} \mathsf{b} \wedge \mathbb{A}\right), \\
\mathscr{S}^{\text{source}}_{M} &= -\left(\int_{\partial_2 M} \mathbb{B} \wedge \alpha + \int_{\partial_1 M} \beta \wedge \mathbb{A}\right).
\end{aligned}$$

Proof Assume we have chosen non-singular extensions $\widetilde{\mathbb{A}}$, $\widetilde{\mathbb{B}}$ and split $\mathsf{A} = \widetilde{\mathbb{A}} + \mathsf{a} + \alpha$, $\mathsf{B} = \widetilde{\mathbb{B}} + \mathsf{b} + \beta$. The action then reads

$$\mathscr{S}^{\mathscr{P}}_{\mathscr{M}} = \int_M (\widetilde{\mathbb{B}} + \mathsf{b} + \beta) \wedge \mathrm{d}(\widetilde{\mathbb{A}} + \mathsf{a} + \alpha) - \int_{\partial_2 M} \iota_2^*((\widetilde{\mathbb{B}} + \mathsf{b} + \beta) \wedge (\widetilde{\mathbb{A}} + \mathsf{a} + \alpha))$$

where ι_2 denotes the inclusion $\partial_2 M \hookrightarrow M$. We can assume the supports of $\widetilde{\mathbb{B}}$ and $\widetilde{\mathbb{A}}$ are disjoint. Furthermore, we have that $\iota_2^* \mathsf{b} = \iota_2^* \beta = \iota_2^* \widetilde{\mathbb{A}} = 0$ and $\mathrm{d}\mathsf{a} = \mathrm{d}\mathsf{b} = 0$. We then get

$$\mathscr{S}^{\mathscr{P}}_{\mathscr{M}} = \int \mathsf{b} \wedge \mathrm{d}\widetilde{\mathbb{A}} + \beta \wedge \mathrm{d}\widetilde{\mathbb{A}} + \widetilde{\mathbb{B}} \wedge \mathrm{d}\alpha + \mathsf{b} \wedge \mathrm{d}\alpha + \beta \wedge \mathrm{d}\alpha - \left(\int_{\partial_2 M} \widetilde{\mathbb{B}} \wedge \mathsf{a} + \widetilde{\mathbb{B}} \wedge \alpha\right)$$

The integral of $\mathsf{b} \wedge \mathrm{d}\alpha$ vanishes by integration by parts since b is closed and $\mathsf{b} \wedge \mathrm{d}\alpha$ is zero restricted to ∂M. Now integrate the $\widetilde{\mathbb{A}}$ terms by parts, resulting in

$$\mathscr{S}^{\mathscr{P}}_{\mathscr{M}} = \int_M \mathrm{d}\beta \wedge \widetilde{\mathbb{A}} + \widetilde{\mathbb{B}} \wedge \mathrm{d}\alpha + \beta \wedge \mathrm{d}\alpha - \left(\int_{\partial_1 M} \mathsf{b} \wedge \widetilde{\mathbb{A}} + \beta\widetilde{\mathbb{A}}\right) - \left(\int_{\partial_2 M} \widetilde{\mathbb{B}} \wedge \mathsf{a} + \widetilde{\mathbb{B}} \wedge \alpha\right)$$

Sending $\widetilde{\mathbb{A}}, \widetilde{\mathbb{B}}$ to singular extensions proves the claim as the first two terms will vanish.

9.2.6.4 The State

We now would like to compute the state

$$\widehat{\psi}_M(\mathbb{A}, \mathbb{B}, \mathsf{a}, \mathsf{b}) = \int_{(\alpha,\beta)\in\mathscr{L}} e^{\frac{i}{\hbar}\mathscr{S}_M^{\mathscr{P}}(\mathbb{A}+\mathsf{a}+\alpha, \mathbb{B}+\mathsf{b}+\beta)}\mathscr{D}\alpha\mathscr{D}\beta$$
$$\in \widehat{\mathscr{H}_M^{\mathscr{P}}} \subset \mathrm{Fun}(\mathscr{B}_{\partial M}^{\mathscr{P}})\hat{\otimes}C^\infty(\mathscr{V}_M). \tag{9.6}$$

as a formal Gaussian integral. Applying decomposition (9.5) of the action, and the general theory of performing such Gaussian integrals in quantum field theory (see [19, 20]), we need to understand the integral

$$T_M := \int_{\mathscr{L}} e^{\frac{i}{\hbar}\widehat{\mathscr{S}}_M}\mathscr{D}\alpha\mathscr{D}\beta. \tag{9.7}$$

as a regularised determinant of the inverse of the operator d in the quadratic part of the action. This is not an easy task (see [16, 23]), but for our purposes it is enough to say that T_M is a number independent of the choice of $\mathscr{L}$ (but that can depend on our choice of representatives of cohomology). The integral (9.6) can then be expressed in terms of the so-called *propagator*[8]

$$\eta(x_1, x_2) = \frac{-1}{T_M}\frac{1}{i\hbar}\int_{\mathscr{L}} e^{\frac{i}{\hbar}\widehat{\mathscr{S}}_M}\alpha(x_1)\beta(x_2)\mathscr{D}\alpha\mathscr{D}\beta. \tag{9.8}$$

Namely,

$$\widehat{\psi}_M(\mathbb{A}, \mathbb{B}, \mathsf{a}, \mathsf{b}) = T_M e^{\frac{i}{\hbar}S^{\mathrm{eff}}(\mathbb{A},\mathbb{B},\mathsf{a},\mathsf{b})}, \tag{9.9}$$

with

$$S^{\mathrm{eff}}(\mathbb{A}, \mathbb{B}, \mathsf{a}, \mathsf{b}) = -\left(\int_{\partial_2 M}\mathbb{B}\wedge\mathsf{a} - \int_{\partial_1 M}\mathsf{b}\wedge\mathbb{A}\right) - \int_{\partial_2 M\times\partial_1 M}\pi_1^*\mathbb{A}\wedge\eta\wedge\pi_2^*\mathbb{B}. \tag{9.10}$$

9.2.6.5 The Propagator

The propagator η is a $(d-1)$-form on the configuration space $C_2^0(M) = \{(x_1, x_2) \in M\times M : x_1 \neq x_2\}$ that vanishes for $x_2 \in \partial_1 M$ or $x_1 \in \partial_2 M$. It is determined by our choice of gauge fixing Lagrangian. It has two important properties:

- Its differential satisfies

$$\mathrm{d}\eta = \sum_i (-1)^{\deg\chi_i}\pi_1^*\chi_i\pi_2^*\chi^i. \tag{9.11}$$

[8] Also known to physicists as 2-point function or - slightly abusing language - Green's function.

- For any $x \in M$, if we fix a chart $\phi\colon U \to \mathbb{R}^3$ satisfying $\phi(x) = 0$, then

$$\lim_{\varepsilon\to 0}\int_{y\in\partial B_\varepsilon(0)} \eta(\phi^{-1}(y), x) = 1 = -\lim_{\varepsilon\to 0}\int_{y\in\partial B_\varepsilon(0)} \eta(x, \phi^{-1}(y)). \tag{9.12}$$

A choice of such a propagator (and representatives of cohomology) also leads to the definition of a gauge-fixing Lagrangian. For computations with Feynman diagrams it is often desirable to have a propagator satisfying also

- $$\int_{y\in M} \eta(x, y)\chi_i(y) = \int_{x\in M} \chi^i(x)\eta(x, y) = 0, \tag{9.13}$$

- $$\int_{y\in M} \eta(x, y)\eta(y, z) = \int_{x\in M} \eta(z, x)\eta(x, y) = 0. \tag{9.14}$$

These properties do not automatically follow from the definition but they can always be satisfied by picking a suitable $\mathscr{L}$ (see Sect. 4 in [8] for a discussion on manifolds without boundary, arguments there can be adapted to the case with boundary using machinery in [10]).

9.2.6.6 mQME

In the case of abelian BF theory, the quantisation of the boundary is simply the "standard" or "canonical" quantisation. It is obtained by the following recipe: In the boundary action, on $\partial_1 M$ we have to replace every occurence of $\widehat{\mathsf{B}}$ by $(-i\hbar\frac{\delta}{\delta\mathbb{A}})$, on $\partial_2 M$, $\widehat{\mathsf{A}}$ has to replaced by $(-i\hbar\frac{\delta}{\delta\mathbb{B}})$. Here we have to integrate by parts to do so. The result is

$$\Omega^{\mathscr{P}}_{\partial M} = (-i\hbar)\left(\int_{\partial_1 M} \mathrm{d}\mathbb{A}\frac{\delta}{\delta\mathbb{A}} + \int_{\partial_2 M} \mathrm{d}\mathbb{B}\frac{\delta}{\delta\mathbb{B}}\right). \tag{9.15}$$

Claim 2 *The state $\widehat{\psi}_M$ defined by* (9.9) *satisfies the mQME* (9.4)

$$(\hbar^2\widehat{\Delta}^{\mathscr{P}}_M + \widehat{\Omega}^{\mathscr{P}}_M)\widehat{\psi}_M = 0. \tag{9.16}$$

Proof Since the effective action S^{eff} given in (9.10) is only linear in coordinates on $\mathcal{V}_M$, it is immediate that $\Delta S^{\text{eff}} = 0$. In this case $(\hbar^2\Delta + \Omega)e^{\frac{i}{\hbar}S^{\text{eff}}} = -\frac{1}{2}(S^{\text{eff}}, S^{\text{eff}})e^{\frac{i}{\hbar}S^{\text{eff}}} + \Omega e^{\frac{i}{\hbar}S^{\text{eff}}}$. Only the first two terms in the action depend on the residual fields and hence contribute to the BV bracket. Also, only the bracket of b with a is nontrivial, so we have

$$\frac{1}{2}(\mathscr{S}^{\text{eff}},\mathscr{S}^{\text{eff}}) = \left(\int_{\partial_2 M}\mathbb{B}\wedge \mathsf{a}, \int_{\partial_1 M}\mathsf{b}\wedge\mathbb{A}\right) = \sum_{i,j}\left(\int_{\partial_2 M}\mathbb{B}\wedge z^i\chi_i, \int_{\partial_1 M} z_j^+\chi^j\wedge\mathbb{A}\right)$$
$$= \sum_i (-1)^{\deg z^i}\int_{\partial_2 M}\mathbb{B}\wedge\chi_i\int_{\partial_1 M}\chi^j\wedge\mathbb{A},$$

since $(z^i, z_j^+) = (-1)^{\deg z^i}\Delta(z^i z_j^+) = (-1)^{\deg z^i}$. On the other hand,

$$\begin{aligned}\Omega e^{\frac{i}{\hbar}S^{\text{eff}}} &= \left(\left(\int_{\partial_1 M}\mathrm{d}\mathbb{A}\frac{\delta}{\delta\mathbb{A}} + \int_{\partial_2 M}\mathrm{d}\mathbb{B}\frac{\delta}{\delta\mathbb{B}}\right)\mathscr{S}^{\text{eff}}\right)e^{\frac{i}{\hbar}\mathscr{S}^{\text{eff}}}\\ &= \left(\int_{\partial_2 M\times\partial_1 M}\pi_1^*\mathbb{A}\wedge\mathrm{d}\eta\wedge\pi_2^*\mathbb{B}\right)e^{\frac{i}{\hbar}\mathscr{S}^{\text{eff}}}\\ &= \sum_i(-1)^{\deg\chi^i+1}\int_{\partial_2 M}\mathbb{B}\wedge\chi_i\int_{\partial_1 M}\chi^j\wedge\mathbb{A},\end{aligned}$$

where we integrated by parts and used property (9.11). Now the claim follows from the fact that $\deg z^i = 1 - \deg\chi^i$.

9.2.6.7 Dependence of the State on the Gauge-Fixing.

Clearly, the state defined in (9.9) depends on the choice of the gauge-fixing. However, one can show (and, by finite-dimensional arguments, this is supposed to hold in any quantum BV-BFV theory) that, upon deformations of the gauge fixing, the state changes as

$$\frac{d}{dt}\widehat{\psi} = (\hbar^2\widehat{\Delta}_M + \widehat{\Omega}_M^{\mathscr{P}})\widehat{\zeta} \tag{9.17}$$

for some $\widehat{\zeta}\in\widehat{\mathscr{H}_M^{\mathscr{P}}}$.

9.2.6.8 Gluing

Suppose we have two manifolds M_1 and M_2 that share a boundary component Σ. Then we can glue them together along Σ to obtain a new manifold $M = M_1\cup_\Sigma M_2$. The state $\widehat{\psi}_M$ can now be computed from the states $\widehat{\psi}_{M_1}$ and $\widehat{\psi}_{M_2}$ in the following way: Fix polarisations such that $\Sigma\subseteq\partial_1 M_1$ on M_1 and $\Sigma\subseteq\partial_2 M_2$ on M_2. Denote by $\mathbb{A}^\Sigma$ coordinates on $\Omega^\bullet(\Sigma)[1]\subseteq\mathscr{B}^{\mathscr{P}}_{\partial M_1}$ and by $\mathbb{B}^\Sigma$ coordinates on $\Omega^\bullet(\Sigma)[1]\subseteq\mathscr{B}^{\mathscr{P}}_{\partial M_2}$. Then we define $\widetilde{\psi}_M$ by

$$\widetilde{\psi}_M = \int_{\mathbb{A}^\Sigma,\mathbb{B}^\Sigma} e^{\frac{i}{\hbar}\int_\Sigma\mathbb{B}^\Sigma\mathbb{A}^\Sigma}\widehat{\psi}_{M_1}\widehat{\psi}_{M_2}.$$

Again, this integration is defined by a variant of Wick's theorem[9]: The integral of a term in the product of the states is nonzero if we can contract every $\mathbb{A}^{\Sigma}$ with to a $\mathbb{B}^{\Sigma}$. In this case, we sum over all possibilities to do so, and every contraction of a $\mathbb{A}^{\Sigma}(x)$ with a $\mathbb{B}^{\Sigma}(y)$ yields a $\delta^{(2)}_{\partial M}(x, y)$.

One also has to take care of the residual fields: This glued state will usually depend on a non-minimal amount of residual fields, and one can pass to the minimal amount of residual fields by a BV pushforward, yielding the "correct" state $\widehat{\psi}_M$.

9.2.6.9 BF-like Theories

As above, we call "BF-like" those theories whose action can be decomposed as $\mathcal{S}_{BF} + \mathcal{S}_{\text{int}}$. It is useful to also allow for the free part to consist of several copies of abelian BF theories. One way to do this is to change the space of fields to $\mathcal{F}_M = (\Omega^\bullet(M) \otimes V[1]) \oplus (\Omega^\bullet(M) \otimes V^*[1])$ with action

$$\mathcal{S}_{M,0} = \int_M \langle \mathsf{B}, \mathrm{d}\mathsf{A} \rangle$$

where V is a finite-dimensional vector space and $\langle \cdot, \cdot \rangle$ denotes the pairing between V and V^*. The above discussion goes through. The only thing that changes in the gauge fixing is that we should replace η by $\tilde{\eta} = \eta \otimes \mathrm{id}_V \in \Omega(C_2^0(M)) \otimes (V \otimes V^*)$, so that in any basis ξ_i of V with dual basis ξ^i it is given by

$$\tilde{\eta}(x_1, x_2) = \sum_{i,j} \eta(x_1, x_2) \delta^i_j \xi_i \otimes \xi^j.$$

9.3 Chern–Simons Theory as a BF-like Theory

9.3.1 *Split BV Chern–Simons Theory*

Let $\mathfrak{g}$ be a Lie algebra with an non-degenerate ad-invariant pairing $\langle \cdot, \cdot \rangle : \mathfrak{g} \times \mathfrak{g} \to \mathbb{R}$, i.e. we have for all $x, y, z \in \mathfrak{g}$ that $\langle x, [y, z] \rangle = \langle [x, y], z \rangle$. Let M be a 3-manifold, and $\mathsf{C} \in \Omega^\bullet(M) \otimes \mathfrak{g}[1]$. Then the BV Chern–Simons action is [9]

$$S[\mathsf{C}] = \int_M \frac{1}{2} \langle \mathsf{C}, \mathrm{d}\mathsf{C} \rangle + \frac{1}{6} \langle \mathsf{C}, [\mathsf{C}, \mathsf{C}] \rangle,$$

where for homogeneous elements $A \otimes v, B \otimes w \in \Omega^\bullet(M) \otimes \mathfrak{g}$ the bracket and the pairing are defined by

---

[9]In the sense that we compute it formally as a Gaussian (or rather, Fresnel) integral.

$$[A \otimes v, B \otimes w] = A \wedge B \otimes [v, w]$$

and

$$\langle A \otimes v, B \otimes w \rangle = \langle v, w \rangle A \wedge B$$

respectively. Now assume that the Lie Algebra $\mathfrak{g}$ admits a splitting $g = V \oplus W$ into maximally isotropic subspaces, i.e. the pairing restricts to 0 on V and W and $\dim V = \dim W = \frac{\dim \mathfrak{g}}{2}$. Then we can identify $W \cong V^*$ via the pairing and decompose $\mathsf{C} = \mathsf{A} + \mathsf{B}$, where $\mathsf{A} \in \Omega^\bullet(M) \otimes V[1]$and $\mathsf{B} \in \Omega^\bullet(M) \otimes W[1]$. The action decomposes into a "free" or "kinetic" part

$$\begin{aligned} S_{free} &= \int_M \frac{1}{2}\langle \mathsf{C}, \mathrm{d}\mathsf{C} \rangle = \int_M \frac{1}{2}\langle \mathsf{A} + \mathsf{B}, \mathrm{d}\mathsf{A} + \mathrm{d}\mathsf{B} \rangle \\ &= \int_M \frac{1}{2}\langle \mathsf{A}, \mathrm{d}\mathsf{B} \rangle + \frac{1}{2}\langle \mathsf{B}, \mathrm{d}\mathsf{A} \rangle = \int_M \langle \mathsf{B}, \mathrm{d}\mathsf{A} \rangle \end{aligned}$$

(where $\langle \mathsf{A}, \mathrm{d}\mathsf{A} \rangle = 0 = \langle \mathsf{B}, \mathrm{d}\mathsf{B} \rangle$ by isotropy and we integrate by parts) and an "interaction" term

$$\mathscr{V}\langle \mathsf{A}, \mathsf{B} \rangle = \frac{1}{6}\langle \mathsf{A} + \mathsf{B}, [\mathsf{A} + \mathsf{B}, \mathsf{A} + \mathsf{B}] \rangle.$$

Hence, the theory is "BF-like".

9.3.2 Perturbative Expansion

Let M be a 3-manifold, possibly with boundary. We want to compute the state $\widehat{\psi}_M$. As described above for the BF example, we choose a decomposition of the boundary $\partial M = \partial_1 M \sqcup \partial_2 M$ and get a polarisation on the space of boundary fields such that $\mathscr{B}^{\mathscr{P}}_{\partial M} = \mathscr{B}_1 \times \mathscr{B}_2 \ni (\mathbb{A}, \mathbb{B})$. Decomposing $\mathsf{A} = \mathbb{A} + \mathsf{a} + \alpha$, $\mathsf{B} = \mathbb{B} + \mathsf{b} + \beta$, we can decompose the action as explained in Remark 10:

$$\mathscr{S}^{\mathscr{P}}_M = \widehat{\mathscr{S}}_{M,0} + \widehat{\mathscr{S}}_{M,\text{pert}} + \mathscr{S}^{\text{back}}_M + \mathscr{S}^{\text{source}}_M,$$

where

$$\begin{aligned} \widehat{\mathscr{S}}_{M,0} &= \int_M \langle \beta, \mathrm{d}\alpha \rangle, \\ \widehat{\mathscr{S}}_{M,\text{pert}} &= \int_M \mathscr{V}(\widehat{A}, \widehat{B}), \\ \mathscr{S}^{\text{back}}_M &= -\left(\int_{\partial_2 M} \langle \mathbb{B}, \mathsf{a} \rangle + \int_{\partial_1 M} \langle \mathsf{b}, \mathbb{A} \rangle \right), \end{aligned}$$

$$\mathscr{S}_M^{\text{source}} = -\left(\int_{\partial_2 M} \langle \mathbb{B}, \alpha \rangle + \int_{\partial_1 M} \langle \beta, \mathbb{A} \rangle\right).$$

The state is given by

$$\widehat{\psi}_M = \widehat{\psi}_M(\mathbb{A}, \mathbb{B}, \mathsf{a}, \mathsf{b}) = \int_{\mathscr{L}} e^{\frac{i}{\hbar}\mathscr{S}_M^{\mathscr{P}}},$$

where $\mathscr{L} \ni (\alpha, \beta)$, the gauge-fixing Lagrangian, is the same as for abelian BF theory (cf. Remark 8). Therefore it does not depend on the boundary and background fields. By virtue of the above decomposition, we can rewrite this as

$$\widehat{\psi}_M(\mathbb{A}, \mathbb{B}, \mathsf{a}, \mathsf{b}) = e^{\frac{i}{\hbar}\mathbb{S}_M^{\text{back}}} \int_{\mathscr{L}} e^{\frac{i}{\hbar}\widehat{\mathbb{S}}_{M,0}} e^{\frac{i}{\hbar}\widehat{\mathbb{S}}_{M,\text{pert}}} e^{\frac{i}{\hbar}\mathbb{S}_M^{\text{source}}}.$$

To do a perturbative (power series) expansion,[10] expand the exponentials

$$\begin{aligned}
&\widehat{\psi}_M(\mathbb{A}, \mathbb{B}, \mathsf{a}, \mathsf{b}) = \\
&= \sum_k \frac{1}{k!}\left(-\frac{i}{\hbar}\right)^k \left(\int_{\partial_2 M} \langle \mathbb{B}, \mathsf{a} \rangle + \int_{\partial_1 M} \langle \mathsf{b}\mathbb{A} \rangle\right)^k \int_{\mathscr{L}} e^{i\widehat{\mathbb{S}}_{M,0}} \sum_l \frac{1}{l!}\left(\frac{i}{\hbar}\right)^l \left(\int_M \mathscr{V}(\widehat{A}, \widehat{B})\right)^l \\
&\times \sum_m \frac{1}{m!}\left(-\frac{i}{\hbar}\right)^m \left(\int_{\partial_2 M} \langle \mathbb{B}, \alpha \rangle + \int_{\partial_1 M} \langle \beta, \mathbb{A} \rangle\right)^m \\
&= \sum_{k,l,m} \frac{1}{k!l!m!}(-1)^{k+m}\left(\frac{i}{\hbar}\right)^{k+l+m} \left(\int_{\partial_2 M} \langle \mathbb{B}, \mathsf{a} \rangle + \int_{\partial_1 M} \langle \mathsf{b}, \mathbb{A} \rangle\right)^k \times \\
&\int_{\mathscr{L}} e^{i\widehat{\mathbb{S}}_{M,0}} \left(\int_M \mathscr{V}(\widehat{\mathsf{A}}, \widehat{\mathsf{B}})\right)^l \left(\int_{\partial_2 M} \langle \mathbb{B}, \alpha \rangle + \int_{\partial_1 M} \langle \beta, \mathbb{A} \rangle\right)^m \\
&= \sum_{l,k,m} \frac{1}{k!l!m!}(-1)^{k+m}\left(\frac{i}{\hbar}\right)^{k+l+m} \left(\int_{\partial_2 M} \langle \mathbb{B}, \mathsf{a} \rangle + \int_{\partial_1 M} \langle \mathsf{b}, \mathbb{A} \rangle\right)^k \times \\
&\int_{\mathscr{L}} e^{i\widehat{\mathbb{S}}_{M,0}} \left(\int_M \frac{1}{6}\langle \widehat{\mathsf{A}} + \widehat{\mathsf{B}}, [\widehat{\mathsf{A}} + \widehat{\mathsf{B}}, \widehat{\mathsf{A}} + \widehat{\mathsf{B}}] \rangle\right)^l \left(\int_{\partial_2 M} \langle \mathbb{B}, \alpha \rangle + \int_{\partial_1 M} \langle \beta, \mathbb{A} \rangle\right)^m.
\end{aligned}$$

Now we choose a basis ξ_i of V and let ξ^i be the corresponding dual basis of W. We expand our fields[11] $\mathsf{A} = \mathsf{A}^i\xi_i$, $\mathsf{B} = \mathsf{B}_i\xi^i$ and also their decompositions accordingly, i.e. $\alpha = \alpha^i\xi_i$, and so on. We then get e.g. $\langle \mathsf{B}, d\mathsf{A} \rangle = \mathsf{B}_i d\mathsf{A}^i$. We now want to expand the perturbation term in this basis. For this purpose we make use of the fact that $\langle X, [Y, Z] \rangle = \langle Z, [X, Y] \rangle = \langle Y, [Z, X] \rangle$ for any $X, Y, Z \in \Omega^\bullet(M) \otimes \mathfrak{g}[1]$, so we can decompose the interaction term as

[10] Actually, a semiclassical expansion around the classical solution given by the trivial connection.

[11] From now on, we will make use of Einstein summation (sums over repeated indices are implied).

$$\mathscr{V}(\widehat{\mathsf{A}},\widehat{\mathsf{B}}) = \frac{1}{6}\langle\widehat{\mathsf{A}},[\widehat{\mathsf{A}},\widehat{\mathsf{A}}]\rangle + \frac{1}{2}\langle\widehat{\mathsf{B}},[\widehat{\mathsf{A}},\widehat{\mathsf{A}}]\rangle + \frac{1}{2}\langle\widehat{\mathsf{A}},[\widehat{\mathsf{B}},\widehat{\mathsf{B}}]\rangle + \frac{1}{6}\langle\widehat{\mathsf{B}},[\widehat{\mathsf{B}},\widehat{\mathsf{B}}]\rangle.$$

Now we make the following simplifying assumption on $\mathfrak{g}$.

Assumption 1 The splitting $\mathfrak{g} = V \oplus W$ is actually a splitting into *Lie subalgebras*, i.e. $(\mathfrak{g}, V, W)$ is a Manin triple.

By isotropy of the subspaces, this implies that the terms $\langle\widehat{\mathsf{A}},[\widehat{\mathsf{A}},\widehat{\mathsf{A}}]\rangle$ and $\langle\widehat{\mathsf{B}},[\widehat{\mathsf{B}},\widehat{\mathsf{B}}]\rangle$ vanish. Splitting $\widehat{\mathsf{A}} = \mathsf{a} + \alpha$, $\widehat{\mathsf{B}} = \mathsf{b} + \beta$, we expand the perturbation term in terms of the type $\langle\gamma_1,[\gamma_2,\gamma_3]\rangle$, where $\gamma_i \in \{\mathsf{a},\alpha,\mathsf{b},\beta\}$. These we can express as

$$\sum_{i,j,k} f_{ijk}\gamma_1^i\gamma_2^j\gamma_3^k,$$

where f_{ijk} are the structure constans of $\mathfrak{g}$ in the basis $\xi_1,\ldots\xi_n,\xi^1,\ldots\xi^n$. Integration over $\mathscr{L}$ can then be performed using Wick's theorem. Let η be an abelian BF propagator on M as discussed above. We exchange integrals over M, $\partial_i M$ and $\mathscr{L}$ and get an integrand which is a sum of products of forms γ. By the Wick theorem, the integral vanishes except for the case where there are precisely as many α's as β's, in which case

$$\int_{\mathscr{L}} e^{i\widehat{\mathbb{S}}_{M,0}}\alpha^{j_1}(x_1)\cdots\alpha^{j_n}(x_n)\beta^{k_1}(y_1)\cdots\beta^{k_n}(y_n) =$$
$$= T_M(-i\hbar)^n\sum_{\sigma\in S_n}\delta^{j_1k_{\sigma(1)}}\eta(x_1,y_{\sigma(1)})\cdots\delta^{j_nk_{\sigma(n)}}\eta(x_n,y_{\sigma(n)}),$$

where $T_M = \int_{\mathscr{L}} e^{i\widehat{\mathbb{S}}_{M,0}}$.

9.3.3 Feynman Graphs and Rules

After integration over $\mathscr{L}$, we can label the terms in the perturbative expansion by graphs as follows. Fix $k, l, m \in \mathbb{N}_0$. We consider graphs Γ with three types of vertices:

- *Boundary background vertices*: There are k of these distributed on ∂M. They are labelled by $\mathbb{B}\mathsf{a}$ if they lie on $\partial_2 M$ and $\mathsf{b}\mathbb{A}$ if they lie on $\partial_1 M$.
- *Boundary source vertices*: There are m boundary source vertices distributed on ∂M. They are labelled by $\mathbb{B}\alpha$ on $\partial_2 M$ and $\mathbb{A}\beta$ on $\partial_1 M$. Vertices on $\partial_2 M$ have an arrow tail originating from them, whereas vertices on $\partial_1 M$ have an arrowhead pointing towards them.
- *Internal interaction vertices*: There are l internal vertices. They come with three half-edges which are labelled by γ_i's in $\{\mathsf{a},\alpha,\mathsf{b},\beta\}$. These half-edges are either marked as leaves if they are labelled by a background, as an arrow tail if they are labelled by α, or an arrowhead if they are labelled by β.

If it is possible to connect every arrow tail α to an arrowhead β (possibly at the same vertex), then the graph resulting from this procedure is called an *admissible graph*. To such a graph we can associate a functional on the space of boundary fields as follows:

- For every background boundary vertex, multiply by $(-i/\hbar)$ times the label and integrate over the corresponding boundary point.
- For every internal vertex multiply by $(-i/\hbar)$ times the correct structure constants (specified by the half-edge labels) and integrate over M.
- For every leaf, multiply by the corresponding background field evaluated at the point.
- For every arrow between vertices in different positions $i \neq j$, with tail labelled by α^k and head β_l, multiply by a propagator $(-i\hbar)\delta_l^k \eta(x_i, y_j)$.
- For every short loop (also called tadpole), i.e. an arrow issueing and ending at the same vertex i, with tail labelled by α^k and head β_l, multiply by $(-i\hbar)\delta_l^k \alpha(x_i)$, where $\alpha \in \Omega^2(M)$ is a so-called "tadpole form".[12]
- For every source boundary vertex, we multiply by $(-i/\hbar)$ times the corresponding boundary field and integrate over the corresponding boundary point.

We denote the result by $\widehat{\psi}_\Gamma$. Denoting the set of all admissible graphs for k, l, m by $\Lambda_{k,l,m}$, we get

$$\widehat{\psi}_M(\mathbb{A}, \mathbb{B}, \mathsf{a}, \mathsf{b}) = T_M \sum_{k,l,m} \sum_{\Gamma \in \Lambda_{k,l,m}} \widehat{\psi}_\Gamma.$$

Remark 11 We can factor out the non-interacting diagram parts (background boundary vertices and source boundary vertices connecting to other source boundary vertices). This will yield a prefactor of $e^{\frac{i}{\hbar}\mathcal{S}_0^{\text{eff}}}$ where $\mathcal{S}_0^{\text{eff}}$ is the free effective action

$$\mathcal{S}_0^{\text{eff}} = -\left(\int_{\partial_2 M} \langle \mathbb{B}, \mathsf{a} \rangle + \int_{\partial_1 M} \langle \mathsf{b}, \mathbb{A} \rangle\right) - \int_{\partial_2 M \times \partial_1 M} \pi_1^* \mathbb{B}_i \eta \mathbb{A}^i \tag{9.18}$$

i.e. the effective action of the unperturbed theory.

The remaining interaction diagrams have $l \geq 1$ internal vertices and $m \leq 3l$ boundary vertices. Denoting the set of admissible interaction diagrams by $\Lambda_{l,m}^{int}$, the above expression becomes

$$\widehat{\psi}_M(\mathbb{A}, \mathbb{B}, \mathsf{a}, \mathsf{b}) = T_M e^{\frac{i}{\hbar}\mathcal{S}_0^{\text{eff}}} \left(1 + \sum_{l=1}^{\infty} \sum_{m=0}^{3l} \sum_{\Gamma \in \Lambda_{l,m}^{int}} \widehat{\psi}_\Gamma\right).$$

Our goal is now to give an asymptotic expansion of the state of the form

[12] These contributions can be ignored if the Lie algebra is unimodular (i.e. the structure constants satisfy $f_{ik}^i = 0$) or the Euler characteristic of M is 0. We will restrict ourselves to these cases.

$$\widehat{\psi}_M(\mathbb{A}, \mathbb{B}, \mathsf{a}, \mathsf{b}) = T_M e^{\frac{i}{\hbar}\mathscr{S}_M^{\text{eff}}} \sum_{j \geq 1} \hbar^j R_j,$$

where $\mathscr{S}_M^{\text{eff}}$ is the so-called *tree effective action*, i.e. the sum of all diagrams whose underlying graphs are trees, and R_j denotes the sum of all diagrams that contain at least one loop.

9.4 Split Chern–Simons Theory on the Solid Torus

In this section we compute a first approximation for the state on the solid torus $K := D \times S^1$ with boundary $\partial M = S^1 \times S^1 =: \mathbb{T}^2$. Here we think of $D = \{z \in \mathbb{C}, |z| \leq 1\}$ as the closed unit disk in the complex plane. This is not just a simple exercise: Note that since the quantum BV-BFV formalism allows also for the gluing of states, given a state on the solid torus one can compute it also for any manifold that can be glued together from tori (namely, all lens spaces).

Since the boundary $\mathbb{T}^2$ is connected, there are only two possible choices for $\partial_1 M$ and $\partial_2 M$, we choose $\partial_1 M := \partial M$ and $\partial_2 M := \emptyset$. In a future paper we plan to do a similar computation for handlebodies, and due to Heegard decomposition this would lead to state for general 3-folds. This leads to the following space of backgrounds:

$$\begin{aligned}\mathscr{V}_M &= H^\bullet_{D1}(M)[1] \otimes V \oplus H^\bullet_{D2}(M)[1] \otimes W = H^\bullet(M, \partial M)[1] \otimes V \oplus H^\bullet(M) \otimes W \\ &\cong (H^\bullet(D, \partial D) \otimes H^\bullet(S^1)) \otimes V \oplus H^\bullet(S^1))[1] \otimes W.\end{aligned}$$

Let μ be a normalised generator of $H^\bullet(D, \partial D)$, i.e. $\int_D \mu = 1$. Denoting t the coordinate on S^1, we get that $\chi_1 = \mu dt$, $\chi_2 = \mu$ is a basis of $H^\bullet_{D1}(M)[1]$, with dual basis $\chi^1 = 1$, $\chi^2 = dt$ of $H^\bullet_{D2}(M)[1]$. We can then expand

$$\begin{aligned}\mathsf{a}^i &= z^{1i}\mu dt + z^{2i}\mu, \\ \mathsf{b}_i &= z^+_{1i}1 + z^+_{2i}dt.\end{aligned}$$

The canonical BV Laplacian on $\mathscr{V}_M$ is then given by

$$\Delta_{\mathscr{V}_M} = -\left(\frac{\partial}{\partial z^1}\frac{\partial}{\partial z^+_1} + \frac{\partial}{\partial z^2}\frac{\partial}{\partial z^+_2}\right).$$

9.4.1 *Effective Action on the Solid Torus*

Assume as above that $\mathfrak{g} = V \oplus W$ is a Manin triple, i.e.

- $V \cong W^*$ as vector spaces
- V, W Lie algebras.

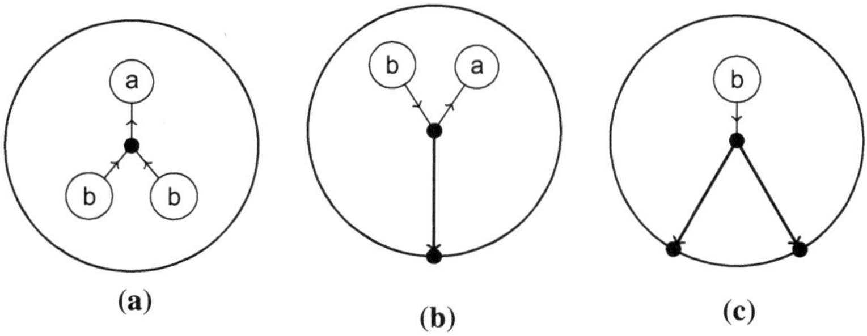

Fig. 9.1 Graphs in the solid torus (depicted in a cross-section) with 1 interaction vertex. A bullet denotes a point we integrate over, a long *arrow* denotes a propagator

Let us introduce bases $\xi_1, \dots, \xi_n$ of V, $\xi^1, \dots, \xi^n$ of W such that $\langle \xi_i, \xi^j \rangle = \delta_i^j$ and structure constants in these bases: $[\xi_i, \xi_j]_V = f_{ij}^k \xi_k$, $[\xi^i, \xi^j]_W = g_k^{ij} \xi^k$. We can then also decompose the fields

$$\mathsf{B} = \mathsf{B}_i \xi^i = \mathsf{b}_i \xi^i + \beta_i \xi^i + \mathbb{B}_i \xi^i,$$
$$\mathsf{A} = \mathsf{A}^i \xi_i = \mathsf{a}^i \xi_i + \alpha^i \xi_i + \mathbb{A}^i \xi_i.$$

The fact we have a Manin triple means that in terms of the structure constants we have

$$f_{ij}^k g_k^{lm} = f_{ik}^l g_j^{km} - f_{jk}^l g_i^{km} + f_{ik}^m g_j^{lk} - f_{jk}^m g_i^{lk}. \tag{9.19}$$

We now want to compute an approximation to the tree effective action by considering tree diagrams that have at most two interaction vertices and at most two boundary vertices.

We will proceed by the number of interaction vertices. There is only a single connected diagram with no interaction vertices, consisting of a single point on the boundary. It yields the free effective action (9.18) for $\partial_2 M = \emptyset$, namely

$$S_0^{\text{eff}} = -\int_{\partial_1 M} \mathsf{b}_k \mathbb{A}^k.$$

9.4.1.1 1-Point Contribution

Let us continue with diagrams containing a single interaction vertex. It is now important that the solid torus has zero Euler characteristic, so we do not need to consider tadpoles. Since there can be no arrows issuing from $\partial_1 M$, diagrams with a half-edge labelled by β at the interaction point are not admissible. Also notice that $\mathsf{a} \wedge \mathsf{a} = 0$ (it is a 4-form on a 3-manifold). In the end, there are only three contributing diagrams (see Fig. 9.1):

(a) The single interaction vertex with three leaves labelled by a, b and b, corresponding to

$$S_1^{\text{eff}} := \frac{1}{2}\int_M \langle \mathsf{a}, [\mathsf{b}, \mathsf{b}]\rangle.$$

We should explain some notation. We denote by $C_{m,n}(M, \partial M)$ (a suitable compactification of) the configuration space of m points in the bulk and n in the boundary. It comes with natural projections

$$\pi_i : C_{m,n}(M, \partial M) \to \begin{cases} M & i \leq m \\ \partial M & i \geq m \end{cases}$$

and

$$\pi_{ij} : C_{i,j}(M, \partial M) \to \begin{cases} C_2(M) & i, j \leq m \\ C_{1,1}(M, \partial M) & i \leq m, j \leq n \\ C_2(\partial M) & i, j \geq m \end{cases}.$$

By writing γ_i resp. γ_{ij} we mean the pullback of γ under the corresponding projection.

(b) The single interaction vertex with two leaves labelled b and a and an arrow connecting to a boundary source vertex $\beta\mathbb{A}$. It evaluates to

$$S_2^{\text{eff}} := -\int_{C_{1,1}(M,\partial_1 M)} f^i_{jk}\mathsf{b}_{1,i}\mathsf{a}^j_1\eta_{12}\mathbb{A}^k_2.$$

(c) The single interaction vertex with a leaf labelled by b and two arrows connecting to two different boundary source vertices. This evaluates to

$$S_{\text{eff},3} := \frac{1}{2}\int_{C_{1,2}(M,\partial_1 M)} f^i_{jk}b_{1,i}\eta_{12}\eta_{13}\mathbb{A}^j_2\mathbb{A}^k_3.$$

9.4.1.2 2-Point Contribution

Now we consider tree diagrams with two interaction vertices. Since the diagrams have to be connected, there has to be at least one arrow between the vertices. Since we are only considering trees, there is exactly one arrow between them. Also, we are considering only diagrams that have at most two boundary vertices. The diagrams in Fig. 9.2 below show the admissible graphs in the relevant degrees. (admissible graphs with no boundary vertices all evaluate to 0 because of property 9.13) We will discuss the results below.

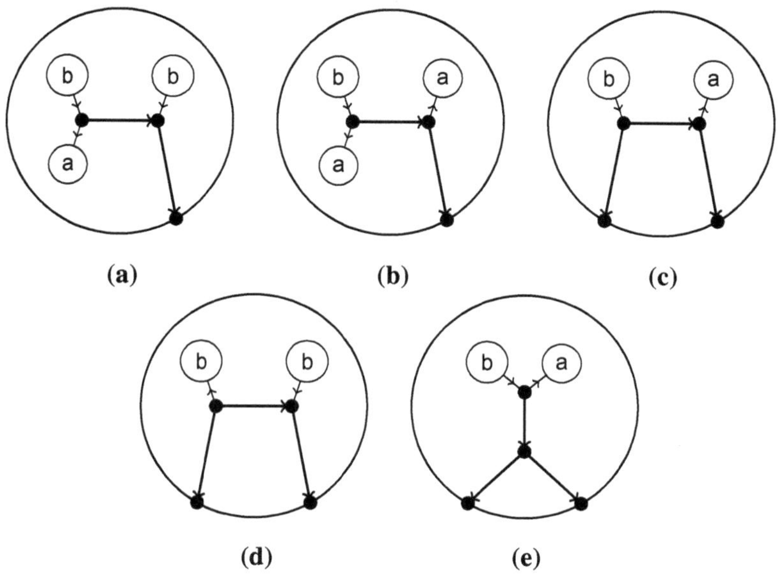

Fig. 9.2 Graphs with 2 interaction vertices. A bullet denotes a point we integrate over, long arrow denotes a propagator

9.4.1.3 Performing Integration over M

We now want to perform the integration over the bulk points. There are two possibilities to proceed:

1. One constructs an explicit propagator on M and computes the integrals analytically.
2. One analyses how the resulting form on the boundary behaves under de Rham differential and integration of points, and picks a form which is a product of propagators and representatives of cohomology on the boundary that has the same properties. Since only these properties enter into the proof of the mQME, this produces a valid state. We will discuss this procedure and the question of uniqueness in more depth in a future paper.

With the second approach, choosing a propagator satisfying also (9.13) and (9.14), one can see that the only non-vanishing contributions from two-point diagrams come from diagrams Fig. 9.2c and e. Denoting the results by S_4^{eff} and S_5^{eff} respectively, we obtain

$$
\begin{aligned}
S_0^{\text{eff}} &= -z^+_{1,k}\int_{\partial_1 M}\mathbb{A}^k - z^+_{2,k}\int_{\partial_1 M} dt\mathbb{A}^k,\\
S_1^{\text{eff}} &= \frac{1}{2}g_i^{jk}(z^{1i}z^+_{1j}z^+_{1k} + 2z^{2i}z^+_{1j}z^+_{2k}),\\
S_2^{\text{eff}} &= f^i_{jk}z^+_{1i}z^{2j}\int_{\partial_1 M} d\theta\mathbb{A}^k + f^i_{jk}(z^+_{1i}z^{1j} - z^+_{2i}z^{2j})\int_{\partial_1 M} dtd\theta\mathbb{A}^k,\\
S_3^{\text{eff}} &= \frac{1}{2}f^i_{jk}z^+_{1i}\int_{C_2(\partial_1 M)} \eta^T_{12}\mathbb{A}^j_1\mathbb{A}^k_2,\\
&\quad + \frac{1}{2}f^i_{jk}z^+_{2i}\int_{C_2(\partial_1 M)} \eta^T_{12}\frac{dt_1 + dt_2}{2}\mathbb{A}^j_1\mathbb{A}^k_2,\\
S_4^{\text{eff}} &= f^i_{jk}f^j_{lm}z^+_{1i}z^{2l}\int_{C_2(\partial_1 M)} d\theta_1\eta^T_{12}\mathbb{A}^k_1\mathbb{A}^m_2\\
&\quad + f^i_{jk}f^j_{lm}(z^+_{1i}z^{1l} - z^+_{2i}z^{2l})\int_{C_2(\partial_1 M)} dt_1d\theta_1\eta^T_{12}\mathbb{A}^k_1\mathbb{A}^m_2,\\
S_5^{\text{eff}} &= f^i_{jk}f^k_{lm}z^+_{1i}z^{2j}\int_{C_2(\partial_1 M)} d\theta_1\eta^T_{12}\mathbb{A}^l_1\mathbb{A}^m_2\\
&\quad + f^i_{jk}f^k_{lm}(z^+_{1i}z^{1j} - z^+_{2i}z^{2j})\int_{C_2(\partial_1 M)} dt_1d\theta_1\eta^T_{12}\mathbb{A}^l_1\mathbb{A}^m_2,
\end{aligned}
$$

where t denotes the parallel (longitudinal) and θ the meridian coordinate on the boundary torus (i.e. in the solid torus $[d\theta] = 0$), and η^T is a propagator for abelian BF theory on the boundary torus.

9.4.2 mQME

Our goal in this section is to prove the modified Quantum Master Equation

$$(\hbar^2\Delta + \Omega)e^{\frac{i}{\hbar}S_{\text{eff}}} = 0,$$

ignoring terms of nonzero order in $\hbar$, more than two boundary vertices or more than second power in the interaction. Here Ω is given by the standard quantisation of

$$S^\partial = \int_{\partial M}\langle \mathsf{B}, \mathsf{dA}\rangle + \frac{1}{2}\langle \mathsf{B}, [\mathsf{A}, \mathsf{A}]\rangle + \frac{1}{2}\langle \mathsf{A}, [\mathsf{B}, \mathsf{B}]\rangle,$$

which (on the solid torus) is

$$\Omega_{\text{st}} = -i\hbar\int_{\partial_1 M}\mathrm{d}\mathbb{A}^k\frac{\delta}{\delta\mathbb{A}^k} + \frac{1}{2}g_a^{bc}\int_{\partial_1 M} -\hbar^2\mathbb{A}^a\frac{\delta}{\delta\mathbb{A}^b}\frac{\delta}{\delta\mathbb{A}^c} - \frac{i\hbar}{2}f^a_{bc}\int_{\partial_1 M}\mathbb{A}^b\mathbb{A}^c\frac{\delta}{\delta\mathbb{A}^a}.$$

Remark 12 The second term containing two derivatives yields possibly singular results when applied to a single term in the effective action. Therefore the two derivatives are allowed to act only on different terms in a product of terms of the effective action. With this regularisation one can also check that $\Omega_{st}^2 = 0$.

One can check that $\Delta S_{\text{eff}} = 0$ and therefore $(\hbar^2 \Delta + \Omega) e^{\frac{i}{\hbar} S^{\text{eff}}} = -\frac{1}{2}(S^{\text{eff}}, S^{\text{eff}}) e^{\frac{i}{\hbar} S^{\text{eff}}} + \Omega e^{\frac{i}{\hbar} S^{\text{eff}}}$. So we should check that $\frac{1}{2}(S^{\text{eff}}, S^{\text{eff}}) e^{\frac{i}{\hbar} S^{\text{eff}}} = \Omega e^{\frac{i}{\hbar} S^{\text{eff}}}$ up to higher order corrections.

9.4.2.1 BV Bracket

Let us compute first $(S^{\text{eff}}, S^{\text{eff}})$. Abbreviating $S_i^{\text{eff}} =: S_i$, we get that $(S^{\text{eff}}, S^{\text{eff}}) = \sum_i (S_i, S_i) + 2\sum_{i<j}(S_i, S_j)$.

We have that $(z_{1i}^+, z^{1j}) = \delta_{ij} = -(z_{2i}^+, z^{2j})$, and all other brackets vanish.

Since S_0 and S_3 only contain z^+ variables, we get that $(S_0, S_0) = (S_3, S_3) = (S_0, S_3) = 0$. Also, (S_2, S_3) contains three boundary fields, so we neglect it. The same is true for any bracket of S_4 with the rest, except (S_1, S_4), which is third power in the structure constants. So the only contributing brackets are (S_0, S_1), (S_0, S_2), (S_1, S_1), (S_1, S_2), (S_1, S_3) and (S_2, S_2).

9.4.2.2 Ω Part

Now let us compute $\Omega_{\text{st}} e^{\frac{i}{\hbar} S^{\text{eff}}}$. At first, we will consider only contributions of order 0 in $\hbar$ and less than two $\mathbb{A}'s$. Let us split Ω into the following 3 terms:

$$\Omega_0 := -\mathrm{i}\hbar \int_{\partial_1 M} \mathrm{d}\mathbb{A}^k \frac{\delta}{\delta \mathbb{A}^k},$$

$$\Omega_1 := -\frac{\mathrm{i}\hbar}{2} f_{bc}^a \int_{\partial_1 M} \mathbb{A}^b \mathbb{A}^c \frac{\delta}{\delta \mathbb{A}^a},$$

$$\Omega_2 := -\frac{\hbar^2}{2} g_a^{bc} \int_{\partial_1 M} \mathbb{A}^a \frac{\delta}{\delta \mathbb{A}^b} \frac{\delta}{\delta \mathbb{A}^c}.$$

By the usual rules of derivatives we will have

$$\Omega_{st} e^{\frac{i}{\hbar} S^{\text{eff}}} = \left((\Omega_0 + \Omega_1) \frac{i}{\hbar} S^{\text{eff}} + \Omega_2 \left(\frac{i}{\hbar}\right)^2 \frac{1}{2} (S^{\text{eff}})^2 \right) e^{\frac{i}{\hbar} S^{\text{eff}}}.$$

Let us look at the linear term first. Notice that $\Omega_0(S_0) = \Omega_0(S_1) = \Omega_0(S_2) = 0$, since we can integrate by parts, and the forms appearing in these integrals are closed. Also, since we are ignoring terms with more than two boundary fields, and $\Omega_1(S_1) = 0$, we only need to consider $\Omega_1(S_0)$ and $\Omega_1(S_2)$. Now we need to consider $\Omega_2 \left(\frac{i}{\hbar}\right)^2 \frac{1}{2!} (S^{\text{eff}})^2$.

Since Ω_2 removes one $\mathbb{A}$, but adds one power in the interaction, we have to consider terms in $(S^{\text{eff}})^2$ with two or three $\mathbb{A}$'s and at most first power in the interaction. One can easily check that the only products to consider are S_0^2, $S_0 S_2$ and $S_0 S_3$.

9.4.2.3 Proving the mQME

Proposition 1 *To prove the mQME in the chosen degrees one can equivalently prove that*

$$(S_0, S_1) + (S_0, S_2) + \frac{1}{2}(S_1, S_1) + (S_1, S_2) + (S_1, S_3) + \frac{1}{2}(S_2, S_2) =$$
$$= \frac{i}{\hbar}\left(\Omega_0(S_3) + \Omega_0(S_4) + \Omega_0(S_5) + \Omega_1(S_0) + \Omega_1(S_2)\right)$$
$$+ \frac{1}{2}\left(\frac{i}{\hbar}\right)^2 \Omega_2(S_0^2 + 2S_0S_2 + 2S_0S_3).$$

This can be shown using a direct computation, which we summarise as follows.

Lemma 7 *The following identities hold:*

(i) $(S_0, S_1) = \frac{1}{2}\left(\frac{i}{\hbar}\right)^2 \Omega_2(S_0^2)$,
(ii) $(S_1, S_1) = 0$,
(iii) $(S_0, S_2) = \frac{i}{\hbar}\left(\Omega_0(S_3) + \Omega_1(S_0)\right)$,
(iv) $(S_1, S_2) = \left(\frac{i}{\hbar}\right)^2 \Omega_2(S_0 S_2)$,
(v) $(S_1, S_3) = \left(\frac{i}{\hbar}\right)^2 \Omega_2(S_0 S_3)$,
(vi) $(S_2, S_2) = \frac{i}{\hbar}\left(\Omega_0(S_4) + \Omega_0(S_5) + \Omega_1(S_2)\right)$.

Corollary 8 *The state defined by* $\widehat{\psi} = e^{\frac{i}{\hbar}S^{\text{eff}}}$ *satisfies the mQME on the solid torus at zeroth order in* $\hbar$*, considering terms with at most two boundary fields and at most second order in the interaction.*

9.4.3 Change of Data

Now we will analyse how the state behaves under an infinitesimal change of gauge-fixing, i.e. the representatives of cohomology and the propagator. Such a change can be described by the action of a vector field X on M on these forms by the Lie derivative

$$\dot{\chi}_i = L_X \chi_i, \dot{\chi}^i = L_X \chi^i, \dot{\eta} = L_X \eta$$

(we will always write X to mean the vector field $(X, \dots, X) \in TM \oplus \cdots \oplus TM \cong T(M \times \cdots \times M)$). Clearly we have

$$\frac{d}{dt}\widehat{\psi} = \frac{i}{\hbar}\frac{d}{dt}(S^{\mathrm{eff}})e^{\frac{i}{\hbar}S^{\mathrm{eff}}}.$$

Proposition 2 *If we expand S^{eff} as a sum of terms of the form*

$$S^{\mathit{eff}} = \sum \int_{C_n(\partial_1 M)} \gamma \pi_1^* \mathbb{A} \cdots \pi_n^* \mathbb{A},$$

then its time derivative is given by

$$\frac{d}{dt}(S^{\mathit{eff}}) = \sum \int_{C_n(\partial_1 M)} (L_{X^\partial}\gamma)\pi_1^* \mathbb{A} \cdots \pi_n^* \mathbb{A},$$

where X^∂ denotes restriction of X to the boundary.

Proof S^{eff} is a sum of terms of the form

$$\int_{C_{m,n}(M,\partial_1 M)} \widehat{\gamma} \pi_1^* \mathbb{A} \cdots \pi_n^* \mathbb{A},$$

where $\widehat{\gamma}$ is a product of background fields and propagators on M. Since L_X is a derivation, we have $\frac{d}{dt}\widehat{\gamma} = L_X \widehat{\gamma}$. But the Lie derivative commutes with the integration over the bulk vertices, so we have proved the statement.

We are now going to define a state ζ such that

$$(\hbar^2 \Delta + \Omega)(\widehat{\psi}\zeta) = \frac{d}{dt}\widehat{\psi}$$

(as in (9.17)) for our example on the torus. Namely, we define $\gamma_i \in \Omega^{k_i}(C_{n_i}(\partial_1 M))$ by

$$S^{\mathrm{eff}} = \sum_i F_i(f, g, z, z^+)_{j_1 \cdots j_{n_i}} \int_{C_{n_i}(\partial_1 M)} \gamma_i \pi_1^* \mathbb{A}^{j_1} \cdots \pi_{n_i}^* \mathbb{A}^{j_{n_i}}.$$

Then ζ is defined by

$$\zeta = \sum_i F_i(f, g, z, z^+)_{j_1 \cdots j_{n_i}} \int_{C_{n_i}(\partial_1 M)} (\iota_{X^\partial}\gamma_i)\pi_1^* \mathbb{A}^{j_1} \cdots \pi_{n_i}^* \mathbb{A}^{j_{n_i}},$$

i.e. we replace every differential form γ_i by its contraction with X.

Proposition 3 *For the change of data described above and the effective action described in the last paragraph, we have that*

$$(\hbar^2 \Delta + \Omega)(\widehat{\psi}\zeta) = \frac{d}{dt}\widehat{\psi}$$

at zeroth order in $\hbar$*, considering only terms of at most two boundary fields and at most second power in the interaction.*

Proof (*Sketch of the proof*) We have that

$$\Delta((\widehat{\psi}\zeta)) = \Delta(\widehat{\psi})\zeta \pm \widehat{\psi}\Delta(\zeta) \pm (\psi, \zeta) = \Delta(\widehat{\psi})\zeta \pm (\psi, \zeta),$$

since $\Delta(\zeta) = 0$. On the other hand, using that Ω_0 and Ω_1 are first-order differential operators and Ω_2 is a second-order differential operator,

$$\begin{aligned}\Omega(\widehat{\psi}\zeta) &= \Omega_0(\widehat{\psi}\zeta) + \Omega_1(\widehat{\psi}\zeta) + \Omega_2(\widehat{\psi}\zeta) \\ &= \Omega_0(\widehat{\psi})\zeta + \widehat{\psi}\Omega_0(\zeta) + \Omega_1(\widehat{\psi})\zeta + \widehat{\psi}\Omega_1(\zeta) + \Omega_2(\widehat{\psi})\zeta + \widehat{\psi}\Omega_2(\zeta) + (\widehat{\psi}\zeta)' \\ &= \Omega(\widehat{\psi})\zeta + \widehat{\psi}\Omega(\zeta) + (\widehat{\psi}\zeta)',\end{aligned}$$

where $(\widehat{\psi}\zeta)'$ denotes the term where one derivative in Ω_2 acts on $\widehat{\psi}$ and the other acts on ζ. By the mQME, terms where Δ and Ω act on ψ only cancel. Let us first consider the term where Ω acts on ζ only. After integrating by parts, $\Omega_0(\zeta)$ replaces $\iota_{X^\partial}\gamma_i$ by $\mathrm{d}\iota_{X^\partial}\gamma_i$, plus contributions from the boundary of the configuration space. As in the proof of the mQME, those are cancelled by $\Omega_1(\zeta)$. Since Ω_2 can only act on products of terms, $\Omega_2(\zeta) = 0$. Next, notice that by properties of BV brackets and derivatives we have

$$(\psi, \zeta) = (S^{\text{eff}}, \zeta)\psi \qquad \text{and} \qquad (\psi\zeta)' = (S^{\text{eff}}\zeta)'\psi.$$

We are left to prove that $(S^{\text{eff}}, \zeta) + (S^{\text{eff}}\zeta)'$ produces all the terms of the form $\iota_{X^\partial}\mathrm{d}\gamma$, then the result follows from Proposition 2 and Cartan's magic formula. We summarise this as follows.

Lemma 9 *Let* S_i *be the parts of the effective action as above. Denote by* $\iota_{X^\partial}S_i$, $\mathrm{d}S_i$ *the operation of replacing all differential forms* γ *appearing in* S_i *by* $\iota_{X^\partial}\gamma$ *or* $\mathrm{d}\gamma$ *respectively. Then the following identities hold:*

$$\Omega_2(S_0\iota_{X^\partial}S_0) = (S_1, \iota_{X^\partial}S_0), \tag{9.20}$$

$$\Omega_2(S_0\iota_{X^\partial}S_2) + \Omega_2(S_2\iota_{X^\partial}S_0) = (S_1, \iota_{X^\partial}S_2), \tag{9.21}$$

$$\Omega_2(S_0\iota_{X^\partial}S_3) + \Omega_2(S_3\iota_{X^\partial}S_0) = (S_1, \iota_{X^\partial}S_3), \tag{9.22}$$

$$(S_2, \iota_{X^\partial}S_0) + (S_0, \iota_{X^\partial}S_2) = \iota_{X^\partial}\mathrm{d}S_3, \tag{9.23}$$

$$(S_2, \iota_{X^\partial}S_2) = \iota_{X^\partial}\mathrm{d}S_4 + \iota_{X^\partial}\mathrm{d}S_5. \tag{9.24}$$

As in the proof of the mQME, these are all the relevant brackets and products for our choice of degrees. Since S_3 and S_4 are the only terms with differential forms that are not closed, all the terms we need are produced and we conclude the statement. □

9.5 Conclusions and Outlook

We have shown that the BV-BFV formalism can be applied to split Chern–Simons theory and produces a non-trivial example. Using the method applied in Sect. 9.4.1.3 it is possible to make statements about the effective action to all orders. Furthermore, the structure of the identities in Lemmas 7 and 9 seems to hint to the structure of the effective action being governed by the mQME alone, i.e. to the fact that one can recover the state in the perturbed theory from the state in the unperturbed theory requiring only that the mQME is satisfied. A natural question to consider would be: to what extent one can make such a statement rigorous, and in what generality one can prove it.

In another direction, the next step is to use the state on the solid torus to compute the Chern–Simons theta invariants of lens spaces via the gluing operation. The relatively simple expression for the effective action in terms of a propagator and the cohomology on the boundary should also allow for an extension to higher genus handlebodies and other background flat connections, and thereby the computation of the Chern–Simons invariants for all 3-manifolds.

Acknowledgements A.S.C. and K.W. acknowledge partial support of SNF Grants No. 200020-149150/1 and PDFMP2_137103. This research was (partly) supported by the NCCR SwissMAP, funded by the Swiss National Science Foundation, and by the COST Action MP1405 QSPACE, supported by COST (European Cooperation in Science and Technology). P. M. acknowledges partial support of RFBR Grant No. 13-01-12405-ofi-m.

References

1. M. Atiyah, Topological quantum field theories. *Publications mathématiques de l'IHÉS*, **68** (1), 175–186 (1988). ISSN 1618-1913
2. I.A. Batalin, G.A. Vilkovisky, Quantization of gauge theories with linearly dependent generators. Phys. Rev. D **28**(10), 2567–2582 (1983). ISSN 0556-2821
3. I.A. Batalin, E.S. Fradkin, A generalized canonical formalism and quantization of reducible gauge theories. Phys. Lett. B **122**(2), 157–164 (1983). ISSN 0370-2693
4. I.A. Batalin, G.A. Vilkovisky, Gauge algebra and quantization. Phys. Lett. B **102**(1), 27–31 (1981)
5. N. Beisert, Quantum Field Theory II, Lecture Notes, 2014, Retrieved from http://edu.itp.phys.ethz.ch/fs13/qft2/
6. S. Cappell, D. DeTurck, H. Gluck, E.Y. Miller, Cohomology of Harmonic Forms on Riemannian Manifolds with Boundary (2007)
7. A.S. Cattaneo, G. Felder, A path integral approach to the Kontsevich quantization formula. Commun. Math. Phys. **212**, 591–611 (2000)

8. A.S. Cattaneo, P. Mnev, Remarks on Chern-Simons invariants. Commun. Math. Phys. **293**, 803–836 (2008)
9. A.S. Cattaneo, P. Mnev, N. Reshetikhin, Classical BV theories on manifolds with boundary. Commun. Math. Phys. **332**(2), 535–603 (2014)
10. A.S. Cattaneo, P. Mnev, N. Reshetikhin. Perturbative Quantum Gauge Theories on Manifolds with Boundary (2015)
11. P. Clavier, V.D. Nguyen, Batalin-Vilkovisky formalism as a theory of integration for polyvectors, in *Quantization, Geometry and Noncommutative Structures in Mathematics and Physics*, ed. by A. Cardona, P. Morales, H. Ocampo, A. Reyes Lega, S. Paycha (Springer, Berlin, 2017)
12. D. Fiorenza, An Introduction to the Batalin-Vilkovisky Formalism (2003)
13. M. Henneaux, C. Teitelboim, *Quantization of Gauge Systems* (Princeton University Press, Princeton, 1994)
14. R. Kirby, P. Melvin, The 3-manifold invariants of Witten and Reshetikhin-Turaev for sl(2, c). Invent Math. **105**(1), 473–545 (1991)
15. P. Mnev, Discrete BF Theory (2008)
16. P. Mnev, Lecture Notes on Torsions (2014)
17. T. Ohtsuki, *Quantum Invariants: A Study of Knots, 3-manifolds, and Their Sets*. K & E Series on Knots and Everything (World Scientific, Singapore, 2002)
18. E.M. Peskin, D.V. Schroeder, *An Introduction to Quantum Field Theory* (Westview Press, 1995)
19. M. Polyak, Feynman diagrams for pedestrians and mathematicians. Proc. Symp. Pure Math. **73**, 15–42 (2005)
20. N. Reshetikhin, Lectures on quantization of gauge systems, in *Proceedings of the Summer School "New paths towards quantum gravity", Holbaek, Denmark*, Lecture Notes in Physics, vol. 807, ed. by B. Booss-Bavnbek, B. D'Esposito, M. Lesch (Springer, Berlin, 2010), pp. 3–58
21. N. Reshetikhin, V.G. Turaev, Invariants of 3-manifolds via link polynomials and quantum groups. Invent. Math. **103**(1), 547–597 (1991)
22. A. Schwarz, Topological Quantum Field Theories (2007)
23. A.S. Schwarz, The partition function of degenerate quadratic functional and Ray-Singer invariants. Lett. Math. Phys. **2**(3), 247–252 (1978)
24. F. Schätz, Invariance of the bfv complex. Pacific J. Math. **248**(2), 453–474 (2010)
25. G.B. Segal, The definition of conformal field theory, in *Differential Geometrical Methods in Theoretical Physics (Como, 1987)*. NATO Adv. Sci. Inst. Ser. C Math. Phys. Sci., vol. 250 (Kluwer Acad. Publ., Dordrecht, 1988), pp. 165–171
26. D. Tong, Lectures on Quantum Field Theory (2006), Retrieved from http://www.damtp.cam.ac.uk/user/tong/qft.html
27. V.G. Turaev, *Quantum Invariants of Knots and 3-manifolds* (De Gruyter Studies in Mathematics, New York, 2010)

Chapter 10
Weighted Direct Product of Spectral Triples

Kevin Falk

Abstract We present conditions on a family of spectral triples for its direct product to be a spectral triple again. We also exhibit an example of such construction, involving Toeplitz operators with polynomial symbols and acting on weighted Bergman spaces over the unit ball of $\mathbb{C}^n$.

10.1 Introduction and Motivation

The main idea of Connes's non-commutative geometry is to characterize the geometry of a space in the language of algebras [2]. We know, for instance, that a compact Hausdorff space can be equivalently seen as the *commutative* C^*-algebra of continuous functions living on it. By analogy, a *non-commutative* algebra would correspond to a space of quantum nature: a *non-commutative space*. More precisely, the algebraic description of a space is based on the notion of *unital spectral triple*, consisting of the data $(\mathscr{A}, \mathscr{H}, \mathscr{D})$, where $\mathscr{A}$ is an involutive unital *-algebra $\mathscr{A}$ represented faithfully on a Hilbert space $\mathscr{H}$, and $\mathscr{D}$ is a selfadjoint operator acting on $\mathscr{H}$ with compact resolvent and such that for any $a \in \mathscr{A}$, $\pi(a)\text{dom}(\mathscr{D}) \subset \text{dom}(\mathscr{D})$ and the extended operator of $[\,\mathscr{D}, \pi(a)\,]$ is bounded. When $\mathscr{A}$ is not unital, consider its unitization $\mathscr{A} \oplus \mathbb{C}$ and replace the compactness of the resolvent by the compactness of $\pi(a)(\mathscr{D} - \lambda)^{-1}$ for any $a \in \mathscr{A}$ and $\lambda \notin \text{Spec}(\mathscr{D})$: The induced triple is then called *non-unital*. Among the various geometric entities which are encoded in the spectrum of $\mathscr{D}$, we are interested in the so-called *spectral dimension*, defined as the quantity $d := \inf\{s \in \mathbb{R}\,,\ \text{Tr}\ |\mathscr{D}|^{-s} < +\infty\}$.

One of the simplest examples of spectral triple concerns the unit circle S^1, where $\mathscr{A} = C^\infty(S^1)$ is the algebra of smooth functions over the circle, $\mathscr{H}$ is the Hilbert

K. Falk (✉)
Centre de Physique Théorique, CNRS, CPT, UMR 7332, Aix-Marseille Université, 13288 Marseille, France
e-mail: kevin.falk@cpt.univ-mrs.fr

K. Falk
et Université de Toulon, La Garde, France

A. Cardona et al. (eds.), *Quantization, Geometry and Noncommutative Structures in Mathematics and Physics*, Mathematical Physics Studies,
DOI 10.1007/978-3-319-65427-0_10

space completion of $\mathscr{A}$ itself endowed with the flat metric on the circle, and $\mathscr{D} = -i\frac{d}{dx}$, whose spectrum is given by $\{2\pi k\,,\, k \in \mathbb{Z}\}$. This spectral triple has spectral dimension 1 since

$$\mathrm{Tr}(\mathscr{D})^{-s} = \sum_{k\in\mathbb{Z}\setminus\{0\}} |2\pi k|^{-s} = \tfrac{2}{(2\pi)^s} \sum_{k\in\mathbb{N}\setminus\{0\}} k^{-s}$$

is finite if and only if $s > 1$.

Different classes of spectral triples based on algebras generated by Toeplitz operators acting on weighted Bergman spaces over a strictly pseudoconvex domain Ω of $\mathbb{C}^n$ are presented in [4]. A spectral triple related to the so-called Berezin–Toeplitz quantization was constructed [4, Sect. 6] and the interesting fact for us is that this spectral triple happens to be decomposable into a direct product of a family of spectral triples.

We focus here on the reciprocal, which is more subtle to treat: Given a countable family of spectral triples $(\mathscr{A}_m, \mathscr{H}_m, \mathscr{D}_m)_{m\in\mathbb{N}}$, representing a family of (not necessarily commutative) spaces, what are the conditions for the direct product over $m \in \mathbb{N}$ (the precise definition is given below) to be a spectral triple again? The main difficulty lies in the fact that the behavior of the spectral triples as m tends to infinity (boundedness of the representations, of the commutator between the algebras $\mathscr{A}_m$ and the operators $\mathscr{D}_m$, etc.) is hard to control in general.

We first present the notion of *weighted direct product* of spectral triples and then give an explicit example (also related to Toeplitz operators) and compute the corresponding spectral dimension.

Remark 10.1 The following results have been presented in [5] and related to non-commutative versions of self-similar sets.

10.2 Weighted Direct Product of Spectral Triples

As the following result states, a family of spectral triples must verify some (quite restrictive) conditions if we want the direct product to be a spectral triple again.

Lemma 10.1 *Let $(\mathscr{H}_m)_{m\in\mathbb{N}}$ be a family of Hilbert spaces, $(\mathscr{D}_m)_{m\in\mathbb{N}}$ be a family of unbounded selfadjoint operators with corresponding dense domains* $(\mathrm{dom}(\mathscr{D}_m) \subset \mathscr{H}_m)_{m\in\mathbb{N}}$ *and* $(\beta_m)_{m\in\mathbb{N}} \in (\mathbb{R}\setminus\{0\})^{\mathbb{N}}$.

Let $\mathscr{D}^{\oplus} := \bigoplus_{m\in\mathbb{N}} \beta_m \mathscr{D}_m$ *with domain* $\mathrm{dom}(\mathscr{D}^{\oplus}) := \big\{ \bigoplus_{m=0}^{N} \phi_m \in \mathscr{H}^{\oplus}\,,\, N \in \mathbb{N}\,,\, \phi_m \in \mathrm{dom}(\mathscr{D}_m) \big\}$.

Then $\mathscr{D}^{\oplus}$ is essentially selfadjoint.

Proof Let $\phi^{\oplus} := \bigoplus_{m\in\mathbb{N}} \phi_m \in \mathscr{H}^{\oplus}$. For any $m \in \mathbb{N}$, the operator $\mathscr{D}_m$ is densely defined so there is a sequence $(\phi_{mj})_{j\in\mathbb{N}}$ of elements in $\mathrm{dom}(\mathscr{D}_m)$ converging to ϕ_m as $j \to \infty$. Thus for any fixed $(m, j) \in \mathbb{N}^2$, there is $M_{mj} \in \mathbb{N}$ such that

$\| \phi_m - \phi_{m,M_{mj}+k} \|^2_{\mathscr{H}_m} < 2^{-j}$ for any $k \in \mathbb{N}$. Define for any $j \in \mathbb{N}$, the vector $\psi_j^\oplus := \bigoplus_{m=0}^{j} \phi_{m,M_{mj}} \in \mathrm{dom}(\mathscr{D}^\oplus)$. For any $j \in \mathbb{N}$, $\psi_j^\oplus \in \mathrm{dom}(\mathscr{D}^\oplus)$ and

$$\begin{aligned} \| \phi^\oplus - \psi_j^\oplus \|^2_{\mathscr{H}^\oplus} &= \sum_{m=0}^{j} \| \phi_m - \phi_{m,M_{mj}+k} \|^2_{\mathscr{H}_m} + \sum_{m>j} \| \phi_m \|^2_{\mathscr{H}_m} \\ &< j\, 2^{-j} + \sum_{m>j} \| \phi_m \|^2_{\mathscr{H}_m} \underset{j\to+\infty}{\longrightarrow} 0\,. \end{aligned}$$

Thus for any $\varepsilon > 0$, there exists $N \in \mathbb{N}$ such that $\| \phi^\oplus - \psi_N^\oplus \|_{\mathscr{H}^\oplus} < \varepsilon$, which shows that $\mathscr{D}^\oplus$ is densely defined.

Using the same reasoning and the fact that for any $m \in \mathbb{N}$, $\mathrm{Ran}(\beta_m \mathscr{D}_m \pm i) = \mathscr{H}_m$ (since $\beta_m \mathscr{D}_m$ is selfadjoint), it can be shown that for any $\phi^\oplus \in \mathscr{H}^\oplus$ and $\varepsilon > 0$, there is $N \in \mathbb{N}$ and $\psi_N^\oplus \in \mathrm{dom}(\mathscr{D}^\oplus)$ defined as above and such that $\| \phi^\oplus - (\mathscr{D}^\oplus \pm i)\psi_N^\oplus \|_{\mathscr{H}^\oplus} < \varepsilon$, thus $\mathrm{Ran}(\mathscr{D}^\oplus \pm i)$ is dense in $\mathscr{H}^\oplus$.

The operator $\mathscr{D}^\oplus$ is also symmetric since for any $\phi^\oplus := \sum_{m=0}^{N} \phi_m$ and $\phi'^\oplus := \sum_{m=0}^{N'} \phi'_m$ in $\mathrm{dom}(\mathscr{D}^\oplus)$,

$$\begin{aligned} \langle\, \mathscr{D}^\oplus \phi^\oplus \,,\, \phi'^\oplus \,\rangle_{\mathscr{H}^\oplus} &= \sum_{m=0}^{\min(N,N')} \langle\, \beta_m \mathscr{D}_m \phi_m \,,\, \phi'_m \,\rangle_{\mathscr{H}_m} = \\ &= \sum_{m=0}^{\min(N,N')} \langle\, \phi_m \,,\, \beta_m \mathscr{D}_m \phi'_m \,\rangle_{\mathscr{H}_m} = \langle\, \phi^\oplus \,,\, \mathscr{D}^\oplus \phi'^\oplus \,\rangle_{\mathscr{H}^\oplus}\,, \end{aligned}$$

which shows that $\mathscr{D}^\oplus$ is essentially selfadjoint (see [7, Chap. VIII.2, Corollary p.257]).

Proposition 10.1 *Let $(\mathscr{A}_m, \mathscr{H}_m, \mathscr{D}_m)_{m\in\mathbb{N}}$ be a family of (not necessarily unital) spectral triples, with corresponding representations $(\pi_m)_{m\in\mathbb{N}}$, and denote $\| \,.\, \|_m$ the norm on $\pi_m(\mathscr{A}_m)$.*

Define the following objects:

- $\mathscr{H}^\oplus := \bigoplus_{m\in\mathbb{N}} \mathscr{H}_m$,
- $\mathscr{D}^\oplus := \bigoplus_{m\in\mathbb{N}} \beta_m\, \mathscr{D}_m$ *and* $\overline{\mathscr{D}^\oplus}$ *as above, both acting on* $\mathscr{H}^\oplus$,
- $$\begin{aligned} \mathscr{A}^\oplus := \{(a_m)_{m\in\mathbb{N}} \in \prod_{m\in\mathbb{N}} \mathscr{A}_m \,:\, &\sup_{m\in\mathbb{N}} \| \,\pi_m(a_m)\, \|_m < +\infty, \\ &\sup_{m\in\mathbb{N}} \| \,[\,\beta_m \mathscr{D}_m,\, \pi_m(a_m)\,]\, \|_m < +\infty\,, \text{ and} \\ &\| \,\pi_m(a_m)(1 + \beta_m^2 \mathscr{D}_m^2)^{-1/2}\, \|_m \underset{m\to+\infty}{\longrightarrow} 0\}, \end{aligned}$$
- $\pi^\oplus(a^\oplus) := \bigoplus_{m\in\mathbb{N}} \pi_m(a_m)$, *for* $a^\oplus \in \mathscr{A}^\oplus$.

Then $(\mathscr{A}^\oplus, \mathscr{H}^\oplus, \overline{\mathscr{D}^\oplus})$ is a (not necessarily unital) spectral triple.

Proof For two elements $a^{\oplus} = (a_m)_{m\in\mathbb{N}}$ and $b^{\oplus} = (b_m)_{m\in\mathbb{N}}$ in $\mathscr{A}^{\oplus}$, we have:

$$\sup_{m\in\mathbb{N}} \| \pi_m(a_m b_m) \|_m \leq \sup_{m\in\mathbb{N}} \| \pi_m(a_m) \|_m \sup_{m\in\mathbb{N}} \| \pi_m(b_m) \|_m < +\infty\,, \qquad \text{and}$$
$$\sup_{m\in\mathbb{N}} \| [\, \beta_m \mathscr{D}_m,\, \pi_m(a_m b_m) \,] \|_m \leq \sup_{m\in\mathbb{N}} \| \pi_m(a_m) \|_m \sup_{m\in\mathbb{N}} \| [\, \beta_m \mathscr{D}_m,\, \pi_m(b_m) \,] \|_m$$
$$+ \sup_{m\in\mathbb{N}} \| [\, \beta_m \mathscr{D}_m,\, \pi_m(a_m) \,] \|_m \sup_{m\in\mathbb{N}} \| \pi_m(b_m) \|_m < +\infty\,,$$

hence $\mathscr{A}^{\oplus}$ is an algebra with involution $* : a^{\oplus} = (a_m)_{m\in\mathbb{N}} \mapsto (a^{\oplus})^* := (a_m^*)_{m\in\mathbb{N}}$.

For $a^{\oplus} \in \mathscr{A}^{\oplus}$, we have

$$\pi^{\oplus}(a^{\oplus}) \left(1 + (\mathscr{D}^{\oplus})^2\right)^{-1/2} = \bigoplus_{m\in\mathbb{N}} \pi_m(a_m)\, (1 + \beta_m^2\, \mathscr{D}_m^2)^{-1/2}.$$

Combining the fact that for any $m \in \mathbb{N}$, the summand $\pi_m(a_m)\,(1 + \beta_m^2\, \mathscr{D}_m^2)^{-1/2}$ is compact, and the third condition in the definition of $\mathscr{A}^{\oplus}$, we conclude that $\pi^{\oplus}(a^{\oplus}) \left(1 + (\mathscr{D}^{\oplus})^2\right)^{-1/2}$ is compact.

From Lemma 10.1, $\mathscr{D}^{\oplus}$ is essentially selfadjoint with selfadjoint extension $\overline{\mathscr{D}^{\oplus}}$.

For $a^{\oplus} \in \mathscr{A}^{\oplus}$ and $\phi_N^{\oplus} := \bigoplus_{m=0}^{N} \phi_{N,m} \in \mathrm{dom}(\mathscr{D}^{\oplus})$, for some $N \in \mathbb{N}$, we have

$$\pi^{\oplus}(a^{\oplus}) \phi_N^{\oplus} = \bigoplus_{m=0}^{N} \pi_m(a_m) \phi_{N,m}$$

and each summand on the right-hand side belongs to $\mathrm{dom}(\mathscr{D}_m)$ since $(\mathscr{A}_m, \mathscr{H}_m, \mathscr{D}_m)$ is a spectral triple for any $m \in \mathbb{N}$. Thus $\pi^{\oplus}(a^{\oplus})$ maps $\mathrm{dom}(\mathscr{D}^{\oplus})$ into itself for any $a^{\oplus} \in \mathscr{A}^{\oplus}$.

Moreover, for any $a^{\oplus} \in \mathscr{A}^{\oplus}$ and $\phi_N^{\oplus} := \bigoplus_{m=0}^{N} \phi_{N,m} \in \mathrm{dom}(\mathscr{D}^{\oplus})$ of norm 1, we have

$$\| [\, \mathscr{D}^{\oplus},\, \pi^{\oplus}(a^{\oplus}) \,] \phi_N^{\oplus} \| = \sup_{m=0,\dots,N} \| [\, \beta_m \mathscr{D}_m,\, \pi_m(a_m) \,] \phi_{N,m} \|$$
$$\leq \sup_{m\in\mathbb{N}} \| \, [\, \beta_m \mathscr{D}_m,\, \pi_m(a_m) \,] \, \|_m < +\infty\,,$$

so $[\, \mathscr{D}^{\oplus},\, \pi^{\oplus}(a^{\oplus}) \,]$ is bounded on $\mathrm{dom}(\mathscr{D}^{\oplus})$. Moreover, since $\mathrm{dom}(\mathscr{D}^{\oplus})$ is a core for $\overline{\mathscr{D}^{\oplus}}$, we conclude that for any $a^{\oplus} \in \mathscr{A}^{\oplus}$, $\pi^{\oplus}(a^{\oplus})\big(\mathrm{dom}(\overline{\mathscr{D}^{\oplus}})\big) \subset \mathrm{dom}(\overline{\mathscr{D}^{\oplus}})$, thus $[\, \overline{\mathscr{D}^{\oplus}},\, \pi^{\oplus}(a^{\oplus}) \,]$ extends to a bounded operator on $\mathscr{H}^{\oplus}$ (see [6, Proposition A.1]).

The previously defined spectral triple $(\mathscr{A}^{\oplus}, \mathscr{H}^{\oplus}, \overline{\mathscr{D}^{\oplus}})$ is called the *weighted direct product* of the sequence of quintuples $(\mathscr{A}_m, \mathscr{H}_m, \mathscr{D}_m, \pi_m, \beta_m)_{m\in\mathbb{N}}$, where $(\mathscr{A}_m, \mathscr{H}_m, \mathscr{D}_m)$ is a spectral triple for any $m \in \mathbb{N}$, with corresponding representations π_m and weights $\beta_m \in \mathbb{R}\backslash\{0\}$.

The first two conditions in the definition of $\mathscr{A}^\oplus$ correspond to the boundedness of both the representation $\pi^\oplus$ and the commutator $[\overline{\mathscr{D}^\oplus}, \pi^\oplus(\mathscr{A}^\oplus)]$ for the norm $\| \,.\, \|^\oplus := \sup_{m\in\mathbb{N}} \| \,.\, \|_m$ on $\pi^\oplus(\mathscr{A}^\oplus)$. The third condition is equivalent to the compactness of the resolvent of $\mathscr{D}^\oplus$ (recall that an operator $\bigoplus_{m\in\mathbb{N}} A_m$ is compact if and only if A_m is compact for any $m \in \mathbb{N}$ and $\| A_m \|_m \to 0$ as $m \to \infty$; see [3, Exercice II.4.13]). As a consequence, the sequence $(\mathscr{D}_m)_{m\in\mathbb{N}}$ is such that $\sum_{m\in\mathbb{N}} \dim(\text{Ker } \mathscr{D}_m) < \infty$. In particular, if we take the same $\mathscr{D}_m = \mathscr{D}_0$ at each level $m \in \mathbb{N}$, the latter must be invertible.

The parameter $(\beta_m)_{m\in\mathbb{N}}$ has been introduced in order to control the behavior of the sequence $(\mathscr{D}_m)_{m\in\mathbb{N}}$ as m tends to infinity. This can be avoided by putting some constraints directly on the operators $\mathscr{D}_m$, but this restricts the set of summable families of spectral triples. Indeed, when $\mathscr{D}^\oplus := \bigoplus_{m\in\mathbb{N}} \mathscr{D}_0$, with $\mathscr{D}_0$ invertible, then the resolvent of $\mathscr{D}^\oplus$ is not compact and the use of a sequence $(\beta_m)_{m\in\mathbb{N}}$ is necessary for the direct product.

An alternative would consist in rescaling the norm $\| \,.\, \|_m$ at each level by multiplying it by the term β_m and set $\mathscr{D}^\oplus$ as the simple direct product of all $\mathscr{D}_m$.

10.3 Example of Weighted Direct Product with Toeplitz Operators

Denote $\mathscr{B}^n$ the unit open ball of $\mathbb{C}^n$, with the unit sphere $\partial\mathscr{B}^n$ as boundary. Let r be the defining function $r : z \in \overline{\mathscr{B}^n} \mapsto r(z) := |z|^2 - 1$,[1] and consider a family of weights on $\mathscr{B}^n$ of the form $w_m := (-r)^m$, with $m \in \mathbb{N}$. The corresponding *weighted Bergman space* over $\mathscr{B}^n$ is given by

$$A^2_m(\mathscr{B}^n) := \{\phi \in L^2(\mathscr{B}^n, w_m d\mu)\,,\, \phi \text{ holomorphic in } \mathscr{B}^n\}\,,$$

where $d\mu$ is the usual Lebesgue measure over $\mathscr{B}^n$. Denote $\boldsymbol{\Pi}_m$ the orthogonal projection from $L^2(\mathscr{B}^n)$ onto $A^2_m(\mathscr{B}^n)$. The *Toeplitz operator* corresponding to $f \in C^\infty(\overline{\mathscr{B}^n})$ is defined as

$$\mathbf{T}^{(m)}_f : \phi \in A^2_m(\mathscr{B}^n) \mapsto \boldsymbol{\Pi}_m(f\phi) \in A^2_m(\mathscr{B}^n)\,. \tag{10.1}$$

For instance, since $r < 0$ on $\mathscr{B}^n$, $\mathbf{T}^{(m)}_{-r}$ is a selfadjoint positive definite Toeplitz operator on $A^2_m(\mathscr{B}^n)$, admitting an unbounded selfadjoint inverse $(\mathbf{T}^{(m)}_{-r})^{-1}$, densely defined on $A^2_m(\mathscr{B}^n)$. For a general Toeplitz operator $\mathbf{T}^{(m)}_f$, $f \in C^\infty(\overline{\mathscr{B}^n})$, we have the relations

$$\| \mathbf{T}^{(m)}_f \| \leq \| f \|_\infty\,, \text{ and } \quad (\mathbf{T}^{(m)}_f)^* = \mathbf{T}^{(m)}_{\bar{f}}\,.$$

---

[1] r is a defining function for $\mathscr{B}^n$ means that it verifies $r|_{\mathscr{B}^n} < 0$, $r|_{\partial\mathscr{B}^n} = 0$ and $dr|_{\partial\mathscr{B}^n} \neq 0$.

Since in general the product of two Toeplitz operators is not Toeplitz anymore, we will consider the $*$-algebra *generated* by the Toeplitz operators (10.1), the involution given by the usual Hilbert space adjoint.

The following result is a particular case of [4, Proposition 5.4]:

Proposition 10.2 *Let $\mathscr{A}_m$ be the algebra generated by the Toeplitz operators $\mathbf{T}_f^{(m)}$, $f \in C^\infty(\overline{\mathscr{B}^n})$, with the identity representation on $\mathscr{H}_m := \mathrm{A}_m^2(\mathscr{B}^n)$, and $\mathscr{D}_m := (\mathbf{T}_{-r}^{(m)})^{-1}$.*

Then $(\mathscr{A}_m, \mathscr{H}_m, \mathscr{D}_m)$ is a unital spectral triple of spectral dimension $n = \dim_{\mathbb{C}} \mathscr{B}^n$.

From a geometric point of view, we get a non-commutative version of the unit ball of $\mathbb{C}^n$, of same dimension as the usual one.

For a general multiindex $\alpha \in \mathbb{N}^n$ and $z \in \mathbb{C}^n$, we use the notation

$$z^\alpha := z_1^{\alpha_1} z_2^{\alpha_2} \dots z_n^{\alpha_n},$$

and define $1_j := (0, \dots, 0, 1, 0, \dots, 0) \in \mathbb{N}^n$ (the 1 being at the jth place), for $j = 1, \dots, n$.

We introduce the operators $\mathscr{R} := \sum_{j=1}^n \mathscr{R}_j$ and $\overline{\mathscr{R}} := \sum_{j=1}^n \overline{\mathscr{R}}_j$ with $\mathscr{R}_j := z_j \partial_{z_j}$ and $\overline{\mathscr{R}}_j := \bar{z}_j \, \partial_{\bar{z}_j}$, acting on $C^\infty(\overline{\mathscr{B}^n})$. Let $\mathrm{Pol}(\mathbb{B}^n)$ be the set of polynomials on $\mathbb{B}^n$ in z and $\bar{z}$.

Proposition 10.3 *For $p(z) = \sum_{|\alpha| \le d, |\beta| \le d'} p_{\alpha\beta} \, z^\alpha \, \bar{z}^\beta \in \mathrm{Pol}(\mathbb{B}^n)$, and setting for short $\mathbf{T}_p = \mathbf{T}_p^{(m)}$, we get*

$$[\, \mathbf{T}_{-r}^{-1}, \, \mathbf{T}_p \,] = \tfrac{1}{m+1} \, \mathbf{T}_{(\mathscr{R} - \overline{\mathscr{R}}) \, p} \, , \quad \textit{on } A_m^2(\mathbb{B}^n) \, .$$

Proof By [8, (2.9)], a standard orthonormal basis of $A_m^2(\mathbb{B}^n)$ is given by

$$v_\alpha(z) = \left(\tfrac{\Gamma(|\alpha|+m+n+1)}{\Gamma(m+n+1) \, \alpha!} \right)^{1/2} z^\alpha \, .$$

Using the shift operators $S_j : v_\alpha \mapsto v_{\alpha + 1_j}$, $j = 1, \dots, n$, we have the relations

$$\mathbf{T}_{z_j} = S_j \left(\tfrac{\mathscr{R}_j + 1}{\mathscr{R} + m + n + 1} \right)^{1/2} , \; \mathscr{R}_j S_j = S_j (\mathscr{R}_j + 1), \; S_j^* S_j = 1, \quad \text{for } j = 1, \dots, n, \text{ and}$$

$$\begin{aligned} \mathbf{T}_{-r}^{-1} &= \left(1 - \sum_{j=1}^n \mathbf{T}_{|z_j|^2} \right)^{-1} = \left(1 - \sum_{j=1}^n (\mathbf{T}_{z_j})^* \mathbf{T}_{z_j} \right)^{-1} = \left(1 - \sum_{j=1}^n \tfrac{\mathscr{R}_j + 1}{\mathscr{R} + m + n + 1} \right)^{-1} \\ &= \tfrac{1}{m+1} (\mathscr{R} + m + n + 1) \, . \end{aligned}$$

Hence we get

$$\begin{aligned}
[\mathbf{T}_{-r}^{-1}, \mathbf{T}_{z_j}] &= \tfrac{1}{m+1}\Big((\mathscr{R}+m+n+1)\,\boldsymbol{S}_j\left(\tfrac{\mathscr{R}_j+1}{\mathscr{R}+m+n+1}\right)^{1/2} \\
&\qquad - \boldsymbol{S}_j\left(\tfrac{\mathscr{R}_j+1}{\mathscr{R}+m+n+1}\right)^{1/2}(\mathscr{R}+m+n+1)\Big) \\
&= \tfrac{1}{m+1}\,\boldsymbol{S}_j\left(\tfrac{\mathscr{R}_j+1}{\mathscr{R}+m+n+1}\right)^{1/2}\big(\mathscr{R}+m+n+2-(\mathscr{R}+m+n+1)\big) \\
&= \tfrac{1}{m+1}\,\mathbf{T}_{z_j}\,.
\end{aligned}$$

From this last equality, combined with the fact that $[\mathbf{T}_{z_j}, \mathbf{T}_{z_k}] = 0$, for $j, k = 1, \dots, n$, we get by iteration of the formula $[A, BC] = B[A, C] + [A, B]C$ for $\alpha, \beta \in \mathbb{N}^n$

$$[\mathbf{T}_r^{-1}, \prod_{j=1}^{n}\mathbf{T}_{z_j}^{\alpha_j}] = \tfrac{|\alpha|}{m+1}\prod_{j=1}^{n}\mathbf{T}_{z_j}^{\alpha_j}, \text{ and similarly } [\mathbf{T}_{-r}^{-1}, \prod_{j=1}^{n}(\mathbf{T}_{z_j}^{*})^{\beta_j}] = -\tfrac{|\beta|}{m+1}\prod_{j=1}^{n}(\mathbf{T}_{z_j}^{*})^{\beta_j}.$$

Hence, the relation $\mathbf{T}_{z^\alpha\bar{z}^\beta} = \big(\prod_{j=1}^{n}(\mathbf{T}_{z_j}^{*})^{\beta_j}\big)\big(\prod_{j=1}^{n}\mathbf{T}_{z_j}^{\alpha_j}\big)$ yields to

$$\begin{aligned}
[\mathbf{T}_{-r}^{-1}, \mathbf{T}_p] &= \sum_{|\alpha|\le d, |\beta|\le d'}[\mathbf{T}_{-r}^{-1}, \mathbf{T}_{z^\alpha\bar{z}^\beta}] = \sum_{|\alpha|\le d, |\beta|\le d'}[\mathbf{T}_{-r}^{-1}, \big(\prod_{j=1}^{n}(\mathbf{T}_{z_j}^{*})^{\beta_j}\big)\big(\prod_{j=1}^{n}\mathbf{T}_{z_j}^{\alpha_j}\big)] \\
&= \sum_{|\alpha|\le d, |\beta|\le d'} p_{\alpha\beta}\Big(\big(\prod_{j=1}^{n}(\mathbf{T}_{z_j}^{*})^{\beta_j}\big)[\mathbf{T}_{-r}^{-1}, \prod_{j=1}^{n}\mathbf{T}_{z_j}^{\alpha_j}] + [\mathbf{T}_{-r}^{-1}, \prod_{j=1}^{n}(\mathbf{T}_{z_j}^{*})^{\beta_j}]\prod_{j=1}^{n}\mathbf{T}_{z_j}^{\alpha_j}\Big) \\
&= \tfrac{1}{m+1}\sum_{|\alpha|\le d, |\beta|\le d'} p_{\alpha\beta}\,(|\alpha|-|\beta|)\big(\prod_{j=1}^{n}(\mathbf{T}_{z_j}^{*})^{\beta_j}\big)\big(\prod_{j=1}^{n}\mathbf{T}_{z_j}^{\alpha_j}\big) \\
&= \tfrac{1}{m+1}\sum_{|\alpha|\le d, |\beta|\le d'} p_{\alpha\beta}\,(|\alpha|-|\beta|)\,\mathbf{T}_{z^\alpha\bar{z}^\beta} \\
&= \tfrac{1}{m+1}\,\mathbf{T}_{(\mathscr{R}-\overline{\mathscr{R}})\,p}\,.
\end{aligned}$$

□

Remark 10.2 The previous result is restricted to polynomials only. Indeed, we cannot apply the Stone–Weierstrass theorem to extend the result for general smooth functions over $\overline{\mathscr{B}^n}$ since $f \mapsto [\mathbf{T}_{-r}^{-1}, \mathbf{T}_f]$ is not continuous on A_m^2 for the norm $\|\,.\,\|_\infty$.

Theorem 10.1 *For $m \in \mathbb{N}$, let*

- $\mathscr{H}_m := A_m^2(\mathbb{B}^n)$,
- $\mathscr{D}_m := (\mathbf{T}_{-r}^{(m)})^{-1}$,
- $\mathscr{A}_m$ *be the* **-algebra generated by Toeplitz operators* $\mathbf{T}_p^{(m)}$ *acting on* $\mathscr{H}_m$, *with* $p \in \mathrm{Pol}(\mathbb{B}^n)$,
- π_m *be the identity representation on* $\mathscr{H}_m$,
- $\|\,.\,\|_m$ *be the usual norm of operators,*
- $\beta_m := m + 1$.

If we define $\mathscr{H}^{\oplus}$, $\mathscr{D}^{\oplus}$, $\pi^{\oplus}$ as in Proposition 10.1 and $\mathscr{A}'^{\oplus}$ as the algebra generated by elements of the form $(\mathbf{T}_p^{(m)})_{m\in\mathbb{N}}$, with $p \in \mathrm{Pol}(\mathbb{B}^n)$ (i.e., keeping the same polynomial at all levels $m \in \mathbb{N}$), then $(\mathscr{A}'^{\oplus}, \mathscr{H}^{\oplus}, \overline{\mathscr{D}^{\oplus}})$ is the weighted direct product of the sequence $(\mathscr{A}_m'^{\oplus}, \mathscr{H}_m^{\oplus}, \mathscr{D}_m)_{m\in\mathbb{N}}$.

Proof First, we know from Proposition 10.2 that for any $m \in \mathbb{N}$, $(\mathscr{A}_m, \mathscr{H}_m, \mathscr{D}_m)$ defines a unital spectral triple of dimension n. Let us show that $\mathscr{A}'^{\oplus}$ is a subalgebra of $\mathscr{A}^{\oplus}$ of Proposition 10.1: If $(a_m)_{m\in\mathbb{N}} = (\mathbf{T}_p^{(m)})_{m\in\mathbb{N}}$ of $\mathscr{A}'^{\oplus}$, $p \in \mathrm{Pol}(\mathbb{B}^n)$, is a generator, the conditions are satisfied since

$$\begin{aligned}
&\sup_{m\in\mathbb{N}} \| \pi_m(a_m) \|_m \leq \| p \|_\infty < +\infty\,,\\
&\sup_{m\in\mathbb{N}} \| [\, \beta_m \mathscr{D}_m, \pi_m(a_m) \,] \|_m = \sup_{m\in\mathbb{N}} \tfrac{m+1}{m+1} \| \mathbf{T}^{(m)}_{(\mathscr{R}-\overline{\mathscr{R}})p} \| \leq \| (\mathscr{R} - \overline{\mathscr{R}})\, p \|_\infty\\
&\qquad\qquad\qquad\qquad\qquad < +\infty\,, \text{ from Proposition 10.3, and}\\
&\| \pi_m(a_m)(1 + \beta_m^2\, \mathscr{D}_m^2)^{-1/2} \|_m = \| \pi_m(a_m)(1 + \beta_m^2\, (\mathbf{T}_{-r}^{(m)})^{-2})^{-1/2} \|_m\\
&\qquad\qquad \leq |\beta_m|^{-1} \| \pi_m(a_m) \|_m \| \mathbf{T}_{-r}^{(m)} \|_m\\
&\qquad\qquad \leq |\beta_m|^{-1} \sup_{m\in\mathbb{N}} \| \pi_m(a_m) \|_m \; \| r \|_\infty \underset{m\to+\infty}{\longrightarrow} 0\,.
\end{aligned}$$

These inequalities remain valid for a general element of $\mathscr{A}'^{\oplus}$, which consists, at each level $m \in \mathbb{N}$, of the same finite sum of finite products of Toeplitz operators acting on A_m^2. Since $\mathscr{A}'^{\oplus}$ forms a *-algebra, we conclude that it is a *-subalgebra of $\mathscr{A}^{\oplus}$ and from Proposition 10.1, it follows that $(\mathscr{A}'^{\oplus}, \mathscr{H}^{\oplus}, \overline{\mathscr{D}^{\oplus}})$ is a spectral triple.

We now compute its dimension. For $s \in \mathbb{R}$, we have

$$\begin{aligned}
\mathrm{Tr}\, |\overline{\mathscr{D}^{\oplus}}|^{-s} &= \sum_{m\in\mathbb{N}} \beta_m^{-s}\, \mathrm{Tr}\, (\mathbf{T}_{-r}^{(m)})^{s} = \sum_{m\in\mathbb{N}} \left(\tfrac{\beta_m}{m+1}\right)^{-s} \mathrm{Tr}\, (\mathscr{R} + m + n + 1)^{-s}\\
&= \sum_{m\in\mathbb{N}} \mathrm{Tr}\, (\mathscr{R} + m + n + 1)^{-s}\,.
\end{aligned}$$

Denote $\lambda_k(m) := k + m + n + 1$ the eigenvalues of $\mathscr{R} + m + n + 1$, and $M_k := \binom{k+n-1}{n-1}$ the corresponding multiplicities, then

$$\mathrm{Tr}\, (\mathscr{R} + m + n + 1)^{-s} = \sum_{k=0}^{\infty} M_k\, \lambda_k(m)^{-s} =: I_m(s)\,. \tag{10.2}$$

We know from Proposition 10.2 that $I_m(s)$ is finite for $s > n$. So when $s = n + \varepsilon$, $\varepsilon > 0$, we can estimate the asymptotic behavior of this quantity as $m \to \infty$ using forthcoming Lemma 10.2, and so $\mathrm{Tr}\, |\overline{\mathscr{D}^{\oplus}}|^{-(n+\varepsilon)}$ is finite when

$$\sum_{m\in\mathbb{N}} \left(\tfrac{m+1}{\beta_m}\right)^{n+\varepsilon} m^{-\varepsilon} < +\infty\,, \tag{10.3}$$

i.e., when $\varepsilon > 1$, which leads to the result.

Lemma 10.2 *With the notations of* (10.2)*, we have* $I_m(n+\varepsilon) \underset{m\to\infty}{\sim} (n-1)!\, \frac{\Gamma(\varepsilon)}{\Gamma(\varepsilon+n)} m^{-\varepsilon}$.

Proof Recall that the Pochhammer symbol $(x)_{(n)} := x(x-1)\dots(x-n+1)$ and the Stirling numbers of the first kind $s(n,k)$ are related by

$$(x)_{(n)} = \sum_{k=0}^{n} s(n,k)\, x^k\,.$$

First, if $n \geq 2$, we can express M_k as a polynomial in $\lambda_k(m)$:

$$\begin{aligned} M_k &= \tfrac{1}{(n-1)!}\,(k+n-1)(k+n-2)\cdots(k+1) = \tfrac{1}{(n-1)!}\,\big(\lambda_k(m)-(m+2)\big)_{(n-1)} \\ &= \tfrac{1}{(n-1)!}\sum_{j=1}^{n-1} s(n-1,j)\,\big(\lambda_k(m)-(m+2)\big)^j = \tfrac{1}{(n-1)!}\sum_{l=0}^{n-1} c_l(m)\,\lambda_k(m)^l, \end{aligned}$$

where $c_l(m) := \sum_{i=l}^{n-1} \binom{i}{l} s(n-1,i)\big(-(m+2)\big)^{i-l}$ for any $l = 0,\cdots,n-1$. Thus we have

$$I_m(n+\varepsilon) = \tfrac{1}{(n-1)!}\sum_{k=0}^{\infty} g_m(k)\,, \quad \text{where} \quad g_m(k) := \sum_{l=0}^{n-1} c_l(m)\,\lambda_k(m)^{l-(n+\varepsilon)}.$$

We use the Euler–Maclaurin formula on g_m (see, for instance, [1, Chap. VI.7 (37)]):

$$\begin{aligned} \sum_{k\in\mathbb{N}} g_m(k) &= \int_0^{+\infty} g_m(x)\,dx + \tfrac{1}{2}\big(f(0) + \lim_{x\to+\infty} g_m(x)\big) \\ &\quad + \sum_{j=2}^{N} \tfrac{B_j}{j!} \lim_{x\to+\infty} \big(\partial_x^{j-1} g_m(x) - \partial_x^{j-1} g_m(0)\big) + R(m) \\ &=: T_1(m) + T_2(m) + T_3(m) + R(m), \end{aligned}$$

where B_j is the jth Bernoulli number and

$$R(m) := \tfrac{(-1)^{N+1}}{N!} \int_0^{\infty} \partial_x^N g_m(x)\, b_N(x - \lfloor x \rfloor)\, dx,$$

and b_N being the Nth-Bernoulli polynomial. We get

$$\begin{aligned}
T_1(m) &= \int_0^\infty g_m(x)\,dx = \sum_{l=0}^{n-1} c_l(m) \int_0^\infty (x+m+n+1)^{l-(n+\varepsilon)}\,dx \\
&= -\sum_{l=0}^{n-1} c_l(m)\,\tfrac{1}{l-(n+\varepsilon)+1}\,(m+n+1)^{l-(n+\varepsilon)+1} \\
&= -\sum_{l=0}^{n-1}\sum_{i=l}^{n-1} \tbinom{i}{l} s(n-1,i)\big(-(m+2)\big)^{i-l}\tfrac{1}{l-(n+\varepsilon)+1}(m+n+1)^{l-(n+\varepsilon)+1} \\
&\underset{m\to\infty}{\sim} \sum_{l=0}^{n-1} \tbinom{n-1}{l}\,\tfrac{(-1)^{n-l}}{l-(n+\varepsilon)+1}\,m^{-\varepsilon} = (n-1)!\,\tfrac{\Gamma(\varepsilon)}{\Gamma(\varepsilon+n)}\,m^{-\varepsilon}\,, \quad \text{and} \\
T_2(m) &= \tfrac{1}{2}g(0) = \tfrac{1}{2}\sum_{l=0}^{n-1}\sum_{i=l}^{n-1} s(n-1,i)\big(-(m+2)\big)^{i-l}\,(m+n+1)^{l-(n+\varepsilon)} \\
&= O(m^{-(\varepsilon+1)})\,.
\end{aligned}$$

Since, for $j \geq 2$,

$$\partial_x^{j-1} g_m(x) = \sum_{l=0}^{n-1} c_l(m)(l-(n+\varepsilon))_{(j-1)}\,(x+m+n-1)^{l-(n+\varepsilon)-(j-1)}, \tag{10.4}$$

we get, for $N \geq 2$, $T_3(m) = O(m^{-(\varepsilon+2)})$. We have the following upper bound for the remainder (obtained by computing the Fourier series of the Bernoulli polynomial b_N)

$$|R(m)| \leq \tfrac{2}{(2\pi)^N}\,\zeta(N)\int_0^{+\infty} |g_m^{(N)}|(x)\,dx\,,$$

which gives, using (10.4), and after integration over x for $N \geq 2$,

$$\begin{aligned}
|R(m)| &\leq \tfrac{2}{(2\pi)^N}\,\zeta(N)\sum_{l=0}^{n-1} |c_l(m)||(l-n+\varepsilon)_{(N)}|\frac{(m+n+1)^{l-(n+\varepsilon)-N+1}}{(N-1+n+\varepsilon-l)} \\
&= O(m^{-(\varepsilon+N-1)}) = O(m^{-(\varepsilon+2)})\,.
\end{aligned}$$

□

Remark 10.3

(i) A possible extension of Theorem 10.1, in which any $(a_m)_{m\in\mathbb{N}} \in \mathscr{A}'^{\oplus}$ is defined as the copy of the same element on each level $m \in \mathbb{N}$, consists in replacing a finite number of a_m by arbitrary elements of $\mathscr{A}_m$.

Thus the representation of an element $a^{\oplus}$ of this new algebra $\mathscr{A}''^{\oplus}$ is of the form

$$\pi''^{\oplus}(a^{\oplus}) = \bigoplus_{m \leq N} \pi_m(a_m) \oplus \bigoplus_{m>N} \sum_{i=1}^{p} \prod_{j=1}^{q_i} \mathbf{T}_{p_{ij}}^{(m)},$$

for some integer N, some arbitrary $a_m \in \mathscr{A}_m$, $m \leq N$, and fixed family of polynomials p_{ij} in $\mathrm{Pol}(\mathbb{B}^n)$, $i = 1, \ldots, p$, $j = 1, \ldots, q_i$.

(ii) In Theorem 10.1, we can consider a more general sequence $(\beta_m)_{m \in \mathbb{N}}$ such that $\beta_m \sim m^{\delta}$, as m tends to infinity, for $0 < \delta \leq 1$ (the upper bound comes from the boundedness of the commutator between the representation of an element of the algebra and $\overline{\mathscr{D}^{\oplus}}$). Then, the conclusions of Proposition 10.1 remain valid but the dimension changes: In this case, (10.3) is true when $\varepsilon > \big(1 + n(1-\delta)\big)/\delta$. As a consequence, making δ varying in $(0, 1]$ leads to a dimension lying in $[n+1, +\infty)$.

References

1. N. Bourbaki, *Éléments de mathématique, première partie: structures fondamentales de l'analyse*, Fonctions d'une variable réelle (Springer, Berlin, 2007)
2. A. Connes, *Noncommutative Geometry* (Academic Press, Dublin, 1994)
3. J.B. Conway, *A Course in Functional Analysis*, 2nd edn. (Springer, New York, 1997)
4. M. Engliš, K. Falk, B. Iochum, Spectral triples and Toeplitz operators, arXiv:1402:3061
5. K. Falk, Examples of infinite direct sums of spectral triples, arXiv:1602.07458
6. A.L. Paterson, Contractive spectral triples for crossed products. Math. Scand. **114**, 275–298 (2014)
7. M. Reed, B. Simons, *Methods of Modern Mathematical Physics I.* (Academic Press, Dublin, 1980)
8. K. Zhu, *Spaces of Holomorphic Functions in the Unit Ball*, Graduate Texts in Mathematics (Springer, New York, 2005)

Index

A. Cardona et al. (eds.), *Quantization, Geometry and Noncommutative Structures in Mathematics and Physics*, Mathematical Physics Studies,
DOI 10.1007/978-3-319-65427-0

MIX
Papier aus verantwortungsvollen Quellen
Paper from responsible sources
FSC® C105338

If you have any concerns about our products,
you can contact us on
ProductSafety@springernature.com

In case Publisher is established outside the EU,
the EU authorized representative is:
Springer Nature Customer Service Center GmbH
Europaplatz 3, 69115 Heidelberg, Germany

Printed by Libri Plureos GmbH
in Hamburg, Germany